Lecture Notes in Computer Science   6168

Commenced Publication in 1973
Founding and Former Series Editors:
Gerhard Goos, Juris Hartmanis, and Jan van Leeuwen

Ron Steinfeld   Philip Hawkes (Eds.)

Information Security and Privacy

15th Australasian Conference, ACISP 2010
Sydney, Australia, July 5-7, 2010
Proceedings

 Springer

Volume Editors

Ron Steinfeld
Macquarie University, Department of Computing
North Ryde, NSW 2109, Australia
E-mail: rons@science.mq.edu.au

Philip Hawkes
Qualcomm Incorporated
Suite 301, Level 3, 77 King Street, Sydney, NSW 2000, Australia
E-mail: phawkes@qualcomm.com

Library of Congress Control Number: 2010929205

CR Subject Classification (1998): E.3, K.6.5, D.4.6, C.2, J.1, G.2.1

LNCS Sublibrary: SL 4 – Security and Cryptology

ISSN	0302-9743
ISBN-10	3-642-14080-7 Springer Berlin Heidelberg New York
ISBN-13	978-3-642-14080-8 Springer Berlin Heidelberg New York

springer.com

© Springer-Verlag Berlin Heidelberg 2010
Printed in Germany

Typesetting: Camera-ready by author, data conversion by Scientific Publishing Services, Chennai, India
Printed on acid-free paper   06/3180

Preface

The annual Australasian Conference on Information Security and Privacy is the premier Australian academic conference in its field, showcasing research from around the globe on a range of topics. ACISP 2010 was held during July 5-7, 2010, at Macquarie University in Sydney, Australia.

There were 97 paper submissions for the conference. These submission were reviewed by the Program Committee and a number of other individuals, whose names can be found overleaf. The Program Committee then selected 24 papers for presentation at the conference. These papers are contained in these proceedings.

In addition to the peer-reviewed papers, two invited speakers presented talks at the conference: Craig Gentry (IBM, USA); and Stephan Overbeek (Shearwater Solutions, Australia). We would like to express our gratitude to Craig and Stephan for contributing their knowledge and insight, and thus expanding the horizons of the conference delegates.

We would like to thank the authors of all of submission for offering their research for publication in ACISP 2010. We extend our sincere thanks to the Program Committee and other reviewers for the high-quality reviews and in-depth discussion. The Program Committee made use of the iChair electronic submission and reviewing software written by Thomas Baignères and Matthieu Finiasz at EPFL, LASEC. We would like to express our thanks to Springer for continuing to support the ACISP conference and for help in the conference proceedings production. We also thank the Organizing Committee, led by the ACISP 2010 General Chair Josef Pieprzyk, for their contribution to the conference.

Finally, we would like to thank our sponsors, iRobot, and our hosts, Qualcomm Inc. and the Centre for Advanced Computing - Algorithms and Cryptography (ACAC) at Macquarie University.

July 2010   Ron Steinfeld
                                                                             Philip Hawkes

Organization

General Chair

Josef Pieprzyk   Macquarie University, Australia

Program Co-chairs

Ron Steinfeld   Macquarie University, Australia
Philip Hawkes   Qualcomm Incorporated, Australia

Program Committee

Michel Abdalla	École Normale Supérieure, France
Masayuki Abe	NTT, Japan
Magnus Almgren	Chalmers University of Technology, Sweden
Joonsang Baek	Institute for Infocomm Research, Singapore
Feng Bao	Institute for Infocomm Research, Singapore
Lynn Batten	Deakin University, Australia
Alex Biryukov	University of Luxembourg, Luxembourg
Colin Boyd	Queensland University of Technology, Australia
Joo Yeon Cho	Nokia A/S, Denmark
Carlos Cid	Royal Holloway, University of London, UK
Andrew Clark	Queensland University of Technology, Australia
Nicolas Courtois	University College London, UK
Yvo Desmedt	University College London, UK and RCIS, AIST, Japan
Christophe Doche	Macquarie University, Australia
Ulrich Flegel	SAP Research, Germany
Steven Galbraith	University of Auckland, New Zealand
Juan Gonzalez Nieto	Queensland University of Technology, Australia
Maria Isabel Gonzalez Vasco	Universidad Rey Juan Carlos, Spain
Peter Gutmann	University of Auckland, New Zealand
Svein Knapskog	Norwegian University of Science and Technology, Norway
Xuejia Lai	Shanghai Jiao Tong University, China
Mark Manulis	TU Darmstadt, Germany
Keith Martin	Royal Holloway, University of London, UK
Mitsuru Matsui	Mitsubishi Electric, Japan
Krystian Matuiesewicz	Technical University of Denmark, Denmark
Chris Mitchell	Royal Holloway, University of London, UK
Atsuko Miyaji	JAIST, Japan

Yi Mu	University of Wollongong, Australia
C. Pandu Rangan	IIT, Madras, India
Vincent Rijmen	KU Leuven, Belgium and TU Graz, Austria
Rei Safavi Naini	University of Calgary, Canada
Palash Sarkar	Indian Statistical Institute, India
Berry Schoenmakers	TU Eindhoven, The Netherlands
Jennifer Seberry	University of Wollongong, Australia
Damien Stehlé	CNRS, France and University of Sydney, Australia and Macquarie University, Australia
Willy Susilo	University of Wollongong, Australia
Serge Vaudenay	EPFL, Switzerland
Damien Vergnaud	École Normale Supérieure, France
Huaxiong Wang	Nanyang Technological University, Singapore and Macquarie University, Australia
Duncan Wong	City University of Hong Kong, Hong Kong
Kan Yasuda	NTT, Japan
Yuliang Zheng	University of North Carolina at Charlotte, USA

External Reviewers

Hadi Ahmadi	Jialin Huang	Kaisa Nyberg
Man Ho Au	Qiong Huang	Tomas Olovsson
Gleb Beliakov	Ralf Hund	Kazumasa Omote
Simon R. Blackburn	Shaoquan Jiang	Khaled Ouafi
Thomas Bläsing	Charanjit Jutla	Vijayakrishnan
Jens-Matthias Bohli	Dmitry Khovratovich	Pasupathinathan
Ignacio Cascudo	Andreas Larsson	Serdar Pehlivanoglu
Rafik Chaabouni	Pho Le	Henning Rogge
Xiaofeng Chen	Gregor Leander	Markus Rueckert
Zhengjie Cheng	Benoit Libert	Minoru Saeki
Cline Chevalier	Tingting Lin	Yu Sasaki
Sebastiaan de Hoogh	Joseph K. Liu	Takashi Satoh
Georg Fuchsbauer	Shengli Liu	Jacob Schuldt
Jun Furukawa	Yi Lu	Sharmila Deva Selvi
Martin Gagne	Yiyuan Luo	Pouyan Sepehrdad
Choudary Gorantla	Takahiro Matsuda	Siamak F. Shahandashti
Vipul Goyal	Florian Mendel	Masaaki Shirase
Jian Guo	Ivan Morel	Magnus Själander
Fuchun Guo	Sumio Morioka	Benjamin Smith
Hua Guo	Sean Murphy	Boyeon Song
Gerhard Hancke	Jorge Nakahara Jr	Suriadi Suriadi
Javier Herranz	Kris Narayan	Daisuke Suzuki
Fumitaka Hoshino	Andrew Novocin	Christophe Tartary
Xinyi Huang	Attrapadung Nuttapong	Alan Tickle

Table of Contents

Symmetric Key Encryption

Cryptanalysis of a Generalized Unbalanced Feistel Network Structure... 1
Ruilin Li, Bing Sun, Chao Li, and Longjiang Qu

Improved Algebraic Cryptanalysis of QUAD, Bivium and Trivium via
Graph Partitioning on Equation Systems............................ 19
Kenneth Koon-Ho Wong and Gregory V. Bard

On Multidimensional Linear Cryptanalysis 37
Phuong Ha Nguyen, Lei Wei, Huaxiong Wang, and San Ling

Side-Channel Analysis of the K2 Stream Cipher..................... 53
Matt Henricksen, Wun She Yap, Chee Hoo Yian,
Shinsaku Kiyomoto, and Toshiaki Tanaka

On Unbiased Linear Approximations 74
Jonathan Etrog and Matthew J.B. Robshaw

Hash Functions

Distinguishers for the Compression Function and Output
Transformation of Hamsi-256 87
Jean-Philippe Aumasson, Emilia Käsper, Lars Ramkilde Knudsen,
Krystian Matusiewicz, Rune Ødegård, Thomas Peyrin, and
Martin Schläffer

Second-Preimage Analysis of Reduced SHA-1 104
Christian Rechberger

Some Observations on Indifferentiability 117
Ewan Fleischmann, Michael Gorski, and Stefan Lucks

Public Key Cryptography

Adaptive and Composable Non-committing Encryptions 135
Huafei Zhu, Tadashi Araragi, Takashi Nishide, and Kouichi Sakurai

Relations among Notions of Complete Non-malleability:
Indistinguishability Characterisation and Efficient Construction
without Random Oracles... 145
Manuel Barbosa and Pooya Farshim

Strong Knowledge Extractors for Public-Key Encryption Schemes 164
Manuel Barbosa and Pooya Farshim

A Multi-trapdoor Commitment Scheme from the RSA Assumption   182
    Ryo Nishimaki, Eiichiro Fujisaki, and Keisuke Tanaka

Identity-Based Chameleon Hash Scheme without Key Exposure   200
    Xiaofeng Chen, Fangguo Zhang, Willy Susilo, Haibo Tian,
    Jin Li, and Kwangjo Kim

The Security Model of Unidirectional Proxy Re-Signature with Private
Re-Signature Key   216
    Jun Shao, Min Feng, Bin Zhu, Zhenfu Cao, and Peng Liu

Security Estimates for Quadratic Field Based Cryptosystems   233
    Jean-François Biasse, Michael J. Jacobson Jr., and Alan K. Silvester

Solving Generalized Small Inverse Problems   248
    Noboru Kunihiro

Protocols

One-Time-Password-Authenticated Key Exchange   264
    Kenneth G. Paterson and Douglas Stebila

Predicate-Based Key Exchange   282
    James Birkett and Douglas Stebila

Attribute-Based Authenticated Key Exchange   300
    M. Choudary Gorantla, Colin Boyd, and
    Juan Manuel González Nieto

Optimally Tight Security Proofs for Hash-Then-Publish
Time-Stamping   318
    Ahto Buldas and Margus Niitsoo

Additive Combinatorics and Discrete Logarithm Based Range
Protocols   336
    Rafik Chaabouni, Helger Lipmaa, and Abhi Shelat

Proof-of-Knowledge of Representation of Committed Value and Its
Applications   352
    Man Ho Au, Willy Susilo, and Yi Mu

Network Security

Pattern Recognition Techniques for the Classification of Malware
Packers   370
    Li Sun, Steven Versteeg, Serdar Boztaş, and Trevor Yann

Repelling Sybil-Type Attacks in Wireless Ad Hoc Systems   391
    Marek Klonowski, Michał Koza, and Mirosław Kutyłowski

Author Index   403

Cryptanalysis of a Generalized Unbalanced Feistel Network Structure[*]

Ruilin Li[1], Bing Sun[1], Chao Li[1,2], and Longjiang Qu[1,3]

[1] Department of Mathematics and System Science, Science College,
National University of Defense Technology, Changsha, 410073, China
`securitylrl@gmail.com, happy_come@163.com`
[2] State Key Laboratory of Information Security, Institute of Software,
Chinese Academy of Sciences, Beijing, 100190, China
`lichao_nudt@sina.com`
[3] National Mobile Communications Research Laboratory,
Southeast University, Nanjing, 210096, China
`ljqu_happy@hotmail.com`

Abstract. This paper reevaluates the security of GF-NLFSR, a new kind of generalized unbalanced Feistel network structure that was proposed at ACISP 2009. We show that GF-NLFSR itself reveals a very slow diffusion rate, which could lead to several distinguishing attacks. For GF-NLFSR containing n sub-blocks, we find an n^2-round integral distinguisher by algebraic methods and further use this integral to construct an $(n^2 + n - 2)$-round impossible differential distinguisher. Compared with the original $(3n - 1)$-round integral and $(2n - 1)$-round impossible differential, ours are significantly better.

Another contribution of this paper is to introduce a kind of non-surjective attack by analyzing a variant structure of GF-NLFSR, whose provable security against differential and linear cryptanalysis can also be provided. The advantage of the proposed non-surjective attack is that traditional non-surjective attack is only applicable to Feistel ciphers with non-surjective (non-uniform) round functions, while ours could be applied to block ciphers with bijective ones. Moreover, its data complexity is $\mathcal{O}(l)$ with l the block length.

Keywords: block ciphers, generalized unbalanced Feistel network, integral attack, impossible differential attack, non-surjective attack.

1 Introduction

Differential cryptanalysis (DC) [6] and linear cryptanalysis (LC) [23] are the two most powerful known attacks on block ciphers since 1990s. For a new block

---

[*] The work in this paper is supported by the Natural Science Foundation of China (No: 60803156), the open research fund of State Key Laboratory of Information Security (No: 01-07) and the open research fund of National Mobile Communications Research Laboratory of Southeast University (No: W200807).

R. Steinfeld and P. Hawkes (Eds.): ACISP 2010, LNCS 6168, pp. 1–18, 2010.

cipher algorithm, designers must guarantee that it can resist these two attacks. However, even the security against DC and LC can be proved, the algorithm may suffer other attacks, such as truncated differential attack [13], higher-order differential attack [13,18], impossible differential attack [4,14], boomerang attack [27], amplified boomerang attack [16], rectangle attack [5], integral attack [15], interpolation attack [12], non-surjective attack [24], algebraic attack [8], related-key attack [3], slide attack [1] and so on. Among these methods, integral attack and impossible differential attack are of special importance. Take the well-known 128-bit version block cipher Rijndael as an example, six rounds is sufficient for resisting DC and LC. However, by integral attack or impossible differential attack, one can break six, seven, even eight rounds [9,11,20,29].

Integral cryptanalysis [15], which is especially well-suited for analyzing ciphers with primarily bijective components, was proposed by Knudsen et al.. In fact, it is a more generalization of Square attack [9], Saturation attack [19] and Multiset attack [2] proposed by Daemen et al., Lucks, and Biryukov et al., respectively. These methods exploit the simultaneous relationship between many encryptions, in contrast to differential cryptanalysis, where only pairs of encryptions are considered. Consequently, integral cryptanalysis applies to a lot of ciphers which are not vulnerable to DC and LC. These features have made integral an increasingly popular tool in recent cryptanalysis work.

The concept of using impossible differentials (differentials with probability 0) to retrieve the secret key of block ciphers was firstly introduced by Knudsen [14] against the DEAL cipher and further by Biham et al. [4] to attack Skipjack. Unlike differential cryptanalysis which recovers the right key through the obvious advantage of a high probability differential (differential characteristic), impossible differential cryptanalysis is a sieving attack that excludes all the wrong candidate keys using impossible differentials. Since its emergence, impossible differential cryptanalysis has been applied to attack many well-known block ciphers [20,21,28,29].

Non-surjective attack [24] was introduced by Rijmen et al. and it is applicable to Feistel ciphers with non-surjective, or more generally, non-uniform round functions such as CAST and LOKI 91. If the round function of Feistel ciphers is non-surjective (non-uniform), then by analyzing the statistical bias of some expression derived from the round function, one can apply a key recovery attack. However, if the round function is a surjective (uniform) one, it is impossible to apply this kind of non-surjective attack.

At ACISP 2009, Choy et al. proposed a new block cipher structure called n-cell GF-NLFSR [7], which is a kind of generalized unbalanced Feistel network [26] containing n sub-blocks. The advantages of this structure are that it allows parallel computations for encryption and that it can provide provable security against DC and LC, given that the round function is bijective. Meanwhile, the designers show the existence of a $(3n - 1)$-round integral distinguisher and a $(2n - 1)$-round impossible differential distinguisher. In the same paper, a new block cipher Four-Cell is designed as an application of the theoretical model of 4-cell GF-NLFSR.

Main Contribution. (1) We demonstrate that GF-NLFSR itself reveals a very slow diffusion rate, which could lead to several distinguishing attacks. We especially apply *algebraic methods* to find integral distinguishers in n-cell GF-NLFSR. In this method, plaintexts of special forms as well as their indeterminate states are treated as polynomial functions over finite fields, and in many cases, more precise information among these states could be obtained, which would lead to a better distinguisher.

Our cryptanalytic results show that, for n-cell GF-NLFSR, there exists an n^2-round integral distinguisher, which could be extended to an $(n^2 + n - 2)$-round higher-order integral distinguisher. Furthermore, by studying the relationship between integral and truncated differential, an $(n^2 + n - 2)$-round impossible differential distinguisher could be constructed. These distinguishers are significantly better than the original ones.

(2) We introduce a kind of *non-surjective attack* by analyzing a variant structure of GF-NLFSR, whose provable security against DC and LC can also be provided. The advantage of the proposed attack is that traditional non-surjective attack is only applicable to Feistel ciphers with non-surjective (non-uniform) round functions, while ours could be applied to block ciphers with bijective ones. Moreover, its data complexity is $\mathcal{O}(l)$ with l the block length.

Outline. We begin with a brief description of n-cell GF-NLFSR in Section 2. Encryption properties of n-cell GF-NLFSR by every n rounds are studied in Section 3. The existence of n^2-round integral distinguisher and $(n^2 + n - 2)$-round impossible differential distinguisher are shown in Section 4 and Section 5, respectively. Section 6 presents a kind of non-surjective attack by analyzing a variant structure of GF-NLFSR. Section 7 contains results of the experiment with the proposed non-surjective attack on a toy cipher, and finally Section 8 is the conclusion.

2 Description of n-Cell GF-NLFSR

As shown in Fig. 1, assume the input, output and round key to the i-th round of n-cell GF-NLFSR are $(x_0^{(i)}, x_1^{(i)}, \ldots, x_{n-1}^{(i)}) \in \mathbb{F}_{2^b}^n$, $(x_0^{(i+1)}, x_1^{(i+1)}, \ldots, x_{n-1}^{(i+1)}) \in \mathbb{F}_{2^b}^n$, and $K_i = (k_i, k_i')$, then the round transformation can be described as follow:

$$(x_0^{(i)}, x_1^{(i)}, \ldots, x_{n-2}^{(i)}, x_{n-1}^{(i)}) \mapsto (x_0^{(i+1)}, x_1^{(i+1)}, \ldots, x_{n-2}^{(i+1)}, x_{n-1}^{(i+1)}),$$

where

$$\begin{cases} x_l^{(i+1)} = x_{l+1}^{(i)}, & \text{if } l = 0, 1, \ldots, n-2 \\ x_{n-1}^{(i+1)} = F(x_0^{(i)}, K_i) \oplus x_1^{(i)} \oplus x_2^{(i)} \oplus \ldots \oplus x_{n-1}^{(i)} \end{cases}$$

and $F(\cdot, K_i) \triangleq F_{K_i}(\cdot)$ is a permutation on $\mathbb{F}_{2^b}$.

From [7], this kind of generalized unbalanced Feistel network can provide its provable security against DC and LC, which is summarized in the following proposition.

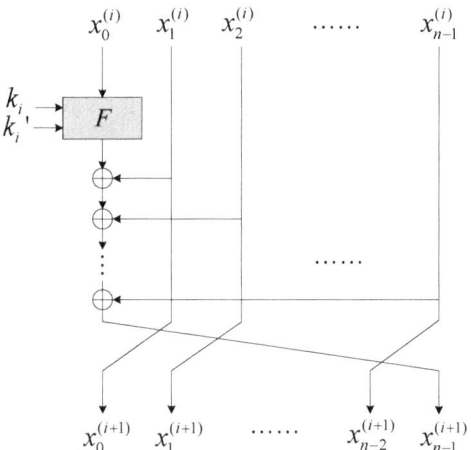

Fig. 1. The i-th round transformation of n-cell GF-NLFSR

Proposition 1. [7] *Let the round function of n-cell GF-NLFSR $F : \mathbb{F}_{2^b} \times \mathbb{F}_{2^b} \times \Omega \to \mathbb{F}_{2^b}$ be of the form $F(x, k_i, k_i') = f(x \oplus k_i, k_i')$, where $f : \mathbb{F}_{2^b} \times \Omega \to \mathbb{F}_{2^b}$ is bijective for all fixed $k_i' \in \Omega$. If the maximum differential (linear hull) probability of f satisfies $DP(LP)_{max}(f) \leq p(q)$, then the differential (linear hull) probability of the $(n+1)$-round encryption is upper bounded by $p^2(q^2)$.*

3   Encryption Property of n-Cell GF-NLFSR

In this section, we study the encryption property of n-cell GF-NLFSR by every n rounds. From now on, the round function $F_{K_i}(x)$ is treated as a permutation polynomial over $\mathbb{F}_{2^b}$.

Firstly, according to the definition of n-cell GF-NLFSR, the following result could be obtained.

Proposition 2. *Let $(x_0, x_1, \ldots, x_{n-1})$ be the input of the i-th round of n-cell GF-NLFSR, and $(y_0, y_1, \ldots, y_{n-1})$ be the output of the $(i+n-1)$-th round, then*

$$\begin{cases} y_0 = F_{K_i}(x_0) \oplus x_1 \oplus x_2 \oplus \ldots \oplus x_{n-1} \\ y_m = F_{K_{i+m-1}}(x_{m-1}) \oplus F_{K_{i+m}}(x_m) \oplus x_m, & if \ 1 \leq m \leq n-1 \end{cases}$$

and

$$\bigoplus_{j=0}^{n-1} y_j = F_{K_{i+n-1}}(x_{n-1}).$$

Proposition 2 can be verified directly by the encryption procedure of n-cell GF-NLFSR, based on which we could deduce the following proposition.

Proposition 3. *Let the input of n-cell GF-NLFSR be $(x, c_1, \ldots, c_{n-1})$, where x is a variable and each c_i is some constant with $1 \leq i \leq n-1$, let the output of the r-th round be $\left(y_0^{(r)}(x), y_1^{(r)}(x), \ldots, y_{n-1}^{(r)}(x) \right)$, and $1 \leq m \leq n-1$, then*

(1) $y_i^{(m \times n)}(x)$ is a permutation polynomial over $\mathbb{F}_{2^b}$ if $i = m$,

(2) $y_i^{(m \times n)}(x)$ is a constant if $i > m$.

Table 1 is the encryption results of every n rounds of n-cell GF-NLFSR when plaintexts are of the form $(x, c_1, \ldots, c_{n-1})$ as described in Proposition 3. Note that the first column denotes the round number, and each of the other columns represents the corresponding output sub-block. The letter C denotes some constant which could be different from each other. $P_m(x)$ is some permutation polynomial over $\mathbb{F}_{2^b}$ with $1 \leq m \leq n-1$, and those blank cells (elements under the diagonal) indicate that their behaviors are unknown.

An immediate conclusion, from Proposition 3 and Table 1, is that the diffusion rate of n-cell GF-NLFSR is very slow, since the input variable x needs at least $(n-1) \times n$ rounds to influence the last (rightmost) sub-block of the output.

Table 1. Output of every n rounds of n-cell GF-NLFSR

0	x	C	C	$\ldots$	C	$\ldots$ C	C
n	$P_1(x)$	C		$\ldots$	C	$\ldots$ C	C
$\vdots$			$\ddots$		$\vdots$	$\vdots$	$\vdots$
$(m-1) \times n$				$P_{m-1}(x)$	C	$\ldots$ C	C
$m \times n$					$P_m(x)$	$\ldots$ C	C
$\vdots$						$\ddots$ $\vdots$	$\vdots$
$(n-2) \times n$						$P_{n-2}(x)$	C
$(n-1) \times n$							$P_{n-1}(x)$

4 Integral Distinguisher of GF-NLFSR

4.1 Preliminaries

To apply integral cryptanalysis, one should first find an integral distinguisher of the reduced-round cipher, then apply the key recovery attack. In this section, we show how to construct an n^2-round integral distinguisher of n-cell GF-NLFSR by using algebraic techniques.

Firstly, recall that most traditional methods in finding integral distinguishers are based on the so-called *empirical methods*. They firstly treat each part of plaintexts with special forms as active or passive state (see definitions below), then study the property (active, passive or balanced) of its corresponding intermediate state after passing through several encryption rounds.

Definition 1. *A set $\{a_i | a_i \in \mathbb{F}_{2^b}, 0 \leq i \leq 2^b - 1\}$ is active, if for any $0 \leq i < j \leq 2^b - 1$, $a_i \neq a_j$. We use $\mathbf{A}$ to denote the active set.*

Definition 2. *A set $\{a_i | a_i \in \mathbb{F}_{2^b}, 0 \leq i \leq 2^b - 1\}$ is* passive *or constant, if for any $0 < i \leq 2^b - 1$, $a_i = a_0$. We use* **C** *to denote the passive set.*

Definition 3. *A set $\{a_i | a_i \in \mathbb{F}_{2^b}, 0 \leq i \leq 2^b - 1\}$ is* balanced, *if the XOR-sum of all element of the set is 0, that is $\oplus_{i=0}^{2^b-1} a_i = 0$. We use* **B** *to denote the balanced set.*

Moreover, three principles are widely used when applying empirical methods: (1) An active set remains active after passing a bijective transform. (2) The linear combination of several active/balanced sets is a balanced set. (3) The property of a balanced set after passing a nonlinear transformation is generally unknown.

Obviously, the third one is the bottleneck of empirical methods, thus if one could determine the property of a balanced set after it passes a nonlinear transformation, integral distinguisher with more rounds can be constructed.

4.2 n^2-Round Integral Distinguisher of n-Cell GF-NLFSR

By using the empirical method, the designers presented the following $(3n - 1)$-round integral distinguisher:

$$(A, C, C, \ldots, C) \rightarrow (C, ?, ?, \ldots, ?),$$

where A is active in $\mathbb{F}_{2^b}$, C is constant in $\mathbb{F}_{2^b}$, and ? is unknown.

Now we describe the newly constructed n^2-round integral in the following theorem, the proof is based on algebraic methods. See Appendix B for a 16-round integral distinguisher of 4-cell GF-NLFSR as an example.

Theorem 1. *There is an n^2-round integral distinguisher of n-cell GF-NLFSR:*

$$(A, C, \ldots, C) \rightarrow (S_0, S_1, \ldots, S_{n-1}),$$

where A is active, C is constant and $(S_0 \oplus S_1 \oplus \ldots \oplus S_{n-1})$ is active.

Proof. Let the input of n-cell GF-NLFSR be $(x, c_1, \ldots, c_{n-1})$ and the output of the $((n-1) \times n)$-th round be

$$\left(y_0^{((n-1) \times n)}(x), y_1^{((n-1) \times n)}(x), \ldots, y_{n-1}^{((n-1) \times n)}(x) \right),$$

then $y_{n-1}^{((n-1) \times n)}(x)$ is a permutation polynomial by Proposition 3.

Assume the output of the n^2-round is

$$\left(y_0^{(n^2)}(x), y_1^{(n^2)}(x), \ldots, y_{n-1}^{(n^2)}(x) \right),$$

according to Proposition 2,

$$y_0^{(n^2)}(x) \oplus y_1^{(n^2)}(x) \oplus \ldots \oplus y_{n-1}^{(n^2)}(x) = F_{K_{n^2}}\left(y_{n-1}^{((n-1) \times n)}(x) \right).$$

Since $y_{n-1}^{((n-1) \times n)}(x)$ is a permutation polynomial, so is $F_{K_{n^2}}\left(y_{n-1}^{((n-1) \times n)}(x) \right)$, which ends the proof. □

From the idea of higher-order integral [15], the above n^2-round integral can be extended to an $(n^2 + n - 2)$-round higher-order one.

Theorem 2. *There is an $(n^2 + n - 2)$-round higher-order integral distinguisher of n-cell GF-NLFSR:*

$$(A_0, A_1, \ldots, A_{n-2}, C) \rightarrow (S_0, S_1, \ldots, S_{n-1}),$$

where $(A_0, A_1, \ldots, A_{n-2})$ is active in $\mathbb{F}_{2^b}^{n-1}$, C is constant and $(S_0 \oplus S_1 \oplus \ldots \oplus S_{n-1})$ is balanced.

Proof. First, according to bijective property of the encryption structure of n-cell NLFSR, if the input is $(x_0, x_1, \ldots, x_{n-2}, c)$, where $(x_0, x_1, \cdots, x_{n-1})$ is active in $\mathbb{F}_{2^b}^{n-1}$, $c \in \mathbb{F}_{2^b}$ is constant, after $n - 2$ rounds encryption, the intermediate state must be $(y_0, c, y_2, \ldots, y_{n-1})$, where $(y_0, y_2, \ldots, y_{n-1})$ is active in $\mathbb{F}_{2^b}^{n-1}$.

Next, let's focus on the set containing these $2^{(n-1)b}$ intermediate states after $n - 2$ rounds encryption. Fix $(y_2, y_3, \ldots, y_{n-1}) \in \mathbb{F}_{2^b}^{n-2}$, we thus get a structure with 2^b elements, which is the input of the n^2-round integral distinguisher as shown in Theorem 1(From now on, we call this structure a Λ set).

Now, these $2^{(n-1)b}$ intermediate states can be divided into $2^{(n-2)b}$ indistinguishable Λ sets. When each Λ set passes through the n^2 rounds encryption, the XOR sum of the n sub-blocks of outputs is active (thus balanced) in $\mathbb{F}_{2^b}$. Consequently, the XOR sum of the n sub-blocks of outputs for these $2^{(n-2)b}$ indistinguishable Λ sets is balanced. Let $E_j^{(i)}(\cdot)$ denote the j-th sub-block after i rounds encryption of the input, then we can explain the higher-order integral distinguisher as follows:

$$\bigoplus_{x_0, x_1, \ldots, x_{n-2}} \bigoplus_{j=0}^{n-1} E_j^{(n^2+n-2)}(x_0, x_1, \ldots, x_{n-2}, c)$$

$$= \bigoplus_{y_0, y_2, \ldots, y_{n-1}} \bigoplus_{j=0}^{n-1} E_j^{(n^2)}(y_0, c, y_2, \ldots, y_{n-1})$$

$$= \bigoplus_{y_2, \ldots, y_{n-1}} \left(\bigoplus_{y_0} \bigoplus_{j=0}^{n-1} E_j^{(n^2)}(y_0, c, y_2, \ldots, y_{n-1}) \right)$$

$$= \bigoplus_{y_2, \ldots, y_{n-1}} 0$$

$$= 0 \qquad \qquad \square$$

5 Impossible Differential of GF-NLFSR

By using the $\mathcal{U}$-method [17], the designers of n-cell GF-NLFSR found a $(2n - 1)$-round impossible differential: $(0, 0, 0, \ldots, \alpha) \nrightarrow (\psi, \psi, 0, \ldots, 0)$, where $\alpha \neq 0$, $\psi \neq 0$. In this section, we show how to construct an $(n^2 + n - 2)$-round impossible differential by studying the relationship between integral and truncated differential as described in the following theorem:

Theorem 3. *The n^2-round integral distinguisher of Theorem 1 corresponds to the following n^2-round truncated differential with probability 1:*

$$(\delta, 0, \ldots, 0) \rightarrow (\delta_0, \delta_1, \ldots, \delta_{n-1}),$$

where $\delta \neq 0$ and $\delta_0 \oplus \delta_1 \oplus \ldots \oplus \delta_{n-1} \neq 0$.

Proof. Let the input of the n-cell GF-NLFSR be $(x, c_1, c_2, \ldots, c_{n-1})$, after n^2 rounds, the output is $(q_0(x), q_1(x), \ldots, q_{n-1}(x))$, then according to Proposition 2, $q_0(x) \oplus q_1(x) \oplus \ldots \oplus q_{n-1}(x) \triangleq q(x) \in \mathbb{F}_{2^b}[x]$ is a permutation polynomial.

Assume two inputs are $(x_1, c_1, c_2, \ldots, c_{n-1})$ and $(x_2, c_1, c_2, \ldots, c_{n-1})$ with $x_1 \neq x_2$, thus $q(x_1) \neq q(x_2)$. Now the input difference is $(\delta, 0, \ldots, 0)$ with $\delta = x_1 \oplus x_2 \neq 0$, and the output difference is $(\delta_0, \delta_1, \ldots, \delta_{n-1})$, satisfying $\delta_0 \oplus \delta_1 \oplus \ldots \oplus \delta_{n-1} = q(x_1) \oplus q(x_2) \neq 0$. □

Theorem 4. *There exists an $(n^2 + n - 2)$-round impossible differential in n-cell GF-NLFSR of the following form:*

$$(\delta, 0, \ldots, 0) \nrightarrow (\psi, \psi, 0, \ldots, 0),$$

where $\delta \neq 0$ and $\psi \neq 0$.

Proof. From encrypt direction, the n^2-round truncated differential $(\delta, 0, \ldots, 0) \rightarrow (\delta_0, \delta_1, \ldots, \delta_{n-1})$ is with probability 1, where $\delta \neq 0$ and $\delta_0 \oplus \delta_1 \oplus \ldots \oplus \delta_{n-1} \neq 0$. From decrypt direction, the $(n-2)$-round truncated differential $(\psi, \psi, 0, \ldots, 0) \rightarrow (0, \ldots, 0, \psi, \psi)$ is with probability 1. Since $\psi \oplus \psi = 0$, we find a contradiction. □

Remark. Wu *et al.* [30] independently found the same $(n^2 + n - 2)$-round impossible differential through a more direct approach. By using the 18-round impossible differential when $n = 4$, they presented a key recovery attack on the full round block cipher Four-Cell. Due to these new distinguishers and the full round attack, the designers have modified Four-Cell to Four-Cell$^+$ for better protection against the integral and impossible differential attacks.

6   A Kind of Non-surjective Attack

Our goal for introducing this kind of attack is that traditional non-surjective attack is only applicable to Feistel ciphers with non-surjective (non-uniform) round functions, while ours could be applied to block ciphers with bijective ones. Moreover, its data complexity is $\mathcal{O}(l)$ with l the block length.

To this end, we describe a variant structure of n-cell GF-NLFSR, denoted as n-cell VGF-NLFSR. As shown in Fig. 2, the main difference between these two structures is the round function. In n-cell VGF-NLFSR, the round function is $F(x \oplus K_i)$ with F bijective. One can easily demonstrate that the provable security against DC and LC for n-cell VGF-NLFSR can be provided using the same technique in [7]. Furthermore, Proposition 2 and 3 also suit for n-cell VGF-NLFSR, thus there exist the same n^2-round integral and $(n^2 + n - 2)$-round impossible differential as in n-cell GF-NLFSR.

Now, we introduce the non-surjective attack by analyzing VGF-NLFSR in the following two subsections.

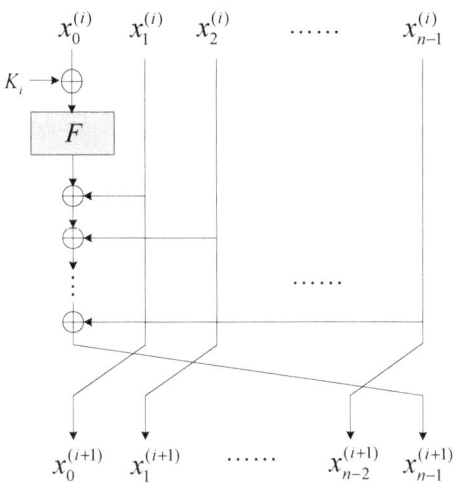

Fig. 2. The i-th round transformation of n-cell VGF-NLFSR

6.1   Description of the Non-surjective Distinguisher

Let the input of n-cell VGF-NLFSR be $(x, c_1, \ldots, c_{n-1})$, according to Proposition 2 and Proposition 3, $y_{n-1}^{((n-2) \times n)}$ is a constant, say C, and

$$\bigoplus_{j=0}^{n-1} y_j^{(n^2-n)} = F(C \oplus K_{n^2-n}) \triangleq C'.$$

Thus

$$y_0^{(n^2-n)} = C' \oplus \bigoplus_{j=1}^{n-1} y_j^{(n^2-n)}.$$

Assume the output of the n^2-th round is $(q_0(x), q_1(x), \ldots, q_{n-1}(x))$, from Proposition 2, we have

$$q_0(x) = F\left(y_0^{(n^2-n)} \oplus K_{n^2-n+1}\right) \oplus \bigoplus_{j=1}^{n-1} y_j^{(n^2-n)}.$$

Let $t = y_0^{(n^2-n)} \oplus K_{n^2-n+1}$, then

$$q_0(x) = F(t) \oplus t \oplus K_{n^2+n-1} \oplus C'$$
$$= F(t) \oplus t \oplus C^*,$$

where $C^* = K_{n^2+n-1} \oplus C'$ represents some unknown constant.

Let $f(t) = F(t) \oplus t$, and define $\mathcal{D}_f = \{y | y = f(t), t \in \mathbb{F}_{2^b}\}$. From the above fact, we have the following n^2-round distinguisher:

Theorem 5. *Let the input to n-cell VGF-NLFSR be $(x, c_1, \ldots, c_{n-1})$, where c_i is constant, and the output of the n^2-th round be $(q_0(x), q_1(x), \ldots, q_{n-1}(x))$, then there exists some constant $C^* \in \mathbb{F}_{2^b}$, such that for any $x \in \mathbb{F}_{2^b}$, $q_0(x) \oplus C^* \in \mathcal{D}_f$.*

Consider the distinguisher in Theorem 5, in this situation, the input to the $(n^2 + 1)$-th round function F is $q'(x) = q_0(x) \oplus K_{n^2+1}$, let $c^* = C^* \oplus K_{n^2+1}$, then $q'(x) \oplus c^* = q_0(x) \oplus C^*$. In other words, for all $x \in \mathbb{F}_{2^b}$, there exists some constant c^*, such that $q'(x) \oplus c^* \in \mathcal{D}_f$. Thus we could get the following theorem:

Theorem 6. *Let the input of n-cell VGF-NLFSR be $(x, c_1, \ldots, c_{n-1})$, where c_i is constant, and the input of the $(n^2 + 1)$-th round function F be $q'(x)$, then there exists some constant $c^* \in \mathbb{F}_{2^b}$, such that for any $x \in \mathbb{F}_{2^b}$, $q'(x) \oplus c^* \in \mathcal{D}_f$.*

One should note that if $\mathcal{D}_f = \mathbb{F}_{2^b}$, then both $F(x)$ and $F(x) \oplus x$ are permutations on $\mathbb{F}_{2^b}$, which indicates that $F(x)$ is an *orthormorphic permutation* [22]. Since the number of all orthormorphic permutations is small, in general, for a randomly chosen permutation $F(x)$, $f(x) = F(x) \oplus x$ can be seen as a random function (as the Davies-Meyer construction in hash function), thus $\mathcal{D}_f \subsetneq \mathbb{F}_{2^b}$. From now on, we will call the above distinguisher a *non-surjective distinguisher*, since the range of the function f is only a subset of $\mathbb{F}_{2^b}$.

6.2   Description of the Non-surjective Attack

By using the non-surjective distinguisher, one can attack $(n^2 + n')$-round n-cell VGF-NLFSR by Algorithm 1, where $n' > 1$.

Algorithm 1. Non-surjective attack on n-cell VGF-NLFSR	
Step 1	Compute and store $\mathcal{D}_f$.
Step 2	Given t plaintexts $(x_i, c_1, \ldots, c_{n-1})$, obtain the corresponding $(n^2 + n')$-round ciphertexts, $i = 1, \ldots, t$.
Step 3	Guess the last $(n' - 1)$ round-keys $rk = (rk_1, rk_2, \ldots, rk_{n'-1})$, decrypt the ciphertext to get the input of the $(n^2 + 1)$-round function F, denoted by $q'_{rk}(x_i)$.
Step 4	For all x_i in Step 2, test whether there exists some constant c^* satisfying $q'_{rk}(x_i) \oplus c^* \in \mathcal{D}_f$. If not, the guessed round-keys rk must be wrong.
Step 5	If necessary, repeat Step 2 $\sim$ Step 5 to further filter the wrong round keys until only one left.

In order to estimate the complexity of the above attack, we need the following two lemmas and their proofs can be found in Appendix A.

Lemma 1. *Given $A \subseteq \mathbb{F}_{2^b}$, $|A|$ denotes the number of different elements in A. For a randomly chosen set $X \subseteq \mathbb{F}_{2^b}(|X| \leq |A|)$, let p be the probability that there exists some constant $c \in \mathbb{F}_{2^b}$, such that $X \oplus c = \{x \oplus c | x \in X\} \subseteq A$, then*

$$p \leq 2^b \times \frac{|A|}{2^b} \times \frac{|A| - 1}{2^b - 1} \times \ldots \times \frac{|A| - (|X| - 1)}{2^b - (|X| - 1)}.$$

Lemma 2. *Let $f(x)$ be a random function from $\mathbb{F}_q$ to $\mathbb{F}_q$, $\mathcal{D}_f = \{f(x)|x \in \mathbb{F}_q\}$, let $\epsilon = E(|\mathcal{D}_f|)$ and $\sigma^2 = V(|\mathcal{D}_f|)$ be the expectation and variance of $|\mathcal{D}_f|$, respectively, then*

(i) $\displaystyle \lim_{q \to \infty} \frac{\epsilon}{q} = 1 - \frac{1}{e} \approx 0.632,$

(ii) $\displaystyle \lim_{q \to \infty} \frac{\sigma^2}{q} = \frac{e-2}{e^2} \approx 0.097.$

From Lemma 1, for a randomly chosen $X \subseteq \mathbb{F}_{2^b}$, if $|X| \ll |A|$, the upper bound of p can be well approximated by $2^b \times \left(|A|/2^b\right)^{|X|}$.

From Lemma 2, when q is large, the *Chebyshev Inequality* [27] indicates

$$\mathbf{Pr}\left(||\mathcal{D}_f| - \epsilon| \le l\sigma\right) \ge 1 - \frac{1}{l^2}.$$

If we choose $q = 2^b$ and $l = 10$, then for a randomly chosen f,

$$\mathbf{Pr}\left(0.63 \times 2^b - 3 \times 2^{b/2} \le |\mathcal{D}_f| \le 0.63 \times 2^b + 3 \times 2^{b/2}\right) \ge 0.99.$$

Thus we can estimate with high probability that $|\mathcal{D}_f|$ is less than $0.63 \times 2^b + 3 \times 2^{b/2}$. Moreover, when b is large, $|\mathcal{D}_f|$ can be approximated by 0.63×2^b.

Now, the data, time and space complexity of the proposed non-surjective attack can be analyzed as follows:

Data Complexity. Firstly, we note that when applying integral attack to n-cell VGF-NLFSR, one must choose at least a structure of all possible $(x, c_1, \ldots, c_{n-1})$, where $c_i's$ are constants. While for the non-surjective attack, only a fraction of them are needed.

Assume the number of chosen plaintexts as $(x, c_1, \ldots, c_{n-1})$ is t, let $\mathcal{T}$ denote the set of their corresponding ciphertexts, $\mathcal{T}_{rk}$ denote the set of the input to the $(n^2 + 1)$-round F function from decrypting the ciphertexts in $\mathcal{T}$ by guessing the last $n' - 1$ round keys rk.

The crucial step in Algorithm 1 is to check whether there exists a constant $c^* \in \mathbb{F}_{2^b}$ such that $\mathcal{T}_{rk} \oplus c^* \subseteq \mathcal{D}_f$. Assume wrong key values can pass such test with probability P_{err}, then from Lemma 1,

$$P_{err} \le (2^{(n'-1)b} - 1) \times 2^b \times \binom{|\mathcal{D}_f|}{t} / \binom{2^b}{t} \triangleq P_t,$$

thus in order to identify the right keys for the last $n' - 1$ rounds, P_{err} must be small enough. If b is large, and $t \ll |\mathcal{D}_f|$,

$$P_t \approx 2^{n'b} \times \left(|\mathcal{D}_f|/2^b\right)^t \approx 2^{n'b} \times 0.63^t.$$

Let $P_t = 2^{-\lambda}$, where the parameter λ is related to the success probability, and can be deduced by experiments, then $P_{err} \le P_t = 2^{-\lambda}$, which indicates that the probability that wrong key values can pass the test in Step 4 is less than $2^{-\lambda}$.

From $2^{n'b} \times 0.63^t = 2^{-\lambda}$, we get $t \approx \frac{3}{2}n'b + \frac{3}{2}\lambda$. Thus the data complexity of the above non-surjective attack is $\mathcal{O}(b)$.

To sum up, for attacking $(n^2 + n')$-round n-cell VGF-NLFSR, the data complexity is about $\frac{3}{2}n'b + \frac{3}{2}\lambda$.

Time Complexity. As explained before, Step 4 of Algorithm 1 needs to verify whether there exists a constant $c^* \in \mathbb{F}_{2^b}$, s.t. $\mathcal{T}_{rk} \oplus c^* \subseteq \mathcal{D}_f$ for each possible r_k. Assume for each possible c^*, the time complexity for testing whether $\mathcal{T}_{rk} \oplus c^* \subseteq \mathcal{D}_f$ is equivalent to u encryptions, then the time complexity is about

$$\left(\frac{3}{2}n'b + \frac{3}{2}\lambda\right) \times (2^{(n'-1)b}) \times 0.63 \times 2^b \times u \approx (n'b + \lambda) \times 2^{n'b} \times u,$$

thus a good algorithm for testing whether one set is included in another is required.

Space Complexity. Since one must store $\mathcal{D}_f$ to apply the non-surjective attack, the space complexity is about 0.63×2^b.

7   Experiments with the Proposed Non-surjective Attack

This section describes a 32-bit toy cipher based on 4-cell VGF-NLFSR, where the round function is defined by $F(x, k) = S(x \oplus k)$ with S as the S-box of AES. It is well known that the differential (linear hull) probability of the S-box of AES is upper bounded by 2^{-6}, thus the differential (linear hull) probability for five rounds is upper bounded by $(2^{-6})^2 = 2^{-12}$. Now we can see that the differential (linear) characteristic probability for 15 rounds is at most $(2^{-12})^3 = 2^{36} \leq 2^{-32}$, that is to say such toy cipher with more than 15 rounds is practically secure against DC and LC.

As an example, we use the method in Section 6 to mount a non-surjective attack on the 18-round toy cipher. In this case, $b = 8$ and $|\mathcal{D}_f| = 163 \approx 0.63 \times 2^8$. Table 2 lists our experimental results. For each $\lambda = 2, 4, 6, 8, 10$, t_λ denotes the number of chosen plaintexts and p_λ denotes the success probability, where the "success" means the adversary can uniquely recover the right 18-th round key. For each chosen parameter λ, we do the non-surjective attack 1000 times, and in each time the plaintext as well as the encryption key are randomly generated. The success probabilities are $0.474, 0.758, 0.873, 0.965, 0.992$.

One could also apply the integral attack to the 18-round toy cipher, however, to get a high success probability, its data complexity is about $2 \times 2^8 = 2^9$.

Table 2. Experiments with the non-surjective attack on the 18-round toy cipher

parameter λ	chosen plaintexts $t_\lambda = 3b + 1.5\lambda$	success probability p_λ
2	27	0.474
4	30	0.758
6	33	0.873
8	36	0.965
10	39	0.992

8 Conclusion

This paper presents several security analysis on GF-NLFSR. Although such structure allows parallel computations for encryption and can even provide its provable security against DC and LC, the structure itself reveals a very slow diffusion rate, which could lead to several distinguishing attacks.

For n-cell GF-NLFSR, our cryptanalytic results show that there exists an n^2-round integral distinguisher, which could be extended to an $(n^2 + n - 2)$-round higher-order one. Based on this n^2-round integral distinguisher, an $(n^2 + n - 2)$-round impossible differential is constructed. These results are significantly better than the original ones and thus imply that the security of n-cell GF-NLFSR must be carefully reevaluated.

Besides, a kind of non-surjective attack is proposed, which is different in essence with the one introduced by Rijmen $et\ al.$, since traditional non-surjective attack is only applicable to Feistel ciphers with non-surjective (non-uniform) round functions while ours can be applied to block ciphers with round functions being bijective. To demonstrate this, we describe a variant structure of n-cell GF-NLFSR, whose round function is defined by $F(x \oplus K)$. The provable security against DC and LC can also be provided for this variant structure, however, by using the proposed non-surjective attack, an efficient key recovery attack with very low data complexity could be mounted. Some experimental results are given for this non-surjective attack on a toy cipher based on the S-box of AES.

It is interesting that whether this kind of non-surjective attack can be applied to other block ciphers.

Acknowledgments. The authors wish to thank the anonymous reviewers of ACISP 2010 for their valuable suggestions and comments.

References

1. Biryukov, A., Wagner, D.: Slide Attack. In: Knudsen, L.R. (ed.) FSE 1999. LNCS, vol. 1636, pp. 245–259. Springer, Heidelberg (1999)
2. Biryukov, A., Shamir, A.: Structural Cryptanalysis of SASAS. In: Pfitzmann, B. (ed.) EUROCRYPT 2001. LNCS, vol. 2045, pp. 394–405. Springer, Heidelberg (2001)
3. Biham, E.: New Types of Cryptanalytic Attacks Using Related Keys. In: Helleseth, T. (ed.) EUROCRYPT 1993. LNCS, vol. 765, pp. 398–409. Springer, Heidelberg (1994)
4. Biham, E., Biryukov, A., Shamir, A.: Cryptanalysis of Skipjack Reduced to 31 Rounds Using Impossible Differentials. In: Stern, J. (ed.) EUROCRYPT 1999. LNCS, vol. 1592, pp. 12–23. Springer, Heidelberg (1999)
5. Biham, E., Dunkelman, O., Keller, N.: The Rectangle Attack- Rectangling the Serpent. In: Pfitzmann, B. (ed.) EUROCRYPT 2001. LNCS, vol. 2045, pp. 340–357. Springer, Heidelberg (2001)
6. Biham, E., Shamir, A.: Differential Cryptanalysis of DES-like Cryptosystems. Journal of Cryptology 3, 3–72 (1991)

7. Choy, J., Chew, G., Khoo, K., Yap, H.: Cryptographic Properties and Application of a Generalized Unbalanced Feistel Network Structure. In: Boyd, C., González Nieto, J. (eds.) ACISP 2009. LNCS, vol. 5594, pp. 73–89. Springer, Heidelberg (2009)

8. Courtois, N., Pieprzyk, J.: Cryptanalysis of Block Ciphers with Overdefined Systems of Equations. In: Zheng, Y. (ed.) ASIACRYPT 2002. LNCS, vol. 2501, pp. 267–287. Springer, Heidelberg (2002)

9. Daemen, J., Knudsen, L., Rijmen, V.: The Block Cipher Square. In: Biham, E. (ed.) FSE 1997. LNCS, vol. 1267, pp. 149–165. Springer, Heidelberg (1997)

10. Feller, W.: An Introduction to Probability Theory and Its Applications, 3rd edn. Wiley, New York (1968)

11. Ferguson, N., Kelsey, J., Lucks, S., Schneier, B., Stay, M., Wagner, D., Whiting, D.: Improved Cryptanalysis of Rijndael. In: Schneier, B. (ed.) FSE 2000. LNCS, vol. 1978, pp. 213–230. Springer, Heidelberg (2001)

12. Jackobsen, T., Knudsen, L.: The Interpolation Attack on Block Cipher. In: Biham, E. (ed.) FSE 1997. LNCS, vol. 1267, pp. 28–40. Springer, Heidelberg (1997)

13. Knudsen, L.: Truncated and High Order Differentials. In: Preneel, B. (ed.) FSE 1994. LNCS, vol. 1008, pp. 196–211. Springer, Heidelberg (1995)

14. Knudsen, L.: DEAL – A 128-bit Block Cipher. Technical Report 151, Department of Informatics, University of Bergen, Bergen, Norway (February 1998)

15. Knudsen, L., Wagner, D.: Integral Cryptanalysis. In: Daemen, J., Rijmen, V. (eds.) FSE 2002. LNCS, vol. 2365, pp. 112–127. Springer, Heidelberg (2002)

16. Kelsey, J., Kohno, T., Schneier, B.: Amplified Boomerang Attacks against Reduced-round MARS and Serpent. In: Schneier, B. (ed.) FSE 2000. LNCS, vol. 1978, pp. 75–93. Springer, Heidelberg (2001)

17. Kim, J., Hong, S., Sung, J., Lee, S., Lim, J., Sung, S.: Impossible Differential Cryptanalysis for Block Cipher Structures. In: Johansson, T., Maitra, S. (eds.) INDOCRYPT 2003. LNCS, vol. 2904, pp. 82–96. Springer, Heidelberg (2003)

18. Lai, X.: High Order Derivatives and Differential Cryptanalysis. In: Communications and Cryptography, pp. 227–233 (1994)

19. Lucks, S.: The Saturation Attack — A Bait for Twofish. In: Matsui, M. (ed.) FSE 2001. LNCS, vol. 2355, pp. 1–15. Springer, Heidelberg (2002)

20. Lu, J., Dunkelman, O., Keller, N., Kim, J.: New Impossible Differential Attacks on AES. In: Chowdhury, D.R., Rijmen, V., Das, A. (eds.) INDOCRYPT 2008. LNCS, vol. 5365, pp. 279–293. Springer, Heidelberg (2008)

21. Lu, J., Kim, J., Keller, N., Dunkelman, O.: Improving the Efficiency of Impossible Differential Cryptanalysis of Reduced Camellia and MISTY1. In: Malkin, T.G. (ed.) CT-RSA 2008. LNCS, vol. 4964, pp. 370–386. Springer, Heidelberg (2008)

22. Mittenthal, L.: Block Substitutions Using Orthomorphic Mappings. Advances in Applied Mathematics 16(1), 59–71 (1995)

23. Matsui, M.: Linear Cryptanalysis Method for DES Cipher. In: Helleseth, T. (ed.) EUROCRYPT 1993. LNCS, vol. 765, pp. 386–397. Springer, Heidelberg (1994)

24. Rijmen, V., Preneel, B., De Win, E.: On Weaknesses of Non-surjective Round Functions. Designs, Codes, and Cryptography 12, 253–266 (1997)

25. Roberts, F., Tesman, B.: Applied Combinatorics, 2nd edn. Pearson Education, London (2005)

26. Schneier, B., Kelsey, J.: Unbalanced Feistel Networks and Block Cipher Design. In: Gollmann, D. (ed.) FSE 1996. LNCS, vol. 1039, pp. 121–144. Springer, Heidelberg (1996)

27. Wanger, D.: The Boomerang Attack. In: Knudsen, L.R. (ed.) FSE 1999. LNCS, vol. 1636, pp. 156–170. Springer, Heidelberg (1999)

28. Wu, W., Zhang, W., Feng, D.: Impossible differential cryptanalysis of Reduced-Round ARIA and Camellia. Journal of Compute Science and Technology 22(3), 449–456 (2007)
29. Zhang, W., Wu, W., Feng, D.: New Results on Impossible Differential Cryptanalysis of Reduced AES. In: Nam, K.-H., Rhee, G. (eds.) ICISC 2007. LNCS, vol. 4817, pp. 239–250. Springer, Heidelberg (2007)
30. Wu, W., Zhang, L., Zhang, L., Zhang, W.: Security Analysis of the GF-NLFSR Structure and Four-Cell Block Cipher. In: ICICS 2009. LNCS, vol. 5927, pp. 17–31. Springer, Heidelberg (2009)

A   Proofs of Lemma 1 and Lemma 2

1. Proof of Lemma 1

First note that the number of different sets chosen from $\mathbb{F}_{2^b}$ with $|X|$ elements is $\binom{2^b}{|X|}$. Consider the subset $A \subseteq \mathbb{F}_{2^b}$, the number of different sets chosen from A with $|X|$ elements is $\binom{|A|}{|X|}$. Now for every fixed $c \in \mathbb{F}_{2^b}$, the probability p_c that $X \oplus c \subseteq A$ is upper bound by $\binom{|A|}{|X|}/\binom{2^b}{|X|}$. Thus we have

$$p = \sum_{c \in \mathbb{F}_{2^b}} p_c \leq 2^b \times \binom{|A|}{|X|} / \binom{2^b}{|X|}. \qquad \square$$

2. Proof of Lemma 2

Lemma 2 can be extended to a more general situation, where $\mathbb{F}_q$ can be replaced by any set with n elements and we will prove this more general conclusion. Note that the result of (i) can also be found in [24], however, by using their technique, one could not get the result of (ii). So, we introduce a formal method and prove these two results in a unified approach.

Given a set S, $|S| = n$, let f be a random function from S to S and $\mathcal{D}_f = \{f(a)|a \in S\} \subseteq S$.

(i) By the definition of expectation,

$$\epsilon = \sum_f \frac{1}{n^n} \times |\mathcal{D}_f| = \frac{1}{n^n} \times \sum_f |\mathcal{D}_f|. \qquad (1)$$

From the "Principle of Inclusive and Exclusive" [25], we have

$$\sum_f |\mathcal{D}_f| = \sum_{t=1}^{n} t \cdot \binom{n}{t} \cdot \sum_{i=0}^{t-1} \binom{t}{t-i} \cdot (-1)^i \cdot (t-i)^n$$

$$= \sum_{t=1}^{n} t \cdot \binom{n}{t} \cdot \sum_{u=1}^{t} \binom{t}{u} \cdot (-1)^{t-u} \cdot u^n \text{ (where } u = t - i)$$

$$= \sum_{u=1}^{n} u^n \cdot \sum_{t=u}^{n} t \cdot \binom{n}{t} \cdot \binom{t}{u} \cdot (-1)^{t-u}$$

$$= \sum_{u=1}^{n} u^n \cdot \sum_{k=0}^{n-u} (k+u) \cdot \binom{n}{k+u} \cdot \binom{k+u}{u} \cdot (-1)^k \text{ (where } k = t - u)$$

$$= \sum_{u=1}^{n} u^n \cdot \sum_{k=0}^{n-u} (k+u) \cdot \binom{n}{u} \cdot \binom{n-u}{k} \cdot (-1)^k$$

$$= \sum_{u=1}^{n} u^n \cdot \binom{n}{u} \cdot \sum_{k=0}^{n-u} (k+u) \cdot \binom{n-u}{k} \cdot (-1)^k$$

$$\triangleq A + B, \tag{2}$$

where

$$A = \sum_{u=1}^{n} u^n \cdot \binom{n}{u} \cdot \sum_{k=0}^{n-u} k \cdot \binom{n-u}{k} \cdot (-1)^k$$

$$= \sum_{u=1}^{n-1} u^n \cdot \binom{n}{u} \cdot \sum_{k=1}^{n-u} k \cdot \binom{n-u}{k} \cdot (-1)^k$$

$$= \sum_{u=1}^{n-1} u^n \cdot \binom{n}{u} \cdot \sum_{k=1}^{n-u} (n-u) \cdot \binom{n-u-1}{k-1} \cdot (-1)^k$$

$$= -\sum_{u=1}^{n-1} u^n \cdot \binom{n}{u} \cdot \sum_{k'=0}^{n-u-1} (n-u) \cdot \binom{n-u-1}{k'} \cdot (-1)^{k'}$$

$$= -n \cdot (n-1)^n,$$

and

$$B = \sum_{u=1}^{n} u^{n+1} \cdot \binom{n}{u} \cdot \sum_{k=0}^{n-u} \binom{n-u}{k} \cdot (-1)^k = n^{n+1}.$$

From (1) and (2), we get

$$\epsilon = \frac{A+B}{n^n} = \frac{1}{n^n} \times \left(n^{n+1} - n \cdot (n-1)^n \right) = n - n \cdot (1 - 1/n)^n.$$

Thus

$$\lim_{n \to \infty} \frac{\epsilon}{n} = \lim_{n \to \infty} \left(1 - \left(1 - \frac{1}{n} \right)^n \right) = 1 - \frac{1}{e}.$$

(ii) By the definition of variance,

$$\sigma^2 = \sum_f \frac{1}{n^n} \times (|\mathcal{D}_f| - \epsilon)^2 = \frac{1}{n^n} \times \sum_f (|\mathcal{D}_f| - \epsilon)^2. \tag{3}$$

From the result of (i),

$$\sum_f \left(|\mathcal{D}_f| - \epsilon \right)^2$$

$$= \sum_{t=1}^{n} \left(t - n\left(1 - \left(1 - \frac{1}{n}\right)^n\right) \right)^2 \cdot \binom{n}{t} \cdot \sum_{i=0}^{t-1} \binom{t}{t-i} \cdot (-1)^i \cdot (t-i)^n$$

$$= \sum_{t=1}^{n} \left(t^2 - 2nt\left(1 - \left(1 - \frac{1}{n}\right)^n\right) + \left(1 - \left(1 - \frac{1}{n}\right)^n\right)^2 \cdot n^2 \right)$$

$$\cdot \binom{n}{t} \cdot \sum_{i=0}^{t-1} \binom{t}{t-i} \cdot (-1)^i \cdot (t-i)^n$$

$$\triangleq A + B + C, \tag{4}$$

where

$$A = \sum_{t=1}^{n} t^2 \cdot \binom{n}{t} \cdot \sum_{i=0}^{t-1} \binom{t}{t-i} \cdot (-1)^i \cdot (t-i)^n,$$

$$B = -2n\left(1 - \left(1 - \frac{1}{n}\right)^n\right) \cdot \sum_{t=1}^{n} t \cdot \binom{n}{t} \cdot \sum_{i=0}^{t-1} \binom{t}{t-i} \cdot (-1)^i \cdot (t-i)^n,$$

$$C = \left(1 - \left(1 - \frac{1}{n}\right)^n\right)^2 \cdot n^2 \cdot \sum_{t=1}^{n} \cdot \binom{n}{t} \cdot \sum_{i=0}^{t-1} \binom{t}{t-i} \cdot (-1)^i \cdot (t-i)^n.$$

Using the same technique as in the proof of (i), after careful calculation,

$$A = n^{n+2} - 2n(n-1)^{n+1} + \left(2(n-2)^n \binom{n}{2} - n(n-1)^n\right),$$

$$B = -2n\left(1 - \left(1 - \frac{1}{n}\right)^n\right) \cdot (n^{n+1} - n(n-1)^n),$$

$$C = \left(1 - \left(1 - \frac{1}{n}\right)^n\right)^2 \cdot n^2 \cdot n^n.$$

From (3) and (4), we get

$$\sigma^2 = \frac{A + B + C}{n^n}.$$

Thus

$$\lim_{n \to \infty} \frac{\sigma^2}{n} = \lim_{n \to \infty} \frac{A + B + C}{n^{n+1}} = \frac{e - 2}{e^2}.$$

□

B   16-Round Integral Distinguisher of 4-Cell GF-NLFSR

0	x	C_1	C_2	C_3
1	C_1	C_2	C_3	$y \oplus C_4$
2	C_2	C_3	$y \oplus C_4$	$y \oplus C_5$
3	C_3	$y \oplus C_4$	$y \oplus C_5$	C_6
4	$y \oplus C_4$	$y \oplus C_5$	C_6	C_7
5	$y \oplus C_5$	C_6	C_7	$y \oplus z \oplus C_8$
6	C_6	C_7	$y \oplus z \oplus C_8$	$y \oplus z \oplus w \oplus C_9$
7	C_7	$y \oplus z \oplus C_8$	$y \oplus z \oplus w \oplus C_9$	$w \oplus C_{10}$
8	$y \oplus z \oplus C_8$	$y \oplus z \oplus w \oplus C_9$	$w \oplus C_{10}$	C_{11}
9	$y \oplus z \oplus w \oplus C_9$	$w \oplus C_{10}$	C_{11}	$t_1 \oplus y \oplus z \oplus C_{12}$
10	$w \oplus C_{10}$	C_{11}	$t_1 \oplus y \oplus z \oplus C_{12}$	$t_2 \oplus t_1 \oplus y \oplus z \oplus w \oplus C_{13}$
11	C_{11}	$t_1 \oplus y \oplus z \oplus C_{12}$	$t_2 \oplus t_1 \oplus y \oplus z \oplus w \oplus C_{13}$	$t_2 \oplus w \oplus u \oplus C_{14}$
12	$t_1 \oplus y \oplus z \oplus C_{12}$	$t_2 \oplus t_1 \oplus y \oplus z \oplus w \oplus C_{13}$	$t_2 \oplus w \oplus u \oplus C_{14}$	$u \oplus C_{15}$
13	$t_2 \oplus t_1 \oplus y \oplus z \oplus w \oplus C_{13}$	$t_2 \oplus w \oplus u \oplus C_{14}$	$u \oplus C_{15}$	$t_3 \oplus t_1 \oplus y \oplus z \oplus C_{16}$
14	$t_2 \oplus w \oplus u \oplus C_{14}$	$u \oplus C_{15}$	$t_3 \oplus t_1 \oplus y \oplus z \oplus C_{16}$	$t_4 \oplus t_3 \oplus t_2 \oplus t_1 \oplus y \oplus z \oplus w \oplus C_{17}$
15	$u \oplus C_{15}$	$t_3 \oplus t_1 \oplus y \oplus z \oplus C_{16}$	$t_4 \oplus t_3 \oplus t_2 \oplus t_1 \oplus y \oplus z \oplus w \oplus C_{17}$	$t_5 \oplus t_4 \oplus t_2 \oplus w \oplus u \oplus C_{18}$
16	$t_3 \oplus t_1 \oplus y \oplus z \oplus C_{16}$	$t_4 \oplus t_3 \oplus t_2 \oplus t_1 \oplus y \oplus z \oplus w \oplus C_{17}$	$t_5 \oplus t_4 \oplus t_2 \oplus w \oplus u \oplus C_{18}$	$t_5 \oplus u \oplus v \oplus C_{19}$

The parameters in the above 16-round integral distinguisher are as follows: C_i, $4 \leq i \leq 19$ is passive (constant) in $\mathbb{F}_{2^b}$, x, y, z, w, u, v are active in $\mathbb{F}_{2^b}$, and t_j, $1 \leq j \leq 5$ is some unknown intermediate value in $\mathbb{F}_{2^b}$. It can be easily verified that

$$(t_3 \oplus t_1 \oplus y \oplus z \oplus C_{16}) \oplus (t_4 \oplus t_3 \oplus t_2 \oplus t_1 \oplus y \oplus z \oplus w \oplus C_{17}) \oplus (t_5 \oplus t_4 \oplus t_2 \oplus w \oplus u \oplus C_{18}) \oplus (t_5 \oplus u \oplus v \oplus C_{19})$$

$$= v \oplus C_{16} \oplus C_{17} \oplus C_{18} \oplus C_{19}.$$

Improved Algebraic Cryptanalysis of QUAD, Bivium and Trivium via Graph Partitioning on Equation Systems

Kenneth Koon-Ho Wong[1] and Gregory V. Bard[2]

[1] Information Security Institute, Queensland University of Technology,
Brisbane QLD 4000, Australia
kk.wong@qut.edu.au

[2] Mathematics Department, Fordham University, The Bronx NY 10458, USA
bard@fordham.edu

Abstract. We present a novel approach for preprocessing systems of polynomial equations via graph partitioning. The variable-sharing graph of a system of polynomial equations is defined. If such graph is disconnected, then the corresponding system of equations can be split into smaller ones that can be solved individually. This can provide a tremendous speed-up in computing the solution to the system, but is unlikely to occur either randomly or in applications. However, by deleting certain vertices on the graph, the variable-sharing graph could be disconnected in a balanced fashion, and in turn the system of polynomial equations would be separated into smaller systems of near-equal sizes. In graph theory terms, this process is equivalent to finding balanced vertex partitions with minimum-weight vertex separators. The techniques of finding these vertex partitions are discussed, and experiments are performed to evaluate its practicality for general graphs and systems of polynomial equations. Applications of this approach in algebraic cryptanalysis on symmetric ciphers are presented: For the QUAD family of stream ciphers, we show how a malicious party can manufacture conforming systems that can be easily broken. For the stream ciphers Bivium and Trivium, we achieve significant speedups in algebraic attacks against them, mainly in a partial key guess scenario. In each of these cases, the systems of polynomial equations involved are well-suited to our graph partitioning method. These results may open a new avenue for evaluating the security of symmetric ciphers against algebraic attacks.

1 Introduction

There has been a long history of the use of graph theory in solving systems of equations. Graph partitioning techniques are applied to processes such as re-ordering variables in matrices to reduce fill-in for sparse systems [19, Ch. 7] and partitioning a finite element mesh across nodes in parallel computations [42]. These techniques primarily focus on linear systems over the real or complex numbers. In this paper, we apply similar graph theory techniques to systems of

R. Steinfeld and P. Hawkes (Eds.): ACISP 2010, LNCS 6168, pp. 19–36, 2010.
© Springer-Verlag Berlin Heidelberg 2010

multivariate polynomial equations, and develop methods of partitioning these systems into ones of smaller sizes via their "variable-sharing" graphs. These techniques are intended to work over any field, finite or infinite, but are particularly suited to GF(2) for use in algebraic cryptanalysis of symmetric ciphers. In most algebraic cryptanalysis, the symmetric ciphers are described by systems of polynomial equations over GF(2) or its algebraic extensions. The graph theory methods introduced in this paper can be used to improve the efficiency of solving these systems of equations, which would translate to a reduction of the security of these ciphers. This will be exemplified with the QUAD [9], Bivium [48] and Trivium [20] stream ciphers.

Computing the solution to a system of multivariate polynomial equations is an NP-hard problem [5, Ch. 3.9]. A variety of solution techniques have been developed for solving these polynomial systems over finite fields, such as linearization and XL [18], Gröbner bases, and resultants [6, Ch. 12], as well as recent ones such as SAT-solvers [7], Vielhaber's AIDA [50], Raddum-Semaev method [47], and the triangulation algorithm [35]. Over the real and complex numbers, numerical techniques are also known, but require the field to be ordered and complete—GF(2) is neither. The graph partitioning method introduced in this paper could be a novel addition to the variety of methods available, principally as a preprocessor.

From a multivariate polynomial system of equations, a variable-sharing graph is constructed with a vertex for each variable in the system, and an edge between two vertices if and only if those variables appear together in any equation in the system. Clearly, if the graph is disconnected, the system can be split into two separate systems of smaller sizes, and they can be solved for individually. However, even if the graph is connected, we show that it may be possible to disconnect the graph by eliminating a few variables by, for example, guessing their values when computing over a small finite field, and thereby splitting the remaining system. This suggests a divide-and-conquer approach to solving systems of equations. When the polynomial terms in the system of equations are very sparse, we show that the system can usually be reduced to a set of smaller systems, whose solutions can be computed individually in much less time. It should be noted that for large finite fields, and infinite fields as well, the technique of resultants can be used to achieve similar objectives [53].

In order for a partition of a system to be productive, the minimum number of variables should be eliminated, and the two subsystems must be approximately equal in size. This ensures that the benefit of partitioning the system is maximised. These conditions lead to the problem of finding a balanced vertex partition with a minimum-weight vertex separator on its variable-sharing graph, which is an NP-hard problem [31, 43]. Nevertheless, heuristic algorithms can often find near-optimal partitions efficiently [32].

In this paper, we offer two cryptographic applications of vertex partitioning arising from the algebraic cryptanalysis of stream ciphers, where both achieve positive results. First, we describe a method whereby a manufacturer of a sparse implementation of QUAD [9], a provably-secure infinite family of stream ciphers, could "poison" the polynomial system in the cipher, and thereby enable messages transmitted with it to be read by the manufacturer. Second, we present an algebraic cryptanalysis of Trivium [20], a profiled stream cipher in the eSTREAM project, as well as its reduced versions Bivium-A and Bivium-B, and discuss the implications of graph partitioning methods on solving the corresponding systems of equations. Improvements to partial key guess attacks against Trivium and Bivium are observed.

Section 2 introduces the necessary background in graph theory and graph partitioning. Section 3 shows how a system of polynomial equations can be split into ones of smaller sizes using graph partitioning methods. Section 4 provides results for some partitioning experiments and analyses the feasibility of equation solving via graph partitioning methods. Section 5 presents the applications of graph partitioning methods on the algebraic cryptanalysis of QUAD, Bivium and Trivium. Conclusions will be drawn in Section 6. In Appendix A, we discuss the possibility for vertex connectivities of variable-sharing graphs becoming a security criterion for symmetric ciphers.

2   Preliminaries

Let $G = (V, E)$ be a graph with vertex set V and edge set E. Two vertices $v_i, v_j \in V$ are connected if there is a path from v_i to v_j through edges in E. A disconnected graph is a graph where there exists at least one pair of vertices that is not connected, or if the graph has only one vertex. A graph $G_1 = (V_1, E_1)$ with vertex set $V_1 \subseteq V$ and edge set $E_1 \subseteq E$ is called a subgraph of G. Given a graph G, subgraphs of G can be obtained by removing vertices and edges from G. Let $G = (V, E)$ be a graph with k vertices and l edges, such that $V = \{v_1, v_2, \ldots, v_{k-1}, v_k\}$, $E = \{(v_{i_1}, v_{j_1}), (v_{i_2}, v_{j_2}), \ldots, (v_{i_l}, v_{j_l})\}$. Removing a vertex v_k from V forms a subgraph $G_1 = (V_1, E_1)$ with $V_1 = \{v_1, v_2, \ldots, v_{k-1}\}$ and $E_1 = \{(v_i, v_j) \in E \mid v_k \notin \{v_i, v_j\}\}$. We call G_1 the subgraph of G induced by the vertex set $(V - \{v_k\})$.

Let $G_1 = (V_1, E_1)$ and $G_2 = (V_2, E_2)$ be two subgraphs of G. G_1, G_2 are considered disjoint if no vertices in G_1 are connected to vertices in G_2. Clearly, the condition $V_1 \cap V_2 = \emptyset$ is necessary but insufficient.

2.1   Graph Connectivity

The goal of partitioning a graph is to make the graph disconnected by removing some of its vertices or edges. The number of vertices or edges that needs to be removed to disconnect a graph are its vertex- or edge-connectivities respectively.

Definition 2.1. *The* vertex connectivity $\kappa(G)$ *of a graph G is the minimum number of vertices that must be removed to disconnect G.*

Definition 2.2. *The* edge connectivity $\lambda(G)$ *of G is the minimum number of edges that must be removed to disconnect G.*

Clearly, a disjoint graph has vertex connectivity zero. On the other extreme, a complete graph K_n, where all n vertices are connected to each other, has vertex connectivity $(n - 1)$. The removal of all but one vertex from K_n results in a graph consisting of a single vertex, which is considered to be disconnected.

2.2   Graph Partitioning

The process of removing vertices or edges to disconnect a graph is called vertex partitioning or edge partitioning respectively. All non-empty graphs admit trivial vertex and edge partitions, where all connections to a single vertex are removed. This is obviously not useful for most applications. In this paper, we only consider balanced partitions with minimum-weight separators, in which a graph is separated into subgraphs of roughly equal sizes by removing as few vertices or edges as possible. More specifically, our primary focus is on balanced vertex partitions.

Definition 2.3. *Let $G = (V, E)$ be a graph. A vertex partition (V_1, C, V_2) of G is a partition of V into mutually exclusive and collectively exhaustive sets of vertices V_1, C, V_2, where V_1, V_2 are non-empty, and where no edges exist between vertices in V_1 and vertices in V_2. The removal of C causes the subgraphs induced by V_1 and V_2 to be disjoint, hence C is called the vertex separator.*

For a balanced vertex partition, we require V_1 and V_2 to be of similar size. For a minimum-weight separator, we also require that C be small. This is to ensure that the vertex partition obtained is useful for applications.

Definition 2.4. *Let $G = (V, E)$ be a graph, and (V_1, C, V_2) be a vertex partition of G with vertex separator C. If $\max(|V_1|, |V_2|) \leq \alpha|V|$, then G is said to have an α-vertex separator.*

The problem of finding α-vertex separators is known to be NP-hard [31, 43].

Definition 2.5. *Let $G = (V, E)$ be a graph. If (V_1, C, V_2) is a vertex partition of G, then define*

$$\beta = \frac{\max(|V_1|, |V_2|)}{|V_1| + |V_2|} = \frac{\max(|V_1|, |V_2|)}{|V| - |C|} = \frac{\alpha|V|}{|V| - |C|}$$

to be the balance of the vertex partition. Note further if $|C| \ll |V|$ then $\alpha \approx \beta$.

Suppose the balance of a vertex partition of G into (V_1, C, V_2) is β, then the partition also satisfies $\max(|V_1|, |V_2|) = \beta(|V_1| + |V_2|) \leq \beta|V|$, and hence the G has a β-vertex separator. Therefore, theorems that apply to α-vertex separators would also apply to vertex partitions with balance β. See [43] for more details of α-vertex separators. Several theorems governing the existence of α-vertex separators have been shown in [3, 27, 38, 41].

Figure 1 presents examples of balanced and unbalanced partitions, and their respective β values. The vertex separators C are circled, with the partitioned vertices V_1, V_2 outside. The removal of the vertices in the separators disconnects the graphs. For a balanced partition, β should be close to $1/2$.

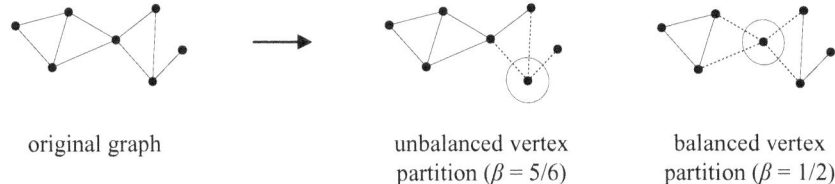

| original graph | unbalanced vertex partition ($\beta = 5/6$) | balanced vertex partition ($\beta = 1/2$) |

Fig. 1. Balanced and Unbalanced Vertex Partitions

2.3 Partitioning Algorithms and Software

While balanced partitioning is an NP-hard problem, a variety of heuristic algorithms have been found to be very efficient in finding near-optimal partitions.

One efficient scheme for balanced graph partitioning is called multilevel partitioning. Suppose a graph G_0 is to be partitioned. Firstly, G_0 "coarsened" progressively into simpler graphs $G_1, G_2, \ldots, G_r$ by contracting adjacent vertices. The process of choosing vertices for contraction is called matching. After reaching a graph G_r with the desired level of simplicity, a partitioning is performed. The result is then progressively refined back through the chain of graphs $G_{r-1}, G_{r-2}, \ldots, G_0$. At each refining step, a contracted vertices are expanded and partitioned. The output is then a partition of G_0. Details of multilevel partitioning can be found in [30, 32]. Examples of partitioning and refinement algorithms include the ones by Kerighan-Lin [34] and Fiduccia-Mattheyses [25].

Balanced edge partitioning is widely used in scientific and engineering applications, such as electric circuit design [49], parallel matrix computations [37], and finite element analysis [42]. Software packages are readily available for computing balanced edge partitions using a variety of algorithms [8, 11, 26, 29, 44, 45, 51]. On the other hand, balanced vertex partitioning has fewer applications, one of which being variable reordering in linear systems [19]. We are not aware of publicly available software that could be used for directly computing balanced vertex partitions with minimum-weight vertex separators.

This is also true for multilevel vertex-partitioning algorithms. Therefore, we have chosen to compute vertex partitionings through the use of the multilevel edge-partitioning software Metis [33] for our study. The Matlab interface Meshpart [28] to Metis is used to access the algorithms. It also contains a routine to convert an edge partition found by Metis to a vertex partition. We have also implemented an alternative greedy algorithm for this task. Both are used for the experiments in Section 4 and for the algebraic cryptanalysis of Trivium in Section 5.2.

Unless otherwise stated, from here on we will only consider the problem of balanced vertex partitioning with minimum-weight vertex separators (sometimes simply referred to as vertex partitioning or partitioning) and its applications to solving systems of multivariate polynomial equations.

3   Partitioning Polynomial Systems

In this section, our method for partitioning systems of multivariate polynomal equations by finding balanced vertex partitions of their variable-sharing graphs is described.

Definition 3.1. *Let F be the polynomial system*

$$f_1(x_1, x_2, \ldots, x_n) = 0$$
$$f_2(x_1, x_2, \ldots, x_n) = 0$$
$$\vdots$$
$$f_m(x_1, x_2, \ldots, x_n) = 0$$

of m polynomial equations in the variables $x_1, x_2, \ldots, x_n$. The variable-sharing graph $G = (V, E)$ of F is obtained by creating a vertex $v_i \in V$ for each variable x_i, and creating an edge $(v_i, v_j) \in E$ if two variables x_i, x_j appear together (with non-zero coefficient) in any polynomial f_k.

Example 3.1. Suppose we have the following quadratic system of equations over GF(2), where the variables $x_1, x_2, \ldots, x_5$ are known to take values in GF(2).

$$x_1 x_3 + x_1 + x_5 = 1$$
$$x_2 x_4 + x_4 x_5 = 0$$
$$x_1 x_5 + x_3 x_5 = 1 \tag{1}$$
$$x_2 x_5 + x_2 + x_4 = 0$$
$$x_2 + x_4 x_5 = 1$$

The corresponding variable-sharing graph G and a balanced vertex partition is shown in Figure **??**.

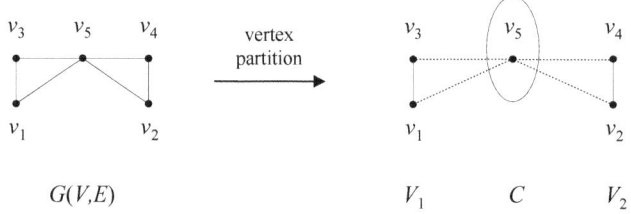

$$G(V,E) \qquad\qquad V_1 \qquad C \qquad V_2$$

Fig. 2. Variable-sharing graph of the quadratic system (1) and a vertex partition

The quadratic system can then be partitioned into two systems of equations with the common variable x_5 as follows.

$$
\begin{array}{ll}
x_1 x_3 + x_1 + x_5 = 1 & x_2 x_4 + x_4 x_5 = 0 \\
x_1 x_5 + x_3 x_5 = 1 & x_2 x_5 + x_2 + x_4 = 0 \\
 & x_2 + x_4 x_5 = 1
\end{array}
\tag{2}
$$

Since $x_5 \in \mathrm{GF}(2)$, we can substitute all possible values of x_5 into (1) and compute solutions to the reduced systems to give

$$x_5 = 0 \Rightarrow \text{no solution}$$
$$x_5 = 1 \Rightarrow (x_1, x_3) = (0, 1), (x_2, x_4) = (1, 0)$$
$$\Rightarrow \boldsymbol{x} = (0, 1, 1, 0, 1)$$

The solution obtained is the same as if we had directly computed the solution to the full system of equations. However, the systems have been reduced to having less than half the number of variables compared to the the original, at the cost of applying guesses to one variable.

This method of guessing and solving will be used for the algebraic cryptanalysis of the Trivium stream cipher in Section 5. For simplicity, from here on we might use the terms for variables and vertices interchangeably to denote the variables in the polynomial systems of equations and their corresponding vertices in the variable-sharing graphs, provided there are no ambiguities.

4   Graph Partitioning Experiments

To evaluate the practicality of partitioning large systems of equations, experiments have been performed on random graphs of different sizes resembling typical variable-sharing graphs. These experiments were run on a Pentium M 1.4 GHz CPU with 1 GB of RAM using the Meshpart [28] Matlab interface to the Metis [33] partitioning software.

Definition 4.1. *Let $G = (V, E)$ be a graph. The* degree $\deg(v)$ *of a vertex $v \in V$ is the number of edges $e \in E$ incident upon (connecting to) v.*

Definition 4.2. *The* density $\rho(G)$ *of a graph $G = (V, E)$ is the ratio of the number of edges $|E|$ in G to the maximum possible number $\frac{1}{2}|V|(|V| - 1)$ of edges in G.*

In each experiment, random graphs $G = (V, E)$ are generated, each with prescribed number of vertices $|V|$, number of edges $|E|$, and average degree d of its vertices. Their densities ρ are also computed. For each graph, a vertex partition is performed to give (V_1, C, V_2), where C is the vertex separator. The balance measure β is then computed, and the time required is also noted. Some experimental results are shown in Table 1.

It can be observed from Table 1 that the graph of β is likely to be correlated with the average degree d of the graphs. Small vertex separators can be obtained when the number of edges is a small factor of the number of vertices. At $d = 16$, the value of β is near its upper bound of 1, which means that those partitions are unlikely to be useful. Since the maximum number of edges for a graph of size n is $O(n^2)$, the edge density must be smaller with a larger graph for practical partitions. This is a reasonable assumption for polynomial systems, since certain sparse systems have only a small number of variables in each equation, regardless

Table 1. Vertex Partitioning Experiments

$|V|$	$|E|$	ρ	d	$|C|$	$|V_1|$	$|V_2|$	β	Time
64	64	0.0308	2	5	31	28	0.5254	61.26 ms
64	128	0.0615	4	15	30	19	0.6122	63.06 ms
64	256	0.1231	8	26	28	10	0.7368	80.95 ms
64	512	0.2462	16	32	3	29	0.9063	67.36 ms
128	128	0.0155	2	7	64	57	0.5289	64.80 ms
128	256	0.0310	4	28	60	40	0.6000	66.73 ms
128	512	0.0620	8	55	45	28	0.6164	63.27 ms
128	1024	0.1240	16	62	63	3	0.9545	83.51 ms
1024	1024	0.0020	2	51	508	465	0.5221	74.58 ms
1024	2048	0.0039	4	222	482	320	0.6010	90.05 ms
1024	4096	0.0078	8	418	355	251	0.5858	113.66 ms
1024	8192	0.0156	16	509	511	4	0.9922	168.55 ms
4096	4096	0.0005	2	183	2039	1874	0.5211	122.48 ms
4096	8192	0.0010	4	877	1903	1316	0.5912	175.20 ms
4096	16384	0.0020	8	1697	1539	860	0.6415	289.24 ms
4096	32768	0.0039	16	2037	2047	12	0.9942	548.75 ms

of the total number of variables in the system. This fact is true for the case of Trivium in Section 5. One could say that the weight or length (i.e. the number of terms) of the polynomials should be short for useful cuts to be guaranteed.

It is also noted that the time required to compute vertex partitions are quite short for the practical graph sizes considered. Therefore, we can safely assume that the time complexity of the partitioning algorithm is negligible compared to that required to solve the partitioned equation systems, which in the worst case is exponential in the number of variables and the maximum degree.

5   Applications to Algebraic Cryptanalysis

A bit-based stream cipher can be thought of as an internal state $s \in \mathrm{GF}(2)^n$ and two maps, $g : \mathrm{GF}(2)^n \to \mathrm{GF}(2)^n$ and $f : \mathrm{GF}(2)^n \to \mathrm{GF}(2)$. At each clock tick, $s_{t+1} = g(s_t)$, and so g can be called the state-update function. Also at each clock tick $f(s_t) = z_t$ is outputted, and this is called the keystream. Given plaintext bits $p_1, p_2, \ldots$, the stream cipher encrypts them using the keystream into ciphertext bits $c_1, c_2, \ldots$ via $c_t = p_t + z_t$, with the addition being over $\mathrm{GF}(2)$.

At the start of an encryption, a key-initialisation phase would take place, whereby a secret key k and a known initialisation vector IV are used to set s to its secret initial state s_0. The cipher then begins its keystream generation phase, and outputs a series of keystream bits $z_1, z_2, \ldots$, as explained above.

By the Universal Mapping Theorem [6, Th. 72], since f and g are maps from finite sets to finite sets, we know that they can be written as polynomial systems of equations over any field, but $\mathrm{GF}(2)$ is most useful to us. Since stream ciphers are traditionally designed for implementation in digital circuits, where there are economic motivations to keep the gate-count low, f and g can often be represented by very simple polynomial functions. Then, if both p_t and c_t are known for enough timesteps, one can write a system of equations based on $z_t = p_t + c_t$ using f and g. This forms the foundation of algebraic cryptanalysis.

To perform an algebraic cryptanalysis of the stream cipher, the cipher is first described as a system of equations. Its variables usually correspond to the bits in the key k or the initial state s_0. If the variables are from k, solving the system is called "key recovery", and the cipher is immediately broken. If the variables are from s_0, solving the system is called "state recovery", and the key could be derived from the solution, whose difficulty depends on the specific cipher design.

Every attack on every cipher has its nuances, and so above description is necessarily vague. For an overview of algebraic cryptanalysis, see [6]. For techniques of algebraic cryptanalysis on specific types of ciphers, see [17, 16, 15, 2, 54]. Some uses of graph theory for algebraic attacks can also be found in [48, 52]. In this section, two applications of our equation partitioning to algebraic cryptanalysis are presented. Firstly, we show a malicious use of the stream cipher QUAD [9]. Then, we describe and perform an algebraic cryptanalysis to the stream cipher Trivium [20] and its variants Bivium-A and Bivium-B [48]. We discuss only the equations arising from the cipher, and refer the reader to the respective references of these ciphers for their design and implementation details.

5.1   QUAD

The stream-cipher family QUAD is given in [9]. The security of QUAD is based on the Multivariate Quadratic (MQ) problem. The heart of the cipher is a random system of kn quadratic equations in n variables over a finite field $GF(q)$. Usually, we have $q = 2$, but implementations with $q = 2^s$ have also been discussed [55]. This system of equations is not secret, but publicly known, and there are criteria for these equations, such as those relating to rank, which we omit here. In a different context, QUAD has been analyzed in [55, 4], and [6, Ch. 5.2].

Equations of QUAD. The authors of QUAD recommend $k = 2$ and $n \geq 160$, so it is assumed that we have a randomly generated system of $2n = 320$ equations in $n = 160$ unknowns. The system is to be drawn uniformly from all those possible, which is to say that the coefficients can be thought of as generated by fair coins.

Each quadratic equation is a map $GF(2)^n \rightarrow GF(2)$, so the first set of n equations form a map $GF(2)^n \rightarrow GF(2)^n$ called f_1, and the second set of n equations also form a map of the same dimensions called f_2. The internal state is a vector s of 160 bits. The first 160 equations are evaluated at s, and the resulting vector $f_1(s_t) = s_{t+1}$ becomes the new state. The second 160 equations are evaluated to become the output of that timestep $z_t = f_2(s_t)$. The vector z_t is added to the next n bits of the plaintext p_t over $GF(2)$, and is transmitted as the ciphertext $c_t = p_t + z_t$. (Each bit is added independently, without carries.) There is also an elaborate setup stage which maps the secret key and an initialization vector to the initial state s_0.

Finding a pre-image under the maps f_1, f_2 i.e. finding s_i given s_{i+1} and z_i, is equivalent to solving a quadratic system of $2n$ equations in n unknowns, and is NP-hard [5, Ch. 3.9]. This is further complicated by the fact that the adversary would not have s_{i+1}, but rather only $z_i + p_i$.

Given a known-plaintext scenario, where the attacker knows both the plaintext $p_1, p_2 \ldots, p_n$ and ciphertext $c_1, c_2, \ldots, c_n$, one can write the following system of equations.

$$
\begin{aligned}
c_1 + p_1 &= z_1 = f_2(s_1) \\
c_2 + p_2 &= z_2 = f_2(s_2) = f_2(f_1(s_1)) \\
&\vdots = \vdots \\
c_t + p_t &= z_t = f_2(s_t) = f_2(\underbrace{f_1(f_1(f_1(\cdots f_1(s_1)\cdots))))}_{t-1 \text{ times}}
\end{aligned}
$$

The interesting fact here is that $f_2(f_1(f_1(\cdots f_1(s_1)\cdots))))$ and higher iterates might be quite dense even if f_1 is sparse. The authors of QUAD have excellent security arguments when the polynomial system is generated by fair coins. In this case, the variable-sharing graph of the cipher will have density close to that of a complete graph, and our graph partition method will have little use. However, it will have on average 6440.5 monomials per equation or roughly 2 million in the system, which would require a large gate count or would be slow in software. Thus, in their conference presentation, the authors of QUAD mention that a slightly sparse f might still be secure against algebraic attacks, because of the repeated iterations and the general difficulty of the MQ problem. Nevertheless, in this case, if a sparse system can be chosen such that it contains a small balanced vertex separator, then the cipher can be made insecure by a malicious attacker as follows.

Poisoned Equations and QUAD. One could imagine the following scenario, which is inspired by Jacques Patarin's system "Oil and Vinegar" [36]. A malicious manufacturer does not generate the system at random, but rather creates a system that is sparse and has vertex connectivity of 20, for some vertex partition with $\beta \approx 0.6$. Our experiments in Section 4 show that this is a feasible partition. The malicious manufacturer would claim that the system is sparse for efficiency reasons and it might have a considerably faster encryption throughput than a QUAD system with quadratic equations generated by fair coins.

Some separator of 20 vertices divides the variable sharing graph into roughly 56 and 84 vertices. This means that an attacker would need only to know the plaintext and ciphertext of one 160-bit sequence, and solve the equation

$$
f_2(\underbrace{f_1(f_1(\cdots f_1(f_1(s_1))\cdots)))}_{i-1 \text{ times}} = p_t + c_t \tag{3}
$$

For any guess of the key, this would be solving 56 equations in 56 unknowns and 84 equations in 84 unknowns. Such a problem is certainly trivial for a SAT-solver, as shown in [7], [5, Ch. 3] and [6, Ch. 7]. Only 2^{20} such systems would need to be solved, and with a massive parallel network, such as BOINC [1], this would be feasible [12], although experiments would be required for confirmation.

Remedy to Poisoned Systems for QUAD. While finding a balanced vertex partition of a graph G is NP-hard, calculating the vertex connectivity $\kappa(G)$ is easier. If $\kappa(G) > 80$, for example, then there is no vertex partition, balanced or otherwise, with fewer than 80 vertices in the vertex separator. Then, by calculating $\kappa(G)$, a manufacturer of QUAD could prove that they are not poisoning the quadratic system. There are also techniques to generate functions with verifiable randomness [14], which could be used to construct polynomial systems of equations for QUAD, such that they are provably not poisoned.

5.2   Trivium

Trivium [20] is a bit-based stream cipher in the eSTREAM project portfolio for hardware implementation with an 80-bit key, 80-bit initialization vector, and a 288-bit internal state. As at the end of the eSTREAM project, after three phases of expert and community reviews, no feasible attacks faster than an exhaustive key search on the full implementation of Trivium were found. However, Trivium without key initialisation, as well as its reduced versions Bivium-A and Bivium-B with a 177-bit internal state, admit attacks faster than exhaustive key search. Cryptanalytic results on Trivium and Bivium have been presented in [10, 21, 22, 23, 39, 40, 46, 50].

Equation Construction. The equations governing keystream generation from the initial state s_0 can be found in [20] for Trivium and [48] for Bivium. In the algebraic cryptanalysis presented in this paper, we do not consider the initialisation phase from the key k and initialisation vector IV, and hence we are performing state recovery of the cipher.

Trivium can be described as a system of 288 multivariate polynomial equations in 288 variables, but we found that this is too dense for partitioning to be useful. Instead, we use the system of quadratic equations presented in [48], which contains more variables, but is very sparse. The quadratic system of Trivium consists of 954 sparse quadratic equations in 954 variables, and observed keystream from 288 clocks. Similarly, the polynomial system of Bivium-A and Bivium-B consists of 399 sparse quadratic equations in 399 variables, and observed keystream from 177 clocks. We attempt to solve these equations via partitioning.

Equation Partitioning. The sparse quadratic equations for Trivium and Bivium are constructed as per [48], and their variable-sharing graphs are then computed. Figure 3 shows the adjacency matrix for the variable-sharing graph of Trivium. The sparsity of this matrix appears promising for a reasonable partition. Graphs for Bivium are of similar sparsity.

Partitioning these variable-sharing graphs $G = (V, E)$ into vertex sets V_1, V_2 and vertex separator C with [33] as in Section 4 gives the results shown in Table 2. From these results, it seems that both of the Bivium ciphers admit very balanced partitions, whereas Trivium did not. Nevertheless, using our alternative

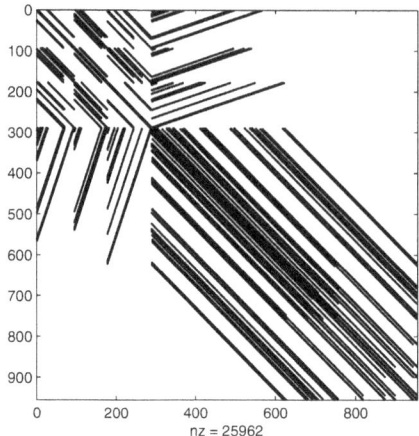

Fig. 3. Graph Adjacency Matrix of Trivium Equations

Table 2. Partitioning Equations of Bivium-A, Bivium-B and Trivium

Cipher	State Size	Number of Variables	$|C|$	$|V_1|$	$|V_2|$	β
Bivium-A	177	399	96	156	147	0.5149
Bivium-B	177	399	14	128	127	0.5020
Trivium	288	954	288	476	190	0.7147

greedy algorithm for converting edge partitions from the Metis software to vertex partitions, we were able to find a balanced partition for Trivium with $|C| = 295$ and $\beta \approx 0.5$, at the cost of having a larger vertex separator.

The sizes of the vertex separators C are the number of variables that must be eliminated to separate the systems into two. In algebraic attacks, this corresponds to the number of variables whose values are to be discovered or guessed at a complexity of $2^{|C|}$. The process of guessing certain bits in order to find a solution is called partial key guessing. If the guessed bits are correct, then solving the remaining system would lead to the solution.

For Trivium, the separator size is at least the internal state size, so a partition on the equation system is not useful, as guessing the variables in the separator is as costly as an exhaustive search on the initial state. For Bivium and Bivium-A, the separator sizes are less than the internal state size, but larger than the key size of 80-bits. This means that the time complexity of partial key guessing on all bits of the separators would be higher than that of a brute-force search on the key, but lower than a brute-force search on the initial state.

Partial Key Guessing and Perforated Systems. However, we can attempt to guess fewer bits than the size of the separator C. The remaining system would not be separated, but it can still be solved. Since it is close to being partitioned,

Table 3. Partial Key Guessing on Trivium and Bivium

Cipher	All Guesses in $|C|$	n	m	q	Time	Memory
Bivium-A	No	24	422	193	26 s	42 MB
Bivium-A	No	20	421	200	195 s	234 MB
Bivium-A	No	18	417	203	2558 s	843 MB
Bivium-A	Yes	18	417	195	80 s	127 MB
Bivium-A	Yes	16	415	201	1101 s	751 MB
Bivium-A	Yes	14	413	202	2023 s	1200 MB
Bivium-B	No	80	479	143	392 s	1044 MB
Bivium-B	No	78	477	146	740 s	1044 MB
Bivium-B	No	76	475	141	1213 s	1044 MB
Bivium-B	Yes	70	469	132	12 s	62 MB
Bivium-B	Yes	66	465	136	623 s	546 MB
Bivium-B	Yes	62	461	141	3066 s	1569 MB
Trivium	No	280	1333	329	13 s	80 MB
Trivium	No	272	1224	343	155 s	554 MB
Trivium	No	264	1217	344	594 s	1569 MB
Trivium	Yes	178	1130	499	18 s	596 MB
Trivium	Yes	176	1127	499	4511 s	1875 MB
Trivium	Yes	174	1126	501	10543 s	3150 MB

we will call such systems "perforated". We have discovered by experiments that partial key guesses on subsets of bits in C provide significant advantages over those on random bits, in that the reduced polynomials systems are much easier to solve. The experiments were performed using Magma 2.12 [13] with its implementation of the Gröbner basis algorithm F_4 [24] for solving the reduced polynomial systems. The results are shown in Table 3, where n is the number of bits guessed, m is the number of equations resulting from the guess, with q of them being quadratic. Correct guesses are always used to reduce the polynomial systems, which means that the time and memory use presented are for solving the entire system arriving at a unique solution. All values are averaged over 10 individual runs.

The experimental results show that the time required for partial key guessing on n bits is reduced significantly if those bits are taken from the separator. This means that, by finding partitions to the system of equations, we have reduced the resistance of these ciphers to algebraic cryptanalysis, since a feasible partial-key-guess attack can potentially be launched on fewer bits with this extra information. For example, with Bivium-B, the time to compute a solution by guessing 78 bits randomly is roughly equivalent to that by guessing 66 bits in the separator. Hence, the time complexity for an attack on Bivium-B is reduced from $2^{78}T_B$ to $2^{66}T_B$ with the use of the separator, where T_B denotes the time complexity required to compute a solution to a reduced system of Bivium-B. For Trivium, the improvement is even more pronounced. The time complexity could be reduced from $2^{280}T_C$ to $2^{178}T_C$, which T_C denotes the time complexity required to compute a solution to a reduced system of Trivium.

In an actual algebraic attack, many of the guesses will result in inconsistent equations with no solutions, which can be checked and discarded easily. This means that the time required to process a guess is at most T_B or T_C. A full attack attempt was launched on Bivium-A with a partial key guess on 20 bits in its separator. About $200,000$ guesses of out the possible 2^{20} were made, with each guess taking on average about 0.15 seconds to process. This is much faster than the 45 seconds required from the experimental results to process a correct guess to completion.

Although the time complexity for the algebraic attack on Trivium is significantly reduced through our partitioning method, it is still much higher than that of exhaustive key search, which is 2^{80}. On the other hand, this method applied on Bivium-A and Bivium-B may be faster than exhaustive key search, depending on the time complexity of solving the reduced equation systems, which varies with the equation solving technique employed. Experiments with different techniques would be needed for a sound conclusion to be drawn.

A Bit-Leakage Attack. There is another scenario whereby the graph partitioning would provide an advantage to algebraic cryptanalysis. Suppose by some means, accidental or deliberate, some bits of the internal state of a cipher could be leaked to an attacker. This would occur in a side-channel attack setting. If the attacker could control which bits are leaked, then the best choices would be those variables in the separator. If all bits in the separator are leaked, then the equation system is immediately split into two, and the time complexity of solving for the remaining bits is significantly reduced. If only some bits in the separator are leaked, we have shown in the earlier experiments that this leads to faster attacks than if an equal amount of random bits are leaked. This also means that if bits can be leaked from the separator, fewer of them would be needed before the system of equations can be solved in a reasonable time, compared to the case where bits are leaked randomly.

6   Conclusions

In this paper, the concept of a variable-sharing graph of a system of polynomial equations was defined. It has been shown that this concept can be used to break systems of polynomial equations into pieces, which can be solved separately, provided that the graph has a vertex partition satisfying various requirements: namely that the vertex separator should be small, and the partition should be balanced. We also presented methods for finding the partition, and methods for using the partition to solve polynomial systems of equations over GF(2).It has been shown that balanced vertex partitions are feasible to obtain for some sparse systems of polynomial equations. Experiments on random graphs of reasonable size and sparsity, resembling variable-sharing graphs of equation systems, have been performed.

The practicality of this partitioning technique has been demonstrated in the algebraic cryptanalysis of the stream cipher Trivium and its reduced versions, where we have found balanced partitions of useful sizes. These partitions provide information for launching more effective algebraic attacks with partial key guessing, and improves the attack time by at least a few orders of magnitude. Furthermore, we show how the partitioning technique can be used to poison the provably secure stream cipher QUAD, so that a malicious manufacturer can recover the keystream much more efficiently.

As discussed earlier, this paper has provided a novel technique for preprocessing large sparse systems of equations, which could be used together with popular techniques such as Gröbner basis methods to significantly reduce the time for computing solutions to these systems. It has also been shown that this technique provides improvements to algebraic cryptanalysis, and further research into this area is warranted, since there may be security implications for other ciphers that are susceptible to this technique.

References

[1] BOINC: Berkeley Open Infrastructure for Network Computing, http://boinc.berkeley.edu/
[2] Al-Hinai, S., Batten, L., Colbert, B., Wong, K.K.H.: Algebraic attacks on clock-controlled stream ciphers. In: Batten, L.M., Safavi-Naini, R. (eds.) ACISP 2006. LNCS, vol. 4058, pp. 1–16. Springer, Heidelberg (2006)
[3] Alon, N., Semour, P., Thomas, R.: A separator theorem for graphs with an excluded minor and its applications. Journal of the American Mathematical Society 3(4), 801–808 (1990)
[4] Arditti, D., Berbain, C., Billet, O., Gilbert, H., Patarin, J.: QUAD: Overview and recent developments. In: Biham, E., Handschuh, H., Lucks, S., Rijmen, V. (eds.) Symmetric Cryptography. Dagstuhl Seminar Proceedings, vol. 07021. Internationales Begegnungs- und Forschungszentrum fuer Informatik (IBFI), Schloss Dagstuhl, Germany (2007)
[5] Bard, G.V.: Algorithms for solving linear and polynomial systems of equations over finite fields with applications to cryptanalysis. Ph.D. thesis, Department of Applied Mathematics and Scientific Computation, University of Maryland, College Park (August 2007), http://www.math.umd.edu/~bardg/bard_thesis.pdf
[6] Bard, G.V.: Algebraic Cryptanalysis. Springer, Heidelberg (2009)
[7] Bard, G.V., Courtois, N., Jefferson, C.: Efficient methods for conversion and solution of sparse systems of low-degree multivariate polynomials over GF(2) via SAT-Solvers. Cryptology ePrint Archive, Report 2007/024 (2007), http://eprint.iacr.org/2007/024.pdf
[8] Baños, R., Gil, C., Ortega, J., Montoya, F.G.: Multilevel heuristic algorithm for graph partitioning. In: Raidl, G.R., Cagnoni, S., Cardalda, J.J.R., Corne, D.W., Gottlieb, J., Guillot, A., Hart, E., Johnson, C.G., Marchiori, E., Meyer, J.-A., Middendorf, M. (eds.) EvoIASP 2003, EvoWorkshops 2003, EvoSTIM 2003, EvoROB/EvoRobot 2003, EvoCOP 2003, EvoBIO 2003, and EvoMUSART 2003. LNCS, vol. 2611, pp. 143–153. Springer, Heidelberg (2003)
[9] Berbain, C., Gilbert, H., Patarin, J.: QUAD: A practical stream cipher with provable security. In: Vaudenay, S. (ed.) EUROCRYPT 2006. LNCS, vol. 4004, pp. 109–128. Springer, Heidelberg (2006)

[10] Bernstein, D.: Response to slid pairs in Salsa20 and Trivium. Tech. rep., The University of Illinois, Chicago (2008), http://cr.yp.to/snuffle/reslid-20080925.pdf

[11] Berry, J., Dean, N., Goldberg, M., Shannon, G., Skiena, S.: Graph computation with LINK. Software: Practice and Experience 30, 1285–1302 (2000)

[12] Black, M., Bard, G.: SAT over BOINC: Satisfiability solving over a volunteer grid. Draft Article (2010) (Submitted for Publication), http://www.math.umd.edu/~bardg/publications.html

[13] Bosma, W., Cannon, J., Playoust, C.: The MAGMA algebra system. I. The user language. Journal of Symbolic Computation 24(3-4), 235–265 (1997)

[14] Chase, M., Lysyanskaya, A.: Simulatable $vrfs$ with applications to multi-theorem nizk. In: Menezes, A. (ed.) CRYPTO 2007. LNCS, vol. 4622, pp. 303–322. Springer, Heidelberg (2007)

[15] Cho, J.Y., Pieprzyk, J.: Algebraic attacks on SOBER-t32 and SOBER-t16 without stuttering. In: Roy, B., Meier, W. (eds.) FSE 2004. LNCS, vol. 3017, pp. 49–64. Springer, Heidelberg (2004)

[16] Courtois, N.: Algebraic attacks on combiners with memory and several outputs. In: Park, C.-s., Chee, S. (eds.) ICISC 2004. LNCS, vol. 3506, pp. 3–20. Springer, Heidelberg (2005)

[17] Courtois, N., Meier, W.: Algebraic attacks on stream cipher with linear feedback. In: Biham, E. (ed.) EUROCRYPT 2003. LNCS, vol. 2656. Springer, Heidelberg (2003)

[18] Courtois, N., Shamir, A., Patarin, J., Klimov, A.: Efficient algorithms for solving overdefined systems of multivariate polynomial equations. In: Preneel, B. (ed.) EUROCRYPT 2000. LNCS, vol. 1807, pp. 392–407. Springer, Heidelberg (2000)

[19] Davis, T.A.: Direct methods for sparse linear systems, Fundamentals of Algorithms, vol. 2. SIAM, Philadelphia (2006)

[20] De Cannière, C., Preneel, B.: Trivium specifications. Tech. rep., Katholieke Universiteit Leuven (2007), http://www.ecrypt.eu.org/stream/p3ciphers/trivium/trivium_p3.pdf

[21] Dinur, I., Shamir, A.: Cube attacks on tweakable black box polynomials. In: Joux, A. (ed.) EUROCRYPT 2009. LNCS, vol. 5479, pp. 278–299. Springer, Heidelberg (2010)

[22] Eén, N., Sörensson, N.: Minisat — a SAT solver with conflict-clause minimization. In: Bacchus, F., Walsh, T. (eds.) SAT 2005. LNCS, vol. 3569, pp. 61–75. Springer, Heidelberg (2005)

[23] Eibach, T., Pilz, E., Völkel, G.: Attacking Bivium using SAT solvers. In: Büning, H. K., Zhao, X. (eds.) SAT 2008. LNCS, vol. 4996, pp. 63–76. Springer, Heidelberg (2008)

[24] Faugère, J.C.: A new efficient algorithm for computer Gröbner bases (f_4). Journal of Pure and Applied Algebra 139, 61–88 (1999)

[25] Fiduccia, C., Mattheyses, R.: A linear time heuristic for improving network partitions. In: 19th ACM/IEEE Design Automation Conference, pp. 175–181 (1982)

[26] Fremuth-Paeger, C.: Goblin: A graph object library for network programming problems (2007), http://goblin2.sourceforge.net/

[27] Gilbert, J.R., Hutchinson, J.P., Tarjan, R.E.: A separation theorem for graphs of bounded genus. Journal of Algorithms 5, 391–407 (1984)

[28] Gilbert, J.R., Teng, S.H.: Meshpart: Matlab mesh partitioning and graph separator toolbox (2002), http://www.cerfacs.fr/algor/Softs/MESHPART

[29] Hendrickson, B., Leland, R.: The Chaco user's guide: Version 2.0. Tech. Rep. SAND94-2692, Sandia National Laboratories (1994)

[30] Hendrickson, B., Leland, R.: A multilevel algorithm for partitioning graphs. In: 1995 ACM/IEEE Supercomputing Conference. ACM, New York (1995)

[31] Johnson, D.S.: The NP-completeness column: An on-going guide. J. Algorithms 8, 438–448 (1987)

[32] Karypis, G., Kumar, V.: A fast and high quality multilevel scheme for partitioning irregular graphs. SIAM Journal on Scientific Computing 20(1), 359–392 (1999)

[33] Karypis, G., et al.: Metis — Serial graph partitioning and fill-reducing matrix ordering (1998), http://glaros.dtc.umn.edu/gkhome/views/metis/

[34] Kernighan, B., Lin, S.: An efficient heuristic procedure for partitioning graphics. Bell Systems Technical Journal 49, 291–307 (1970)

[35] Khovratovich, D., Biryukov, D., Nikolic, I.: Speeding up collision search for byte-oriented hash functions. In: Halevi, S. (ed.) Advances in Cryptology - CRYPTO 2009. LNCS, vol. 5677, pp. 164–181. Springer, Heidelberg (2009)

[36] Kipnis, A., Patarin, J., Goubin, L.: Unbalanced oil and vinegar signature schemes. In: Stern, J. (ed.) EUROCRYPT 1999. LNCS, vol. 1592, pp. 206–222. Springer, Heidelberg (1999)

[37] Kumar, V., Grama, A., Gupta, A., Karypis, G.: Introduction to Parallel Computing: Design and Analysis of Algorithms. Benjamin/Cummings Publishing Company, Redwood City (1994)

[38] Lipton, R.J., Tarjan, R.E.: A separator theorem for planar graphs. SIAM Journal on Applied Mathematics 36(2), 177–189 (1979)

[39] Maximov, A., Biryukov, A.: Two trivial attacks on Trivium. In: Adams, C.M., Miri, A., Wiener, M.J. (eds.) SAC 2007. LNCS, vol. 4876, pp. 36–55. Springer, Heidelberg (2007), http://eprint.iacr.org/2007/021

[40] McDonald, C., Charnes, C., Pieprzyk, J.: An algebraic analysis of Trivium ciphers based on the boolean satisfiability problem. In: Presented at the International Conference on Boolean Functions: Cryptography and Applications, BFCA2008 (2008), Cryptology ePrint Archive, Report 2007/129 (2007), http://eprint.iacr.org/2007/129

[41] Menger, K.: Zur allgemeinen Kurventheorie. Fundamenta Mathematicae 10, 96–115 (1927)

[42] Miller, G.L., Teng, S.H., Thurston, W., Vavasis., S.A.: Automatic mesh partitioning. In: George, A., Gilbert, J., Liu, J. (eds.) Graph Theory and Sparse Matrix Computation. The IMA Volumes in Mathematics and its Application, vol. 56, pp. 57–84. Springer, Heidelberg (1993)

[43] Müller, R., Wagner, D.: α-vertex separator is NP-hard even for 3-regular graphs. J. Computing 46, 343–353 (1991)

[44] Pellegrini, F., Roman, J.: SCOTCH: A software package for static mapping by dual recursive bipartitioning of process and architecture graphs. In: Liddell, H., Colbrook, A., Hertzberger, B., Sloot, P.M.A. (eds.) HPCN-Europe 1996. LNCS, vol. 1067, pp. 493–498. Springer, Heidelberg (1996)

[45] Preis, R., Diekmann, R.: The PARTY partitioning-library, user guide - version 1.1. Tech. Rep. tr-rsfb-96-024, University of Paderborn (1996)

[46] Priemuth-Schmid, D., Biryukov, A.: Slid pairs in Salsa20 and Trivium. In: Chowdhury, D.R., Rijmen, V., Das, A. (eds.) INDOCRYPT 2008. LNCS, vol. 5365, pp. 1–14. Springer, Heidelberg (2008)

[47] Raddum, H., Semaev, I.: New technique for solving sparse equation systems. Cryptology ePrint Archive, Report 2006/475 (2006), http://eprint.iacr.org/2006/475

[48] Raddum, H.: Cryptanalytic results on Trivium. Tech. Rep. 2006/039, The eS-TREAM Project (March 27, 2006),
http://www.ecrypt.eu.org/stream/papersdir/2006/039.ps
[49] Schweikert, D.G., Kernighan, B.W.: A proper model for the partitioning of electrical circuits. In: 9th workshop on Design automation, pp. 57–92. ACM, New York (1972)
[50] Vielhaber, M.: Breaking One. Fivium by AIDA an algebraic IV differential attack. Cryptology ePrint Archive, Report 2007/413 (2007),
http://eprint.iacr.org/2007/413
[51] Walshaw, C., Cross, M.: JOSTLE: Parallel Multilevel Graph-Partitioning Software - An Overview. Tech. rep., Civil-Comp Ltd. (2007)
[52] Wong, K.K.H.: Application of Finite Field Computation to Cryptology: Extension Field Arithmetic in Public Key Systems and Algebraic Attacks on Stream Ciphers. PhD Thesis, Information Security Institute, Queensland University of Technology (2008)
[53] Wong, K.K.H., Bard, G., Lewis, R.: Partitioning multivariate polynomial equations via vertex separators for algebraic cryptanalysis and mathematical applications. Draft Article (2008),
http://www.math.umd.edu/~bardg/publications.html
[54] Wong, K.K.H., Colbert, B., Batten, L., Al-Hinai, S.: Algebraic attacks on clock-controlled cascade ciphers. In: Barua, R., Lange, T. (eds.) INDOCRYPT 2006. LNCS, vol. 4329, pp. 32–47. Springer, Heidelberg (2006)
[55] Yang, B.Y., Chen, O.C.H., Bernstein, D.J., Chen, J.M.: Analysis of QUAD. In: Biryukov, A. (ed.) FSE 2007. LNCS, vol. 4593, pp. 290–308. Springer, Heidelberg (2007)

A   New Criterion for Symmetric Ciphers?

In this paper, we have shown that our graph partitioning method for solving equation systems describing ciphers works particularly well when these systems are sparse. This, in turn, means that the vertex connectivities of their variable-sharing graphs are low. Therefore, for maximum protection against an algebraic attack of this kind, a cipher should be designed such that its variable-sharing graph is close to being a complete graph, and hence does not admit any useful balanced vertex partition. The vertex connectivity could become the measure of this criterion, since it can be computed efficiently. This measure can be used to predict the usefulness of the graph partitioning method, which is primarily determined by the size of the vertex separator and the balance of the partition. Furthermore, the vertex connectivity can be computed efficiently. However, care must be taken to account for the effects of variable relabelling techniques, which would alter the variable-sharing graph and hence its vertex connectivity. Further research would be required to refine this possible new criterion for use in practice.

On Multidimensional Linear Cryptanalysis

Phuong Ha Nguyen, Lei Wei, Huaxiong Wang, and San Ling

Division of Mathematical Sciences,
School of Physical and Mathematical Sciences,
Nanyang Technological University, Singapore
{ng0007ha,wei0005,hxwang,lingsan}@ntu.edu.sg

Abstract. Matsui's Algorithms 1 and 2 with multiple approximations have been studied over 16 years. In CRYPTO'04, Biryukov *et al.* proposed a formal framework based on m statistically independent approximations. Started by Hermelin *et al.* in ACISP'08, a different approach was taken by studying m-dimensional combined approximations from m base approximations. Known as multidimensional linear cryptanalysis, the requirement for statistical independence is relaxed. In this paper we study the multidimensional Alg. 1 of Hermelin *et al.*. We derive the formula for N, the number of samples required for the attack and we improve the algorithm by reducing time complexity of the distillation phase from $2^m N$ to $2m2^m + mN$, and that of the analysis phase from 2^{2m} to $3m2^m$. We apply the results on 4- and 9-round Serpent and show that Hermelin *et al.* actually provided a formal model for the hypothesis of Biryukov *et al.* in practice, and this model is now much more practical with our improvements.

1 Introduction

Linear cryptanalysis [12] was formally introduced in 1993 by Matsui, who suggested 2 algorithms to exploit linear approximations of block ciphers. Consider a block cipher $E_k(\cdot)$ with a linear approximation: $g = uX \oplus vY \oplus cK$ and $Pr(g = 0) = 1/2 + \epsilon$, where u, v, and c are the selection patterns for the plaintext, ciphertext and the extended key, respectively, X is the plaintext and Y is ciphertext, K is extended key of secret key k, and ϵ is the bias of linear approximation.

Algorithm 1 in [12] is a known-plaintext attack and it requires a pool of sufficiently many random plaintext-ciphertext pairs. If successful, the parity cK can be recovered and N – the number of random samples as (X, Y) pairs needed – is proportional to c/ϵ^2, where c is a constant that depends on the success probability. In 1994 Matsui provided the first experimental cryptanalysis of DES [13], using two linear approximations derived from the best 14-round expression.

From then researchers started the quest for better attacks by using multiple linear approximations to improve the basic form of Alg. 1 and 2 in [12]. In the same year, Kaliski and Robshaw [11] combined m linear approximations $g_1, ..., g_m$, with biases of $\epsilon_1, \ldots, \epsilon_m$ respectively. The linear approximations are required to have the same selection pattern c for the extended key. N can achieve

R. Steinfeld and P. Hawkes (Eds.): ACISP 2010, LNCS 6168, pp. 37–52, 2010.

an m-fold reduction while keeping the same success rate, with N proportional to $1/\sum_{i=1}^{m}\epsilon_i^2$. Due to the restriction on the key mask, it recovers at most 1-bit parity – $c \cdot K$. Intuitively speaking, asking 1 parity bit from m approximations simultaneously would require much fewer samples than from a single approximation, under the same success rate.

In 2004 Biryukov et al. [2] show a statistical framework for using multiple approximations in both Alg 1 and Alg 2. In their generalization to Alg. 1, m statistically independent linear approximation $g_1, \ldots, g_m$ are used, with biases of $\epsilon_1, \ldots, \epsilon_m$ respectively. The attack is able to recover at most m parity bits if successful, with N proportional to $1/\sum_{i=1}^{m}\epsilon_i^2$. In practice, m' linearly dependent masks are used with $m' > m$. It is expected that the performance in this case is strictly better than that of m independent approximations, although no explicit estimation of N was given. They confirmed the reduction in data complexity by using $m' = 86$ approximations for 8-round DES, the 86 approximations only give 10 linearly independent key masks. In 2007, this reduction in N by using multiple approximations is further confirmed by experiments of Collard et al. [7], with $m' = 64$ for 4-round Serpent from 10 linearly independent text masks. Collard et al. also observed that the gain increases 8 times faster with 64 approximations than with 10 approximations. Both [2] and [7] did not analyze why such an advantage is present. Some intuitive guesses were given by Collard et al. [7], referring to linear hulls, error correcting codes, etc. They also noticed that in practice, m and m' cannot be too large due to computation limits.

Significant reduction in data complexity can be achieved by the methods in [2]. However, in practice, it is generally not easy to verify whether the m approximations are statistically independent. Instead, linear independence is used as a criterion for the m approximations, in both [2] and [7]. In the meantime, it is also helpful to doubt whether there is a better attack if the m approximations are linearly independent in text masks but statistically correlated, as shown by the experimental results in [2] and [7]. Notably, these experiments work with linearly dependent text masks.

Hermelin et al. [10] introduced a multidimensional framework, in which Matsui's Alg. 1 is generalized to m-dimensions to exploit correlations between the m approximations, to achieve a higher capacity. Experiments on 4-round Serpent have shown that this method reduces N compared with a similar attack in [7]. In this framework, the requirement for statistical independence of approximations is relieved. Instead, the approximations only need to have linearly independent text masks. This resembles the experiment scenarios in [2] and [7]. In fact, in [10] it is shown clearly that the statistical independence assumption of [2] does not hold.

In [10], the formula for N is derived as the amount of data needed to tackle the $|\mathcal{Z}|$-ary hypothesis testing problem, where $\mathcal{Z}$ is the set of key classes. It does not reflect how it depends on the approximations $g_1, \cdots, g_m$ and it is not efficiently computable. The reduction in N compared with [2] was observed empirically, rather than theoretically. Our first contribution is the provision with proof of a much simpler theoretical formula for the number of samples required, to complement this theoretical framework in [10]. This formula gives us insights on how much can be achieved with multidimensional linear cryptanalysis. We can now easily estimate N given $g_1, \ldots, g_m$, hence the attack complexity. The simplicity of the formula eases cryptanalysis greatly since to compute N with the original formula in [10] (for N_{key}) requires a lot of computation when m is large.

A major obstacle for the method in [10] to be useful is that the number of approximations m is much limited due to complexity bottlenecks in the distillation phase ($2^m N$) and the analysis phase (2^{2m}) of the online stage. This limitation hinders the application of multidimensional Alg. 1 in practice when more approximations have to be used. Our second contribution is that we present a method (Method-A) for the distillation phase to speed up the computation for distributions to $2m2^m + mN$, using ideas from [6], and we develop Method-B to the analysis phase which reduces the complexity to $3m2^m$. We arrive at an improved algorithm for the multidimensional generalization of Matsui's Alg. 1 in [10], better than previous generalizations to a number of degrees. Most importantly, the algorithm is practical, with strong support from theoretical work in [10]. We show applications to 4- and 9-round Serpent as examples.

The paper is organized as follows. Section 2 contains notations and some basic notions. Section 3 is about the statistical model and algorithm in [10]. We show briefly how the distribution of an m-dimensional vectorial Boolean function g can be constructed from m given approximations $g_1, \ldots, g_m$ as Boolean functions. Section 4 contains our contributions. After the proof for N, we analyze the algorithm of [10] step-wise, to introduce our improvements, followed by the descriptions for each improvement. Section 5 describes the improved algorithm in detail. Section 6 gives the application results on Serpent and comparisons with previous cryptanalysis results [1] and [4]. In Section 7, we conclude and discuss the implication of our work.

2   Notations and Background

We follow the notations used in [10]. Denote the space of m-dimensional binary vectors by V_m or $V_m := GF(2)^m$. The inner product of 2 vectors $a = (a_1, \ldots, a_m), b = (b_1, \ldots, b_m), a, b \in V_m$ is $ab = \bigoplus_{i=1}^m a_i b_i$.

The function $f : V_m \to V_1$ is called a Boolean function and $f : V_n \to V_m, f = (f_1, \ldots, f_m)$ is called a vectorial Boolean function, where each f_i is a Boolean function for all $i = 1, \cdots, m$.

Let X be a random variable (r.v.) in V_m. Let $p_\eta = Pr(X = \eta)$, with $\eta \in V_m$. Then $p = (p_0, p_1, \ldots, p_{2^m-1})$ is the probability distribution (p.d) of r.v. X. If we associate with a vectorial Boolean function $f : V_n \to V_m$ an r.v. $Y := f(X)$, where X is uniformly distributed in V_m, then the p.d. of Y is $p(f) := (p_0(f), \ldots, p_{2^m-1}(f))$ where $p_\eta(f) = Pr(f(X) = \eta)$, for all $\eta \in V_m$. Two Boolean functions f and g are called statistically independent if their associated r.v.'s $f(X)$ and $g(Y)$ are statistically independent, with X, Y uniform in V_n.

The correlation between a binary r.v. X and 0 is $\rho = Pr(X = 0) - Pr(X = 1) = 2\epsilon$, where ϵ is the bias of the r.v. X. Let $g : V_m \to V_1$ be a Boolean function. Its correlation with 0 is defined as

$$\rho = 2^{-m}(\#\{\eta \in V_m | g(\eta) = 0\}) - \#\{\eta \in V_m | g(\eta) = 1\}) = 2Pr(g(X) = 0) - 1,$$

where X is uniformly distributed in V_m.

Definition 1. *Let $p = (p_0, \ldots, p_M)$ and $q = (q_0, \ldots, q_M)$ be two p.d.'s. Then their (mutual) capacity is*

$$C(p\|q) = \sum_{\eta=0}^{M} \frac{(p_\eta - q_\eta)^2}{q_\eta}. \tag{1}$$

Definition 2. *The relative entropy or the Kullback-Leibler (KL) distance between two distributions $p = (p_0, \ldots, p_M)$ and $q = (q_0, \ldots, q_M)$ is defined as*

$$D(q\|p) = \sum_{\eta=0}^{M} q_\eta \log \frac{q_\eta}{p_\eta}.$$

In [6], Collard *et al.* presented the following theorems concerning circulant matrices.

Theorem 1. *A circulant $\mathbf{S}$ of level k and type $(m, n, o, \ldots, r)$ is diagonalizable by the unitary matrix $\mathbf{F} = \mathbf{F}_m \otimes \mathbf{F}_n \otimes \mathbf{F}_o \otimes \cdots \otimes \mathbf{F}_r$*

$$\mathbf{S} = \mathbf{F}^* diag(\lambda)\mathbf{F},$$

where λ is the vector of eigenvalues of $\mathbf{S}$, the symbol $\otimes$ is the Kronecker product and $\mathbf{F}_n$ is the Fourier matrix of size $n \times n$ defined by:

$$\mathbf{F}_n(i, j) = \frac{1}{\sqrt{n}} \omega^{ij}, \ (0 \le i, j \le n - 1)$$

with

$$\omega = e^{\frac{2\pi\sqrt{-1}}{n}}.$$

Theorem 2. *The eigenvalues vector λ of a circulant matrix $\mathbf{S}$ of level k and type $(m, n, o, \ldots, r)$ can be computed with the following matrix-vector product:*

$$\lambda = \mathbf{F}\mathbf{S}(:, 1)\sqrt{mno \ldots r}$$

where $\mathbf{S}(:, 1)$ means we take the first column of $\mathbf{S}$.

We recall important results on the Fast Fourier Transform [8], Fast Walsh-Hadamard Transform [14] and Parseval's theorem. Given an M-dimensional vector $\mathbf{E} = (E_1, \ldots, E_M)$ and a matrix $\mathbf{F}^{M \times M}$, we have M-dimensional vector

$$\mathbf{D} = \mathbf{F}\mathbf{E}^T,$$

where $\mathbf{E}^T$ is the transpose of $\mathbf{E}$ and $\mathbf{F}$ is a Hadamard matrix if $\mathbf{F}(i,j) = (-1)^{ij}$, for all $i, j = 0, \cdots, M - 1$. If matrix $\mathbf{F}$ is either Fourier or Hadamard, vector $\mathbf{D}$ can be computed with complexity $\mathcal{O}(M \log M)$ instead of $\mathcal{O}(M^2)$ by Fast Fourier Transform or Fast Walsh-Hadamard Transform, respectively. We recall basic facts due to Parseval:

Let $f : V_m \to \mathcal{R}$ where $\mathcal{R}$ is the real field, and $a \in V_m$. We define $f_1(a) = \sum_{b \in V_m} (-1)^{ab} f(b)$, $A = \sum_{a \in V_m} f^2(a)$ and $A_1 = \sum_{a \in V_m} f_1^2(a)$. Then

$$2^m A = A_1$$

or

$$2^m \left(\sum_{a \in V_m} f^2(a) \right) = \sum_{a \in V_m} f_1^2(a). \tag{2}$$

3   Statistical Model and Algorithm of Hermelin *et al.*

3.1   Constructing Multidimensional Probability Distribution

Let $f : V_l \to V_n$ be a vectorial Boolean function and binary vectors $w_i \in V_n, u_i \in V_l, i = 1, \ldots, m$ be selection patterns such that pairs of input and output masks (u_i, w_i) are linearly independent. Define the functions g_i as

$$g_i(\eta) = w_i f(\eta) \oplus u_i \eta, \quad \forall \eta \in V_l, i = 1, \cdots, m$$

and g_i has correlation ρ_i, $i = 1, \cdots, m$. Then $\rho_1, \ldots, \rho_m$ are called the base-correlations, and $g_1, \ldots, g_m$ are the base approximations of f. Let $g=(g_1, \ldots, g_m)$ be an m-dimensional vectorial Boolean function, and matrices $W = (w_1, \ldots, w_m)$ and $U = (u_1, \ldots, u_m)$ contain the output and input masks for each of the g_i, then we find the p.d. $p = (p_0, \ldots, p_{2^m-1})$ of

$$g(\eta) = W f(\eta) \oplus U \eta.$$

Lemma 1. *[10] Let $g = (g_1, \ldots, g_m) : V_l \to V_m$ be a vectorial Boolean function and $p = (p_0, \ldots, p_{2^m-1})$ its p.d. Then*

$$2^l p_\eta = 2^{-m} \sum_{a \in V_m} \sum_{b \in V_l} (-1)^{a(g(b) \oplus \eta)}.$$

Define

$$\rho(a) = 2^{-l} \sum_{b \in V_l} (-1)^{ag(b)} = Pr(ag(X) = 0) - Pr(ag(X) = 1),$$

where X is an r.v. uniformly distributed in V_l.

Corollary 1. *Let $g : V_n \to V_m$ be a Boolean function with p.d. p and correlations $\rho(a)$ of the combined approximations ag, for all $a \in V_m$. Then for $\eta \in V_m$,*

$$p_\eta = 2^{-m} \sum_{a \in V_m} (-1)^{a\eta} \rho(a). \tag{3}$$

3.2   Multidimensional Generalization of Matsui's Alg. 1

We describe the core idea of the multidimensional algorithm 1 in [10]. Let there be m linear approximations $g_i := u_i X \oplus v_i Y \oplus c_i K$, $(i = 1, \cdots, m)$. Their corresponding correlations are

$$\rho_i := 2Pr(u_i X \oplus v_i Y \oplus c_i K = 0) - 1$$

where the masks c_i for the extended key K are linearly independent. In addition, the pairs of input and output masks (u_i, v_i) are linearly independent.

Define $g := (g_1, \ldots, g_m)$ with p.d. p, and $h := (h_1, \ldots, h_m)$, where $h_i = u_i X \oplus v_i Y$. We call h an experimental function, as we use two of its probability distributions, namely, the theoretical p.d. q and the empirical p.d. $\hat{q}$. In the attack, q is approximated by $\hat{q}$, which is computed from N samples. Let $w_i = c_i K$, $i = 1, \cdots, m$ be the parity bits of the extended key K. As the c_i's are linearly independent, $w = (w_1, \ldots, w_m)$ defines a key class of K. Thus we have

$$g = h \oplus w.$$

Hence, the p.d. q of experimental function h is a permutation of p. Since $\{c_i\}$ are linearly independent, with the 2^m possible parity vectors w, the key space K can be classified into 2^m classes. Let w^* be the correct key class, as p^{w^*} is the permutation of p corresponding to w^*, we have $q = p^{w^*}$. By [3], given w, the relationship between p^w and p is

$$p_\eta^w = \sum_{a \in V_m} (-1)^{a(\eta \oplus w)} \rho(a) = p_{\eta \oplus w}, \quad \forall \eta \in V_m. \tag{4}$$

The KL distance is then used to determine the correct key class w^*, by constructing a hypothesis testing problem of finding the closest distribution p^w with $\hat{q}$ among the 2^m possibilities of w. In [10] the following theorem is described.

Theorem 3. *Let us have an $|\mathcal{Z}|$-ary hypothesis problem, with $|\mathcal{Z}|$ hypotheses H_w stating that the data originates from p^w, where $w \in \mathcal{Z}$ corresponds to the key. The hypothesis for which the Kullback-Leibler distance $D(\hat{q}||p^w)$ is smallest is selected. Given some success probability P_{sc}, the lower bound N for the amount of data required to give the smallest value of the statistic when the correct key is used, is given by*

$$N \approx \frac{4 \log_2 |\mathcal{Z}|}{min_{w \neq 0} C(p^0, p^w)}. \tag{5}$$

Now we analyze the multidimensional algorithm 1 step by step.

Algorithm of Hermelin *et al.* [10]

Input of offline stage: m linear approximations $g_i, i = 1, \cdots, m$ with correlation ρ_i and p, the p.d. of g.

Offline: Compute N based on (5) and p.

Input of online stage: N pairs of (X, Y), with N computed in offline stage.

Online:

    1 Distillation phase: Compute empirical p.d. $\hat{q}$ of h using 2^m counters.

    2 Analysis phase:

    – Construct matrix $\mathbf{T}^{2^m \times 2^m}$, with cell $\mathbf{T}(w, \eta) = \log(\frac{\hat{q}_\eta}{p_\eta^w})$, for all $w, \eta \in V_m$.

    – Compute $D(\hat{q}||p^w)$, for all $w \in V_m$,

$$D = \mathbf{T}\hat{q}^T = (D(\hat{q}||p^0), \ldots, D(\hat{q}||p^{2^m-1})). \tag{6}$$

    3 Sorting phase: Sort the list of w with $D(\hat{q}||p^w)$ in ascending order.

    4 Searching phase: Choose the correct key class as the first element in the sorted list.

Note: in Step 2 the matrix $\mathbf{T}$ does not need to be stored. Storing a single row is sufficient. Other rows can be obtained as (4) by permuting this row when it's required for computation.

4   The Improvements

4.1   A Formula for N, the Number of Samples Required

Given $g = (g_1, \ldots, g_m)$ and $\rho(a)$, for all $a \in V_m$, by (5), the capacity of the attack of [10] is $\min_{w \neq 0} C(p^0||p^w)$. Let $p = p^0$ and $q = p^w$, we derive the following lemma. The proof is given in Appendix A.

Lemma 2.
$$C(p||q) \geq 2 \sum_{\forall a \in V_m \backslash \{0\}} \rho^2(a).$$

Combining with (5) we have the following theorem.

Theorem 4. *The estimation for N in the attack of [10] with m linear approximations g_i, $(i = 1, \ldots, m)$, where the $\{c_i\}$ are linearly independent and (u_i, v_i) are linearly independent, is given by*

$$N \approx \frac{m}{2 \sum_{\forall a \in V_m \backslash \{0\}} \epsilon^2(a)}. \tag{7}$$

With the new formula for N, we can now compare the multidimensional Alg. 1 with previous attacks on their data complexities, as shown in Table 1. The last row contains our formula for N. We can see that the framework provided by Hermelin *et al.* in [10] is able to exploit all the combined approximations systematically, as compared to Biryukov *et al.* in [2].

Table 1. Comparisons of data complexities with different attack frameworks

N	Framework	Comments
$1/\epsilon^2$	[12] Matsui	
$1/\sum_{i=1}^{m}\epsilon_i^2$	[11] Kaliski and Robshaw	
$1/\sum_{i=1}^{m}\epsilon_i^2$	[2] Biryukov *et al.*	
$m/2\sum_{a\neq 0}\epsilon^2(a)$	[10] Hermelin *et al.*	See (7)

4.2   Analysis of the Multidimensional Alg. 1 in [10]

We analyze the multidimensional Alg. 1 of Hermelin *et al.* step by step and introduce our improvements.

Identifying the Bottleneck – Complexity Analysis.   Offline: An estimation of N can be computed from (5), which is slow – to the best of our knowledge, no obvious algorithm is much faster than $3m2^m$ steps.

Online:
  1 Distillation Phase: Using 2^m counters to compute $\hat{q}$ from N samples. The complexity is $2^m N$.
  2 Analysis Phase: Computing D has a time complexity of $O(2^{2m})$.

If we increase m, the number of base approximations $g_1, \ldots, g_m$, we may expect N to decrease [10], but the complexities in the distillation and the analysis phases suffer from exponential increase. The actual number of approximations that can be used is hence limited by the computation resources allowed. It is a trade-off.

Improving the Bottleneck.   To compute the KL distance between a p.d. $\hat{q}$ and p^w, we observe that after expanding $D(\hat{q}||p^w) = \sum_{\eta \in V_m} \hat{q}_\eta \log(\hat{q}_\eta/p_\eta^w) = \sum_{\eta \in V_m} \hat{q}_\eta \log \hat{q}_\eta - \sum_{\eta \in V_m} \hat{q}_\eta \log p_\eta^w$, the term $\sum_{\eta \in V_m} \hat{q}_\eta \log \hat{q}_\eta$ is a constant and hence does not affect the ranking of key classes. We can define $\bar{D}(\hat{q}||p^w) = \sum_{\eta \in V_m} \hat{q}_\eta \log p_\eta^w$ and use $\bar{D}(\cdot||\cdot)$ to rank the key class candidates. The list of w in the sorting phase of the online stage is now sorted by the values of $\bar{D}(\hat{q}||p^w)$ in descending order.
  A new matrix $\bar{\mathbf{T}}$ is constructed as

$$\bar{\mathbf{T}}(w, \eta) = \log(p_\eta^w), \quad \forall w, \eta \in V_m. \tag{8}$$

In Section 4.3 we show that matrix $\bar{\mathbf{T}}$ is a circulant matrix. By Theorem 1 the Fast Fourier Transform algorithm can be applied for fast computation. For $\bar{\mathbf{T}}$, only the first column ($\bar{\mathbf{T}}(w, 0) = \log(p_0^w) = \log(p_w)$, for all $w \in V_m$) needs to be stored. The memory requirement is 2^m. $\bar{\mathbf{T}}$ is used for computing $\bar{D} = \bar{\mathbf{T}}\hat{q}^T$.

  Our Improvements:
  Offline: We present a formula for N based on p, without using (5).
  Online:

  Step 1: We present Method-A to calculate $\hat{q}$ from N samples given. The complexity is $2m2^m + mN$.
  Step 2: We present Method-B to compute $\bar{D}$. The complexity is $3m2^m$.

4.3   Fast Computation of Empirical Distribution – Method A

Method A is for computing the empirical distribution $\hat{q}$ of the experimental function $h = (h_1, \ldots, h_m)$ from N samples $(\mathbf{X}, \mathbf{Y})$. First, the correlations of the combined linear approximations bh, for all $b \in V_m$,

$$\hat{\gamma}(b) = Pr(bh(\mathbf{X}, \mathbf{Y}) = 0) - Pr(bh(\mathbf{X}, \mathbf{Y}) = 1),$$

are calculated. Then the p.d. $\hat{q}$ is computed from $\hat{\gamma}(b)$ by (3).

We show how to compute $\hat{\gamma}(b)$, for all $b = (b_1, \ldots, b_m) \in V_m$ from N samples:

$$\hat{\gamma}(b) = \hat{\gamma}\left(\bigoplus_{i=1}^{m} b_i h_i\right) = \frac{\sum_{j=1}^{N}(-1)^{\bigoplus_{i=1}^{m} b_i h_i(X_j, Y_j)}}{N}$$

$$= \sum_{a \in V_m} (-1)^{\bigoplus_{i=1}^{m} b_i a_i} \frac{T_a}{N} = \sum_{a \in V_m} (-1)^{ba} \frac{T_a}{N},$$

where $a = (a_1, \ldots, a_m)$ and $T_a = \#\{(X_j, Y_j), j = 1, \ldots, N : h_i(X_j, Y_j) = a_i\}$.

Let $S^{2^m \times 2^m}$ be the matrix defined by $S(b, a) = (-1)^{ba}$, and let $E = (\frac{T_0}{N}, \ldots, \frac{T_{2^m-1}}{N})$, and $\hat{\gamma} = (\hat{\gamma}_0, \ldots, \hat{\gamma}_{2^m-1})$, then

$$\hat{\gamma}(b) = \sum_{a \in V_m} S(b, a) E_a, \forall b \in V_m, \text{ or } \hat{\gamma} = SE^T.$$

Since $\mathbf{S}$ is a Hadamard matrix, we can apply the Fast Walsh-Hadamard Transform algorithm for computing $\hat{\gamma}(b)$ for all $b \in V_m$ with complexity $m2^m$, and the storage is $\mathcal{O}(2^m)$. The complexity for computing the counter vector $T = (T_1, \ldots, T_{2^m})$ is mN, by evaluating the m base approximations against each of the N samples.

Construct the vector $R = (R_0, \ldots, R_{2^m-1})$ with $R_b = 2^{-m}\hat{\gamma}(b)$, for all $b \in V_m$. From (3) we have

$$\hat{q}_\eta = \sum_{b \in V_m} (-1)^{\eta b} R_b, \forall \eta \in V_m, \text{ or } \hat{q} = SR^T.$$

The Fast Hadamard Transform is used again to compute $\hat{q}$. Hence, the total complexity is $mN + 2m2^m$ for computing $\hat{q}$ from N samples.

To Summarize Method A

Step 1: Construct the vector E from N samples (X, Y).
Step 2: Compute $\hat{\gamma} = SE^T$, then construct the vector R.
Step 3: Compute $\hat{q} = SR^T$.

4.4   Fast Computation of Kullback-Leibler Distance – Method B

Method-B is used to compute the vector $\bar{D}$ in (6) with modified matrix $\bar{T}$ in (8) based on the idea of circulant matrix in [7]. The proof of the following theorem can be found in Appendix B.

Theorem 5. *The matrix $\bar{T}$ is level-m circulant with type* $\underbrace{(2, 2, \ldots, 2)}_{m\text{-}times}$.

Method-B: From Theorems 1, 2, and 5 we have

$$\bar{\mathbf{T}} = \mathbf{F}^* diag(\lambda)\mathbf{F}, (\mathbf{F}^*\mathbf{F} = \mathbf{I}), \tag{9}$$

$$\lambda = \mathbf{F}\bar{\mathbf{T}}(:, 1)\sqrt{2^m}, \tag{10}$$

and

$$\bar{D} = (\bar{D}(\hat{q}\|p^0), \ldots, \bar{D}(\hat{q}\|p^{2^m-1})) = \bar{\mathbf{T}}\hat{q} = (\mathbf{F}^*(diag(\lambda)(\mathbf{F}\hat{q}))).$$

Applying the Fast Fourier Transform three times for $\mathbf{F}$, $diag(\lambda)$, and $\mathbf{F}^*$ we get all the values $\bar{D}(\hat{q}\|p^w)$, for all $w \in V_m$. The complexity is $3m2^m$.

5   Efficiency of the Improved Algorithm

The improved algorithm works as follows:

Input of offline stage: m linear approximations $g_1, \ldots, g_m$ and p.d. p and $\rho(a)$ of g, for all $a \in V_m$.
Offline stage:

Step 1: Compute N by (7).
Step 2: Compute the eigenvalue vector λ of matrix $\bar{\mathbf{T}}$ based on (10) and p.
Input of online stage : N samples (X, Y).

Online stage :

Step 1 – Distillation: Compute empirical p.d. $\hat{q}$ of h from N by Method-A.
Step 2 – Analysis: From λ and $\hat{q}$, compute vector $\bar{D}$ by Method-B.
Step 3 – Sorting: Sort the key classes w with $\bar{D}(\hat{q}\|p^w)$ in descending order.
Step 4 – Searching: Choose the correct key class as the first element in the sorted list.

The improved results are presented in Table 2 to compare with previous results on the complexities of the distillation phase and the analysis phase, following [10]. In practise mN is always larger than $2m2^m$, and we estimate $\mathcal{O}(2m2^m + mN) = \mathcal{O}(mN)$.

Following the definitions of $N_{s.i.}$ and N_{plain} in [10], $N < N_{s.i.} < N_{plain}$, $m < m'$ (also see Table 1), we have significant improvement in the time complexities

Table 2. Complexity comparison between different algorithms

	Distillation Phase				Analysis Phase			
	Plain	Biryukov	Hermelin	Method-A	Plain	Biryukov	Hermelin	Method-B
Data	$\mathcal{O}(N_{plain})$	$\mathcal{O}(N_{s.i})$	$\mathcal{O}(N)$	$\mathcal{O}(N)$	-	-	-	-
Time	$\mathcal{O}(mN_{plain})$	$\mathcal{O}(m'N_{s.i})$	$\mathcal{O}(2^m N)$	$\mathcal{O}(mN)$	$\mathcal{O}(m2^m)$	$\mathcal{O}(m'2^m)$	$\mathcal{O}(2^{2m})$	$\mathcal{O}(m2^m)$
Mem	$\mathcal{O}(m)$	$\mathcal{O}(m')$	$\mathcal{O}(2^m)$	$\mathcal{O}(2^m)$	$\mathcal{O}(2^m)$	$\mathcal{O}(2^m)$	$\mathcal{O}(2^m)$	$\mathcal{O}(2^m)$

for the distillation phase and the analysis phase. In addition, we have proved that the multidimensional algorithm 1 requires fewer samples than Biryukov *et al.*[2]. The same result was observed in [10] with only empirical evidence.

6   Application to Cryptanalysis

In this section we apply the improved algorithm to reduced-round Serpent and derive the attack complexities. We derive attack scenarios by using 4-round and 9-round linear characteristics used in [10] and [4] to show that the improvement made in this paper can improve previous cryptanalysis by many orders of magnitude.

6.1   Application to the 4-Round Serpent Scenario

In [7] 64 approximations were used to obtain 10 parity bits of 4-round Serpent, from S_4 to S_7. The approximations are modified from the first 4 rounds of the 6-round linear characteristic of [4], with the details described in [5]. As shown in [10], these approximations used are not linearly independent in text masks and key masks. There are 8 of them with correlation in magnitude of 2^{-11} and 56 of 2^{-12}. This gives an overall capacity of $4\sum_{i=0}^{63}\epsilon_i^2 = 2^{-17.54}$ hence estimation of $4m/4\sum_{i=1}^{m'}\epsilon_i^2 = 2^{22.86}$ for N. By selecting a basis of 10 approximations $L_0,\dots,L_9$ from the 64, where L_i is $u_iX \oplus w_0Y \oplus c_iK = 0$, Hermelin *et al.* [10] studied all the approximations generated as in $span\{L_0,\dots,L_9\}$. Of the 1023 combinations, 8, 64 and 128 are with non-negligible correlation in magnitude of 2^{-11}, 2^{-12} and 2^{-13}, respectively. Lemma 2 gives capacity $C(p||q)$ of at least $2\sum_{a\neq 0}\rho^2(a) = 2\cdot(8\cdot 2^{-22} + 64\cdot 2^{-24} + 128\cdot 2^{-26}) = 2^{-16}$ and hence the estimation for N is $4m/C(p||q) = 2^{21.3}$, in perfect correspondence to the experimental results in [10]. With Method-A and Method-B, we can set $m = 16$, yielding an attack with better complexity than the multidimensional Alg. 1 of [10] with $m = 10$.

The approximations $L_0,\dots,L_9$ are derived from the same linear characteristic with input masks $u_0,\dots,u_9$ and the same output mask w_0. We obtain 6 additional ciphertext masks $w_1,\dots,w_6$ and use the following 16 approximations as base approximations: $(u_0,w_0),\dots,(u_9,w_0),(u_1,w_1),\dots,(u_1,w_6)$. An exhaustive check of all combined correlations shows that 32, 384, 1664, 3072 and 2048 of the combined approximations have correlation in magnitude of 2^{-11}, 2^{-12}, 2^{-13}, 2^{-14} and 2^{-15} respectively. This gives a capacity of $2^{-12.8}$, hence we estimate $N \sim 4m/C(p||q) = 2^{18.8}$. We tabulate the comparisons in Table 3. From the table we can conclude that it is clearly advantageous to be able to have a larger m, i.e., when using $m = 16$ instead of $m = 10$, in the case of Serpent. A larger m in this case makes it possible to have a larger number of non-negligible approximations, hence larger capacity, which implies reduced data complexity thus the overall time complexity. Setting $m = 16$ improves the cryptanalysis due to the fact that the number of non-negligible approximations in Serpent is exponential in m. It may appear that the attack is slower with a larger m. However, in fact, it

Table 3. Attack complexities on 4-round Serpent

m	$C(p\|q)$	N	Distillation Phase		Analysis Phase		Memory
			Hermelin *et al.* [10]	This paper	Hermelin *et al.* [10]	This paper	
10	2^{-16}	$2^{21.3}$	$2^{31.3}$	$2^{24.6}$	2^{20}	$2^{14.9}$	2^{10}
16	$2^{-12.8}$	$2^{18.8}$	$2^{34.8}$	$2^{23.2}$	2^{32}	$2^{21.6}$	2^{16}

is the reduction in data complexity N that dominates the time complexity, so we obtain a faster attack. Essentially, this is a trade-off and it is always meaningful to find an appropriate m for a block cipher to produce optimal attack complexity. However, by our experiments, in the case of DES, the number of high probability approximations is much fewer than that of Serpent, so larger values of m do not give better results than fewer approximations. We believe that SPN block ciphers with small S-boxes are more likely to be vulnerable to our attack.

6.2   Multidimensional Linear Cryptanalysis of 9-Round Serpent

We take the 9-round linear characteristic of [4], where the details are described in [5]. This linear characteristic starts from S_3 and ends after the next S_3, with correlation 2^{-49}. The first round has 11 active S-boxes, which results in a correlation of 2^{-11} and the remaining 8 rounds with correlation 2^{-38}. By modification to the input masks, there is a total of 10^{11} masks, with the magnitude of first round correlation from 2^{-11} to 2^{-22}. By picking 44 independent base input masks, we can expect to have around 10^{11} out of the 2^{44} combined approximations giving non-negligible correlations. We can exploit the huge number of approximations by a 44-dimensional attack, with capacity

$$C(p\|q) = 2 \cdot [\sum_{i=0}^{11} \binom{11}{i} \cdot 2^i \cdot 8^{11-i} \cdot ((2^{-1})^i \cdot (2^{-2})^{11-i})^2] \cdot (2^{-38})^2 = 2^{-75}$$

which is 2^{22} times larger than the capacity 2^{-98} for the single approximation scenario. Correspondingly the estimation for N is $4m/C(p\|q) = 2^{82.5}$. The time complexity for the distillation phase is 2^{88} and the analysis phase is 2^{51}. Around 2^{44} memory are needed. In [4], this 9-round linear characteristic is used to break a 10-round and an 11-round Serpent with Matsui's Alg. 2 extended with multiple approximations. It requires at least 2^{99} known-plaintext which is much higher than our estimation. Moreover, the time complexities presented in [4] did not take into account the distillation phase, which should require no lower than mN which is greater than 2^{99}. Hence, it is much higher than our overall time complexity 2^{88}.

Comparing with extensions of Alg. 2, a disadvantage of Alg. 1 is that an r-round linear characteristic can be used to attack an r-round block cipher. However, it is noted in [7] that optimal application of Algorithm 2 with multiple approximations requires accurate estimation of the biases, which can be unreliable if multiple linear characteristics exist with non-negligible probability under the same text mask. Intuitively, Alg. 1 is likely to give more reliable estimation for theoretical cryptanalysis.

7 Conclusions

In this paper, we have presented a new formula for N, in terms of $\rho(a)$ and m, for the $|\mathcal{Z}|$-ary hypothesis testing problem in the multidimensional generalization of Matsui's Algorithm 1 in [10]. The number of known plaintext needed can now be computed from $\rho(a)$ directly, whereas it is harder to compute $\min_{w \neq 0} C(p^0, p^w)$ in the original formula (5).

Method-A and Method-B have been presented to compute the empirical distribution and Kullback-Leibler distance, as improvements to the multidimensional Alg. 1 in [10]. A significant reduction of time complexity in the distillation and analysis phases can be achieved. Breaking these bottlenecks allows many more base approximations to be used.

As shown in the case of 4- and 9-round Serpent, the increase in m brings significant reduction on the data complexity and the time complexity. We expect this improved algorithm to outperform previous multiple linear cryptanalysis. We observed that the framework of Hermelin *et al.* [10], with our improvement, solves the problem of Biryukov *et al.* [2] that the statistical independence assumption cannot in general be guaranteed in practice, and in fact, does not need to be guaranteed, because there exists a large number of combined approximations with non-negligible correlations. Meanwhile, the series of experiments in [4] and [7] provide excellent examples for this argument.

It also implies that, for block cipher designs, bounding the maximum correlation for any single linear characteristic is not sufficent to claim security. Especially for SPN block ciphers with small S-boxes, a single linear trail with multiple active S-boxes in the first or last round can be modified to have many approximations, exponential in the number of active S-boxes of outer rounds. When these approximations are with similar magnitude of correlations, multidimensional linear cryptanalysis as a systematic way to exploit these combined correlations can reduce the attack complexity greatly. It's worthy for designers to have larger security margins or to try to develop specific mechanisms to prevent an attacker from forming an exponential number of valid linear approximations.

Acknowledgements. We thank Joo Yeon Cho for providing the linear approximations used in [10]. This reserach is supported by the Singapore National Research Foundation under Research Grant NRF-CRP2-2007-03 and the Singapore Ministry of Education under Research Grant T206B2204. The first author is supported by the Singapore International Graduate (SINGA) Scholarship.

References

1. Biham, E., Dunkelman, O., Keller, N.: Linear Cryptanalysis of Reduced Round Serpent. In: Matsui, M. (ed.) FSE 2001. LNCS, vol. 2355, pp. 16–27. Springer, Heidelberg (2002)

2. Biryukov, A., De Cannière, C., Quisquater, M.: On Multiple Linear Approximations. In: Franklin, M. K. (ed.) CRYPTO 2004. LNCS, vol. 3152, pp. 1–22. Springer,
   Heidelberg (2004)
3. Cho, J.Y., Hermelin, M., Nyberg, K.: A New Technique for Multidimensional Linear Cryptanalysis with Applications On Reduced Round Serpent. In: Lee, P.J.,
   Cheon, J.H. (eds.) ICISC 2008. LNCS, vol. 5461, pp. 383–398. Springer, Heidelberg (2009)
4. Collard, B., Standaert, F.-X., Quisquater, J.-J.: Improved and Multiple Linear
   Cryptanalysis of Reduced Round Serpent. In: Pei, D., Yung, M., Lin, D., Wu, C.
   (eds.) Inscrypt 2007. LNCS, vol. 4990, pp. 51–65. Springer, Heidelberg (2008)
5. Collard, B., Standaert, F.-X., Quisquater, J.-J.: Improved and Multiple Linear
   Cryptanalysis of Reduced Round Serpent - Description of the Linear Approximations, 2007 (unpublished manuscript)
6. Collard, B., Standaert, F.-X., Quisquater, J.-J.: Improving the Time Complexity
   of Matsui's Linear Cryptanalysis. In: Nam, K.-H., Rhee, G. (eds.) ICISC 2007.
   LNCS, vol. 4817, pp. 77–88. Springer, Heidelberg (2007)
7. Collard, B., Standaert, F.-X., Quisquater, J.-J.: Experiments on the Multiple Linear Cryptanalysis of Reduced Round Serpent. In: Nyberg, K. (ed.) FSE 2008.
   LNCS, vol. 5086, pp. 382–397. Springer, Heidelberg (2008)
8. Cormen, T.H., Stein, C., Rivest, R.L., Leiserson, C.E.: Introduction to Algorithms.
   McGraw-Hill Higher Education, New York (2001)
9. Desmedt, Y.G. (ed.): CRYPTO 1994. LNCS, vol. 839. Springer, Heidelberg (1994)
10. Hermelin, M., Cho, J.Y., Nyberg, K.: Multidimensional Linear Cryptanalysis of
    Reduced Round Serpent. In: Mu, Y., Susilo, W., Seberry, J. (eds.) ACISP 2008.
    LNCS, vol. 5107, pp. 203–215. Springer, Heidelberg (2008)
11. Kaliski Jr., B.S., Robshaw, M.J.B.: Linear Cryptanalysis Using Multiple Approxi-
    mations. In: Desmedt (ed.) [9], pp. 26–39
12. Matsui, M.: Linear Cryptanalysis Method for DES Cipher. In: Helleseth, T. (ed.)
    EUROCRYPT 1993. LNCS, vol. 765, pp. 386–397. Springer, Heidelberg (1994)
13. Matsui, M.: The First Experimental Cryptanalysis of the Data Encryption Stan-
    dard. In: Desmedt [9], pp. 1–11
14. Rao Yarlagadda, R.K., Hershey, J.E.: Hadamard Matrix Analysis and Synthesis:
    with Applications to Communications and Signal/image Processing. Kluwer Aca-
    demic Publishers, Norwell (1997)

Appendix

A   Proof for Lemma 2

Proof. Let $g = (g_1, \ldots, g_m)$ and given $\rho(a)$, for all $a \in V_m$, from (2), (3)

$$2^m \left(\sum p_\eta^2 \right) = \sum \rho^2(a).$$

We have 2 facts: $p^0 = p$ and p^w is a permutation of p. From (1), we replace p^0 and p^w by p, q respectively in (5). Define $u = \max\{q_\eta, \eta \in V_m\} = \max\{p_\eta, \eta \in V_m\}$. Then

$$C(p||q) = \sum_{\eta \in V_m} \frac{(p_\eta - q_\eta)^2}{q_\eta} \geq \frac{1}{u} \sum_{\eta \in V_m} (p_\eta^2 + q_\eta^2 - 2p_\eta q_\eta).$$

We have

$$\frac{q_\eta}{u} \leq 1 \Rightarrow -\frac{q_\eta}{u} \geq -1.$$

Since q is a permutation of p:

$$\sum_{\eta \in V_m} p_\eta = 1 \text{ and } \sum_{\eta \in V_m} p_\eta^2 = \sum_{\eta \in V_m} q_\eta^2.$$

Hence

$$\frac{1}{u} \sum_{\eta \in V_m} (p_\eta^2 + q_\eta^2 - 2p_\eta q_\eta) = \frac{2}{u} \left(\sum_{\eta \in V_m} p_\eta^2 \right) - 2 \sum_{\eta \in V_m} \frac{q_\eta}{u} p_\eta$$

$$\geq \frac{2 \cdot 2^{-m} \sum_{a \in V_m} \rho^2(a)}{u} - 2 \left(\sum_{\eta \in V_m} p_\eta \right)$$

$$= \frac{2 \cdot 2^{-m} \sum_{a \in V_m} \rho^2(a)}{u} - 2.$$

From (3),

$$u = \max\{p_\eta\} \leq 2^{-m} \left(\sum_{a \in V_m} |\rho(a)| \right),$$

so that

$$C(p\|q) \geq \frac{2 \sum_{a \in V_m} \rho^2(a)}{\sum_{a \in V_m} |\rho(a)|} - 2.$$

In practice, since $\rho(0) = 1$ and $\rho(a) \ll 1$, $\forall a \neq 0$, we have $\sum_{a \in V_m} |\rho(a)| \approx 1$. Hence

$$C(p\|q) \geq 2 \sum_{\forall a \in V_m \setminus \{0\}} \rho^2(a).$$

B   Proof for Theorem 5

Proof. The structure of the modified matrix $\bar{T}^{2^m \times 2^m}$ is

$$\begin{pmatrix} \log(p_0^0) & \log(p_1^0) & \cdots & \log(p_{2^m-1}^0) \\ \log(p_0^1) & \log(p_0^1) & \cdots & \log(p_{2^m-1}^1) \\ \vdots & \vdots & \ddots & \vdots \\ \log(p_0^{2^m-1}) & \log(p_1^{2^m-1}) & \cdots & \log(p_{2^m-1}^{2^m-1}) \end{pmatrix}.$$

From the relation between p^w ($\forall w \neq 0$) and $p^0 = p$:

$$p_i^w = \sum_{a \in V_m} (-1)^{a(w \oplus i)} \rho(a) = \sum_{a \in V_m} (-1)^{aj} \rho(a) = p_j,$$

where $j = w \oplus i$.

Hence, $\log(p_i^w) = \log(p_j), j = w \oplus i$. It means that $\bar{T}(w,i) = \bar{T}(0,j) = \bar{T}(0, w \oplus i)$.

Divide $\bar{T}$ into 4 blocks, each with size $(2^{m-1} \times 2^{m-1})$:

$$\begin{pmatrix} \bar{T}_{11} & \bar{T}_{12} \\ \bar{T}_{21} & \bar{T}_{22} \end{pmatrix}.$$

Then for $0 \leq i, j \leq 2^{m-1} - 1$:

- $\bar{T}_{11}(i,j) = \bar{T}(i,j) = \bar{T}(0, i \oplus j)$,
- $\bar{T}_{21}(i,j) = \bar{T}(i + 2^{m-1}, j) = \bar{T}(0, (i + 2^{m-1}) \oplus j) = \bar{T}(0, i \oplus j \oplus 2^{m-1})$,
- $\bar{T}_{12}(i,j) = \bar{T}(i, j + 2^{m-1}) = \bar{T}(0, i \oplus (j + 2^{m-1})) = \bar{T}(0, (i + 2^{m-1}) \oplus j)$,
- $\bar{T}_{22}(i,j) = \bar{T}(i+2^{m-1}, j+2^{m-1}) = \bar{T}(0, (i+2^{m-1}) \oplus (j+2^{m-1})) = \bar{T}(0, i \oplus j)$.

Consequently, $\bar{T}_{11} = \bar{T}_{22}$ and $\bar{T}_{12} = \bar{T}_{21}$, hence $\bar{T}$ is 2-block circulant. We can inductively repeat the same argument to $\bar{T}_{11}$ with $m = m - 1$. Since $\bar{T}_{12} = \bar{T}_{11} \oplus 2^{m-1}$, the structure of $\bar{T}_{12}$ is similar to the circulant structure of $\bar{T}_{11}$. Hence, the matrix $\bar{T}$ is level-m circulant with type $\underbrace{(2, 2, \ldots, 2)}_{m\text{-times}}$.

Side-Channel Analysis of the K2 Stream Cipher

Matt Henricksen[1], Wun She Yap[1], Chee Hoo Yian[1],
Shinsaku Kiyomoto[2], and Toshiaki Tanaka[2]

[1] Institute for Infocomm Research,
A*STAR, Singapore
{mhenricksen,wsyap,chyian}@i2r.a-star.edu.sg
[2] KDDI R&D Laboratories Inc
2-1-15 Ohara, Fujimino-shi, Saitama 356-8502, Japan
{kiyomoto,toshi}@kddilabs.jp

Abstract. In this paper we provide the first side-channel analysis of the
K2 stream cipher. K2 is a fast and secure stream cipher built upon the
strengths of SNOW 2.0. We apply timing attacks, power analysis, and
differential fault analysis to K2. We show that naively implemented K2
is vulnerable to cache-timing attacks, and describe how to implement
efficient countermeasures to protect K2 against side-channel attacks in
hardware and software.

Keywords: Stream Cipher, K2, side-channel, differential fault analysis,
cache timing attacks, power analysis, SNOW 2.0.

1 Introduction

Many methods of cryptanalysis, such as linear cryptanalysis [15] and algebraic
attacks [5], exploit flaws in the design of the cryptographic algorithm. In contrast, side-channel attacks exploit leakage of information during execution of key-
related operations on the cryptographic device. Timing information, power consumption, electromagnetic leaks, or even sound can provide exploitable sources
of leaky state or key information. Side-channel attacks are differentiated from
direct attacks by relying on not only the algorithm design, but also upon its implementation and the characteristics of the hardware on which it is implemented.

Launching side-channel attacks is in most cases likely to require considerable
technical knowledge of the implementation platform. However, potential vulnerabilities can be detected at the design level, and moreover, algorithm designers
can do much to alleviate the vulnerability of their algorithms at the design stage.

Side-channel attacks can be categorized as passive or active, and invasive or
non-invasive. Passive attacks, such as power analysis, timing analysis and cache-
timing attacks are comparatively inexpensive, and rely on observation of the
algorithm as it encrypts or decrypts using secret material. The attacker is likely
to be undetected. Active attacks, such as fault analysis, disrupt the algorithm
to reveal secret material. The assumptions concerning active attacks are usual
stronger than those for passive attacks, and the attacks are usually expensive,

R. Steinfeld and P. Hawkes (Eds.): ACISP 2010, LNCS 6168, pp. 53–73, 2010.

requiring a high initial capital investment plus a moderate amount of investment for each chip attacked (although Skorobogatov and Anderson showed how to conduct optical induction fault attacks for several tens of dollars [17]).

In this paper, we describe the effect of a handful of side-channel attacks - timing attacks, power attacks and fault attacks - on the K2 stream cipher [12], which is based upon the well-known SNOW 2.0 cipher [6]. K2 has additional strength through a reinforced Finite State Machine, and its 'Dynamic Feedback Control' mechanism, meaning that it is more resilient against many attacks, while remaining competitively efficient. We provide the first analysis of K2 with respect to side-channel attacks.

In Section 2, we provide a specification of the K2 stream cipher, and discuss one reasonable way of implementing it. In Section 3 we reiterate an observation on LFSR-based stream ciphers that has impact on almost every side-channel attack that can be applied to K2. In Section 4 we respectively consider both conventional timing and cache-timing attacks applied to K2, especially in light of the successful conventional timing attacks and cache-timing attacks applied to Sosemanuk by Leander, Zenner and Hawkes [14]. In Section 5, we discuss simple and differential power analysis attacks on K2. In Section 6, we survey a differential fault analysis of SNOW 2.0, correct mistakes in that attack, and apply it to K2. In Section 7, we provide concluding notes.

2   Specification

The state of the K2 stream cipher comprises two Feedback-Shift Registers (FSR), respectively *FSR-A* and *FSR-B*, and a finite-state machine *FSM-C*.

FSR-A consists of five 32-bit stages, denoted just after initialization as $r_0, ..., r_4$ while FSR-B consists of eleven 32-bit stages $s_0, ..., s_{10}$. Sometimes, we represent the state of FSR-A as a snapshot at time t using $A[0]...A[4]$, and the state of FSR-B as $B[0]...B[10]$. The FSM-C contains four 32-bit words of memory entitled $L1, L2, R1$ and $R2$. The total state size of K2 is 640 bits.

Update function. FSR-A is autonomous. The contents of its stages are determined by the recurrence $r_{t+5} = \alpha_0 \cdot r_t \oplus r_{t+3}$, where α_0 is the root of a polynomial in $GF(2^4)$. Each element of this field is in turn expressed in terms of the root of a polynomial β in $GF(2^8)$.

FSR-B is driven by FSR-A using Dynamic Feedback Control (DFC). The contents of its stages are determined by the recurrence

$$s_{t+11} = (\alpha_1^{r_{t+2}[30]} \oplus (\alpha_2^{1-r_{t+2}[30]}) - 1)s_t \oplus s_{t+1} \oplus s_{t+6} \oplus \alpha_3^{r_{t+2}[31]} s_{t+8}$$

for 32-bit values α_1, α_2 and α_3. $r^{[i]}$ denotes the ith bit of register stage r.

The contents of the Finite State Machine (FSM) registers is determined as follows:

$$L1_t = Sub(R2_{t-1} \boxplus s_{t+3})$$
$$R1_t = Sub(L2_{t-1} \boxplus s_{t+8})$$
$$L2_t = Sub(L1_{t-1})$$
$$R2_t = Sub(R1_{t-1})$$

Sub is a 32×32-bit bijection composed of four parallel invocations of an 8×8-bit s-box S followed by a linear 32×32 multiplication in $GF(2^8)^4$. The s-box used is the AES s-box and the multiplication is the AES *MixColumn* operation.

Output function. Keystream is generated as the 64-bit word $(z_t^H || z_t^L)$.

$$z_t^H = (s_{t+10} \boxplus L2_t) \oplus L1_t \oplus r_t$$

and

$$z_t^L = (s_t \boxplus R2_t) \oplus R1_t \oplus r_{t+4}$$

Initialization function. The key initialization algorithm must be performed to initialize the cipher with a new key or IV. A key-IV pair can be used to generate at most 2^{58} keystream words, after which the key initialization algorithm must again be invoked. The same key-IV pair cannot be used twice to initialize the stream cipher. Generally initialization will involve the same key and a different IV.

The K2 key initialization schedule has three phases. Phase 1 expands the master key into a series of extended key words, which are used to populate the cipher internal state in phase 2. In phase 3, the cipher's clocking function is invoked twenty-four times to mix the internal state.

Phase 1 uses a 128-, 192- or 256-bit key K to generate an twelve extended key words $EK_i, 0 \leq i \leq 11$. The extended key word EK_i for a 128-bit key $K = K_0 || K_1 || K_2 || K_3$ is generated as:

$$EK_i = \begin{cases} K_i & \text{if } i \in \{0,1,2,3\} \\ EK_{i-4} \oplus Sub((EK_{i-1} \lll 8)) \oplus C_i & \text{if } i \in \{4,8\} \\ EK_{i-4} \oplus EK_{i-1} & \text{if } i \in \{5,6,7,9,10,11\} \end{cases}$$

$\lll$ is rotation on 32-bit words, and constants C_0, C_1 and C_2 are respectively 0x01000000, 0x02000000, and 0x03000000.

Phase 2 initializes the FSRs using the extended key words and raw IV words. The shortest register, FSR-A is populated such that $A[i] = EK_{4-i}$, ie. it receives all of the raw key words in the form of $EK_0...EK_3$. FSR-B is populated with the remaining extended key words and the four IV words. The FSM registers are set to zero.

Phase 3 invokes the update function 24 times to mix the state. The keystream words z^H and z^L are not discarded, but added back into the feedback for FSR-B and FSR-A respectively using exclusive-or.

2.1   Implementation Aspects

Sub implemented using one table or four Sub is generally implemented as a
single primitive with 32-bit inputs and outputs, using four lookups of a s single
table, each based on one byte of the input. ie. for input $x = x_0|x_1|x_2|x_3$:

$$Sub(x) = T[x_0] \oplus (T[x_1] \gg 8) \oplus (T[x_2] \gg 16) \oplus (T[x_3] \gg 24)$$

The table occupies one kilobyte. The number of operations in a *Sub* can be reduced by building the rotation operations into tables. This increases the number
of tables to four, occupying four kilobytes, and *Sub* becomes

$$Sub(x) = T_0[x_0] \oplus T_1[x_1] \oplus T_2[x_2] \oplus T_3[x_3]$$

While this may provide a significant speed boost on machines with poor support
for rotation, we later show there are good reasons of security to avoid using this
technique.

On the α constants. Multiplication by α_0 is implemented as $\alpha(x) = (x \ll 8) \oplus$
$A_0[x \gg 24]$ where A is a $2^8 \times 32$-bit table such that $A_0[i] = (i\beta^{24}, i\beta^3, i\beta^{12}, i\beta^{71})$.
Other α tables have different bases. We usually treat $\alpha_{0..3}$ as lookup tables.
This has implications with respect to side-channel attacks. We define the macro
ALPHA to implement all four α-table lookups as follows:

```
1  #define ALPHA(x, table)   (x << 8) ^ table[(x >> 24) & 0xFF];
```

3   An Observation on LFSR-Based Stream Ciphers

Hoch notes in his differential fault analysis of LFSR-based stream ciphers:

> Given n output bits of the LFSR, such that the corresponding linear
> relations in the initial state bits are independent, we can reconstruct the
> initial state by solving the system of n linear equations in n unknown
> bits over $GF(2)$ [9, page 13]

Leander et al. use the same result in their cache-timing attack of Sosemanuk [14]
to note that clocking its ten word LFSR with 32-bit stages can be represented
by applying an invertible 320×320 matrix M over GF_2, since there exists a
linear bijection $\phi : GF_{2^{32}} \to GF_2^{32}$, and if the operation is linear in $GF_{2^{32}}$ it is
also linear in GF_2. The LFSR can be considered an element in GF_2^{320} via the
isomorphisms $(s_t, ..., s_{t+9}) \to (\psi(s_t), ..., \psi(s_{t+9}))$. Then $\psi(s_{t+1})$ can be written
as $M^t \cdot \psi(s)$, and bits obtained by cache timing can be related to initial state bits
via M. They generalize this attack to all LFSR-based stream ciphers by noting
that for an internal state of n elements of GF_{2^m}, leakage of k internal state bits
per cycles provides sufficient equations to determine the state of the LFSR in
$n \cdot m \cdot k^{-1}$ cycles.

This means that the LFSR of K2, or of any stream cipher, is not robust against leakage of information in a side-channel attack, unless suitable countermeasures are implemented. This observation pinpoints the weakest component of the K2 stream cipher, and it underlies all of our analyses in the following sections. In particular, if FSR-A can be attacked in isolation, it can be virtually stripped away from the cipher. Since FSR-A drives the Dynamic Feedback Control mechanism, it too can be removed. What remains in the 'virtually reduced' cipher is the linear FSR-B and non-linear FSM. The side-channel technique can be repeated on the FSR-B, which also can be removed, and then other techniques applied to solve the state of the 128-bit FSM.

4 Timing Attacks

Timing attacks exploit a side-channel that leaks information through differentials in the time it takes to execute different operations within the algorithm. The amount of time it takes to compute a function depends not only on the number and nature of the operations within the functions, but also upon the inputs passed to them. If key-dependant operations in an algorithm take variable lengths of time to execute according to the value of key bits, then by repeatedly measuring these variations in time, the values of the corresponding key-bits might be deduced. Cache-timing attacks, in which the location of data in memory forms a side channel, are also a form of timing attack, addressed in Section 4.2.

4.1 K2 and Conventional Timing Attacks

Conventional timing attacks are well known and easy to avoid by eschewing the use of conditional branching or operations that have execution times that vary with the value of the operands.

Most of the operations in the K2 cipher keystream generation are constant-time operations with respect to conventional timing attacks: exclusive-or, addition, and table lookups. A naive implementation of Dynamic Feedback Control might involve branching.

$$s_{t+11} = (\alpha_1^{r_{t+2}[30]} \oplus (\alpha_2^{1-r_{t+2}[30]}) \quad 1)s_t \oplus s_{t+1} \oplus s_{t+6} \oplus \alpha_3^{r_{t+2}[31]}s_{t+8}$$

can be implemented as:

```
1  feedback_b  = FSR_B_STAGE(ctx, 1) ^ FSR_B_STAGE(ctx, 6) ^
2      ALPHA(FSB_B_STAGE(ctx, 0),
3          (FSR_A_STAGE(ctx, 1) & DFC_BIT_1) ?
4              alpha_1 : alpha_2);
5
6  feedback_b ^= (FSR_A_STAGE(ctx, 1) & DFC_BIT_2) ?
7      ALPHA(FSR_B_STAGE(ctx, 8), alpha_3):
8      FSR_B_STAGE(ctx, 8));
```

Lines 2-4 of this listing invoke either the α_1 or α_2 table lookup depending on the value of DFC_BIT_1. But this conditional ternary operator is not vulnerable to a timing attack because the number and order of operations in each conditional block is identical. Only the data varies.

The same does hold for the conditional of lines 6-9, since depending on the value of DFC_BIT_2, either an α-table lookup on $B[8]$ is performed or it is not. The remaining operations in both conditional branches are equal, so executing line (7) will usually take longer than executing line (8). Then a timing differential that leaks the value of DFC_BIT_2 occurs as frequently as in every cipher clock. From the observation made in Section 3, periodic leakage of this bit, which is located in FSR-A, will eventually lead the attacker to compute the entire state of that register.

Nevertheless, a countermeasure is readily available at the cost of a slight reduction in throughput. As the bit DFC_BIT_2 is pseudo-random, then α_3 will be executed on average once in every two cycles in the unprotected implementation. In the following countermeasure, it is executed every cycle, and discarded when the value of DFC_BIT_2 is 0, by masking the result of the table with 0 and **0xFFFFFFFF** for respective off- and on values of DFC_BIT_2. If the α_3 table is invoked, then $B[8]$ must be shifted left by one byte, which is easily implemented by shifting by the value of DFC_BIT_2 multiplied by eight. Then lines 6-9 in the above are replaced with:

```
1  feedback_b ^= FSR_B_STAGE(ctx, 8) << (DFC_BIT_2 * 8) ^
2      (alpha_3[FSR_B_STAGE(ctx, 8) >> 24] & (-DFC_BIT_2));
```

Examination of K2's key initialization algorithm reveals no non-constant time operations in phases 1 or 2, and only the aforementioned vulnerability in phase 3.

4.2   K2 and Cache-Timing Attacks

Cache timing attacks rely on measuring the differential in time between accessing data in cache and data in memory. When the CPU requests data to be fetched to its registers, the request is directed to the cache interspersed between the fast CPU and slow main memory. If the data is not present in the cache, then a cache miss occurs. In this case, the data must be imported from the memory at the cost of some latency. The cache subsequently keeps a copy of the data. If the data is again requested, and present in the cache, then a cache hit occurs, and this time because the data need not be fetched from memory, there is less latency in the operation. This differential between time to access data present in the cache, and data present in memory but not in the cache, becomes a side channel.

For reasons of efficiency, data is not imported into the cache on a word-by-word basis. It is imported *line*-by-line. On modern processors, lines are usually collections of 64bytes of adjacent data, and we assume this hereafter. Cache timing measurements can determine indices in table-lookups, such as s-boxes, to the resolution of the cache line. For example, measuring the latency of an s-box lookup into an 8×32 table can provide the high nibble of the index. More generally, the number of bits b leaked by each cache-timing measurement is determined by $b = c - log_2(\gamma/d)$, where the table size is 2^c bytes, d is the number of bytes per each table entry, and γ is size of each cache line in bytes.

In a prime-then-probe attack [19], the attacker fills the cache with data just prior to the victim executing his cryptographic algorithm. Afterwards, the attacker reloads his data, timing the process. Cache hits highlight the cache lines not used by the victim, since the attacker's data has been not displaced. Conversely, this represents a noisy version of the cache lines used by the victim. Therefore the attacker learns some of the indexing information used by the victim's cryptographic algorithm.

Related Work. Zenner [20] studied the eight eSTREAM software finalists from the perspective of cache-timing attacks. He initially had most success attacking HC-256 [18], with a prime-and-probe attack in conjunction with a back-tracking consistency check algorithm [19]. Complexity for the attack was 2^{55} encryptions given precise cache measurements for 6148 chosen rounds. He noted that Dragon [4] is more resistant to cache-timing attacks than might be expected for an algorithm so heavily dependant on s-boxes, since the twelve invocations per cycle of each s-box touch on average 8.6 cache lines. The attacker has no way of knowing which of the $2^{57.7}$ ways of ordering the s-boxes is applicable. Dragon's usage of s-boxes is similar to K2's.

In [14] Leander, Zenner and Hawkes launched a devastating attack on Sosemanuk, and generalized it to other LFSR-based registers. The attack relies on the property in Section 3 and the fact that each lookup table in the LFSR is accessed at most once per cipher clock. Sosemanuk is similar in structure to K2, in that it contains an autonomous LFSR that contributes input to both a FSM and to the keystream generation function. In Sosemanuk, the LFSR consists of ten 32-bit word stages, with a feedback polynomial that includes α and α^{-1} multipliers implemented by way of separate lookup tables. Each time the LFSR is clocked, one access is made to the α table depending on bits 24..31 of one word of the LFSR, and one access is made to the α^{-1} table depending on bits 7..0 of the same word of the LFSR. This has the consequence that in each clock of the LFSR, on machines with 64-byte cache lines, cache timing measurements leak eight bits across two words of the state, namely bits (31..28) of s_t and bits (7..4) of s_{t+3}. After forty words of output, sufficient equations exist to form a system with full rank, and Gaussian elimination can be used to determine the state of the Sosemanuk LFSR. Once the state of the LFSR is known, it is easy to determine the contents of the 64-bit FSM by guessing one of the 32-bit words and using the keystream to determine the other. Determining the FSM dominates the attack so the cipher can be broken in $O(2^{32})$. This is a practical attack, assuming that the cache-timing measurements can be practically and reliably obtained.

Application to K2. K2 has several tables that reside in the cache during keystream generation. These are s-box tables for the finite state machine, and tables that hold coefficients for feedback polynomials for the shift registers.

Attacks based on the α tables. There are four such tables, for $\alpha_0, \alpha_1, \alpha_2$ and α_3. Each table occupies sixteen cache lines on our model machine, leaking four bits of data per displaced cache line during a prime-and-probe attack.

One lookup each to α_0 and α_3 occurs during every cycle. A further lookup occurs, either to α_1 or to α_2, depending on the value $r_{t+2}(31..30)$. Observing whether α_1 and α_2 have been loaded leaks the values of bits $r_{t+2}[30]$ and $r_{t+2}[31]$, in addition to the upper nibbles of r_t, s_t and s_{t+8}, for that cycle. There is a clear linkage between each table and its index. This leakage of DFC information is sufficient to recover the entire state of FSR-A, due to the intra-word diffusion offered by α-multiplication. The absence of an α^{-1} table in K2 means that the rate of leakage of bits due to cache timing measurements is half that of SNOW 2.0 and Sosemanuk, and only four equations are generated per cycle. However, the FSR-A register is very short, and forty rounds are sufficient to completely determine its contents using the following algorithm.

Once the Dynamic Feedback Control bits of FSR-A have been determined, FSR-B can easily be determined. During each clock cycle, the cipher will access table α_1 with probability 0.5 otherwise it will access α_2. There are four different ways to clock FSR-B, and we know with certainty which way to use at each cycle. We focus on table α_1, and ignore the cache measurements made to α_2 and α_3. This helps to ensure that all of the collected equations are linearly independent. At each cycle, we update the matrix used to formulate the equations according to which of the tables are selected (See [14] for details on this algorithm). Then after 88×2 rounds on average, we have sufficient equations from α_1 to form a full-rank system of equations and solve FSR-B.

The dominating factor in the attack is determining the state of the four word FSM. Assume the attacker guesses on $R1$ at time t, then he knows $R2_{t+1} = Sub(R1_t)$. Since the known keystream z_t^L is formulated in terms of known FSR contents s_t and r_{t+4}, known $R1_t$ and unknown $R2_t$, the attacker learns $R2_t$ and likewise from z_{t+1}^L, $R1_{t+1}$. $R1_{t+1} = Sub(L2_t \oplus s_{t+4})$, the last term of which is also known, so $L2_t$ is revealed, and then by the relationship with z_t^H, $L1_t$ also becomes known. The complexity of the entire attack on the 640-bit state is $O(2^{32})$. We have verified this empirically, using simulated cache-timing measurements.

Countermeasure. Leander et al. [14] propose the countermeasure of splitting each alpha table into two smaller tables, such that each fits into one cache line. As each α table is linear, it can be broken into two tables α_H and α_L such that for every $x, 0 \leq x < 16$, $\alpha_H(x) = \alpha(x \ll 4), \alpha_L(x) = \alpha(x)$. Then $\alpha(y), 0 \leq y < 256$ can be computed as $\alpha_H(y \gg 4) \oplus \alpha_L(y\&0\text{xF})$. Each table contains only sixteen elements, so fits into a 64-byte cache line. The performance degradation is minimal, amounting to a few percent.

This solution does not directly apply to K2, because of the use of DFC to select α_1 or α_2. Even with the Leander et al. countermeasure, two of the four cache lines used to hold these tables remain untouched during each cycle and leak the value of DFC_BIT_1. When DFC_BIT_2 = 0, two cache lines also remain untouched.

The solution is to lookup α_3 tables irrespective of the value of DFC_BIT_2 and to discard when necessary; and to further decompose each of α_1 and α_2 into four interleaved four-element tables accessed respectively on bits $(6, 7),(4, 5)$, $(2, 3)$ and $(0, 1)$ of x. In this solution, during each cycle of K2, all six cache-lines holding α-table lookups are accessed, and no information about α-tables or DFC is leaked. In our optimized implementation, there is no degradation in throughput due to this countermeasure. See Appendix A for more information on this technique.

Attacks based on the *Sub* function. The *Sub* table is an 8×32 table that occupies 16 cache lines. To avoid the use of rotation, the table may be expanded into four tables occupying 64 cache lines. If only one table is used, then during one iteration of the keystream generation algorithm, it is invoked sixteen times, touching on average 10.3 cache lines. In the worst case there are $16! = 2^{44.3}$ ways of the ordering the table accesses. If separate tables are used, then on average during each clock 3.6 cache lines are touched for each table. In the worst case there are $2^{18.3}$ ways of ordering the tables. This disparity is a strong motivator for implementing a single table for the *Sub* function rather than four tables.

K2 can be attacked using cache-timing measurements to completely determine FSR-A and FSR-B in isolation. Then the contents of the FSM can be retrieved at time t by guessing on $R1$ for a complexity of $O(2^{32})$. But this overlooks that the FSM also leaks significant information every time a *Sub* function is accessed. In particular, during each cycle, the *Sub* function is accessed four times per cycle, each time leaking sixteen bits of information from each of four registers, since the component bytes to its index access separate tables.

The attacker does not need to guess the association of each displaced cache line with the appropriate nibble of the register. Instead he must just establish the correct value of any one of the FSM variables, a process that has the complexity $4^4 \cdot 2^{16} = 2^{24}$. He deduces the remaining variables using two words of keystream and the relationships between the FSM variables at time t and $t + 1$.

Thus without countermeasures against cache-timing attacks, K2 can be broken with complexity $O(2^{24})$.

If countermeasures are implemented to protect against leakage in FSR-A and FSR-B, the leakage from the FSM can still be used to recover the entire state of the cipher. Assume that the attacker knows how to associate displaced cache lines with *Sub* operations at time t and $t + 1$. He uses cache timing information to guess $R2_t + B[4+t]$ (or $L2_t + B[9+t]$) at time $t = 0$. There are 2^{16} candidates. However, the cache timing information acquired for $L1_{t+1} = Sub(R2_t + B[4+t])$ (or $R1_{t+1}$) filters those candidates by $(2^{-4})^4$, leaving on average only a single candidate left. If the attacker knows the value of $R2_t$ ($L2_t$), then he learns the value of $B[4 + t]$ ($B[9 + t]$).

In that case, the attacker can apply the following algorithm. To learn words $B[4]...B[4 + b]$ and words $B[9]...B[9 + b]$, he needs to acquire cache timing measurements for all of the *Subs* for consecutive cycles $t - 1...t - 1 + b$ (at time $t - 1$, he does not know $R2_t$ or $L2_t$, so cannot learn FSR-B words, but still needs to establish the respective values).

He must make as many as $(4!)^{(b+1)}$ guesses for the association between each *Sub* and its displaced cache line (this is the worst case, where each *Sub* uniquely displaces a cache line in its cycle). Then he can acquire $B[4]...B[4+b-1]$, $B[9]...B[9+b-1]$ corresponding to his guess. If $b=5$, then the attacker knows $B[4]...B[13]$, which is the complete state of FSR-B. The keystream, phrased in terms of the respective known FSR-B and FSM state words provides six non-consecutive words of FSR-A; the remaining words can be solved via its recurrence. Then the attacker uses consistency checking against the keystream to check whether the guess on cache line-Sub associations is correct.

The complexity of the attack for the four-table version of K2 is $(4!)^{4 \cdot (b+1)} \cdot 2^{64+(log_2 b+1)} = 2^{176.6}$ Subs. Profiling reveals Sub to be the dominant operation in the keystream generation algorithm. Since a brute force attack on a 256-bit key involves 28 cycles of the update algorithm, each involving 4 $Subs$, we consider this attack to have the equivalent complexity of checking 2^{170} keys. The complexity of attacking the one-table version increases to $(16!)^{(b+1)} \cdot 2^{64+(log_2 b+1)} = 2^{332}$ $Subs$, much worse than brute-force.

Countermeasure. There is no need for a countermeasure for the one-table version of Sub, if countermeasures for the FSR-leakage has been implemented.

Leakage from s-boxes can be ameliorated by using the forthcoming Intel/AMD AES instructions [8]. The sequence of operations in the Sub function differs from those in the AES round in that it omits the ShiftRow and KeyBytes operations. However the K2 tables are identical to those used in optimized 32-bit implementations of the AES, since the ShiftRow operation is inbuilt into the indices used to lookup those tables, rather than in the tables themselves, and the KeyBytes operation is performed externally. On the Intel platform, placing the Sub input into the most significant word of XMM1, setting XMM2 to 0, then calling AES-ENC produces the correct result for the Sub function. The tables used by these instructions are on-chip and never placed in the cache.

5   Power Analysis

Power analysis is a side-channel attack technique in which the adversary studies the power consumption of a cryptographic hardware device at various stages of its operation. The attack is non-invasive and passive, and usually conducted using a digital sampling oscilloscope.

The power consumption of a device is the sum of the power dissipated by all of its gates, and includes various noise components. The power consumption varies according to the types of operations being executed and secondarily on the values of their operands [1]. Therefore, assuming that noise can be controlled, measuring the power consumption of the device leaks information about those operations [1]. For example, a multiplication operation consumes more power than an addition operation, and writing 'on' bits uses more power than writing 'off' bits.

---

[1] Kocher et al. note that measurement errors and noise sometimes subsume power variations due to operand values [13].

5.1 Simple Power Analysis

In the latter case, this leads to the notion that an attacker can deduce the hamming weight of any operands by observing traces, the set of power measurements made during an execution of the algorithm. This forms the basis of data-dependent Simple Power Analysis (SPA). SPA provides information about cipher state internals using visual inspection of a single sample. This usage model is consistent with the stream cipher requirement that each key-IV pair is used only once during encryption. If the attacker can determine where certain instructions are being executed, he can determine the Hamming weight of each byte or word by measuring the power consumption at the cycle of the instruction that accesses this data.

Another kind of SPA utilizes the dependence of power consumption upon the order of instructions. The attacker studies the power-consumption trace to determine the order of instructions. If the order of instructions is related to values of key bits (ie. conditionals that consist of an expression of key bits), then those key bits might be deduced. This is similar to the principle underlying the conventional timing attack. Since we showed how to implement K2 to be immune to timing attacks, the same advice applies here, and we don't consider operation-based SPA further.

The body of work on SPA is very large although there are no pre-existing results on K2. or In the analysis of K2, we adopt a similar attack model as that of Gierlichs et al [7], in which the adversary observes a perfect Hamming weight or Hamming distance leakage. By measuring the hamming weights of 8- and 32-bit operands, the attacker learns 2.54 and 3.55 bits of information respectively. K2 uses some 32-bit operations that contain 24 bits of entropy. In these cases, the attacker can learn 3.34 bits of entropy from the hamming weight.

Application to K2. In our model, the attacker knows keystream produced by the algorithm using a fixed key and adaptively chosen IVs. He can choose an IV and reset the device using that IV. However, he cannot reuse the same key-IV pair as previously.

There are two fruitful sources of information in the keystream generation algorithm - the α-table lookups and the Sub operations.

Although the operands involved in α-table lookups are 32-bit words, the difference in Hamming weight is upper bounded by a value of eight, due to the implementation of the α multiplication as $\alpha(x) = (x \ll 8) \oplus A[x \gg 24]$. While bit-wise shifting and rotation operations that shift one bit at a time are extremely vulnerable to SPA, we assume that the shifting here occurs either as a native instruction or as a byte permutation. Then the attacker might obtain two separate hamming weights concerning x: the hamming weight of the most significant byte, and the hamming weight of the remaining 24 bits. Then the leakage concerning x is $2.54 + 3.34 = 5.88$ bits. Since there are three such multiplications per cycle, the α lookups leak 17.64 bits of the internal state.

FSM-C involves four *Sub* operations. Each Sub operation involves four 8×32 table lookups and three exclusive-or operations implemented as [2]:

```
1 #define SUB(x)  (aes_t0[x & 0xFF] ^ \
2                   aes_t1[(x >> 8) & 0xFF] ^ \
3                   aes_t2[(x >> 16) & 0xFF] ^ \
4                   aes_t3[(x >> 24)])
```

This is a fertile source for Hamming-weight SPA, since the 32-bit x is broken into bytes, and leaks $2.54 \times 4 = 10.16$ bits of the contents. The recombination of the four thirty-two bit quantities using exclusive-or again leaks $3.55 \times 4 = 14.2$ bits of information, although it is not independent of the information leaked at the input.

There are four such *Sub*s in the FSM, leaking at least $10.16 \times 4 = 40.64$ independent bits of information in addition to the 17.64 bits leaked in the FSRs. The total of 58.28 bits is insignificant compared to the total state space of 640 bits. While SPA can be conducted across several cycles to derive several independent measurements of FSR entropy, the high diffusion in the FSM means that entropy measurements made in the FSM quickly become related to entropy measurements made in FSR-B. Note that information contained in FSR-A is extremely slow to diffuse into FSR-B or the FSM. The converse is also true in keystream generation, when FSR-A is autonomous.

Key initialization algorithm. The initialization algorithm contains three phases. In the first phase, the secret key is transformed into an extended key consisting of eleven 32-bit words. Generally these words are composed by chaining using 32-bit exclusive-or operations, each of which leaks 3.55 bits on operation output but not on the individual key words used. For a 128-bit key, there are two additional operations involving a 32-bit rotation and a *Sub* operation. As indicated previously, the *Sub* operation leaks 10.16 bits on the input, which in the first instance consists bits 96..127 of the secret key. For a 256-bit key, only one *Sub* operation occurs; this also leaks 10.16 bits about bits 224..255 of the secret key.

It is instructive to consider some empirical results on SPA of a 128-bit key during the first phase of key initialization. The attacker monitors the *Sub* operations and is able to deduce the Hamming weights of the individual bytes that compose their inputs and outputs, including EK_7 and EK_3, and consequently EK_6 since $EK_6 = EK_7 \oplus EK_3$. He builds on a byte-by-byte basis candidates for EK_3 and EK_7. If there are c_1 candidates for EK_7 and c_2 candidates for EK_3, then he also has $c_1 \times c_2$ candidates for EK_6. If the attacker can independently measure the hamming weight of EK_6, he is able to filter out 86% (based on the median over twenty thousand trials) of the pool of candidates permitted by the combination of EK_3 and EK_7. In some cases, the three values are unambiguously determined. More information on this experiment is available in Appendix B.

EK_6 and EK_7 do not contain raw key material, so loss of entropy here is not directly utilizable unless measurements are taken from the exclusive-ors on

---

[2] In the four-table implementation; however the leakage for the one-table implementation is identical.

other other key words. Hamming-weight measurements on exclusive-ors is not immediately rewarding since if the Hamming weight of only one quantity is known, this does not restrict the range of Hamming weights of the other two operands. Only EK_3 is directly taken from the master key. The attacker may learn a further $3 \times 3.55 = 10.65$ independent bits of of master key from exclusive-or operations on $EK_{0..2}$. So we estimate that SPA removes 35 bits of entropy from a 128-bit key (24 bits from EK_3, and 10.65 bits from the remaining three master key words). The remaining material cannot easily be brute-forced.

The attacker can make additional measurements, which leaks entropy in excess of what is described here. However the efficient mixing of K2, in particular the optimal MixColumn diffusion components, means that while each additional 32-bit operation leaks 3.55 bits, a significant portion of this is likely to be known through previous measurements. Untangling that which is known from that which is not is non-trivial.

Differential Power Analysis. (DPA) relies on the dependence of power consumption upon data. DPA uses multiple samples produced under the same key but different IVs to reduce algorithmic noise in the observed power traces. In the stream cipher context, this requirement for multiple samples is frequently not practical [3]. An advantage of DPA is that detailed knowledge of the device is not required.

The attack analyses intermediate values of one or several bits, either at one instant of time (first-order DPA) or at some instants of time (higher-order DPA). The attack relies heavily on the accuracy of the power consumption model. If this is wrongly modelled, then key detection is impossible.

DPA works best when involving a non-linear operation in conjunction with known plaintext (ie. IV) [7]. In K2 the IV is not involved until the second phase of key initialization, and it is not involved with non-linear elements until the third. The attacker is principally interested in:

$$L1_{-21} = Sub(Sub(Sub(EK_5))) \boxplus IV_2)$$
$$L1_{-20} = Sub(Sub(Sub(EK_6 \boxplus 0x63)) \boxplus IV_3)$$

Since IV_2 and IV_3 are known constants which can be chosen by the adversary, EK_5 and EK_6 are the targets of key hypotheses that allow the prediction of intermediate values $L1_3$ and $L1_4$. In a DPA attack, correctly guessing EK_5 and EK_6 presents a high correlation coefficient, and the attacker can determine $EK_2 = EK_5 \oplus EK_6$.

Note that IV_0 and IV_1 are not directly involved in the FSM expressions; because of their placement towards the start of FSR-B, they are never selected as input into $R1$ or $R2$.

For this attack on the *Sub*, 32 initial key bits are recovered by considering 2^8 key hypotheses individually for each byte of K_5 and K_6. The complexity of recovering the 128-bit master key is $O(2 \cdot 4 \cdot 2^8) + O(2^{96})$, which is dominated by the brute-force search of the latter term. The complexity of this attack means it is not a threat to the security of K2.

5.2   Countermeasures

In order to generate keystream efficiently, K2 is designed to be implemented as a series of lookups to large tables in combination with a number of arithmetic operations. There are as many as eight $2^8 \times 32$-bit tables. Gierlichs et al. [7] consider that masking boolean operations and small tables is inexpensive, while masking larger (of size greater than 256 byte) tables is more difficult, requiring additional memory and loss of throughput. Rechberger et al. [16] recommend using as few kinds of different operations as possible in ciphers for which affordable countermeasures are required. S-boxes which can be implemented as combinatorial logic are considered easy to mask. Other kinds of s-boxes, such as K2's, are more difficult. The cost of protecting K2 against DPA is therefore expected to be expensive.

6   Fault Analysis

Differential Fault Analysis is an active side-channel attack. In the attack, one or more faults are injected into the algorithm by an adversary by, for example, varying external voltage or an external clock, or shining a crypto-processor with visible or laser light [9]. Such attacks are not necessarily expensive [17].

   We generally assume that the attacker can create transient bit-flipping faults in a particular register (ie. FSR-A, FSR-B, L1, L2, R1 or R2) with precise timing. In the asymmetric model, as typified in EEPROM, one-bits turn to zero-bits with much greater frequency than the converse. In this model, Biham and Shamir were able to show that any symmetric cryptosystem is suspectible to a simple DFA attack [2]. In the symmetric model, the frequency with which one-bits and zero-bits flip is equal. We use the following assumptions:

1. The attacker is able to inject exactly one fault at the chosen position of the internal state of the analysed stream cipher
2. The attacker is able to repeat the fault injection at the chosen position of the internal state of the analysed stream cipher for the same IV-key pair.
3. The attacker is able to obtain both the correct and faulty keystreams for analysis.

The second assumption is controversial in the symmetric stream cipher model.

6.1   Related Work

The predominant body of work on DFA of stream ciphers is by Hoch [10,9]. In his thesis, he presents generic frameworks for attacks on LFSRs with non-linear filters, clock-controlled LFSRs, and less successfully on LFSRs filtered by Finite State Machines. He presents specific attacks on real-life ciphers including SNOW 2.0, which he attacks by inducing single-bit faults at bit j in R1 at times t and $t + 2$. According to Hoch, as

$$z_t = (s_{t+15} \boxplus R1_t) \oplus R2_t \oplus s_t \tag{1}$$

this induces a difference of $\pm 2^j$. Therefore, inducing faults at all positions in j results in recovery of $R1$. Then, by applying fault induction for $t = 0..15$, and utilizing the relationships between $R1$, $R2$ and z, and the LFSR recurrence, the state can be retrieved using 1000 faults. We observed that this can be reduced to 384 faults by using the recurrence relation of the LFSR. Hoch assumes that equation 1 leads to unequivocal discovery of the faulty bit value.

Two measures of difference between the keystreams can be used to derive information about the generated fault.

$$z_t \oplus z'_t = ((s_{t+15} \boxplus R1_t) \oplus R2_t \oplus s_t) \oplus ((s_{t+15} \boxplus R1'_t) \oplus R2_t \oplus s_t) \quad (2)$$

For a single bit fault at position j the pattern of carries from bits $0...j$ are equal, so

$$z_t^{(j)} \oplus z_t'^{(j)} = ((s_{t+15}^{(j)} \oplus R1_t^{(j)}) \oplus R2_t^{(j)} \oplus s_t^{(j)}) \oplus ((s_{t+15}^{(j)} \oplus R1'_t^{(j)}) \oplus R2_t^{(j)} \oplus s_t^{(j)})$$
$$= R1_t^{(j)} \oplus R1'^{(j)}_t$$

Using the exclusive-or difference can identify the location of the bit fault at the least significant bit of the non-zero difference.

Identifying the value of the bit fault requires the following measure of difference:

$$z_t \boxminus z'_t = ((s_{t+15} \boxplus R1_t) \oplus R2_t \oplus s_t) \boxminus ((s_{t+15} \boxplus R1'_t) \oplus R2_t \oplus s_t) \quad (3)$$

At the location of the bit fault:

$$z_t^{(j)} \boxminus z_t'^{(j)} = ((s_{t+15}^{(j)} \oplus R1_t^{(j)}) \oplus R2_t^{(j)} \oplus s_t^{(j)}) \boxminus ((s_{t+15}^{(j)} \oplus R1'^{(j)}_t) \oplus R2_t^{(j)} \oplus s_t^{(j)})$$
$$= (R1_t^{(j)} \oplus x) \boxminus (R1'^{(j)}_t \oplus x)$$

where $x = (s_{t+15}^{(j)} \oplus R2_t^{(j)} \oplus s_t^{(j)})$. If $x = 0$ and $z_t^{(j)} \boxminus z_t'^{(j)} = 1$ then $R1_t^{(j)} = 1$ and $R1_t^{(j)} = 0$; if $x = 1$ the converse applies. However, x is unknown, and the fault value cannot be derived.

We made the observation that the attack can corrected for SNOW 2.0 by utilizing the structure of its embedded AES s-box. Injecting a bit fault into $R1$ at time t also affects the keystream at time $t + 1$, since $R2_{t+1} = Sub(R1_t)$.

$$z_{t+1} \oplus z'_{t+1} = Sub(R1_t) \oplus Sub(R1'_t) \oplus d \oplus d'$$

where $d = (s_{t+16} \boxplus R1_{t+1}) \oplus s_t$, but $d = d'$ since a bit fault in R1 at time t affects none of the terms in d. The attacker is able to guess the value of $R1_t$ and $R1'_t$, since by the properties of the non-linear s-box there are only 2 or 4 candidates per byte. Injecting a second bit fault in the same byte reduces the number of candidates to 1. By varying the byte in which the fault is contained, the attacker is able to determine the value of $R1_t$ in $O(2^{8+4})$. Consequently 8 bit faults are required to identity the value of $R1_t$, and 96 bit faults are required to determine the state of the entire cipher.

6.2   Attacking FSR-A and FSR-B of K2

The construction of K2 allows the contents of FSR-A to be determined in the asymmetric EEPROM model, where '1'-bits flip to 'D0' but not the converse. The location of a random bit fault in FSR-A can be unambiguously detected within five cycles of keystream. If the fault occurs in $A[0]$, it is immediately detectable at time t in $\Delta z_t^H = z_t^H \oplus z'^H_t$. The location of the bit difference in Δz_t^H directly reflects the location of the faulty bit since $A[0]$ is combined into z_t^H using the bijective bitwise operator exclusive-or ($\oplus$). Likewise a bit fault in $A[4]$ is reflected in the same location in Δz_t^L. If no difference is forthcoming in Δz_t^H or Δz_t^L then the bit fault is located in $A[1]$, $A[2]$ or $A[3]$ and the second word of keystream must be examined. If the difference occurs in Δz_{t+1}^L then the location of the difference bit reflects that the bit fault occurred in the corresponding location in $A[3]$. If there is no difference in Δz_{t+1}^H or Δz_{t+1}^L, then the bit fault occurred somewhere between (inclusive of) bits 0 and 29. This can be determined by looking at the corresponding difference bit in z_{t+2}^H. Whether or not the bit fault occurred in bits 31 or 30 of $A[2]$, or in $A[1]$ can be determined by looking at Δz_{t+3}^L. If that word contains no difference then the fault occurred in $A[1]$ at the corresponding location to the bit fault in Δz_{t+1}^H. Otherwise the bit fault occurred in $A[2]$ at the corresponding location given by the difference bit in Δz_{t+2}^H.

If the fault occurs in $r_{t+2}[30]$ then in the next cycle $\Delta s_{t+11} = \alpha_1(s_t) \oplus \alpha_2(s_t)$. Bytes 0 and 2 of Δs_{t+11} have the full complement of potential values, but bytes 1 and 3 each only have 2^6 possible values. Because the keystream word $\Delta z_{t+1}^H = (s_{t+11} \boxplus L2_{t+1}) \oplus (s'_{t+11} \boxplus L2_{t+1})$, where $L2_{t+1}$ has the full range of potential values, it is difficult to know how to utilize this reduction in entropy of Δs_{t+11}.

Assuming that the location of the bit fault can be precisely controlled, then in the asymmetric model the contents of FSR-A can be determined using about 320 bit faults and about 1280 words of keystream, although it is possible to reduce the amount of keystream by inducing multiple bit faults, the location of which can be easily determined, due to the simple way in which FSR-A contributes to the keystream.

In the symmetric model, although the bit fault can be easily located, the value of the faulty bit cannot be determined due to the absence of non-linear elements in FSR-A, and the combined noise of FSR-B and the FSM.

Attacking FSR-B has the same methodology as that for FSR-A.

6.3   Attacking the FSM

The FSM in SNOW 2.0 can be attacked by determining candidates for $R1$ because the relationship $z_{t+1} \oplus z'_{t+1}$ reduces to $Sub(R1) \oplus Sub(R1')$.

However, in K2, if a bit fault is introduced into $R1$, then the relationship between the correct and faulty keystreams at time $t + 1$ is:

$$
\begin{aligned}
z_{t+1}^L \oplus z'^L_{t+1} &= (s_{t+1} \boxplus R2_{t+1}) \oplus R1_t \oplus r_{t+4} \oplus (s'_{t+1} \boxplus R2'_{t+1}) \oplus R1'_t \oplus r'_{t+4} \\
&= (s_{t+1} \boxplus R2_{t+1}) \oplus (s_{t+1} \boxplus R2'_{t+1}) \\
&= (s_{t+1} \boxplus Sub(R1_t)) \oplus (s_{t+1} \boxplus Sub(R1'_t))
\end{aligned}
$$

For K2, it is necessary to know the pattern of the carries induced by s_{t+1}, or to know s_{t+1}. An alternative is to generate enough bit faults over sufficient cycles for s_{t+1} to be equal to zero, in which case there are no carries.

In either case, the attack is not straight-forward. K2 appears to offer good resistant to temporary-bit flipping differential fault analysis because of it switches the order of modular and binary addition in the FSM contribution to keystream, relative to that in SNOW 2.0.

7   Conclusion

The primary weakness of K2 with respect to side channel analysis is that the state of the autonomous LFSR can be determined with certainty after $160 \cdot k$ cycles if one bit is leaked every k cycles. Without countermeasures in place, it can be attacked independently and stripped from the cipher.

Countermeasures against timing attacks are simple. DFC is naturally implemented using branching but by implementing the α-table lookups in conjunction with bit masks and variable offsets, no leakage occurs. This slows the algorithm down since three α-lookups rather than an average of 2.5 must be made per cycle.

Leander, Zenner and Hawkes [14] showed that all word-oriented LFSR-based stream ciphers are vulnerable to cache-timing attacks when implemented without countermeasures. Splitting the linear $\alpha-$ tables and interleaving where necessary, so that each of the respective six caches lines is always touched in each cycle, and using the Intel/AMD AES instructions to implement the Sub operations removes all leakage of K2 by cache-timing attacks.

The K2 keystream generation algorithm is well-defended against power analysis. The Sub operations in conjunction with exclusive-or operations on raw key material leak about 35 bits of a 128-bit master key. Half of the IV material is not directly involved with key material during phase 3 of the key initialization, offering some resistance against DPA. The complexity for the best DPA attack we have discovered so far is $O(2^{96})$.

K2 seems resilient to differential fault analysis because of the use of incompatible operations in combining elements from FSR-A, FSR-B and FSM to produce keystream output. An attack on one component must contend with the combined noise of the remainder. We have not found any effective differential fault attack using single-bit transient flipping.

Table 1. Benchmarks for some implementations of K2

Cipher	Options	Cycles per byte	Megabits per second
K2	optimized with four tables	13.48	1597
K2	optimized with one table	14.66	1467
K2	Leander et al. countermeasure	14.63	1476
K2	beta-tables implementation	28.76	748

The efficiency cost of the countermeasures are shown in Table 1. The implementation platform is an Intel Core Duo 2 2.2 GHz. No special optimization tricks are employed, so the throughput can be improved.

Even without countermeasures, K2 offers reasonable resistance to side-channel attacks. We note that K2 can only be considered broken by cache-timing measurements if the AES is also considered broken. We have also shown that K2 is more resistant to cache-timing and DFA attacks than SNOW 2.0.

References

1. Aigner, M., Oswald, E.: Oswald: Power analysis tutorial. Technical report, Institute for Applied Information Processing and Communication, University of Technology Graz - Seminar (2001)
2. Biham, E., Shamir, A.: Differential fault analysis of secret key cryptosystems. In: Kaliski Jr., B.S. (ed.) CRYPTO 1997. LNCS, vol. 1294, pp. 513–525. Springer, Heidelberg (1997)
3. Chari, S., Rao, J.R., Rohatgi, P.: Template attacks. In: Jr., et al [11], pp. 13–28
4. Chen, K., Henricksen, M., Millan, W., Fuller, J., Simpson, L., Dawson, E., Lee, H., Moon, S.: Dragon: A fast word based stream cipher. In: Park, C.-s., Chee, S. (eds.) ICISC 2004. LNCS, vol. 3506, pp. 33–50. Springer, Heidelberg (2005)
5. Courtois, N., Pieprzyk, J.: Cryptanalysis of block ciphers with overdefined systems of equations. In: Zheng, Y. (ed.) ASIACRYPT 2002. LNCS, vol. 2501, pp. 267–287. Springer, Heidelberg (2002)
6. Ekdahl, P., Johansson, T.: A new version of the stream cipher SNOW. In: Nyberg, K., Heys, H.M. (eds.) SAC 2002. LNCS, vol. 2595, pp. 47–61. Springer, Heidelberg (2003)
7. Gierlichs, B., Batina, L., Clavier, C., Eisenbarth, T., Gouget, A., Handschuh, H., Kasper, T., Lemke-Rust, K., Mangard, S., Moradi, A., Oswald, E.: Susceptibility of eSTREAM candidates towards side channel analysis. In: Proceedings of SASC 2008, Lausanne, Switzerland, February 2008, pp. 123–150 (2008); Special Workshop hosted by the ECRYPT Network of Excellence. Proceedings available at, http://www.ecrypt.eu.org/stvl/sasc2008/
8. Gueron, S.: Intel's new AES instructions for enhanced performance and security. In: Dunkelman, O. (ed.) FSE 2009. LNCS, vol. 5665, pp. 51–66. Springer, Heidelberg (2009)
9. Hoch, J.J.: Fault analysis of stream ciphers. Master's thesis, Weizmann Institute of Science (2005)
10. Hoch, J.J., Shamir, A.: Fault analysis of stream ciphers. In: Joye, M., Quisquater, J.-J. (eds.) CHES 2004. LNCS, vol. 3156, pp. 240–253. Springer, Heidelberg (2004)
11. Kaliski Jr., B.S., Koç, Ç.K., Paar, C. (eds.): CHES 2002. LNCS, vol. 2523. Springer, Heidelberg (2003)
12. Kiyomoto, S., Tanaka, T., Sakurai, K.: K2: A stream cipher algorithm using dynamic feedback control. In: SECRYPT 2007, International conference on Security and Cryptography, Barcelona, Spain, July 28-31 (2007)
13. Kocher, P.C., Jaffe, J., Jun, B.: Differential power analysis. In: Wiener, M. J. (ed.) CRYPTO 1999. LNCS, vol. 1666, pp. 388–397. Springer, Heidelberg (1999)
14. Leander, G., Zenner, E., Hawkes, P.: Cache Timing Analysis of LFSR-based Stream Ciphers. In: Twelfth IMA International Conference on Cryptography and Coding, Royal Agricultural College, Cirencester, UK (December 2009)

15. Matsui, M.: Linear cryptanalysis method for DES cipher. In: Helleseth, T. (ed.) EUROCRYPT 1993. LNCS, vol. 765, pp. 386–397. Springer, Heidelberg (1994)
16. Rechberger, C., Oswald, E.: Stream ciphers and side-channel analysis. In: SASC 2004, The State of the Art of Stream Ciphers, Bruges, Belgium, pp. 320–326 (2004)
17. Skorobogatov, S.P., Anderson, R.J.: Optical fault induction attacks. In: Jr., et al [11], pp. 2–12
18. Wu, H.: The stream cipher HC-256. In: Roy, B. K., Meier, W. (eds.) FSE 2004. LNCS, vol. 3017, pp. 226–244. Springer, Heidelberg (2004)
19. Zenner, E.: A Cache Timing Analysis of HC-256. In: Avanzi, R.M., Keliher, L., Sica, F. (eds.) SAC 2008. LNCS, vol. 5381, pp. 199–213. Springer, Heidelberg (2009)
20. Zenner, E.: Cache timing analysis of eStream finalists. In: Dagstuhl Seminar on Symmetric Cryptography 2009 (January 2009), www.erikzenner.name/docs/2009_Dagstuhl_Talk.pdf

A   Implementation

The following code is a pseudo-optimized version of the K2 keystream generation with countermeasures against timing and cache-timing attacks.

```
#include "k2.h"

ulong alpha_0H[] = {
    0X00000000, 0X31801F63, 0X62C33EC6, 0X534321A5, 0XC4457C4F, 0XF5C5632C,
    0XA6864289, 0X97065DEA, 0X4B8AF89E, 0X7A0AE7FD, 0X2949C658, 0X18C9D93B,
    0X8FCF84D1, 0XBE4F9BB2, 0XED0CBA17, 0XDC8CA574
};

ulong alpha_0L[] = {
    0X00000000, 0XB6086D1A, 0XAF10DA34, 0X1918B72E, 0X9D207768, 0X2B281A72,
    0X3230AD5C, 0X8438C046, 0XF940EED0, 0X4F4883CA, 0X565034E4, 0XE05859FE,
    0X646099B8, 0XD268F4A2, 0XCB70438C, 0X7D782E96
};

/* Entries in this table are interleaved:
 * alpha_2(bits 6, 7); alpha_1(bits 6, 7); alpha_2(bits 4, 5);
 * alpha_1(bits 4, 5); alpha_2(bits 2, 3); alpha_1(bits 2, 3);
 * alpha_2(bits 0, 1); alpha_1(bits 0, 1);
 */
ulong alpha_1L[] = {
    0X00000000, 0X8AA735A6, 0X59036A01, 0XD3A45FA7, 0X00000000, 0X7C2F35B2,
    0XF85E6A49, 0X84715FFB, 0X00000000, 0X84DC5E8F, 0X45F5BC53, 0XC129E2DC,
    0X00000000, 0X1FD646BA, 0X3E818C59, 0X2157CAE3, 0X00000000, 0X2137B1D6,
    0X426E2FE1, 0X63599E37, 0X00000000, 0XDAA387B8, 0X996B235D, 0X43C8A4E5,
    0X00000000, 0X5BF87F93, 0XB6BDFE6B, 0XED4581F8, 0X00000000, 0XA0F5FC2E,
    0X6DC7D55C, 0XCD322972
};

ulong alpha_3H[] = {
    0X00000000, 0XA104F437, 0X27088D6E, 0X860C7959, 0X4E107FDC, 0XEF148BEB,
    0X6918F2B2, 0XC81C0685, 0X9C20FEDD, 0X3D240AEA, 0XBB2873B3, 0X1A2C8784,
    0XD2308101, 0X73347536, 0XF5380C6F, 0X543CF858
};

ulong alpha_3L[] = {
    0X00000000, 0X4559568B, 0X8AB2AC73, 0XCFEBFAF8, 0X71013DE6, 0X34586B6D,
    0XFBB39195, 0XBEEAC71E, 0XE2027AA9, 0XA75B2C22, 0X68B0D6DA, 0X2DE98051,
    0X9303474F, 0XD65A11C4, 0X19B1EB3C, 0X5CE8BDB7
};
```

```
#define CLOCK_INDEX(index, size) index = (index + 1) % size;

#define ALPHA(x, table, offset)  \
    (SHL32(x, 8) & 0xFFFFFF00) ^ table[BYTE3(x) + offset]

#define SUB(x, y) \
    y  = aes_t0[BYTE0(x)] ^ \
    ROL32(aes_t0[BYTE1(x)], 8) ^ \
    ROL32(aes_t0[BYTE2(x)], 16) ^ \
    ROL32(aes_t0[BYTE3(x)], 24);

void K2_update(K2Ctx* ctx, uchar* keystream, const ulong keystream_bytes)
{
    ulong l2, r2;
    ulong a0, b0;
    ulong d, s, t;
    ulong i;
    ulong a1_offset;
    ulong *index = (ulong *)keystream;

    for (i = 0; i < keystream_bytes >> 3; i++) {
        l2 = ctx->L2;
        r2 = ctx->R2;
        a0 = ctx->FSR_A[ctx->FSR_A_index];
        b0 = ctx->FSR_B[ctx->FSR_B_index];

        /* Generate keystream from FSM, FSR-B, FSR-A */
        *(index++) = (FSR_B_STAGE(ctx, 10) + l2) ^ ctx->L1 ^ a0;
        *(index++) = (b0  + r2) ^ ctx->R1 ^ FSR_A_STAGE(ctx, 4);

        /* clock FSR-A */
        CLOCK_INDEX(ctx->FSR_A_index, NUM_A_STAGES)
        FSR_A_STAGE(ctx, 4) = FSR_A_STAGE(ctx, 2) ^ (a0 << 8);
        a0 >>= 24;
        FSR_A_STAGE(ctx, 4) ^= alpha_0H[a0 >> 4] ^ alpha_0L[a0 & 0xF];

        /* update the FSM */
        SUB(ctx->L1, ctx->L2)
        SUB(ctx->R1, ctx->R2)
        SUB(r2 + FSR_B_STAGE(ctx, 4), ctx->L1)
        SUB(l2 + FSR_B_STAGE(ctx, 9), ctx->R1)

        /* clock FSR-B */
        d  = FSR_A_STAGE(ctx, 1);
        a1_offset = (d >> 28) & 4;

        t  = FSR_B_STAGE(ctx, 1) ^
            FSR_B_STAGE(ctx, 6) ^
            /* alpha-1 or alpha-2 depending upon the value of d; if d = 1
             * then the a1_offset = 4; if d = 0, a1_offset = 0, since
             * a1 and a2 entries are interleaved in blocks of 4, with a2
             * entries first */
            (b0 << 8) ^
            alpha_1L[((b0 >> 30) & 0x3) + a1_offset] ^
            alpha_1L[((b0 >> 28) & 0x3) + a1_offset + 8] ^
            alpha_1L[((b0 >> 26) & 0x3) + a1_offset + 16] ^
            alpha_1L[((b0 >> 24) & 0x3) + a1_offset + 24];

        d >>= 31;
        s  = FSR_B_STAGE(ctx, 8) ;
        t ^= s << (d << 3);
        b0 = alpha_3H[s >> 28] ^ alpha_3L[(s >> 24) & 0xF];
        t ^= (b0 & (-d));

        CLOCK_INDEX(ctx->FSR_B_index, NUM_B_STAGES)
        FSR_B_STAGE(ctx, 10) = t  ;
    }
}
```

B   Experimental Observation of Entropy Loss in K2's Key Initialization Algorithm due to SPA

We ran an experiment on twenty thousand random keys for K2 initialization. In the experiment we analysed the entropy loss of the key given that the attacker knows the the Hamming weights of the inputs and outputs of the *Sub* operations in phase 1 of the algorithm. The results are summarized in Table 2. We base the entropy lost on the median of the samples, due to the large standard deviation. We note that there are some cases where the three values are unambiguously determined by Hamming distance.

Table 2. Entropy measurements using SPA on two Subs and one xor

Key	Num of candidates					Entropy	Entropy Lost
	Min	Max	Median	Avg	Std		
EK_3	1	14400	192	476.6	779.2	7.58	24.41
EK_7	1	10386	210	479.9	771.7	7.71	24.29
EK_6	1	5804586	7510	46294	159409	12.87	19.12

On Unbiased Linear Approximations

Jonathan Etrog[*] and Matthew J.B. Robshaw

Orange Labs
38–40 rue du Général Leclerc
92794 Issy les Moulineaux Cedex 9, France
forename.surname@orange-ftgroup.com

Abstract. In this paper we explore the recovery of key information from a block cipher when using unbiased linear approximations of a certain form. In particular we develop a theoretical framework for their treatment and we confirm their behaviour with experiments on reduced-round variants of DES. As an application we show a novel form of linear cryptanalysis using multiple linear approximations which can be used to extract key information when all pre-existing techniques would fail.

Keywords: linear cryptanalysis, multiple approximations, entangled approximations.

1 Introduction

The technique of linear cryptanalysis [12,13] has become a standard tool in symmetric cryptanalysis. When used with block ciphers the essential idea is to find a linear expression or *linear approximation* that links bits of the plaintext, the ciphertext, and the key and which holds with some probability $\frac{1}{2} + s = \frac{1}{2}(1 + \epsilon)$, where the value s is known as the *bias* and $\epsilon = 2s$ is known as the *imbalance* or *correlation* [5]. If the bias or imbalance of the linear approximation is zero, that is $s = \epsilon = 0$, then we say that the linear approximation is *balanced* or *unbiased*.

As is well-known, a linear approximation can be used to recover one bit of key information if the bias or imbalance of the linear approximation is nonzero, that is $s, \epsilon \neq 0$. While elements of doing this were first described in [18], Matsui's *Algorithm 1* [12] describes how to recover this bit of key information given sufficiently many known plaintext-ciphertext pairs. The more complex *Algorithm 2* [12] uses a linear approximation as a reduced-round distinguisher and allows the recovery of more bits of key from more rounds of the cipher, and its effectiveness has been considered in [3,9,16].

The simultaneous use of *multiple linear approximations* to find a single bit of key information was first proposed in [10]. Such an approach can lead to a reduction in the data-complexity of the attack when compared to the use of a single linear approximation. Further papers exploring and extending the use of multiple linear approximations, under different assumptions and different types of approximations, include [1,4,6,7,8,11].

---

[*] Partially supported by the national research project RFIDAP ANR-08-SESU-009-03.

R. Steinfeld and P. Hawkes (Eds.): ACISP 2010, LNCS 6168, pp. 74–86, 2010.

As early as 1994, an interesting experimental phenomenon in the use of multiple linear approximations was described in [11]. There it was observed that four linear approximations could still give an advantage over using three approximations even though the fourth approximation was the linear algebraic sum of the other three. Whilst this issue has been revisited theoretically and experimentally, for instance [1,4], Murphy stressed the importance of considering the dependencies between linear approximations in recent work [14]. This has been confirmed by the framework of Hermelin *et al* using what have been called *multidimensional* techniques [2,6].

In this paper we highlight a striking consequence of this work and we explore the role of unbiased but dependent linear approximations. As shorthand, we will be referring to these approximations as *entangled* approximations (their definition will be given below) and, using [14], we will illustrate how to derive key information from such linear approximations. We will give a full theoretical treatment of this analysis, including error rates and the data requirements for key recovery, and we will also present the results of confirming experiments that have been carried out on reduced-round DES (which is a normal target cipher for experiments in the field).

We stress that this new technique is not intended as a universal replacement for traditional linear cryptanalysis, using either single or multiple approximations. Indeed techniques using biased approximations, when applicable, are likely to offer the best avenue for attack. However the results in this paper are important for two reasons:

1. We show that it is possible to use **only** the counts related to unbiased linear approximations in a linear cryptanalytic attack and still recover key information.
2. We demonstrate a practical situation where biased approximations cannot be used and all current linear cryptanalytic techniques would fail. Nevertheless, the presence of unbiased linear approximations still compromises the cipher.

2   Linear Approximations

Throughout the paper we consider a block cipher that operates on b-bit blocks, and we suppose that all the subkeys for the block cipher are concatenated to give an expanded-key $\mathbf{k}$ for the block cipher. For plaintext $\mathbf{p}$ and ciphertext $\mathbf{c}$, a linear approximation holding with probability $\frac{1}{2}(1 + \epsilon)$ is usually written in the following way

$$\gamma_{\mathbf{p}} \cdot \mathbf{p} \oplus \gamma_{\mathbf{c}} \cdot \mathbf{c} = \gamma_{\mathbf{k}} \cdot \mathbf{k},$$

where $\gamma_{\mathbf{p}}$, $\gamma_{\mathbf{c}}$, and $\gamma_{\mathbf{k}}$ represent bit-masks for the plaintext, ciphertext, and expanded key. For convenience, we rewrite this linear expression by considering the plaintext and ciphertext masks as a combined single $2b$-bit data mask α, and we use $\gamma = \gamma_k$ to give

$$\alpha^T \begin{pmatrix} \mathbf{p} \\ \mathbf{c} \end{pmatrix} = \gamma^T \mathbf{k}.$$

In its most basic form [10], an analysis using multiple linear approximations uses a collection $\alpha_1, \ldots, \alpha_l$ of data (plaintext-ciphertext) masks relating to the same[1] bit $k = \gamma_{\mathbf{k}}^T \mathbf{k}$ of key information to obtain the collection of approximations

$$\alpha_1^T \begin{pmatrix} \mathbf{p} \\ \mathbf{c} \end{pmatrix} = k \text{ with probability } \frac{1}{2}(1 + \epsilon_1),$$

$$\vdots$$

$$\alpha_l^T \begin{pmatrix} \mathbf{p} \\ \mathbf{c} \end{pmatrix} = k \text{ with probability } \frac{1}{2}(1 + \epsilon_l).$$

If we have N plaintext-ciphertext pairs $\begin{pmatrix} \mathbf{p}_1 \\ \mathbf{c}_1 \end{pmatrix}, \ldots, \begin{pmatrix} \mathbf{p}_N \\ \mathbf{c}_N \end{pmatrix}$, then we can estimate the key bit k in the following way. We let $V_1, \ldots, V_l$ be the counts corresponding to the data masks $\alpha_1, \ldots, \alpha_l$, that is

$$V_1 = \# \left\{ \begin{pmatrix} \mathbf{p}_i \\ \mathbf{c}_i \end{pmatrix} \middle| \alpha_1^T \begin{pmatrix} \mathbf{p}_i \\ \mathbf{c}_i \end{pmatrix} = 0 \right\},$$

$$\vdots$$

$$V_l = \# \left\{ \begin{pmatrix} \mathbf{p}_i \\ \mathbf{c}_i \end{pmatrix} \middle| \alpha_l^T \begin{pmatrix} \mathbf{p}_i \\ \mathbf{c}_i \end{pmatrix} = 0 \right\},$$

and we let $Y_1 = V_1 - \frac{N}{2}, \ldots, Y_l = V_l - \frac{N}{2}$ denote the centred counts.

For ease of exposition, we now restrict ourselves to the case of two linear approximations given by data masks α_{10} and α_{01} for which we have access to two centred counters Y_{10} and Y_{01}. Next we assume that the two approximations given by α_{10} and α_{01} are unbiased and yet the approximation given by the data mask $\alpha_{11} = \alpha_{01} + \alpha_{10}$ is biased. Further we assume that for whatever structural or operational reason, we don't have access to the counter Y_{11} that is related to α_{11} (see Section 5 for an example of this).

Considered individually, neither of the two centred counts Y_{10} and Y_{01} of Section 2 can ever give any information about the key bit k. This is reflected in the typical approaches to multiple linear approximations such as those described in [1,10] which would be unable to recover the key bit with any advantage. Furthermore, even multidimensional attacks using the unbiased span approximations α_{10} and α_{01} [2] would not be able to handle this situation since our problem statement explicitly rules out using counts based on α_{11}.

Nevertheless, the purpose of this paper is to highlight the fact that taken together, the pair of centred counts (Y_{10}, Y_{01}) can be used to recover key information. The main result we exploit in using unbiased linear approximations is Theorem 1 of [14]. This essentially states that if the linear approximation corresponding to α_{11} is biased, then the two centred counts Y_{10} and Y_{01} are correlated. We can use this correlation between the two centred counts Y_{10} and Y_{01} for unbiased linear approximations to recover key information. In particular,

---

[1] This condition was first relaxed in [11].

if Y_{10} and Y_{01} have the same sign, then this indicates one particular value for the key bit k is more likely; whereas if Y_{10} and Y_{01} have opposite signs, then this indicates that the other value for the key bit k is more likely.

We now formalise this notion of two unbiased linear approximations to the same key bit using notation from [14].

Definition 1. Suppose α_{10} and α_{01} are the data masks for unbiased linear approximations for the key bit k, that is

$$\alpha_{10}^T \begin{pmatrix} \mathbf{p} \\ \mathbf{c} \end{pmatrix} = k \quad \text{and} \quad \alpha_{01}^T \begin{pmatrix} \mathbf{p} \\ \mathbf{c} \end{pmatrix} = k$$

each with probability $\frac{1}{2}$. If $\alpha_{11} = \alpha_{10} + \alpha_{01}$ is the data mask for a biased linear approximation for key bit k, that is

$$\alpha_{11}^T \begin{pmatrix} \mathbf{p} \\ \mathbf{c} \end{pmatrix} = (\alpha_{10} + \alpha_{01})^T \begin{pmatrix} \mathbf{p} \\ \mathbf{c} \end{pmatrix} = k$$

with probability $\frac{1}{2}(1 + \epsilon)$ for some $\epsilon \neq 0$, then the two unbiased linear approximations for the key bit k based on the data masks α_{10} and α_{01} are *entangled linear approximations*[2] with *entanglement ϵ*.

The goal of our paper. Now our framework has been set we can state the result of the paper. When we can **only** examine the two entangled approximations, we cannot apply typical multiple approximation techniques [1,10] due to the zero capacity of the available approximations. Further we cannot apply multidimensional techniques [6,7] since we don't have access to the full set of counts related to the full set of approximations generated by the base approximations. Nevertheless, using results in this paper, we are still able to recover key information.

3 Key Recovery Using Entangled Approximations

We assume that we have N plaintext-ciphertext pairs for a pair of entangled unbiased linear approximations with entanglement ϵ. We now derive an asymptotically optimal process for recovering the key bit k from the centred counts Y_{10} and Y_{01}. We can rewrite the entanglement condition of Definition 1 as

$$\alpha_{11}^T \begin{pmatrix} \mathbf{p} \\ \mathbf{c} \end{pmatrix} = (\alpha_{10} + \alpha_{01})^T \begin{pmatrix} \mathbf{p} \\ \mathbf{c} \end{pmatrix} = 0 \text{ with probability } \frac{1}{2}\left(1 + (-1)^k \epsilon\right).$$

Section II of [14] gives a bivariate normal distribution for the asymptotic joint distribution of the normalised centred counts

$$Z_{10} = \frac{2}{\sqrt{N}} Y_{10} \quad \text{and} \quad Z_{01} = \frac{2}{\sqrt{N}} Y_{01}$$

---

[2] Note that entanglement is not exactly coincident with statistical dependence since we specify that the base approximations are unbiased.

using central limit theory ideas, so we have

$$\mathbf{Z} = \begin{pmatrix} Z_{10} \\ Z_{01} \end{pmatrix} \sim N \left(\begin{pmatrix} 0 \\ 0 \end{pmatrix} ; \begin{pmatrix} 1 & (-1)^k \epsilon \\ (-1)^k \epsilon & 1 \end{pmatrix} \right).$$

Thus we can write $\mathbf{Z} \sim N(0; \Sigma_k)$, where

$$\Sigma_k = \begin{pmatrix} 1 & (-1)^k \epsilon \\ (-1)^k \epsilon & 1 \end{pmatrix}, \quad \text{so}$$

$$\Sigma_k^{-1} = \frac{1}{1 - \epsilon^2} \begin{pmatrix} 1 & -(-1)^k \epsilon \\ -(-1)^k \epsilon & 1 \end{pmatrix}.$$

We can now give the likelihood function of the key bit value k given data $\mathbf{z} = \begin{pmatrix} z_{10} \\ z_{01} \end{pmatrix}$ for the normalised centred counts as

$$L(k; z) = (2\pi)^{-1} |\Sigma_k|^{-\frac{1}{2}} \exp \left(-\frac{1}{2} \begin{pmatrix} z_{10} & z_{01} \end{pmatrix} \Sigma_k^{-1} \begin{pmatrix} z_{10} \\ z_{01} \end{pmatrix} \right)$$

$$= (2\pi)^{-1} (1 - \epsilon^2)^{-\frac{1}{2}} \exp \left(-\frac{1}{2} \frac{1}{(1 - \epsilon^2)} \left(z_{10}^2 + z_{01}^2 - 2(-1)^k \epsilon z_{10} z_{01} \right) \right).$$

This means that the likelihood ratio for key bit $k = 1$ versus $k = 0$ is given by

$$\Lambda(\mathbf{z}) = \frac{L(1; \mathbf{z})}{L(0; \mathbf{z})} = \frac{\exp \left(-\frac{1}{2} \frac{1}{(1 - \epsilon^2)} \left(z_{10}^2 + z_{01}^2 + 2\epsilon z_{10} z_{01} \right) \right)}{\exp \left(-\frac{1}{2} \frac{1}{(1 - \epsilon^2)} \left(z_{10}^2 + z_{01}^2 - 2\epsilon z_{10} z_{01} \right) \right)}$$

$$= \exp \left(-\frac{2\epsilon}{1 - \epsilon^2} z_{10} z_{01} \right),$$

so we have the following log-likelihood ratio statistic

$$\log \Lambda(z) = -\frac{2\epsilon}{1 - \epsilon^2} z_{10} z_{01}.$$

Thus the log-likelihood ratio statistic is proportional to $P(\mathbf{z}) = z_{10} z_{01}$, or equivalently to $P(\mathbf{y}) = y_{10} y_{01}$ in terms of the unnormalised centred counts. The Neyman-Pearson Lemma [17] therefore shows that the asymptotically optimal test of $k = 0$ versus $k = 1$ is given by the sign of $P(\mathbf{z})$ or $P(\mathbf{y})$. For example, for positive entanglement ($\epsilon > 0$), we choose $k = 0$ if the centred counts Y_{10} and Y_{01} have the same sign and we choose $k = 1$ if the centred counts Y_{10} and Y_{01} have the opposite signs, and we swap these choices for negative entanglement.

3.1   Success Rates in Using Entangled Approximations

We now calculate the success rates in using the process of Section 3 to recover a key bit using a pair of entangled unbiased linear approximations. Without loss of generality, we suppose that the true value of the key bit k is 0 and that the entanglement $\epsilon > 0$. The accuracy probability $a(\epsilon)$ that this process correctly identifies the key bit value k as 0 is then given by $a(\epsilon) = \mathbf{P}(P(\mathbf{z}) > 0) = \mathbf{P}(z_{10}z_{01} > 0) = 2\mathbf{P}(z_{10}, z_{01} > 0)$. Thus this accuracy probability is given by

$$a(\epsilon) = 2 \int_0^\infty \int_0^\infty \frac{1}{2\pi(1 - \epsilon^2)^{\frac{1}{2}}} \exp\left(-\left(\frac{z_{10}^2 + z_{01}^2 - 2\epsilon z_{10}z_{01}}{2(1 - \epsilon^2)}\right)\right) dz_{10}dz_{01},$$

where the integrand is the joint density function of normalised centred count vector $\mathbf{Z}$. The accuracy and error probabilities for certain entanglements ϵ are given in Table 1.

Table 1. Theoretical accuracy and error rates for single key bit recovery using entangled approximations with entanglement ϵ

Entanglement (ϵ)	0.002	0.004	0.01	0.02	0.04	0.1	0.2	0.4
Accuracy Rate	0.5006	0.5013	0.503	0.506	0.513	0.532	0.564	0.631
Error Rate	0.4994	0.4987	0.497	0.494	0.487	0.468	0.436	0.369

We are usually interested in the case where the entanglement ϵ is small. We can write $a(\epsilon) = a(0) + a'(0)\epsilon + \frac{1}{2}a''(0)\epsilon^2 + o(\epsilon^3)$ and calculate:

$$a(0) = 2 \int_0^\infty \int_0^\infty \frac{1}{2\pi} \exp\left(-\frac{1}{2}\left(z_{10}^2 + z_{01}^2\right)\right) dz_{10}dz_{01} = \frac{1}{2},$$

$$a'(0) = 2 \int_0^\infty \int_0^\infty \frac{1}{2\pi} z_{10}z_{01} \exp\left(-\frac{1}{2}\left(z_{10}^2 + z_{01}^2\right)\right) dz_{10}dz_{01} = \frac{1}{\pi},$$

$$a''(0) = 2 \int_0^\infty \int_0^\infty \frac{1}{2\pi}(z_{10}^2 - 1)(z_{01}^2 - 1)\exp\left(-\frac{1}{2}\left(z_{10}^2 + z_{01}^2\right)\right) dz_{10}dz_{01} = 0.$$

For small ϵ, we can therefore express the accuracy probability as

$$a(\epsilon) = \tfrac{1}{2} + \tfrac{\epsilon}{\pi} + o(\epsilon^3),$$

which gives the error probability as $1 - a(\epsilon) \approx \tfrac{1}{2} - \tfrac{\epsilon}{\pi}$.

These success rates for the recovery of a single key bit k from a pair of entangled unbiased linear approximations do not explicitly depend on the number N of plaintext-ciphertext pairs. Thus the expression for success rate $a(\epsilon) = \tfrac{1}{2} + \tfrac{\epsilon}{\pi}$ is valid for any sample size N large enough for the central limit normal approximation to be valid. It would therefore be reasonable to approximate the distribution of the data class count vector as a multivariate normal random variable with, for example, $N = 64$ plaintext-ciphertext pairs. Furthermore, increasing the number of plaintext-ciphertext pairs beyond a point at which the normal approximation is reasonable does not materially improve the accuracy of the process.

3.2   Experimental Confirmation

To illustrate our analysis we consider some experiments using reduced-round versions of the DES [15], using the notation given by [12]. We first consider a reduced-round version of DES with three rounds and the following pair of linear approximations:

$$P_H[7, 18, 24, 29] \oplus C_H[7, 18, 24, 29] = K_1[22] \oplus K_3[22], \text{ and}$$
$$P_L[15] \oplus C_L[15] = K_1[22] \oplus K_3[22].$$

In the notation of Section 3, we have

$$\alpha_{10}^T \begin{pmatrix} \mathbf{p} \\ \mathbf{c} \end{pmatrix} = P_H[7, 18, 24, 29] \oplus C_H[7, 18, 24, 29],$$

$$\alpha_{01}^T \begin{pmatrix} \mathbf{p} \\ \mathbf{c} \end{pmatrix} = P_L[15] \oplus C_L[15], \text{ and}$$

$$\gamma^T \mathbf{k} = k = K_1[22] \oplus K_3[22].$$

By considering the structure of DES, it can be confirmed that these approximations are unbiased. Then considering the data mask $\alpha_{11} = \alpha_{10} + \alpha_{01}$ we obtain

$$\alpha_{11}^T \begin{pmatrix} \mathbf{p} \\ \mathbf{c} \end{pmatrix} = P_H[7, 18, 24, 29] \oplus P_L[15] \oplus C_H[7, 18, 24, 29] \oplus C_L[15].$$

This data mask coincides with Matsui's best three-round linear approximation $\alpha_{11}^T \begin{pmatrix} \mathbf{p} \\ \mathbf{c} \end{pmatrix} = k$ and so it has an has imbalance 0.391 [12]. As a result, the above pair of unbiased linear approximations are entangled linear approximations with entanglement $\epsilon = 0.391$.

The experimental success rates for key bit recovery using the above pair of three-round entangled linear approximations are given in Table 2. This table also gives experimental success rates for key bit recovery for the following pair of four-round entangled linear approximations with entanglement $\epsilon = -0.122$ [12]:

$$P_H[7, 18, 24, 29] \oplus C_L[7, 18, 24, 29] = K_1[22] \oplus K_3[22] \oplus K_4[42, 43, 45, 46],$$
$$P_L[15] \oplus C_H[15] \oplus C_L[27, 28, 30, 31] = K_1[22] \oplus K_3[22] \oplus K_4[42, 43, 45, 46].$$

The structure of DES can be used to confirm that these two approximations are unbiased. The experimental success rates in Table 2 are based on 10,000 trials.

Table 2. Success rates for key bit recovery for three- and four-round DES using entangled approximations. Results are based on 10,000 trials.

Number of plaintext-ciphertext pairs (N)	2^5	2^6	2^7	2^8	2^9	Theoretical prediction
Three-round experiment	0.6153	0.6305	0.6294	0.6301	0.6265	0.6279
Four-round experiment	0.5345	0.5432	0.5440	0.5425	0.5358	0.5388

4 Entangled Approximations and Large Data Sets

We saw in Section 3.1 that we can make an estimate for the key bit k using entangled approximations based on an asymptotic normal approximation, provided that the number N of plaintext-ciphertext pairs exceeds some bound. Suppose now that we have n such estimates for the key bit k, each based on a *packet* of N plaintext-ciphertext pairs. This gives a process for recovering the key from $T = nN$ plaintext-ciphertext pairs; given the results for n packets of size N we should clearly choose the value of the key bit k that is given by the majority of the packets.

We now determine the success rate of such a packet-based process. The probability that an individual packet gives the correct value of the key bit k is $\frac{1}{2} + \frac{\epsilon}{\pi}$. Thus the number of correct key values in a set of n packets has the following distribution

$$\mathrm{Bin}\left(n, \tfrac{1}{2} + \tfrac{\epsilon}{\pi}\right) \approx \mathrm{N}\left(\tfrac{n}{2} + \tfrac{n\epsilon}{\pi} \; ; \; \tfrac{n}{4}\right).$$

Thus the probability that at least half of the n packets give the correct value for the key bit k is $1 - \phi\left(-\frac{2\epsilon\sqrt{n}}{\pi}\right) = \phi\left(\frac{2\epsilon\sqrt{n}}{\pi}\right)$, where ϕ is the cumulative distribution function of a standard normal random variable. Thus over n packets of size N, where N exceeds the lower threshold established in Section 3.1, the success rate is given by $p = \phi\left(\frac{2\epsilon\sqrt{n}}{\pi}\right)$. Alternatively, for a given success rate p, we require $n = \left(\frac{\pi}{2\epsilon}\right)^2 \phi^{-1}(p)$ packets.

4.1 Experimental Confirmation

We illustrate the use of packets of texts with entangled approximations by considering the same entangled approximations for reduced-round versions of DES as we considered in Section 3.2. Table 3 gives the success rates for key bit recovery using the entangled approximations of Section 3.2 when using packets of text. Each packet contains $N = 64$ plaintext-ciphertext pairs, and the success rates are based on 10,000 trials.

Note that if we were to use packets of size one, *i.e.* $N = 1$, then our method would coincide with regular linear cryptanalysis with imbalance ϵ. However the situations of interest to us in this paper, for instance in Section 5, exclude us

Table 3. Success rates for key bit recovery for three- and four-round DES using entangled approximations and packets consisting of 64 plaintext-ciphertext pairs. Results are based on 10,000 trials.

Number of packets (n)	2^3	2^4	2^5	2^6	2^7	2^8	2^9
Three-round experiment	0.7597	0.8527	0.9270	0.9782	0.9975	1.0000	1.0000
Theoretical prediction	0.7585	0.8394	0.9197	0.9764	0.9975	1.0000	1.0000
Four-round experiment	0.5860	0.6223	0.6588	0.7231	0.8123	0.8890	0.9606
Theoretical prediction	0.5866	0.6216	0.6693	0.7321	0.8094	0.8922	0.9601

from this case. For packet sizes $2 \leq N \leq 63$ we don't have a fully-satisfactory theoretical model, but experimental results given in Table 4 confirm the following intuition. For a fixed number of texts, larger packets necessarily implies less packets and we are exploiting the underlying information less efficiently. This translates into reading down a column in Table 4 which illustrates a reduced success rate. For a given packet size, *i.e.* reading across a row in Table 4, more data translates into more packets and hence an increased success rate.

Table 4. Success rates using the entangled approximations over four-round DES when the total number of available texts is split into packets of size N. Results are based on 10,000 trials.

		Total number of texts T						
		2^6	2^7	2^8	2^9	2^{10}	2^{11}	2^{12}
	2^0	0.8346	0.9155	0.9741	0.9970	1.0000	1.0000	1.0000
	2^1	0.6319	0.6869	0.7512	0.8343	0.9149	0.9786	0.9966
↑	2^2	0.6135	0.6456	0.7073	0.7769	0.8656	0.9432	0.9866
N	2^3	0.5840	0.6154	0.6465	0.7184	0.7909	0.8806	0.9490
↓	2^4	0.5488	0.5709	0.6173	0.6694	0.7307	0.8030	0.8871
	2^5	0.5459	0.5577	0.5813	0.6215	0.6695	0.7333	0.8116
	2^6	0.5432	0.5405	0.5553	0.5860	0.6223	0.6588	0.7231

5   Separating Plaintext and Ciphertext

We can imagine situations where the connection between a plaintext and a ciphertext block is not known to the cryptanalyst. In a communication or transmission scenario, blocks might be delivered out of sequence or their ordering might be hidden or affected by some application. In another situation, cryptanalysis of an encrypted database can be practically thwarted when the correspondence between a database query (the plaintext) and its true location in the encrypted column or row (the ciphertext) is unknown. In such situations, existing linear cryptanalysis cannot be applied since analysis requires that a plaintext and its corresponding ciphertext be matched together and treated at the same time.[3] However we can still use the ideas in this paper to recover the key.

To demonstrate this we consider Matsui's best five-round approximation of DES [12]. Define the key bit

$$k = K_1[42, 43, 45, 46] \oplus K_2[22] \oplus K_4[22] \oplus K_5[42, 43, 45, 46],$$

and then Matsui's approximation, for which $\epsilon = 0.038$, is given by

$$P_H[15] \oplus C_H[15] \oplus$$
$$P_L[7, 18, 24, 27, 28, 29, 30, 31] \oplus C_L[7, 18, 24, 27, 28, 29, 30, 31] = k.$$

---

[3] While naïvely one might suggest that all the different plaintexts and ciphertexts be tried in all combinations, it is clear that this would be completely impractical.

Now consider launching an attack when the association between plaintext and ciphertext is lost; the attacker does not know which ciphertext corresponds to which plaintext. Perhaps we can envisage a less drastic situation where the correlation between plaintext and ciphertext is not completely lost and the attacker knows that groups of ciphertext correspond to groups of plaintexts, but inside each group the attacker is unsure which plaintext should be associated with which ciphertext. In either case regular linear cryptanalysis cannot be used since the attacker cannot compute the necessary counters. But using entangled approximations we can recover key information in both situations.

To see this, consider the two linear approximations

$$P_H[15] \oplus P_L[7, 18, 24, 27, 28, 29, 30, 31] = k, \text{ and}$$
$$C_H[15] \oplus C_L[7, 18, 24, 27, 28, 29, 30, 31] = k.$$

The first approximation does not depend on the ciphertext and the second does not depend on the plaintext. Both linear approximations are unbiased, however the two approximations are entangled since they are derived from Matsui's best five-round DES approximation.

Therefore using the techniques in this paper we can recover the key. We can do this when complete correlation is lost, by considering one packet of text as in Section 3.2, or we can use the techniques in Section 4 in the more practical situation where some packet-level correlation between plaintext and ciphertext remains. Table 5 gives the experimental success rates in the first instance, when using a varying number of plaintext-ciphertext pairs, and the success rate for this method is given by $\frac{1}{2} + \frac{\epsilon}{\pi}$. Meanwhile Table 6 confirms the predicted success rates when using the techniques of Section 4 for multiple packets, with each packet consisting of 64 plaintext-ciphertext pairs. The experimental success rates for both tables are based on 10,000 trials.

Table 5. Success rates for key bit recovery for five-round DES using entangled approximations with the separation of plaintext and ciphertext. Results are based on 10,000 trials.

Number of plaintext-ciphertext pairs (N)	2^5	2^6	2^7	2^8	2^9	Theoretical prediction
Five-round experiment	0.5092	0.5024	0.5252	0.5165	0.5127	0.5121

Table 6. Success rates for key bit recovery for five-round DES using entangled approximations with the separation of plaintext and ciphertext and packets consisting of 64 plaintext-ciphertext pairs. Results are based on 10,000 trials.

Number of packets (n)	2^3	2^4	2^5	2^6	2^7	2^8	2^9
Five-round experiment	0.5282	0.5442	0.5501	0.5762	0.6116	0.6464	0.7013
Theoretical prediction	0.5274	0.5388	0.5547	0.5771	0.6084	0.6514	0.7090

5.1   Discussion

It is important to emphasize exactly what the separation of plaintext and ciphertext entails. In both experiments, with a single packet or multiple packets, the
data processing involves a plaintext counter and a ciphertext counter. The two
counters are entirely separate. Thus we have an analysis using multiple linear approximations in which the plaintext and ciphertext are processed independently.

There is, of course, a small cost which can be manifested as either a modest reduction to the success rate or a moderate increase in the amount of text required
for an unchanged success rate (when compared to regular linear cryptanalysis).
To see the increase in text required, we note that regular cryptanalysis with a
single approximation with imbalance ϵ needs

$$\left(\frac{\phi^{-1}(p)}{\epsilon}\right)^2$$

plaintext/ciphertext pairs to achieve a success rate p. Using entangled linear
approximations, with n packets of sufficient large size, the number of packets
needed to achieve a success rate p is

$$n = \left(\frac{\pi\phi^{-1}(p)}{2\epsilon}\right)^2.$$

As was pointed out previously, once we have a sufficiently large packet, the
success rate no longer depends on the size of the packet so packets of size $N = 64$
will be sufficient. We therefore expect to attain the same success rate using
separated plaintext-ciphertext instead of regular linear cryptanalysis when we
have $\frac{\pi^2}{4} \times 2^6 = 2^{7.3}$ times the data. This result is confirmed experimentally for
four-round DES in Table 7.

Table 7. Success rates with entangled approximations and theoretical success rates
with regular linear cryptanalysis over four-round DES

Number of plaintext-ciphertext pairs (T)	$2^{1.7}$	$2^{2.7}$	$2^{3.7}$	$2^{4.7}$
Success rate (regular)	58.67	62.17	66.94	73.23

Number of plaintext-ciphertext pairs (T)	2^{9}	2^{10}	2^{11}	2^{12}
Success rate (entangled)	58.60	62.23	65.88	72.31

Note that we are not proposing that entangled approximations be used to
replace regular linear cryptanalysis. Instead we are showing that there are situations where regular linear cryptanalysis cannot be used but entangled linear
approximations can still give us information about the key.

6   Conclusions

In this paper we have shown that we can recover key information when using only unbiased linear approximations. In Section 5 we demonstrated a practical situation where by using multiple linear approximations in an entirely novel way—two separate approximations involving only plaintext and ciphertext bits respectively—we can extract key information when all existing techniques would fail. However, while much of the underlying analysis has been demonstrated in this paper the practical implications of entanglement still remain to be quantified. One particularly interesting situation would be using these techniques on the constituent components of ciphers. Some other directions for futher work might also include analysis in the direction of Matsui's Algorithm 2 [12], the further development of the statistical models that depend on entanglement, and an extension to linear approximations that depend on more than one key bit.

Acknowledgement

We would particularly like to thank Sean Murphy who launched this work and contributed throughout.

References

1. Biryukov, A., De Cannière, C., Quisquater, M.: On Multiple Linear Approximations. In: Franklin, M. (ed.) CRYPTO 2004. LNCS, vol. 3152, pp. 1–22. Springer, Heidelberg (2004)
2. Cho, J.Y.: Linear Cryptanalysis of Reduced-Round PRESENT. In: Pieprzyk, J. (ed.) CT-RSA 2010. LNCS, vol. 5985, pp. 302–317. Springer, Heidelberg (2010)
3. Collard, B., Standaert, F.-X., Quisquater, J.-J.: Improving the Time Complexity of Matsui's Linear Cryptanalysis. In: Nam, K.-H., Rhee, G. (eds.) ICISC 2007. LNCS, vol. 4817, pp. 77–88. Springer, Heidelberg (2007)
4. Collard, B., Standaert, F.-X., Quisquater, J.-J.: Experiments on the Multiple Linear Cryptanalysis of Reduced-Round Serpent. In: Nyberg, K. (ed.) FSE 2008. LNCS, vol. 5086, pp. 382–397. Springer, Heidelberg (2008)
5. Daemen, J.: Cipher and Hash Function Design. Ph.D. Thesis (March 1995)
6. Hermelin, M., Cho, J.Y., Nyberg, K.: Multidimensional Linear Cryptanalysis of Reduced Round Serpent. In: Mu, Y., Susilo, W., Seberry, J. (eds.) ACISP 2008. LNCS, vol. 5107, pp. 203–215. Springer, Heidelberg (2008)
7. Hermelin, M., Cho, J.Y., Nyberg, K.: A New Technique for Multidimensional Linear Cryptanalysis with Applications on Reduced Round Serpent. In: Yung, M., Liu, P., Lin, D. (eds.) ICISC 2008. LNCS, vol. 5461, pp. 383–398. Springer, Heidelberg (2009)
8. Hermelin, M., Cho, J.Y., Nyberg, K.: Multidimensional Extension of Matsuis Algorithm 2. In: Goos, G., Hartmanis, J., van Leeuwen, J. (eds.) Fast Software Encryption. LNCS, vol. 5665, pp. 209–227. Springer, Heidelberg (2009)

9. Junod, P.: On the complexity of Matsui's attack. In: Vaudenay, S., Youssef, A.M. (eds.) SAC 2001. LNCS, vol. 2259, pp. 199–211. Springer, Heidelberg (2001)
10. Kaliski, B.S., Robshaw, M.J.B.: Linear Cryptanalysis Using Multiple Approximations. In: Desmedt, Y.G. (ed.) CRYPTO 1994. LNCS, vol. 839, pp. 26–39. Springer, Heidelberg (1994)
11. Kaliski, B.S., Robshaw, M.J.B.: Linear Cryptanalysis and FEAL. In: Preneel, B. (ed.) FSE 1994. LNCS, vol. 1008, pp. 249–264. Springer, Heidelberg (1995)
12. Matsui, M.: Linear Cryptanalysis Method for DES Cipher. In: Helleseth, T. (ed.) EUROCRYPT 1993. LNCS, vol. 765, pp. 386–397. Springer, Heidelberg (1994)
13. Matsui, M.: The first experimental cryptanalysis of the Data Encryption Standard. In: Desmedt, Y.G. (ed.) CRYPTO 1994. LNCS, vol. 839, pp. 1–11. Springer, Heidelberg (1994)
14. Murphy, S.: The Independence of Linear Approximations in Symmetric Cryptanalysis. IEEE Transactions on Information Theory 52, 5510–5518 (2006)
15. National Institute of Standards and Technology. FIPS 46-3: Data Encryption Standard (November 1998), http://csrc.nist.gov
16. Selçuk, A.: On Probability of Success in Linear and Differential Cryptanalysis. Journal of Cryptology 21(1), 131–147 (2008)
17. Silvey, S.D.: Statistical Inference. Chapman and Hall, Boca Raton (1975)
18. Tardy-Corfdir, A., Gilbert, H.: A Known Plaintext Attack on FEAL-4 and FEAL-6. In: Feigenbaum, J. (ed.) CRYPTO 1991. LNCS, vol. 576, pp. 172–182. Springer, Heidelberg (1992)
19. Vaudenay, S.: An Experiment on DES Statistical Cryptanalysis. In: Proceedings of the Third ACM Conference on Computer Security, pp. 386–397 (1996)

Distinguishers for the Compression Function and Output Transformation of Hamsi-256

Jean-Philippe Aumasson[1], Emilia Käsper[2], Lars Ramkilde Knudsen[3],
Krystian Matusiewicz[4], Rune Ødegård[5], Thomas Peyrin[6],
and Martin Schläffer[7]

[1] Nagravision SA, Cheseaux, Switzerland
[2] Katholieke Universiteit Leuven, ESAT-COSIC, Belgium
[3] Department of Mathematics, Technical University of Denmark
[4] Institute of Mathematics and Computer Science, Wroclaw University of Technology
[5] Centre for Quantifiable Quality of Service in Communication Systems at the
Norwegian University of Science and Technology
[6] Ingenico, France
[7] IAIK, TU Graz, Austria

Abstract. Hamsi is one of 14 remaining candidates in NIST's Hash
Competition for the future hash standard SHA-3. Until now, little analysis has been published on its resistance to differential cryptanalysis,
the main technique used to attack hash functions. We present a study
of Hamsi's resistance to differential and higher-order differential cryptanalysis, with focus on the 256-bit version of Hamsi. Our main results
are efficient distinguishers and near-collisions for its full (3-round) compression function, and distinguishers for its full (6-round) finalization
function, indicating that Hamsi's building blocks do not behave ideally.

Keywords: hash functions, differential cryptanalysis, SHA-3.

1   Introduction

Hash functions are one of the most ubiquitous primitives in cryptography, with
digital signatures and integrity checks as their main applications. Collision attacks on the deployed standards MD5 and SHA-1 [18, 19, 20, 21] have weakened
the confidence in the MD family of hash functions. Hence, the US Institute of
Standards and Technology (NIST) launched a public competition to develop a
future SHA-3 standard [13].

The hash function Hamsi [8] is one of 64 designs submitted to NIST in fall
2008. Hamsi is also among the 14 submissions selected for the second round
of the competition in July 2009 as one of the few submissions with no major
weaknesses detected thus far. While Hamsi reuses the round components of the
Serpent block cipher [5], its larger block size and different round structure make
existing cryptanalytic results on Serpent hardly useful in its security analysis.

So far, little research has been published on the resistance of Hamsi to common
cryptanalytic attacks: in a work independent from ours, Çalık and Turan studied

R. Steinfeld and P. Hawkes (Eds.): ACISP 2010, LNCS 6168, pp. 87–103, 2010.

differential properties of Hamsi-256, and presented message-recovery and pseudo-second-preimage attacks. Near collisions were studied by Nikolić [12] and Wang et al. [17], as discussed in Section 4.3.

We study the resistance of Hamsi to differential and higher-order differential cryptanalysis, with focus on the 256-bit version Hamsi-256. In Section 3, we show by higher-order analysis that the 3-round compression function of Hamsi-256 does not achieve maximal degree. This is demonstrated by showing that the output of certain related chaining values (with fixed message word) or related message words (with fixed chaining value) sums to zero with a high probability.

In Sections 4 and 5, we focus on differential cryptanalysis and construct high-probability differential paths for the 3-round compression function as well as the full 6-round output transformation. The former gives near-collisions on $(256-25)$ bits of the compression function output, with only six differences in the input chaining value. Section 4 describes a technique for building low-weight, high-probability differential paths for Hamsi. Finally, Section 5 presents differential paths for six rounds of Hamsi-256 that show that the output transformation of Hamsi-256 does not behave ideally.

2   Description of Hamsi-256

This section describes the hash function Hamsi-256, henceforth just called Hamsi. We refer to [8] for a complete specification.

2.1   High-Level Structure

Like most hash functions, Hamsi builds on a finite-domain *compression function*, which is used to process arbitrary-length messages through the use of a *domain extender* (or operation mode). The compression function of Hamsi can be divided into four operations:

Message expansion	$E : \{0,1\}^{32} \to \{0,1\}^{256}$
Concatenation	$C : \{0,1\}^{256} \times \{0,1\}^{256} \to \{0,1\}^{512}$
Non-linear permutations	$P, P_f : \{0,1\}^{512} \to \{0,1\}^{512}$
Truncation	$T : \{0,1\}^{512} \to \{0,1\}^{256}$

The message M to hash is appropriately padded and split into ℓ blocks of 32 bits: $M_1, \ldots, M_\ell$. Each block is iteratively processed by the compression function, which operates on a 512-bit internal state viewed as a 4×4 matrix of 32-bit words.

Figure 1 depicts an iteration of the compression function H (or H_f). Starting from the predefined initial value (IV) h_0, Hamsi iteratively computes the digest h of M as follows:

$$h_i = H(h_{i-1}, M_i) = (T \circ P \circ C(E(M_i), h_{i-1})) \oplus h_{i-1} \quad \text{for} \quad 0 < i < \ell \; ,$$
$$h = H_f(h_{\ell-1}, M_\ell) = (T \circ P_f \circ C(E(M_\ell), h_{\ell-1})) \oplus h_{\ell-1} \; .$$

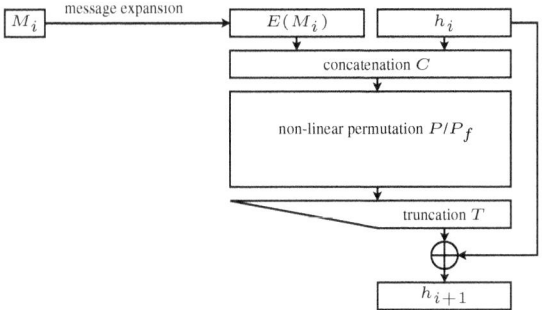

Fig. 1. Domain extension algorithm of Hamsi

2.2   Internals of the Compression Function of Hamsi

Message expansion. The message expansion of Hamsi uses a linear code to expand a 32-bit word into eight words (that is, 256 bits). We write an expanded M_i as $(m_0, \ldots, m_7)$. Thus, the m_j's are defined as the product of a multiplication with the generator matrix of the code:

$$E(M_i) = (m_0, \ldots, m_7) = (M_i \times G) ,$$

where G can be found in [8].

Concatenation. The concatenation function C forms a 512-bit internal state from the 256-bit expanded message $(m_0, \ldots, m_7)$ and the 256-bit incoming chaining value $h_i = (c_0, \ldots, c_7)$ (Figure 2):

$$C(m_0, \ldots, m_7, c_0, \ldots, c_7) = (m_0, m_1, c_0, c_1, c_2, c_3, m_2, m_3, m_4, m_5, c_4, c_5, c_6, c_7, m_6, m_7),$$

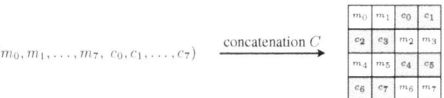

Fig. 2. Concatenation of expanded message words $m_0, \ldots, m_7$ and chaining value words $c_0, \ldots, c_7$ in Hamsi

Truncation. The truncation function T selects eight 32-bit words among the 16 from the internal state to form the new chaining value after feedforward (Figure 3):

$$T(s_0, s_1, s_2, \ldots, s_{14}, s_{15}) = (s_0, s_1, s_2, s_3, s_8, s_9, s_{10}, s_{11}) .$$

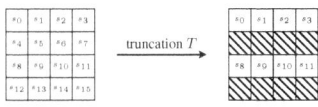

Fig. 3. Truncation selects eight out of 16 words of the internal state

Permutations. Finally, we describe the permutations P and P_f. They only differ in the number of rounds (three for P and six for P_f)[1] and in the round constants. The round function is composed of three layers. First, constants and a counter are XORed to the whole internal state. Then there is a substitution layer, followed by a linear layer.

The substitution layer uses one 4-bit Sbox of the block cipher Serpent [5], in a bitsliced way. That is, four bits, one from each of the four 32-words of the same column in the 4×4 internal state matrix are first extracted and then replaced after application of the Sbox. We denote s_i^j the j-th bit of the internal state word s_i. The substitution layer can be described as follows, for $0 \le j \le 31$ and $0 \le i \le 3$:

$$(s_i^j, s_{i+4}^j, s_{i+8}^j, s_{i+12}^j) := S(s_i^j, s_{i+4}^j, s_{i+8}^j, s_{i+12}^j) ,$$

where S is the 4×4 Sbox given in Table 7 (Appendix A).

The linear diffusion layer applies the Serpent linear transform $L : \{0,1\}^{128} \to \{0,1\}^{128}$ to each of the four diagonals of the state, as follows:

$$(s_0, s_5, s_{10}, s_{15}) := L(s_0, s_5, s_{10}, s_{15})$$
$$(s_1, s_6, s_{11}, s_{12}) := L(s_1, s_6, s_{11}, s_{12})$$
$$(s_2, s_7, s_8, s_{13}) := L(s_2, s_7, s_8, s_{13})$$
$$(s_3, s_4, s_9, s_{14}) := L(s_3, s_4, s_9, s_{14}) .$$

The algorithm below (read column by column) describes the linear transform L on input (a, b, c, d), with $x \lll k$ denoting the left bit rotation of k positions on the word x and $x \ll k$ denoting the left bit shift of k positions on the word x.

$$a := a \lll 13 \qquad\qquad d := d \lll 7$$
$$c := c \lll 3 \qquad\qquad a := a \oplus b \oplus d$$
$$b := a \oplus b \oplus c \qquad\qquad c := (b \ll 7) \oplus c \oplus d$$
$$d := (a \ll 3) \oplus c \oplus d \qquad\qquad a := a \lll 5$$
$$b := b \lll 1 \qquad\qquad c := c \lll 22$$

3   Higher-Order Differential Analysis

This section reports on properties of Hamsi related to higher-order derivatives. After some definitions, we present upper bounds on the algebraic degree of Hamsi's compression function and show how to exploit them to find "k-sums" and "zero-sums". This illustrates the fact that, due to its low algebraic degree, the compression function of Hamsi does not behave ideally.

3.1   Definitions

Higher-order derivatives. Higher-order differential analysis [7,10] of cryptographic algorithms generalizes the notion of differential cryptanalysis by

---

[1] While 6 rounds remains the official parameter, the designer has suggested 8 rounds as a conservative alternative. Our results indicate that moving to 8 rounds may be a necessary precaution.

considering derivatives of order two or more. It is based on the basic observation that for a function f with algebraic degree $s \geq 1$, the degree of a dth-order derivative of f is at most $(s - d)$, where $s \geq d$. Consequently, an sth-order derivative of f is a constant and an $(s+1)$st-order derivative of f is zero, which directly gives a 2^{s+1}-sum for f.

In the following we consider derivatives of functions with domain $\{0,1\}^n$, $n > 1$ and range $\{0,1\}$. Note that a (certain type of) d-th order derivative is then the XOR of 2^d values of the function for the 2^d choices of d input bits.

k-sums. The k-sum problem is, given k lists of random n-bit values (for example, k distinct instances of a compression function $f_1, \ldots, f_k$) , to find one value from each list such that the sum of the k values is zero. The case $k = 2$ is essentially the collision problem.

The k-sum problem can be solved in polynomial time (using the XHASH attack [2]) when $k \geq n$. However, the problem is believed to be hard for small k. The standard method for the k-sum problem with small k is Wagner's "generalized birthday" method, which requires time and space $O(k2^{n/(1+\log k)})$ [16] (see also [3]).

Henceforth, we consider the problem of finding k values whose images by a same function f sum to zero. Note that if f has degree $s < (n-1)$, then a 2^{s+1}-sum can be found by returning the values corresponding to a $(s+1)$st order derivative.

An example of application of k-sums is to forge message authentication codes (MACs). Let H he a hash function and consider the "prefix-MAC" construction defined as $\mathrm{MAC}_K(m) = \mathrm{trunc}\,(H(K\|m))$, where trunc is a function removing some bits of the hash output to combat length extension attacks. Assume we know messages $m_1, \ldots, m_k$ such that the probability

$$\Pr_K \left[\bigoplus_{i=1}^{k} H(K\|m_i) = 0 \right] = p$$

is nonzero. Then by querying MAC_K with $m_1, \ldots, m_{k-1}$ we can determine $\mathrm{MAC}_K(m_k)$ with probability p and thus break the existential forgery of MAC.

This can be generalized to messages whose MAC tags sum to any fixed value, to other MAC constructions, etc. For example, one may fix a message and forge the MAC $H_K(m)$ where K is the IV of H by making related-key queries.

Zero-sums. We define the zero-sum problem as a particular case of the k-sum problem: given a function f, find distinct values that sum to zero such that their images by f also sum to zero.

Both the XHASH attack [2] and Wagner's generalized birthday [16] can be adapted to find zero-sums. These methods are generic, and are probabilistic algorithms whose failure probability can be made exponentially small.

3.2   On the Degree of the Compression Function

Simple bounds. The only nonlinear component of Hamsi's compression function is the layer of 4×4 Sboxes. One round thus has degree three (see [14] for explicit expressions of the Sboxes used), so N rounds have degree at most 3^N, with respect to any choice of variables.

If variables are chosen in $c_0, \ldots, c_3$ only, or in $c_4, \ldots, c_7$ only, then they are all in distinct slices and thus go into distinct Sboxes in the first round. Hence, the first round is linear and after N rounds, the degree is at most 3^{N-1}. This means that the degree is at most 81 after five rounds, and that at least six rounds are necessary to reach maximal degree. In particular, the 3-round compression function has degree at most 9 with respect to choices of 128 variables in distinct slices, which distinguishes it from a randomly chosen function (whose degree would be below 9 with negligible probability).

Case of four variables. If four variables are chosen in the LSB's of $c_0, \ldots, c_3$, after the first application of the Sbox, all the LSB's of a given word depend on the bit varied in the corresponding column. Since only one bit is varied per column, the degree of equations corresponding to LSB's are of degree 1. Then, the linear function $L(a, b, c, d)$ is applied to each column, and we can determine, for a given bit of the state, whether it depends on the single variable of its diagonal. Based on this, we can determine whether a given 4-bit slice depends on 1, 2, 3, or 4 of the variables.

A simple computer-assisted analysis revealed that each slice depends on only one variable. Therefore, the (3-round) compression function of Hamsi always has degree 3 with respect to four variables in the first four LSB's, for any values of the other bits. Ideally, the function should have degree 4 with probability $1/2$, over the choice of the other input bits.

3.3   Finding k-sums for the Compression Function

For randomly chosen 256-bit values, finding 4-sums for the compression function of Hamsi requires an effort of complexity approximately $4 \cdot 2^{256/3} \approx 2^{87}$, using the generalized birthday method. Below we show efficient methods to find 16-, 8-, and 4-sums.

16-sums. Recall the above observation that three rounds have degree at most 3 with respect to a certain choice of four variables. This observation can directly be used to find 16-sums, without any computation. Based on empirical observations, we discovered that we can do better, as presented below.

8-sums. Choose a random value of one 256-bit chaining value, then select seven other chaining values, which are different from the first one only in the LSB's of the first three 32-bit words. Denote these chaining values by $h^0, \ldots, h^7$. Choose a random 32-bit message block M, then compute $\sum_{i=0}^{7} H(h^i, M)$, In 1 000 000 such tests, the above sum was zero in 1458 cases (whereas for a random mapping, the probability to obtain zero is negligible). This indicates that there are 3rd-order derivatives with the value zero (or 8-sums) of a high probability for the compression function of Hamsi. It is very likely that one can identify other 3rd-order derivatives of higher probabilities (our search was limited).

4-sums. We found 2nd-order derivatives with value zero, that is, 4-sums. One example is when one chaining value is the IV of Hamsi specified in [9], and where the three others differ only in two LSB's of the second words; the XOR of the four outputs is the all-zero string (note that the four inputs also sum to zero, thus this is also a zero-sum).

Via an exhaustive search over all 2^{32} message words, we identified 70 messages for which the above four chaining values lead to a 4-sum. We also found 4-sums for the IV given in [8], for 86 values of the 32-bit message block. Although complete analytical justification of these observations remains to be found, the results of these observations strongly differ from what one obtains for a random mapping (for which a 2nd-order derivative is zero with negligible probability).

k-sums for fixed chaining value. Here we report on the case where the chaining value is fixed and where only the message block is varied. The outputs of the compression function in this case has a much higher algebraic degree.

Consider h_0, the IV specified in [9], and 2^{19} values of the 32-bit message block obtained by varying the first and second bytes, and the three least significant bits of the third byte. The remaining bits can be fixed to arbitrary values. Denoting these message words by $m_0, \ldots, m_{2^{19}-1}$, we have:

$$\bigoplus_{i=0}^{2^{19}-1} H(h_0, m_i) = 0 .$$

This observation holds for any initial chaining variable. Here we obtain zero because we perform a 19th-order derivative of a function of degree 18 only. Indeed, in the first round at most two bit variables enter a same Sbox, hence the degree of the first round is 2. Since the two subsequent rounds have degree 3 each, the three rounds have degree $2 \times 3 \times 3 = 18$.

Note that if P_f is replaced by P in Hamsi's domain extender, then the above observation can be used to forge MAC's (cf. Section 3.1), which shows that the extended 6-round output transformation is necessary, and cannot be removed without compromising the security of Hamsi.

3.4   Finding Zero-Sums for the Output Permutation

We describe a dedicated method to find large zero-sums for the 6-round permutation of the finalization function of Hamsi(we stress that it only applies to the internal permutation and not to the finalization as a whole, for it puts no restriction on the initial state). Contrary to Wagner's and the XHASH methods, it is deterministic rather than probabilistic, and needs to evaluate (and to know) only half the function.

In the spirit of [15, §9], we present an "inside-out" technique that exploits the fact that two *halves* of Hamsi's permutation have low algebraic degree. This differs from our method for finding k-sums which exploited the low degree of the full permutation. The attack works as follows:

1. Choose an arbitrary value for the state of Hamsi's permutation after three rounds.
2. Choose 28 distinct bits of the state.
3. Compute the 2^{28} initial states obtained by varying these bits and inverting the first three rounds of the permutation.

We obtain 2^{28} values that sum to zero, since their sum is the 28th-order derivative with respect to three inverse rounds. Their images also sum to zero, since they are the 28th-order derivative with respect to three forwards rounds (although the images are unknown, and need not be computed).

The method works whenever a function can be written as the composition of two low-degree functions. As explained in [4], the proposed technique is slightly more efficient than previous methods, for finding (here) zero-sums of 2^{28} elements.

4   First Order Differential Analysis

In this section, we analyze the differential properties of the Hamsi round transformations and show how to find high-probability differential paths for up to six rounds. Since we use XOR differences in our analysis, the differential propagation is deterministic in the message expansion and in the linear layer based on the L transform. However, the propagation of differences through the Sbox layer is probabilistic and depends on the actual values of the input. To maximize the differential probability of a differential path, we try to minimize the number of active Sboxes during the path search.

4.1   Differential Properties of the Sbox

The differential distribution table (DDT) of the 4-bit Hamsi Sbox S is given in Table 8 (Appendix A). Note that about half the differential transitions are impossible. The probabilities of the non-zero differentials are either 2^{-2} or 2^{-3}. In our approach, besides minimizing the number of active Sboxes, we thus try to minimize the number of probability-2^{-3} differentials.

4.2   Differential Properties of the Linear Transform L

The linear transform L has on average good diffusion properties, that is, a few differences in the input lead to many differences in the output. Additionally, each bit of L contributes to one of the 128 Sboxes in each round. To minimize the number of active Sboxes, we thus need to minimize number of differences in L. The Hamming weight (HW) of a difference is a good heuristic to measure the quality of a differential path. In the following, we first analyze the difference propagation through the linear layer for differences with HW one.

If we introduce a single input difference at bit position i in one input word, the HW of the output differences depends on the position and word of the input difference. In Table 1 and Table 2 give the HW of the output difference for each of the 128 single bit input differences.

We observe that for some specific words and bit positions, the resulting HW can be quite small. This happens if one or more differences are removed by the shift operation. More specifically, the branch number of L is only 3, so certain 1-bit input differences lead to only a 2-bit output difference, and vice versa. Table 1 and Table 2 show the worst case of diffusion, that is, the output HW for a multiple-bit input difference can be upper bounded by summing the corresponding table entries. However, when inserting many differences in several input words, some bit differences might erase each other, thus lowering the overall HW.

Table 1. Hamming weight of output differences if a single difference is introduced at one input word of the 128-bit linear transformation $(a', b', c', d') = L(a, b, c, d)$ of Hamsi in *forward* direction. The total and word-wise Hamming weight of the output difference is given depending on the bit position i and input word of the input difference.

Difference in input word	Position i of input difference	Total HW of output diff.	HW of output diff. in				Conditions (mod 32)
			a'	b'	c'	d'	
a	16,17	3	2	1	-	-	$i + 13 > 28, i + 14 > 24$
	18	4	2	1	1	-	$i + 13 > 28, i + 14 \leq 24$
	11...15	6	3	1	1	1	$i + 13 \leq 28, i + 14 > 24$
	else	7	3	1	2	1	$i + 13 \leq 28, i + 14 \leq 24$
b	24...30	2	1	1	-	-	$i + 1 > 24$
	else	3	1	1	1	-	$i + 1 \leq 24$
c	21...27	6	2	1	2	1	$i + 4 > 24$
	else	7	2	1	3	1	$i + 4 \leq 24$
d		3	1	-	1	1	

Table 2. Hamming weight of input differences if a single difference is introduced at one output word of the 128-bit linear transformation $(a', b', c', d') = L(a, b, c, d)$ of Hamsi in *backward* direction. The total and word-wise Hamming weight of the input difference is given depending on the bit position i and output word of the output difference.

Difference in output word	Position i of output difference	Total HW of input diff.	HW of input diff. in				Conditions (mod 32)
			a	b	c	d	
a'	2...4	2	1	1	-	-	$i + 27 > 28$
	else	3	1	1	-	1	$i + 27 \leq 28$
b'	28...31	3	1	2	-	-	$i > 28, i > 24$
	25...28	4	1	2	-	1	$i \leq 28, i > 24$
	never	6	1	3	1	1	$i > 28, i \leq 24$
	else	7	1	3	1	2	$i \leq 28, i \leq 24$
c'		3	-	1	1	1	
d'	29...31	4	1	-	1	2	$i > 28$
	else	5	1	-	1	3	$i \leq 28$

4.3   Near-Collisions for the Compression Function

Using our observations on the differential properties of Hamsi's Sbox and linear transform, we first searched manually for high-probability paths leading to near-collisions for the compression function, given some difference in the chaining value.

Previous work by Nikolic reported near collisions [12] on $(256 - 25)$ bits with 14 differences in the chaining value; work by Wang et al. reported [17] near collisions on $(256 - 23)$ bits with 16 differences. Below we present near collisions on $(256 - 25)$ bits with only six differences in the chaining value, using the differential path in Table 3.

Table 3. Differential path for three rounds of Hamsi with probability 2^{-26}

It.	Sbox input	Sbox output	Prob.
1	00000000 00000000 00020000 00000002 00004000 00000000 00000000 00000000 00000000 00000000 00020000 00000002 00004000 00000000 00000000 00000000	00000000 00000000 00000000 00000002 00004000 00000000 00000000 00000000 00000000 00000000 00000000 00000000 00000000 00000000 00020000 00000000	8
2	00000000 00000000 00000000 00080000 00000000 00000000 00000000 00000000 00000000 00000000 00000000 00000000 00000000 00000000 00000000 00000000	00000000 00000000 00000000 00000000 00000000 00000000 00000000 00080000 00000000 00000000 00000000 00000000 00000000 00000000 00000000 00080000	3
3	80000000 00000000 02000000 00000000 00000000 00000000 00000000 00100000 00020000 00000000 00010000 00000000 00000000 00000000 00000000 04000000	00000000 00000000 00000000 04100000 80020000 00000000 02010000 04100000 00020000 00000000 00000000 00000000 80000000 00000000 02010000 00000000	15
End	00000000 80400800 00000000 10C130C0 00040105 00000000 04020000 08000000 00020400 A040A0A2 00000000 10004000 00000040 08000000 00820801 00000000		

The differential path in Table 3 is followed with probability 2^{-26} under standard uniformity and independence assumptions. However, for the IV defined in [9] the path is followed with probability 2^{-23}. This is because of the condition put by the two fixed bits in each Sbox. These probabilities were verified experimentally.

Finally, note that the near collisions also result in other 4-sums: for example, for the IV h_0 specified in [9], the IV h_1 obtained by applying the weight-6 initial difference in Table 3, and the message M_1=C33BE456 and M_2=C8D1B855, we have:

1. A near collision between $H(h_0, M_1)$ and $H(h_1, M_1)$.
2. A near collision between $H(h_0, M_2)$ and $H(h_1, M_2)$.
3. A 4-sum $H(h_0, M_1) \oplus H(h_1, M_1) \oplus H(h_0, M_2) \oplus H(h_1, M_2) = 0$.

For inputs of an "ideal" function, the latter equality is unlikely to hold with probability 2^{-23}, but rather with probability close to 2^{-256}.

In the following, we automate our search for high-probability differential paths. Our heuristic algorithm, described in the next section, produced good differential paths for up to six rounds of Hamsi.

4.4   Automated Differential Path Search

As before, we search for differential paths with some difference in the input and output chaining value, and no difference in the input message. The resulting

6-round paths allow us to distinguish the output transform from random, as shown in Sect. 5.4.

Our primary heuristic is to minimise the HW of the differences in each round. To achieve that goal, we start with a very low HW (1 or 2 bit) difference in the middle of the path (at the start of round 3 for a 6-round search) and let the difference spread in both forward and backward directions. Additionally, we try to maximise the transition probabilities and randomize the search.

More precisely, our automated differential path starts from the input of the Sbox layer in round 3, forcing a 1-bit or 2-bit input difference on only one Sbox position i (among the 128 possible bit positions). We then choose one of the best differential transitions through the forward application of the Sbox and apply the linear layer on this new internal state. By best Sbox transitions, we mean the transitions that lead to a low HW after the application of the linear layer. To keep the search complexity feasible, we apply the L-layer to each active S-box separately and use the sum of the HWs as an estimate of the total output HW at the end of each round. Since the path is sparse, the sum of HWs proves to be a good heuristic. We continue picking the best differential transitions for all the active Sbox positions until the end of the fifth round of the output function of Hamsi. As the final output HW of the difference does not influence the path complexity, we optimise for transition probabilities in the last round, and pick the most probable differential Sbox transitions (not the ones minimizing the HW). Finally, we apply the very last linear layer to obtain the full path.

The backward computation is done analogously in the middle rounds, applying the linear layer backward and picking the best backward differential transitions for all active Sboxes. In the first round (the last round when computing backward) we impose additional restrictions in order to fulfill constraints on the message expansion.

As we force no difference in the message input of the compression function, we expect the 256-bit expanded message word to contain no difference at all. Hence, in the first round we only allow Sbox transitions where the difference in the expanded message bits is zero. Note that the probabilities of the first-round transitions do not affect the complexity of the path, as long as they are different from 0. Indeed, in the first round we can use the freedom of the chaining input to fulfill the conditions on the Sboxes and we expect the complexity cost of this first round to be negligible.

In order to increase our chances to obtain a good trail, we randomized the search with several parameters. First, we randomized the first 1-bit or 2-bit perturbation introduction in the output of round 3, as well as its position i among the 128 Sbox locations. Furthermore, we are also randomizing the Sbox transitions when several candidates are equally good. Finally, another improvement has been incorporated in our implementation: after having found a potentially interesting 6-round candidate, we recompute the forward search by allowing more differential transitions through the Sbox. Said in other words, after having placed ourselves in an interesting differential paths subspace, we look in the neighborhood if better ones exist.

Our heuristic search revealed that after three rounds in both backward and forward directions, the diffusion of Hamsi is not sufficient to avoid high-probability differential paths and we can find a differential path with a rather low total HW and good probability. We were able to construct a 6-round differential path with a relatively high probability, which is used to distinguish the the whole Hamsi output transformation in the following section.

5   Non-randomness of the Ouput Transformation

5.1   The Differential Path

The best 6-round path produced by our randomized search program is depicted in Table 4. We can find an input pair (chaining values and messages) conforming to this path with a probability of 2^{-206}. Note that in the first round we have a probability of 2^{-58} for a random message and a random chaining value. However, we can fix a suitable message (see below), and choose a valid chaining value bit-by-bit such that the desired output difference is guaranteed. This means that we can find a conforming input pair to the differential path with a complexity of about 2^{148}.

5.2   First Round and Message Expansion

In the first iteration, active S-boxes impose conditions on the expanded message: for a given non-zero Sbox differential, only one or two pairs of values of the corresponding two expanded message bits are possible. Since we have only 32 degrees of freedom in the message, we need to keep the number of active Sboxes in the first round low. To improve the probability of finding a suitable message candidate, we can vary the differences in the chaining values, whenever several input differences lead to the same output difference of the first Sbox layer. These relaxable differential Sbox transitions are listed in Table 5. In our path, five of the 23 active Sboxes of the first iteration are relaxable. In total, we have only nine Sboxes with two constraints on the message bits; 12 Sboxes with one constraint on the message; and two S-Boxes with a "half" constraint on the message (three of four bit pairs are possible). Therefore, we expect to find $2^{32-2\times 9-12} \cdot \left(\frac{3}{4}\right)^2 \approx 2$ messages satisfying the relaxed first round differential. In practice, we found one such message using the constants of permutation P and three messages using the constants of the output permutation P_f (see the full version of this paper for an example [1]). Note that finding conforming message words can be done in 2^{32} by exhaustive search. The complexity to find chaining values such that the first four rounds of the path are satisfied is about 2^{25}, since we can fulfill the conditions in the first round deterministically.

5.3   Last Round and Truncation

In order to improve the probability of the differential path, we consider truncated differentials in the last application of the Sbox. Namely, we relax the Sbox transitions by fixing some bits in the output difference, while letting the remaining

Table 4. Differential path for six rounds of Hamsi with probability 2^{-148}

It.	Sbox input	Sbox output	Prob.
start		00000000 00000000 84004880 4081C400 2C020018 000045C0 00000000 00000000 00000000 00000000 84024880 4081C400 28020018 000045C0 00000000 00000000	
1	00000000 00000000 84004880 4081C400 2C020018 000045C0 00000000 00000000 00000000 00000000 84024880 4081C400 28020018 000045C0 00000000 00000000	04000000 00000000 04000000 40818000 28020018 000040C0 04020000 00000000 00000018 00004100 00000800 00804000 04020000 000004C0 80024880 00004400	(58)
2	00000000 00000000 00000000 00010000 30000010 00000080 00000000 00000080 30000010 00000080 00000000 00010080 00000000 00000000 00000000 00000000	00000000 00000000 00000000 00010000 30000000 00000000 00000000 00000080 00000010 00000000 00000000 00000000 00000000 00000080 00000000 00000000	17
3	00000000 00000000 00000000 00000000 20000000 00000000 00000000 00000000 20000000 00000000 00000000 00000000 00000000 00000000 00000000 00000000	00000000 00000000 00000000 00000000 20000000 00000000 00000000 00000000 00000000 00000000 00000000 00000000 00000000 00000000 00000000 00000000	3
4	00000000 00000000 00000000 00000008 40000000 00000000 00000000 00000000 00000000 00000000 00000000 00000000 00000000 00000000 00000000 00000000	40000000 00000000 00000000 00000000 40000000 00000000 00000000 00000008 00000000 00000000 00000000 00000000 00000000 00000000 00000000 00000008	5
5	04038000 00000000 00000200 00000010 80000000 00001000 00000000 00000010 00000002 00000000 00000a01 00000000 00000000 00000000 00000000 00200400	80000000 00001000 00000000 00200410 04038002 00001000 00000801 00000000 00000000 00000000 00000000 00000000 84038002 00000000 00000a01 00200400	33
6	08420002 F8022900 00000000 30821140 0903000C 00000000 04001002 00000000 00000000 A0A26145 00041080 12807200 01C0014A 00000000 08051082 10420000	08830144 A0022100 0C051080 10C01000 0181014C 58A04845 0C051082 22406340 01800148 58A04845 08011002 22406340 00400002 58000800 00040080 20020140	90
End	CD9F7546 362513EA 56FE147F 85F6B1E1 8D0682FD F100928A B44C3D06 18A0D101 B8871BEA 70315A82 4819C14B 26257026 A1DD0199 40072022 8329356A A744E830		

bits vary. Since the "a"-bits and "c" bits diffuse faster through the linear layer (see Table 1), we chose to fix these bits in the output of each Sbox. Amongst four different truncated output differences (?0?0, ?0?1, ?1?0 and ?1?1), we chose, for each input, the output difference with the highest probability. Table 6 lists the relaxed input-output transitions for the Sbox. Details of the path used can be found in the full version of the article [1].

Relaxing the Sbox transitions increases the probability of the last round to $2^{-61.8}$, giving a total path complexity $2^{-120.8}$. At the same time, since the "wild card" bits are chosen to have low diffusion, the difference is still fixed in 180 bits of the chaining value. Thus, we obtain a distinguisher by observing the difference in these output bits.

Table 5. Relaxable differential transitions for the first round of the Hamsi Sbox. The first table shows the possible input differences that give the same output if 1, 4 and 5 are the only possible Sbox input differences. The second table shows the same possibilities if 2, 8, and 10 are the only possible Sbox input differences. For each underlined transition two message pairs are possible, while for the other transitions only one message pair is possible.

Desired output	1 a	2 b	3 ab	4 c	5 ac	6 bc	7 abc	8 d	9 ad	10 bd	11 abd	12 cd	13 acd	14 bcd	15 abcd
Possible input					1 5	4 5				1 4	1 4	1 5			

Desired output	1 a	2 b	3 ab	4 c	5 ac	6 bc	7 abc	8 d	9 ad	10 bd	11 abd	12 cd	13 acd	14 bcd	15 abcd
Possible input			2 8		2 8				2 8						2 8

Table 6. Relaxed differential transitions for the last round of the Hamsi Sbox. The table shows the chosen set of output differences for each given input difference. Underlined transitions have probability 2^{-2}, while the other transitions have probability 2^{-3}.

input	1 a	2 b	3 ab	4 c	5 ac	6 bc	7 abc	8 d	9 ad	10 bd	11 abd	12 cd	13 acd	14 bcd	15 abcd
output	12 14	3 9	1 9	10	1 3	2 8	4 12	5 7 13 15	8 10	2 8 10	1 9	11	1 3	7 13	2 10
mask	11?0	?0?1	?001	1010	00?1	?0?0	?100	?1?1	10?0	?0?0	?001	1011	00?1	?1?1	?010

5.4 Distinguishing the Output Transformation

To distinguish the output transformation of Hamsi we use the concept of differential q-multicollision introduced by Biryukov et al. in the cryptanalysis of AES-256 [6] and applied to the SHA-3 candidate SIMD in [11]. Originally, differential q-multicollision have been applied to a block cipher but can be easily adapted to a random function. A differential q-multicollision for a random (compression) function $f(H, M)$ is a set of two differences $\Delta H, \Delta M$ and q pairs $(H_1, M_1), (H_2, M_2), \ldots, (H_q, M_q)$ such that:

$$f(H_1, M_1) \oplus f(H_1 \oplus \Delta H, M_1 \oplus \Delta M) =$$
$$f(H_2, M_2) \oplus f(H_2 \oplus \Delta H, M_2 \oplus \Delta M) =$$
$$\cdots$$
$$f(H_q, M_q) \oplus f(H_q \oplus \Delta H, M_q \oplus \Delta M)$$

The generic complexity to find differential q-multicollision for a random function f with output size n is at least $q \cdot 2^{\frac{q-2}{q+2} \cdot n}$ evaluations of f.

In the case of Hamsi-256, the function f is the output transformation, the message difference ΔM is zero and the output size is $n = 256$. The generic complexity to find differential q-multicollision should be $q \cdot 2^{\frac{q-2}{q+2} \cdot 256}$ and we get for $q = 8$ a generic complexity of $2^{156.1}$. Using our differential path of Section 5.1, we get for $q = 8$ a complexity of $8 \cdot 2^{148} = 2^{151}$. Hence, for $q \geq 8$ we can distinguish the output transfomation of Hamsi from a random function, since we expect to find a q-multicollision approximately 32 times faster than for an ideal transform.

Due to the relaxed conditions, we only fix a truncated difference in 180 output bits and hence, we get $n = 180$. In this case, the generic complexity for $q = 11$ is $q \cdot 2^{\frac{q-2}{q+2} \cdot 180} = 2^{128.1}$. Using the relaxed differential path, we get $q \cdot 2^{120.8} = 2^{124.3}$ and hence, can distinguish the output transfomation of Hamsi from a random function for $q \geq 11$.

6 Conclusion

We investigated the resistance of the 256-bit version of the second round SHA-3 candidate Hamsi against differential and higher-order differential attacks.

Using higher-order analysis, we showed that the 3-round compression function of Hamsi has suboptimal algebraic degree. Using this observation, we provided sets of four related IV's such that the outputs of the compression function obtained with a given fixed message sum to zero. We also presented a set of 2^{19} message words such that the output chaining values, using any fixed IV, sum to zero. The latter result indicates that the compression function of Hamsi, when seen as a function of message words, does not reach the expected maximal degree 27. As an application, we note that the low degree makes the standalone compression function existentially forgeable in the message authentication setting.

Further, we constructed high-probability differential paths for the 3-round compression function to demonstrate a near-collision on $(256 - 25)$ bits with only six differences in the input chaining value. We have also developed a technique for building low-weight, high-probability differential paths for more rounds of Hamsi. Our best differential path for six rounds has probability 2^{-148}, much higher than expected for a random function. Additionally, we gave a truncated differential on 180 output bits with probability $2^{-120.8}$. These are the first results on six rounds of Hamsi, allowing us to distinguish the full output transformation from a random function using differential q-multicollisions.

Although none of our findings directly leads to an attack on the hash algorithm, they indicate that the buildings blocks of Hamsi exhibit nonrandom behavior. We expect our work to serve as a starting point for future analysis of Hamsi.

In order to prevent more serious attacks, we recommend increasing the number of rounds in the output transformation as a precaution. While the current specification does not include performance figures for the 8-round alternative, this change is only expected to noticeably affect the speed of hashing short messages.

Acknowledgements

Emilia Käsper thanks the Computer Laboratory of the University of Cambridge for hosting her.

This work was supported in part by the European Commission through the ICT Programme under Contract ICT-2007-216646 ECRYPT II. Emilia Käsper was also supported by the IAP–Belgian State–Belgian Science Policy BCRYPT and the IBBT (Interdisciplinary institute for BroadBand Technology) of the Flemish Government.

References

1. Aumasson, J.P., Käsper, E., Knudsen, L.R., Matusiewicz, K., Odegaard, R., Peyrin, T., Schlffer, M.: Differential distinguishers for the compression function and output transformation of Hamsi-256. Cryptology ePrint Archive, Report 2010/091 (2010)
2. Bellare, M., Micciancio, D.: A new paradigm for collision-free hashing: Incrementality at reduced cost. In: Fumy, W. (ed.) EUROCRYPT 1997. LNCS, vol. 1233, pp. 163–192. Springer, Heidelberg (1997)
3. Bernstein, D.J.: Better price-performance ratios for generalized birthday attacks. In: SHARCS (2007), http://cr.yp.to/papers.html#genbday
4. Bertoni, G., Daemen, J., Peeters, M., Assche, G.V.: Note on zero-sum distinguishers of keccak-f. NIST mailing list (2010),
http://keccak.noekeon.org/NoteZeroSum.pdf
5. Biham, E., Anderson, R.J., Knudsen, L.R.: Serpent: A new block cipher proposal. In: Vaudenay, S. (ed.) FSE 1998. LNCS, vol. 1372, pp. 222–238. Springer, Heidelberg (1998)
6. Khovratovich, D., Biryukov, A., Nikolić, I.: Distinguisher and related-key attack on the full AES-256. In: Halevi, S. (ed.) CRYPTO 2009. LNCS, vol. 5677, pp. 231–249. Springer, Heidelberg (2009)
7. Knudsen, L.R.: Truncated and higher order differentials. In: Preneel, B. (ed.) FSE 1994. LNCS, vol. 1008, pp. 196–211. Springer, Heidelberg (1995)
8. Kücük, O.: The hash function Hamsi. Submission to NIST (January 2009), http://csrc.nist.gov/groups/ST/hash/sha-3/Round1/documents/HamsiUpdate.zip
9. Kücük, O.: Reference implementation of Hamsi. Submission to NIST (January 2009)
10. Lai, X.: Higher order derivatives and differential cryptanalysis. In: Blahut, R., Costello Jr., D., Maurer, U., Mittelholzer, T. (eds.) Communications and Cryptography, pp. 227–233. Kluwer, Dordrecht (1992)
11. Mendel, F., Nad, T.: A distinguisher for the compression function of simd-512. In: Roy, B.K., Sendrier, N. (eds.) INDOCRYPT 2009. LNCS, vol. 5922, pp. 219–232. Springer, Heidelberg (2009)
12. Nikolić, I.: Near collisions for the compression function of Hamsi-256. CRYPTO rump session (2009),
http://rump2009.cr.yp.to/936779b3afb9b48a404b487d6865091d.pdf
13. NIST: Announcing request for candidate algorithm nominations for a new cryptographic hash algorithm (SHA-3) family. Federal Register Notice. 72(112) (November 2007), http://csrc.nist.gov/groups/ST/hash/documents/FR_Notice_Nov07.pdf

14. Singh, B., Alexander, L., Burman, S.: On algebraic relations of Serpent S-boxes. Cryptology ePrint Archive, Report 2009/038 (2009)
15. Wagner, D.: The boomerang attack. In: Knudsen, L.R. (ed.) FSE 1999. LNCS, vol. 1636, pp. 156–170. Springer, Heidelberg (1999)
16. Wagner, D.: A generalized birthday problem. In: Yung, M. (ed.) CRYPTO 2002. LNCS, vol. 2442, pp. 288–303. Springer, Heidelberg (2002)
17. Wang, M., Wang, X., Jia, K., Wang, W.: New pseudo-near-collision attack on reduced-round of Hamsi-256. Cryptology ePrint Archive, Report 2009/484 (2009)
18. Wang, X., Lai, X., Feng, D., Chen, H., Yu, X.: Cryptanalysis of the hash functions MD4 and RIPEMD. In: Cramer, R. (ed.) EUROCRYPT 2005. LNCS, vol. 3494, pp. 1–18. Springer, Heidelberg (2005)
19. Wang, X., Yin, Y.L., Yu, H.: Finding collisions in the full SHA-1. In: Shoup, V. (ed.) CRYPTO 2005. LNCS, vol. 3621, pp. 17–36. Springer, Heidelberg (2005)
20. Wang, X., Yu, H.: How to break MD5 and other hash functions. In: Cramer, R. (ed.) EUROCRYPT 2005. LNCS, vol. 3494, pp. 19–35. Springer, Heidelberg (2005)
21. Wang, X., Yu, H., Yin, Y.L.: Efficient collision search attacks on SHA-0. In: Shoup, V. (ed.) CRYPTO 2005. LNCS, vol. 3621, pp. 1–16. Springer, Heidelberg (2005)

A   The Sbox of Hamsi

Table 7. The Hamsi Sbox in decimal basis

x	0	1	2	3	4	5	6	7	8	9	10	11	12	13	14	15
$S[x]$	8	6	7	9	3	12	10	15	13	1	14	4	0	11	5	2

Table 8. The differential distribution table (DDT) of the Hamsi Sbox in decimal basis

In \ Out	0	1	2	3	4	5	6	7	8	9	10	11	12	13	14	15
0	16	0	0	0	0	0	0	0	0	0	0	0	0	0	0	0
1	0	0	0	0	0	2	0	2	0	0	2	2	2	0	4	2
2	0	0	0	4	0	4	0	0	0	4	0	0	0	0	0	4
3	0	4	2	0	0	0	2	0	0	2	0	0	2	0	2	2
4	0	0	0	0	0	0	4	0	0	0	4	4	0	4	0	0
5	0	4	0	2	2	2	2	0	2	0	0	0	2	0	0	0
6	0	0	2	2	2	2	0	0	2	2	0	0	0	0	2	2
7	0	0	0	0	4	2	0	2	0	0	2	2	2	0	0	2
8	0	0	0	2	0	2	0	4	0	2	0	0	0	4	0	2
9	0	0	0	2	0	0	0	2	4	2	2	2	2	0	0	0
10	0	0	2	0	2	0	4	0	2	0	4	0	0	0	2	0
11	0	4	0	0	2	0	2	0	2	2	0	0	2	0	0	2
12	0	0	2	0	2	0	0	0	2	0	0	4	0	4	2	0
13	0	4	2	2	0	2	2	0	0	0	0	0	2	0	2	0
14	0	0	2	0	2	0	0	4	2	0	0	0	0	4	2	0
15	0	0	4	2	0	0	0	2	0	2	2	2	2	0	0	0

Second-Preimage Analysis of Reduced SHA-1

Christian Rechberger

[1] Department of Electrical Engineering ESAT/COSIC, Katholieke Universiteit
Leuven. Kasteelpark Arenberg 10, B–3001 Heverlee, Belgium
[2] Interdisciplinary Institute for BroadBand Technology (IBBT), Belgium
christian.rechberger@esat.kuleuven.be

Abstract. Many applications using cryptographic hash functions do not
require collision resistance, but some kind of preimage resistance. That's
also the reason why the widely used SHA-1 continues to be recommended
in all applications except digital signatures after 2010. Recent work on
preimage and second preimage attacks on reduced SHA-1 succeeding up
to 48 out of 80 steps (with results barely below the 2^n time complexity
of brute-force search) suggest that there is plenty of security margin left.

In this paper we show that the security margin is actually somewhat
lower, when only second preimages are the goal. We do this by giving two
examples, using known differential properties of SHA-1. First, we reduce
the complexity of a 2nd-preimage shortcut attack on 34-step SHA-1 from
an impractically high complexity to practical complexity. Next, we show
a property for up to 61 steps of the SHA-1 compression function that violates some variant of a natural second preimage resistance assumption,
adding 13 steps to previously best known results.

Keywords: hash function, cryptanalysis, SHA-1, preimage, second preimage, differential.

1 Introduction and Overview

After the spectacular collision attacks on MD5 and SHA-1 by Wang *et al.* and
follow-up work [7,13,37,40,41,42], implementors reconsider their choices. While
starting a very productive phase of research on design and analysis of cryptographic hash functions, the impact of these results in terms of practical and worry-
ing attacks turned out to be less than anticipated (exceptions are e.g. [18,36,38]).
In addition to collision resistance, another property of hash functions is crucial for
practical security: preimage resistance. Hence, research on preimage attacks and
the security margin of hash functions against those attacks seems well motivated,
especially if those hash functions are in practical use.

1.1 Motivation: Security Margin of SHA-1 against Preimage Style Attacks

SHA-1 continues to get recommended by NIST even after 2010 for applications
that do not require collision resistance [23]. Hence, SHA-1 will globally remain

R. Steinfeld and P. Hawkes (Eds.): ACISP 2010, LNCS 6168, pp. 104–116, 2010.
© Springer-Verlag Berlin Heidelberg 2010

in practical use for a long time. Even though close to practical collision attacks for SHA-1 are described in [6,40], it's resistance against preimage attacks seems very solid.

1.2   The Contribution

Progress in the cryptanalysis of a round-based primitive is often monitored via considering the highest number of rounds for which an attack method violates some assumption about the primitive. For preimage attack, the meet-in-the-middle approach [3,10,15,17,34,35] proved to be successful in doing so. To this end, we devise methods that exhibit non-ideal behavior regarding variants of second preimage resistance for significantly more steps of the SHA-1 compression function (see Sect. 5.2). Another concern is the efficiency of attacks. Also here, we can demonstrate significant efficiency improvements for a step-reduced SHA-1 hash function. Details for this can be found in Section 5.1. As a summary, see Section 1.3 for an overview and a comparison. What is the reason for these improvements? We exclude preimage attacks and specifically use the knowledge of a first preimage to get an advantage as an attacker. The approach we use takes advantage of the existence of differentials with relatively high probability, i.e. it exploits the similar weaknesses that also led to efficient collision search attacks.

1.3   Preview of Our Results on SHA-1

We summarize our results on the second-preimage resistance of SHA-1 hash function and compression function in Table 1 and 2, respectively. There, they are compared with preimage attacks of De Cannière and Rechberger from Crypto 2008 [8], and to preimage attacks from Aoki and Sasaki, from Crypto 2009 [3]. The method in this paper is sensitive to changes of the Boolean function used in the round transformation, hence we distinguish between round-reduced variants that start from step 0, and those that can start anywhere. Note that [8] is not sensitive to the Boolean function used, and hence the number of rounds can not be reduced or extended with a different choice, In case of [3], the impact of the choice of different starting rounds for the reduced variant is more difficult to assess, but likely to be limited. Interestingly, whereas we can cover many steps of the SHA-1 compression function and still show less than ideal properties of it, we fail to do so for the SHA-1 hash function. The efficiency improvement for 34-step SHA-1 however works for both the compression function and the hash function.

1.4   Related Work

This approach was already proposed for MD4 in its basic form by Yu *et al.* [44]. There, a characteristic through all 48 steps of MD4 with probability 2^{-56} was used to state that one in 2^{56} messages is a weak message with respect to a 2nd-preimage attack. Leurent noted [19] that for long messages, this can be turned

Table 1. Comparison of various variants of preimage attacks on the SHA-1 hash function with reduced number of rounds

rounds	complexity time/memory/prob.	type	technique	source
34 (00-33)	$2^{77}/2^{15}/ > 0.5$	2nd-preimage	imp. msg. + P^3graph	[8]
34 (00-33)	$2^{42.42}/$negl./ > 0.5	2nd-preimage	differential	Sect. 5
44 (00-43)	$2^{157}/2^{21}/ > 0.5$	preimage	imp. msg. + P^3graph	[8]
45 (00-44)	$2^{159}/2^{21}/ > 0.5$	2nd-preimage	imp. msg. + P^3graph	[8]
48 (00-47)	$2^{159.3}/2^{40}/ > 0.5$	preimage	MITM	[3]
48 (00-47)	$2^{159.8}/$negl./ > 0.5	preimage	MITM	[3]
48 (00-47)	$2^{159.27}/$negl./ > 0.5	preimage	optimized brute force	[27]

Table 2. Comparison of various variants of preimage attacks on the SHA-1 compression function with reduced number of rounds

rounds	complexity time/memory/prob.	type	technique	source
34 (00-33)	$2^{69}/ - / > 0.5$	preimage	imp. msg.	[8]
34 (00-33)	$2^{42.25}/$negl./ > 0.5	2nd-preimage	differential	Sect. 5
45 (00-44)	$2^{157}/ - / > 0.5$	preimage	imp. msg.	[8]
48 (00-47)	$2^{156.7}/2^{40}/ > 0.5$	preimage	MITM	[3]
48 (00-47)	$2^{157.7}/$negl. / > 0.5	preimage	MITM	[3]
61 (18-79)	$1/$ negl. $/2^{-159.42}$	2nd-preimage	differential	Sect. 5

into an attack actually finding a 2nd-preimage with complexity 2^{56}. Considering second preimage attacks on HMAC when instantiated with concrete hash functions, Kim *et al.* [16] give e.g. results for MD5 up to 33 out of the 64 steps, and for SHA-1 for up to 42 steps.

Relations among various notions of preimage-style resistance requirements are studied in numerous work, e.g. [30,33,39]. Using the notation of [33], we study the aSec property of SHA-1, and show that the SHA-1 compression function is not ideally aSec-secure for up to 61 steps. An example of a construction that explicitly uses the second preimage resistance of a compression function appears in [2].

1.5   Outline of the Paper

We start with a simple definition of second preimage resistance for iterated hash functions in Section 2, followed by a description of SHA-1 in Section 3. The idea of the attack is presented in Section 4. We apply the ideas to step-reduced SHA-1 and show an attack on the compression function and the hash function SHA-1 in Section 5. Finally, we discuss our findings and open problems in Section 6.

2   Definitions

Let an iterated hash function F be built by iterating a compression function $f : \{0,1\}^l \times \{0,1\}^n \to \{0,1\}^n$ as follows:

– Split the message m of arbitrary length into k blocks x_i of size l.
– Set h_0 to a pre-specified IV
– Compute $\forall x_i : h_i = f(h_{i-1}, x_i)$
– Output $F(m) = h_k$

A basic informal definition of second preimage resistance of a hash function is as follows:

Definition 1. *Given $F(\cdot)$, m, it should be hard to find an $m^* \neq m$ such that $F(m^*) = F(m)$. For a hash function with n-bit output size, every guess for an m^* should have success probability of 2^{-n}, and the work to find an m^* should be no less than 2^n.*

Def. 1 applies analogously to a compression function, i.e. with a fixed length instead of arbitrary length input. For a more formal treatment, we refer to [30,33,39].

3   Description of SHA-1

SHA-1 is an iterative hash function that processes up to 2^{55} 512-bit input message blocks and produces a 160-bit hash value. Like many hash functions used today, it is based on the design principle of MD4, pioneered by Rivest [32]. In the following we briefly describe the SHA-1 hash function. It basically consists of two parts: the message expansion and the state update transformation. A detailed description of the hash function is given in [24].

Table 3. Notation

notation	description
$X \oplus Y$	bit-wise XOR of X and Y
$X + Y$	addition of X and Y modulo 2^{32}
X	arbitrary 32-bit word
X^2	pair of words, shortcut for (X, X^*)
M_i	input message word i (32 bits)
W_i	expanded input message word t (32 bits)
$X \lll n$	bit-rotation of X by n positions to the left, $0 \leq n \leq 31$
$X \ggg n$	bit-rotation of X by n positions to the right, $0 \leq n \leq 31$
N	number of steps of the compression function

3.1   Message Expansion

The message expansion of SHA-1 is a linear expansion of the 16 message words (denoted by M_i) to 80 expanded message words W_i.

$$W_i = \begin{cases} M_i, & \text{for } 0 \leq i \leq 15, \\ (W_{i-3} \oplus W_{i-8} \oplus W_{i-14} \oplus W_{i-16}) \lll 1 & \text{for } 16 \leq i \leq 79 . \end{cases} \tag{1}$$

3.2   State Update Transformation

The state update transformation of SHA-1 consists of 4 rounds of 20 steps each. In each step the expanded message word W_i is used to update the 5 chaining variables A_i, B_i, C_i, D_i, E_i as follows:

$$A_{i+1} = E_i + A_i \lll 5 + f(B_i, C_i, D_i) + K_j + W_i$$
$$B_{i+1} = A_i$$
$$C_{i+1} = B_i \ggg 2$$
$$D_{i+1} = C_i$$
$$E_{i+1} = D_i$$

Note that the function f depends on the actual round: round 1 (steps 0 to 19) use f_{IF} and round 3 (steps 40 to 59) use f_{MAJ}. The function f_{XOR} is applied in round 2 (steps 20 to 39) and round 4 (steps 60 to 79). The functions are defined as follows:

$$f_{IF}(B, C, D) = B \wedge C \oplus \overline{B} \wedge D \tag{2}$$
$$f_{MAJ}(B, C, D) = B \wedge C \oplus B \wedge D \oplus C \wedge D \tag{3}$$
$$f_{XOR}(B, C, D) = B \oplus C \oplus D . \tag{4}$$

After the last step of the state update transformation, the chaining variables A_0, B_0, C_0, D_0, E_0 and the output values of the last step $A_{80}, B_{80}, C_{80}, D_{80}, E_{80}$ are combined using word-wise modular addition, resulting in the final value of one iteration (feed forward). The result is the final hash value or the initial value for the next message block.

Note that $B_i = A_{i-1}$, $C_i = A_{i-2} \ggg 2$, $D_i = A_{i-3} \ggg 2$, $E_i = A_{i-4} \ggg 2$. This also implies that the chaining inputs fill all A_j for $-4 \leq j \leq 0$. Thus it suffices to consider the state variable A, which we will for the remainder of this paper.

4   Violating Second Preimage Resistance Properties with Differentials

Assuming the existence of a differential with a certain probability $p > 2^{-n}$, there are two ways to use such a differential in 2nd-preimage attacks. One is to simply use this differential for a single attempt to find a second preimage by being given the first preimage. With $p > 2^{-n}$, this shows less than ideal behavior of the function, even though on average it hardly speeds up the search for an actual second preimage. The second way is to apply this differential in an iterated hash function on individual message blocks, and thereby increasing this probability to actually find a second preimage. In this setting, if the number of message blocks that can be tried is larger than p^{-1} a second preimage can be expected with high probability.

For the description of our approach, we use the framework developed for SHA-1 characteristics by De Cannière and Rechberger [7], and adapt it to the second preimage setting at hand. In the following, we briefly recall those parts that are needed later on.

The expected difference between a particular pair of words X^2 will be denoted by ∇X. For every bit in this pair, we write 'x' if we expect a difference between the same bits of both words, and we write '-' if we do not expect a difference between those two bits.

Let us assume that we are given a complete characteristic for N-step SHA-1, specified by $\nabla A_{-4}, \ldots, \nabla A_N$ and $\nabla W_0, \ldots, \nabla W_{N-1}$, detailing for every bit and every word in the computation, whether or not we expect a difference at a particular bit position. Our goal is to estimate how much effort it would take to, given a message, find another message which follows this characteristic, assuming a simple depth-first search algorithm which tries to determine the pairs of message words L_i^2 one by one starting from L_0^2. In order to estimate the work factor of this algorithm, we will compute the expected number of visited nodes in the search tree. But first another definition, which is needed to estimate the work factor.

Definition 2 ([7]). *The* uncontrolled probability $P_u(i)$ *of a characteristic at step i is the probability that the output A_{i+1}^2 of step i follows the characteristic, given that all input pairs do as well, i.e.,*

$$P_u(i) = P\left(A_{i+1}^2 \in \nabla A_{i+1} \mid A_{i-j}^2 \in \nabla A_{i-j} \text{ for } 0 \leq j < 5, \text{ and } W_i^2 \in \nabla W_i\right).$$

With the definition above, we can now easily express the number of nodes $N_s(i)$ visited at each step of the compression function during the second preimage search.

Taking into account that the average number of children of a node at step i is $P_u(i)$, and that the search stops as soon as step N is reached, we can derive the following recursive relation:

$$N_s(i) = \begin{cases} 1 & \text{if } i = N, \\ N_s(i+1) \cdot P_u^{-1}(i) & \text{if } i < N. \end{cases}$$

The total work factor is then given by

$$N_w = \sum_{i=1}^{N} N_s(i). \tag{5}$$

It is now easy to see that we have two different quantities that define the search for a second preimage. One is the number of step computations N_w, which should be noticeably below $2^n \cdot N$ to be considered an attack. The other one is the number of distinct message blocks N_m that need to be tried during the search:

$$N_m = \prod_{j=1}^{N} P_u(j)^{-1} = N_s(0). \tag{6}$$

Note that N_m could theoretically be above 2^n, while the resulting work factor can still be below an equivalent of 2^n compression function computations. This is because the tree-based model of the search takes *early-stop strategies* into account. However, this only works if in addition to the first preimage, also all intermediate chaining values that lead to the target hash are already available to the attacker. This may be the case in certain settings, but is certainly not a standard assumption for second preimage attacks.

Without this additional assumption on data available to an attacker, the workfactor is in fact

$$N_w = N \cdot \prod_{j=1}^{N} P_u(j)^{-1} \, . \tag{7}$$

We will refer to this as setting 2, and will use setting 1 (and Eq. 5) when we assume the availability of internal chaining inputs.

5   Application to SHA-1

In order to find attacks on the SHA-1 compression function, or the SHA-1 hash function, characteristics need to be found that result in a workfactor N_w which should be noticeably below $2^n \cdot N$. The search algorithms we used are based on methods developed in the early cryptanalysis of SHA-1 regarding collision attacks [4,20,26,31] with the improvement that exact probabilities as described in [7] instead of Hamming weights are used to prune and rank them. More recent characteristic search algorithms (e.g. [7,12,21,43]) which exploit the fact that non-linear propagation of differences with low probability can be useful in collision attacks do not appear to be applicable to the setting considered in this paper. Depending on whether the hash- or the compression function is considered, the chaining input $\nabla A_{-4} \ldots \nabla A_0$ is allowed to have a difference or not.

In order to explain various aspects of the method, we consider two case studies. The first is the SHA-1 hash reduced to the first 34 steps and discussed in Section 5.1. There we show that better attack complexities can be obtained. The second is the SHA-1 compression function reduced to 61 steps and discussed in Section 5.2. There we aim for having results on a higher number of steps.

5.1   Hash Function Attacks: 34-Step SHA-1 as a Case Study

To illustrate the techniques, we consider SHA-1 reduced to the first 34 steps, and walk through the attack reasoning. We aim for a second-preimage attack on the hash functions, i.e., we require from a characteristic that input- and output chaining do not have a difference. The best characteristic we found for our purpose is the same as the one used by Biham *et al.* [4, Tab. 1] for a collision attack, and is also related to those used in Kim *et al.* [16, Tab. 6], and in [28, Tab. 6]. First, we recompute the probabilities $P_u(i)$ of the differential specified by the message difference m', and the chaining output co' (a zero difference). What we are interested in is the probability that, given an m' from a uniform

Table 4. Characteristic with probability $2^{-42.42}$ used for the 34-step (0-33) attack. $P_u(i)$ is written as a log_2, and $N_s(i)$ is written as log_2 as well.

i	∇A_i	∇W_i	$P_u(i)$	$N_s(i)$
-4				
-3				
-2				
-1				
0			1	42.42
1			2	41.42
2			3	39.42
3			2	36.42
4			3	34.42
5			2	31.42
6			2.42	29.42
7			3	27.00
8			4	24.00
9			2	20.00
10			3	18.00
11			4	15.00
12			0	11.00
13			0	11.00
14			1	11.00
15			2	10.00
16			3	8.00
17			0	5.00
18			0	5.00
19			0	5.00
20			1	5.00
21			1	4.00
22			1	3.00
23			1	2.00
24			1	1.00
25			0	0.00
26			0	0.00
27			0	0.00
28			0	0.00
29			0	0.00
30			0	0.00
31			0	0.00
32			0	0.00
33			0	0.00
34				

distribution, $F(m) = F(m \oplus m')$. A good lower bound for this probability is the probability of the particular characteristic as shown in Table 4, which is $2^{-42.42}$.

Taking into account also other, strongly related characteristics with lower probability (see [22,25,28,29] for details), we would arrive at an improved probability of $2^{-42.25}$. The second-preimage finding algorithm hence needs to traverse the first preimage of a length of about $2^{42.25}$ (N_m) message blocks in order to succeed with good probability. The memory requirements for this are negligible as the first preimage can be processed in an on-line manner. In setting 1, when intermediate chaining values are also given, most of the time only the first few step transformations are computed. Hence the computational resources needed in terms of computing step transformations are about an equivalent of $2^{37.87}$ computations of 34-step SHA-1 (N_w according to Eq. 5), taking the early stop technique into account. Without this assumption, the computational effort is hence about $2^{42.25}$ (N_w according to Eq. 7).

Comparisons with results obtained by De Cannière/Rechberger. On one hand, this may be compared with the result from [8], where memory of order 2^{15} and an equivalent of about 2^{77} computations are needed to find a second preimage of 34-step SHA-1 with good probability (a first preimage may be as small as 2^5 message blocks with this approach, but longer first preimages do not help to improve the attack).

Comparison with the generic Kelsey/Schneier 2nd-preimage attack.
On the other hand, this may be compared with the generic method of Kelsey
and Schneier. In [14], Kelsey and Schneier describe a second preimage attack on
iterated hash functions that is independent of the actual compression function.
The approach finds a second preimage for a 2^k-message-block message with
about $k \times 2^{n/2+1} + 2^{n-k+1}$ work. It was then later generalized to also take, among
other aspects, multiple targets into account [1]. Those attacks do not concern our
results on the SHA-1 compression function, but need to be taken into account
when considering the SHA-1 hash function. The new 2nd-preimage result we
described above needs about $2^{42.25}$ message blocks in order to succeed with good
probability, i.e. $k = 42.25$. Using the Kelsey/Schneier approach, the resulting
attack complexity is of order $42.25 \times 2^{160/2+1} + 2^{160-42.25+1} \approx 2^{118.75}$. Hence,
even by neglecting some constants in time complexities comparison, it seems safe
to conclude that the proposed differential based method is considerable faster.

5.2   Compression Function Attacks: 61-Step SHA-1

To further illustrate that the availability of a first preimage helps to improve
upon current preimage attacks on reduced SHA-1, we also seek to increase the
number of steps in which results can be obtained. For this, we relax our requirements on 2nd-preimage attacks in three ways:

1. No practical complexity or probability, better than the ideal 2^{-n} is enough.
2. We do no longer require it to beat the generic Kelsey/Schneier result, i.e.
   the result will only be valid for the compression function rather the hash
   function (as Kelsey/Schneier does not apply there).
3. Any choice of consecutive steps is allowed instead of starting with step 0.

By exploiting all those relaxations, we demonstrate attacks for up to 61 steps,
thereby having reached more steps than in any compression function attack
on SHA-1 before. We used the characteristic given in Table 5. The product of
all uncontrolled probabilities P_u suggests a probability of $2^{-158.42}$. However, this
does not take the feed-forward operation into account. For the previous example,
this was ignored safely, as no probabilistic events happen during the feed forward
operation. As can be seen in Table 5 however, we do have a single bit difference
in the chaining input and chaining output. We do require these differences to
cancel out during the feed forward operation, which happens with probability
$1/2$. Hence a lower bound for the probability to indeed have a second preimage
is $2^{-159.42}$.

   As before, by taking into account also other, strongly related characteristics
with lower probability (see [22,25,28,29] for details), we would arrive at an improved probability of $2^{-159.42+1.61} = 2^{-157.81}$. This probability is above the ideal
2^{-160}, hence exhibiting less than ideal 2nd-preimage resistance.

Comparisons. The best results in terms of number of rounds on the SHA-1
compression function following the impossible message approach [8] is 45 steps.
Also this approach is not able to take advantage of the relaxation of condition

Table 5. Characteristic with probability $2^{-158.42}$ used for the 61-step (18-79) attack. $P_u(i)$ is written as log_2.

i	∇A_i	∇W_i	$P_u(i)$
-4			
-3	------x-		
-2			
-1			
0		x---	1.00
1		x---	0.00
2			0.00
3			0.00
4			0.00
5		-x--	1.00
6	-x---	---x--	1.00
7		-x-- ----x-	2.00
8	--x-	x--x-- ----x-	2.00
9	x---	-x-x-- ---x--x	5.00
10	-x--- x	---x-- -xx-x--	4.00
11		xxx- ---x-	4.00
12	x	xxxx- ----x---x-	5.00
13	x--- x-	-xxx-- -x-x--x	6.00
14	x---	---x-- ---x--	4.00
15	x-	-xx-- ---x----x	4.00
16	x	xx-- ----x--x-	3.00
17		x-- -----xx	3.00
18	x--- x-	xxx-- ---x-x-x-	5.00
19	x-	xx-- ----x-----	3.00
20	x--- x-	xxx-- ---x-x-x-	5.00
21		--x-- -----xx	4.00
22	x--- x	x-- ---xx---	5.00
23		--x-- ------x	5.00
24	x---	xx-- ----x-x-	5.00
25	x-	-xx-- ----x----x	5.00
26	x--- x	-x-- ----xx-x-	7.00
27		--x-- -----xx	6.00
28	x-	xx-- ----x--x-	5.00
29	x-	xxx-- ----x--x-	6.00
30		xxx-- ----x-	4.00
31		----x-	2.00
32	x-	----x-----	2.00
33		x-- ------x	1.42
34	x	x-- ----xx---	3.00
35		x-- ------x	4.00
36	x-	-x-- ---x----x-	4.00
37	x-	xx-- ----x----x-	5.00
38	x-	-x-- ----x----x-	4.00
39			3.00
40	x-	x-- ----x-----	2.00
41		-----xx	2.00
42	x		----x-----	1.00
43		x-- ------x	1.00
44		xx-- -------x-	2.00
45	x-	-x-- -----x-----	2.00
46		-x-- ------x-	2.00
47		x-- -------x	1.00
48	x-	x-- ----x-----	1.00
49		x-- ------x-	1.00
50		x-- -----	0.00
51		x-- -----	0.00
52		x-- -----	0.00
53			0.00
54			0.00
55		----x-	1.00
56	--x-	----x----	1.00
57			1.00
58	--x- x-	----x----	1.00
59		x-- ------x-	1.00
60			0.00
61			

(3) from above. Following the meet-in-the-middle approach, the best result is on 48 steps [3]. There, relaxation of (3) may lead to a slightly better result, but most likely not more than for 1-4 steps.

6   Discussion and Open Problems

Our results on the second preimage resistance of SHA-1 complement earlier analysis regarding its preimage resistance. Both, attacks for more rounds, and more computationally efficient attacks, can be obtained if the existence of a first preimage (especially if it is long) can be assumed. Our results also complement similar results on the iteration mode [1,14]: also there, better second preimage attacks than preimage attacks were obtained. A lesson to be learned from our results are as follows. In the preimage setting, when it comes to squeezing out the most in terms of number of rounds or in terms of attack complexity, the help provided for an attacker by being given an existing preimage is most of the time not used in earlier preimage-style cryptanalysis of the SHA family.

Overall, applications requiring 2nd-preimage resistance of SHA-1 are not endangered by our results. Even though SHA-1 is arguably one of the more interesting cryptanalytic targets, it will be interesting to see this approach considered for other hash functions as well.

Acknowledgements. The work in this paper has been supported in part by the European Commission under contract ICT-2007-216646 (ECRYPT II) and in part by the IAP Programme P6/26 BCRYPT of the Belgian State (Belgian Science Policy).

References

1. Andreeva, E., Bouillaguet, C., Fouque, P.A., Hoch, J.J., Kelsey, J., Shamir, A., Zimmer, S.: Second Preimage Attacks on Dithered Hash Functions. In: Smart, N.P. (ed.) EUROCRYPT 2008. LNCS, vol. 4965, pp. 270–288. Springer, Heidelberg (2008)
2. Andreeva, E., Preneel, B.: A New Three-Property-Secure Hash Function. In: Avanzi, R.M., Keliher, L., Sica, F. (eds.) SAC 2008. LNCS, vol. 5381, pp. 228–244. Springer, Heidelberg (2009)
3. Aoki, K., Sasaki, Y.: Meet-in-the-Middle Preimage Attacks Against Reduced SHA-0 and SHA-1. In: Halevi [11], pp. 70–89
4. Biham, E., Chen, R., Joux, A., Carribault, P., Lemuet, C., Jalby, W.: Collisions of SHA-0 and Reduced SHA-1. In: Cramer [5], pp. 36–57
5. Cramer, R. (ed.): EUROCRYPT 2005. LNCS, vol. 3494. Springer, Heidelberg (2005)
6. De Cannière, C., Mendel, F., Rechberger, C.: Collisions for 70-Step SHA-1: On the Full Cost of Collision Search. In: Adams, C.M., Miri, A., Wiener, M.J. (eds.) SAC 2007. LNCS, vol. 4876, pp. 56–73. Springer, Heidelberg (2007)
7. De Cannière, C., Rechberger, C.: Finding SHA-1 Characteristics: General Results and Applications. In: Lai, X., Chen, K. (eds.) ASIACRYPT 2006. LNCS, vol. 4284, pp. 1–20. Springer, Heidelberg (2006)
8. De Cannière, C., Rechberger, C.: Preimages for Reduced SHA-0 and SHA-1. In: Wagner, D. (ed.) CRYPTO 2008. LNCS, vol. 5157, pp. 179–202. Springer, Heidelberg (2008)
9. Dunkelman, O. (ed.): FSE 2009. LNCS, vol. 5665. Springer, Heidelberg (2009)
10. Guo, J., Ling, S., Rechberger, C., Wang, H.: Advanced Meet-in-the-Middle Preimage Attacks: First Results on Full Tiger, and Improved Results on MD4 and SHA-2. Cryptology ePrint Archive, Report 2010/016 (2010), http://eprint.iacr.org/
11. Halevi, S. (ed.): CRYPTO 2009. LNCS, vol. 5677. Springer, Heidelberg (2009)
12. Hawkes, P., Paddon, M., Rose, G.: Automated Search for Round 1 Differentials for SHA-1: Work in Progress. In: NIST - Second Cryptographic Hash Workshop, August 24-25 (2006)

13. Joux, A., Peyrin, T.: Hash Functions and the (Amplified) Boomerang Attack. In: Menezes, A. (ed.) CRYPTO 2007. LNCS, vol. 4622, pp. 244–263. Springer, Heidelberg (2007)
14. Kelsey, J., Schneier, B.: Second Preimages on n-Bit Hash Functions for Much Less than 2^n Work. In: Cramer [5], pp. 474–490
15. Khovratovich, D., Nikolic, I., Weinmann, R.P.: Meet-in-the-Middle Attacks on SHA-3 Candidates. In: Dunkelman [9], pp. 228–245
16. Kim, J., Biryukov, A., Preneel, B., Hong, S.: On the Security of HMAC and NMAC Based on HAVAL, MD4, MD5, SHA-0 and SHA-1 (Extended Abstract). In: De Prisco, R., Yung, M. (eds.) SCN 2006. LNCS, vol. 4116, pp. 242–256. Springer, Heidelberg (2006)
17. Knudsen, L.R., Mathiassen, J.E., Muller, F., Thomsen, S.S.: Cryptanalysis of MD2. J. Cryptology 23(1), 72–90 (2010)
18. Leurent, G.: Message Freedom in MD4 and MD5 Collisions: Application to APOP. In: Biryukov, A. (ed.) FSE 2007. LNCS, vol. 4593, pp. 309–328. Springer, Heidelberg (2007)
19. Leurent, G.: MD4 is Not One-Way. In: Nyberg, K. (ed.) FSE 2008. LNCS, vol. 5086, pp. 412–428. Springer, Heidelberg (2008)
20. Matusiewicz, K., Pieprzyk, J.: Finding Good Differential Patterns for Attacks on SHA-1. In: Ytrehus, Ø. (ed.) WCC 2005. LNCS, vol. 3969, pp. 164–177. Springer, Heidelberg (2006)
21. McDonald, C., Pieprzyk, J., Hawkes, P.: SHA-1 collisions now 2^{52}. In: Eurocrypt 2009 Rump Session (2009)
22. Mendel, F., Pramstaller, N., Rechberger, C., Rijmen, V.: The Impact of Carries on the Complexity of Collision Attacks on SHA-1. In: Robshaw, M.J.B. (ed.) FSE 2006. LNCS, vol. 4047, pp. 278–292. Springer, Heidelberg (2006)
23. National Institute of Standards and Technology: NIST's Policy on Hash Functions (2008), http://csrc.nist.gov/groups/ST/hash/policy.html
24. National Institute of Standards and Technology (NIST): FIPS-180-2: Secure Hash Standard (August 2002), http://www.itl.nist.gov/fipspubs/
25. Peyrin, T.: Analyse de fonctions de hachage cryptographiques. Ph.D. thesis (2008)
26. Pramstaller, N., Rechberger, C., Rijmen, V.: Exploiting Coding Theory for Collision Attacks on SHA-1. In: Smart, N.P. (ed.) Cryptography and Coding 2005. LNCS, vol. 3796, pp. 78–95. Springer, Heidelberg (2005)
27. Rechberger, C.: Preimage Search for a Class of Block Cipher based Hash Functions with Less Computation (2008) (unpublished manuscript)
28. Rechberger, C., Rijmen, V.: On Authentication with HMAC and Non-random Properties. In: Dietrich, S., Dhamija, R. (eds.) FC 2007 and USEC 2007. LNCS, vol. 4886, pp. 119–133. Springer, Heidelberg (2007)
29. Rechberger, C., Rijmen, V.: New Results on NMAC/HMAC when Instantiated with Popular Hash Functions. Journal = J. UCS 14(3), 347–376 (2008)
30. Reyhanitabar, M.R., Susilo, W., Mu, Y.: Enhanced Target Collision Resistant Hash Functions Revisited. In: Dunkelman [9], pp. 327–344
31. Rijmen, V., Oswald, E.: Update on SHA-1. In: Menezes, A. (ed.) CT-RSA 2005. LNCS, vol. 3376, pp. 58–71. Springer, Heidelberg (2005)
32. Rivest, R.L.: The MD4 Message Digest Algorithm. In: Menezes, A., Vanstone, S.A. (eds.) CRYPTO 1990. LNCS, vol. 537, pp. 303–311. Springer, Heidelberg (1991)
33. Rogaway, P., Shrimpton, T.: Cryptographic Hash-Function Basics: Definitions, Implications, and Separations for Preimage Resistance, Second-Preimage Resistance, and Collision Resistance. In: Roy, B.K., Meier, W. (eds.) FSE 2004. LNCS, vol. 3017, pp. 371–388. Springer, Heidelberg (2004)

34. Sasaki, Y., Aoki, K.: Preimage Attacks on 3, 4, and 5-Pass HAVAL. In: Pieprzyk, J. (ed.) ASIACRYPT 2008. LNCS, vol. 5350, pp. 253–271. Springer, Heidelberg (2008)
35. Sasaki, Y., Aoki, K.: Finding Preimages in Full MD5 Faster Than Exhaustive Search. In: Joux, A. (ed.) EUROCRYPT 2009. LNCS, vol. 5479, pp. 134–152. Springer, Heidelberg (2010)
36. Sasaki, Y., Wang, L., Ohta, K., Kunihiro, N.: Security of MD5 Challenge and Response: Extension of APOP Password Recovery Attack. In: Malkin, T.G. (ed.) CT-RSA 2008. LNCS, vol. 4964, pp. 1–18. Springer, Heidelberg (2008)
37. Stevens, M., Lenstra, A.K., de Weger, B.: Chosen-Prefix Collisions for MD5 and Colliding X.509 Certificates for Different Identities. In: Naor, M. (ed.) EURO-CRYPT 2007. LNCS, vol. 4515, pp. 1–22. Springer, Heidelberg (2007)
38. Stevens, M., Sotirov, A., Appelbaum, J., Lenstra, A., Molnar, D., Osvik, D.A., de Weger, B.: Short chosen-prefix collisions for MD5 and the creation of a rogue CA certificate. In: Halevi [11], pp. 55–69
39. Stinson, D.R.: Some Observations on the Theory of Cryptographic Hash Functions. Des. Codes Cryptography 38(2), 259–277 (2006)
40. Wang, X., Yin, Y.L., Yu, H.: Finding Collisions in the Full SHA-1. In: Shoup, V. (ed.) CRYPTO 2005. LNCS, vol. 3621, pp. 17–36. Springer, Heidelberg (2005)
41. Wang, X., Yu, H.: How to Break MD5 and Other Hash Functions. In: Cramer [5], pp. 19–35
42. Yajima, J., Iwasaki, T., Naito, Y., Sasaki, Y., Shimoyama, T., Peyrin, T., Kunihiro, N., Ohta, K.: A Strict Evaluation on the Number of Conditions for SHA-1 Collision Search (2009)
43. Yajima, J., Sasaki, Y., Naito, Y., Iwasaki, T., Shimoyama, T., Kunihiro, N., Ohta, K.: A New Strategy for Finding a Differential Path of SHA-1. In: Pieprzyk, J., Ghodosi, H., Dawson, E. (eds.) ACISP 2007. LNCS, vol. 4586, pp. 45–58. Springer, Heidelberg (2007)
44. Yu, H., Wang, G., Zhang, G., Wang, X.: The Second-Preimage Attack on MD4. In: Desmedt, Y.G., Wang, H., Mu, Y., Li, Y. (eds.) CANS 2005. LNCS, vol. 3810, pp. 1–12. Springer, Heidelberg (2005)

Some Observations on Indifferentiability

Ewan Fleischmann, Michael Gorski, and Stefan Lucks

Bauhaus-University Weimar, Germany
{ewan.fleischmann,michael.gorski,stefan.lucks}@uni-weimar.de

Abstract. At Crypto 2005, Coron et al. introduced a formalism to study the presence or absence of structural flaws in iterated hash functions. If one cannot differentiate a hash function using ideal primitives from a random oracle, it is considered structurally sound, while the ability to differentiate it from a random oracle indicates a structural weakness. This model was devised as a tool to see subtle real world weaknesses while in the random oracle world. In this paper we take in a practical point of view. We show, using well known examples like NMAC and the Mix-Compress-Mix (MCM) construction, how we can prove a hash construction secure and insecure at the same time in the indifferentiability setting. These constructions do not differ in their implementation but only on an abstract level. Naturally, this gives rise to the question what to conclude for the implemented hash function.

Our results cast doubts about the notion of "indifferentiability from a random oracle" to be a mandatory, practically relevant criterion (as *e.g.*, proposed by Knudsen [17] for the SHA-3 competition) to separate good hash structures from bad ones.

Keywords: hash function, provably secure, indifferentiability framework, ideal world models.

1   Introduction

RANDOM ORACLE METHODOLOGY. A *hash function $H : \{0,1\}^* \to \{0,1\}^n$* is used to compute an n-bit fingerprint from an arbitrarily-sized input. Established security requirements for cryptographic hash functions are collision resistance, preimage and 2nd preimage resistance. But, in an ideal world, most cryptographers expect a good hash function to somehow behave like a *random oracle* [4].

A random oracle is a mathematical abstraction used in cryptographic proofs, hiding away virtually all real world and implementational details. They are typically used when no known implementable function provides the mathematical properties required for the proof – or when it gets too tedious to formalize these properties. From a theoretical point of view, it is clear, that a security proof in the random oracle model is only a heuristic indication of the security of the system when instantiated with a particular hash function. In fact, many recent separation results [2,7,10,13,19,21] illustrated various cryptographic systems secure in the random oracle model but completely insecure for *any* concrete instantiation of the random oracle. Nevertheless, these results do not seem to directly attack any

R. Steinfeld and P. Hawkes (Eds.): ACISP 2010, LNCS 6168, pp. 117–134, 2010.

concrete cryptosystem. In the random oracle model, one proves that the system is at least secure with and "ideal" hash function H. Such formal proof is believed to indicate that there are no structural flaws in the design of the system.

BUILDING A RANDOM ORACLE. In practice, arbitrary length hash functions are built by first heuristically constructing a fixed-length building block, such as a fixed-length *compression function* or a block cipher, and then *iterating* this building block in some manner to extend the input domain arbitrarily.

Current *practical* hash functions, as *e.g.*, SHA-1 [22], SHA-2 [23] or MD5 [26] are all iterated hash functions using a compression function with a fixed-length input, $h : \{0,1\}^{n+l} \to \{0,1\}^n$, and the Merkle-Damgård transformation [9,20] for the full hash function H with arbitrary input sizes. The core idea is to split the message M into l-bit blocks $M_1, \ldots, M_m \in \{0,1\}^l$ (with some padding to ensure that all the blocks are of size l-bit), to define an initial value H_0 and to apply the recurrence $H_i = h(H_{i-1}, M_i)$. The final *chaining variable* H_m is used as the hash output, *i.e.*, $H(M) := H_m$. The main benefit of the MD-transformation is that it preserves collision resistance: if the compression function h is collision resistant, then so is the hash function H.

STRCTURAL FLAWS IN THE HASH FUNCTION. Recent results on the security of the Merkle-Damgård construction [1,14,15,16] indicate that there are some structural weaknesses in the design of the iteration process itself. They can be exploited even if the compression function is ideal, *i.e.*, a fixed input length random oracle. Motivated by the practical need to *"say anything about structural flaws in the design of H itself"*, Coron et al. [8] presented a new notion of security for cryptographic hash functions which is called *indifferentiability*.

In short, if one models the compression function(s) as random oracles with fixed-size inputs, then the iterated hash function composed from these compression functions should be indifferentiable from a random oracle with variably-sized inputs. They propose these as a practically relevant criterion, *e.g.*, to separate practical hash functions with a good structure from those which might suffer from structural flaws, especially in the context of the search for new hash function standard SHA-3 [17,24]. The current paper discusses this issue.

Preliminaries

In this paper, we use notions such as "efficient", "significant" and "negligible" as usual in theoretical cryptography [29], e.g., an algorithm is efficient, if its running time is bounded from above by a polynomial in the security parameters. In the following we will call a hash function *secure* if it is indifferentiable from a random oracle (a formal definition will be given in Section 2), *i.e.*, there exists a simulator so that any efficient distinguisher has negligible advantage in distinguishing the hash function from a random oracle. A hash function is called *insecure* if there exists an efficient distinguisher that can distinguish the hash function for any simulator from a random oracle.

Remark: One purpose of this paper is to inspire a discussion about the practical relevance of the notions *secure* and *insecure*. More precisely: Does *insecure* actually indicate a structural flaw in a hash function whereas *secure* means the absence of them?

Our Contribution

Taking in a practical point of view, we will examine to what extent a structure of a hash function, proven secure using the indifferentiability framework, relates with instantiations satisfying this structure. This perspective is justified by the objective of Coron et. al. [8] to deliver a criteria for the design of practical hash functions that can distinguish between good hash structures and bad hash structures. On a merely abstract level – *i.e.*, if one views a hash function as a sole random oracle – the hash structure is trivially secure. Instantiated as one single collision resistant hash function, it is trivially insecure. We will examine what happens in between these two poles. We will show that one is able to prove one and the same practical hash function secure and insecure at the same time. These hash functions *do not* differ in their implementations but only on an abstract modeling level. Also, we will show how a slight modification to a secure hash function, *e.g.*, concatenation a one way function, can drive it insecure whereas concatenating an easily invertible function apparently preserves its security. We are able to derive some weird features that a secure hash function must offer. Moreover, as we can prove different structures that correspond to one and the same instantiated hash function secure and insecure, we are faced with an open problem what to conclude for the security of the practical hash function.

Section 2 gives an detailed paper outline and further motivates this discussion. Taking the practical point of view as a start, we show how one and the same implementation can be proven secure and insecure in the indifferentiability model. Section 3 introduces the random oracle model and the concept of "indifferentiability from a random oracle" as a security notion for hash functions and compares it to other ideal world security models for hash functions. Sections 4, 5, and Appendix C in the full version of the paper [12] will give proofs for this. In Section 6 we will derive some mandatory design principles for hash functions being secure in the indifferentiability framework. In Section 7 we discuss and conclude.

2   (In)Security in the Indifferentiability World

In the following sections we will examine various constructions that are secure in the indifferentiability framework (details on indifferentiability will follow in Section 3) involving one or more random oracles and show how slight modifications to them (or partial instantiations) drive them insecure (at least in this framework).

In this section we will motivate our research and summarize some of our results in Table 1. Furthermore, we will give a short example in which way our results correlate and with the design of practical hash functions.

Table 1. $\mathcal{RO}$ denotes a random oracle (with fixed or variable length input), $\mathcal{RO}_i$ an injective random 'oracle', $\mathcal{RO}_x$ a random oracle ($\mathcal{RO}_x$ is a fixed or variable input length, injective or not, random oracle), X,Y and Z collision resistant one-way functions (CROWF), W is an easily invertible function.

Section	Secure	Insecure (partial instantiation or modification)	Insecure (extension)	Secure
4	$\mathcal{RO}$	X	$\mathcal{RO} \circ X$ $X \circ \mathcal{RO}$	$\mathcal{RO} \circ W$ $W \circ \mathcal{RO}$
4	$\mathcal{RO} \circ \mathcal{RO}$	$\mathcal{RO} \circ X$ $X \circ \mathcal{RO}$	$\mathcal{RO} \circ \mathcal{RO} \circ X$ $X \circ \mathcal{RO} \circ \mathcal{RO}$	$\mathcal{RO} \circ \mathcal{RO} \circ W$ $W \circ \mathcal{RO} \circ \mathcal{RO}$
5	$\mathcal{RO} \circ MD_{\mathcal{RO}}$ (NMAC)	$\mathcal{RO} \circ MD_Z$ $X \circ MD_{\mathcal{RO}}$	$\mathcal{RO} \circ MD_{\mathcal{RO}} \circ X$ $X \circ \mathcal{RO} \circ MD_{\mathcal{RO}}$	$\mathcal{RO} \circ MD_{\mathcal{RO}} \circ W$ $W \circ \mathcal{RO} \circ MD_{\mathcal{RO}}$
[12] Appendix C	$\mathcal{RO}_i \circ X \circ \mathcal{RO}_i$ (MCM)	$\mathcal{RO}_x \circ X \circ Y$ $Y \circ X \circ \mathcal{RO}_x$ $X \circ \mathcal{RO}_x \circ Y$	$\mathcal{RO}_i \circ X \circ \mathcal{RO}_i \circ Y$ $Y \circ \mathcal{RO}_i \circ X \circ \mathcal{RO}_i$	$\mathcal{RO}_i \circ X \circ \mathcal{RO}_i \circ W$ $W \circ \mathcal{RO}_i \circ X \circ \mathcal{RO}_i$

Our results are even stronger than indicated by Table 1.

Motivational, informal Example. Say we want to design a secure hash function and come up with the idea to design our hash function as a concatenation of a preprocessing function modeled as a random oracle $\mathcal{RO}$ and a collision resistant one way function (CROWF) X. Consequently, our hash function $\mathcal{H}$ for a message M is

$$\mathcal{H}(M) := (\mathcal{RO} \circ X)(M).$$

So we try to proof its security in the indifferentiability framework and come to the conclusion that this hash function is in fact insecure (refer to Theorem 1 (iii)). In the indifferentiability framework we have at least three straightforward approaches to get $\mathcal{H}$ secure:

1. Remove the CROWF X: $\mathcal{H}_1(M) = (\mathcal{RO})(M)$.
2. "Strengthen" X and make it a random oracle: $\mathcal{H}_2(M) = (\mathcal{RO} \circ \mathcal{RO})(M)$.
3. "Weaken" X and make it an easily invertible function W: $\mathcal{H}_3(M) = (\mathcal{RO} \circ W)(M)$.

The hash functions $\mathcal{H}_1, \mathcal{H}_2$ and $\mathcal{H}_3$ can be proven secure in the indifferentiability framework (see Theorem 1 (i) and (ii)).

Indifferentiability was devised as a tool to see subtle real world weaknesses while in the random oracle world. But we can prove $\mathcal{H}$ insecure and $\mathcal{H}_3$ secure. In the real world (*i.e.*, comparing the instantiated hash functions) $\mathcal{H}_3$ is (almost) sure to be substantially weaker than $\mathcal{H}$. Additionally, the hash functions $\mathcal{H}$ and $\mathcal{H}_2$ could be implemented using exact the same lines of code but one is proved to be insecure, the other one seems to be secure. What shall we conclude for the security of our instantiated hash function in the real world?

How can we conclude that $\mathcal{H}$ has some real world weaknesses that $\mathcal{H}_2$ has not. Note that mixing complexity-theoretic and ideal building blocks is common and can *e.g.*, be found in [25].

3 Indifferentiability from a Random Oracle

For hash functions a random oracle serves as a reference model. It offers all the properties a hash function should have. This section gives an overview on all 'known methods' for comparing a hash function with a random oracle: indifferentiability and three weaker models: preimage awarenes, indifferentiability from a public-use random oracle and indistinguishability.

A random oracle, denoted $\mathcal{RO}$, takes as input binary strings of any length and returns for each input a random infinite string, $i.e.$, it is a map $\mathcal{RO} : \mathbb{Z}_2^* \to \mathbb{Z}_2^\infty$, chosen by selecting each bit of $\mathcal{RO}(x)$ uniformly and independently, for every x. As in [8] we will only consider random oracles $\mathcal{RO}$ truncated to a fixed output length $\mathcal{RO} : \mathbb{Z}_2^* \to \mathbb{Z}_2^n$.

Indifferentiability from a Random Oracle. The indifferentiability framework was introduced by Maurer et al. in [19] and is an extension to the classical notion of indistinguishability. Coron et al. [8] applied it to iterated hash function constructions and demonstrated for several iterated hash function constructions that they are indifferentiable from a random oracle if the compression function is a fixed input length (FIL) random oracle. Here, we give a brief introduction on these topics. For a more in-depth treatment, we refer to the original papers. In the context of iterated hashing, the adversary – called distinguisher D – shall distinguish between two systems as illustrated in Figure 1.

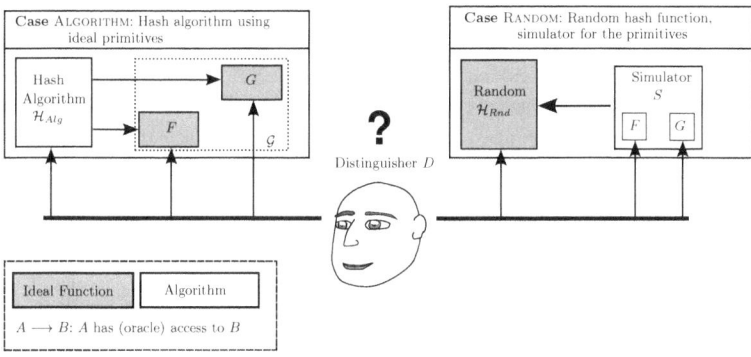

Fig. 1. Defining $\mathcal{H}_{Alg}$ being indifferentiable from a random oracle $\mathcal{H}_{Rnd} := \mathcal{RO}$

The system at the left (Case ALGORITHM) is the hash algorithm $\mathcal{H}_{Alg}$ using some ideal components ($i.e.$, FIL random oracles) contained in the set $\mathcal{G}$. The adversary can make queries to $\mathcal{H}_{Alg}$ as well as to the functions contained in $\mathcal{G}$. The system at the right consists of a random oracle (with truncated output) $\mathcal{H}_{Rnd} := \mathcal{RO}$ providing the same interface as the system on the left. To be indifferentiable to the system at the left, the system at the right (Case RANDOM) also needs a subsystem offering the same interface to the adversary as the ideal compression functions contained in $\mathcal{G}$. A simulator S is needed and its task is to simulate the ideal compression functions so that no distinguisher can tell whether it is interacting with the system at the left or with the one at the right. The output of S should look consistent with what the distinguisher can obtain from the random oracle $\mathcal{H}_{Rnd}$. In order to achieve that, the simulator can query the random oracle $\mathcal{H}_{Rnd}$. Note that the simulator does not see the distinguisher's queries to the random oracle. Formally, the indifferentiability of $\mathcal{H}_{Alg}$ from a random oracle $\mathcal{H}_{Rnd}$ is satisfied if:

Definition 1. *[8] A Turing machine $\mathcal{H}_{Alg}$ with oracle access to a set of ideal primitives contained in the set $\mathcal{G}$ is said to be (t_D, t_S, q, ϵ) indifferentiable from an ideal primitive $\mathcal{H}_{Rnd}$, if there exists a simulator S, such that for any distinguisher D it holds that:*

$$|Pr[D^{\mathcal{H}_{Alg}, \mathcal{G}} = 1] - Pr[D^{\mathcal{H}_{Rnd}, S} = 1]| < \epsilon.$$

The simulator has oracle access to $\mathcal{H}_{Rnd}$ and runs in time at most t_S. The distinguisher runs in time at most t_D and makes at most q queries. Similarly, $\mathcal{H}_{Alg}$ is said to be indifferentiable from $\mathcal{H}_{Rnd}$ if ϵ is a negligible function of the security parameter k.

Now, it is shown in [19] that if $\mathcal{H}_{Alg}$ is indifferentiable from a random oracle, then $\mathcal{H}_{Alg}$ can replace the random oracle in any cryptosystem, and the resulting cryptosystem is at least as secure in the ideal compression function model (*i.e.*, case ALGORITHM) as in the random oracle model (*i.e.*, case RANDOM).

'Non-Optimal' Ideal World Models. At EUROCRYPT'09 Dodis et. al. [11] have presented two ideal world security models that are strictly weaker than indifferentiability: *preimage awareness* and *indifferentiability from a public-use random oracle*. But both model a hash function fairly inadequate. A function that is preimage aware is not guaranteed to be secure against such trivial attacks as, *e.g.*, the Merkle-Damgård extension attack. And a function that is indifferentiability from a public-use random oracle has to 'publish' any oracle query and might be only of limited use in the context of some signature schemes.

If a hash function is indistinguishable from a random oracle, an attacker that can query $\mathcal{H}_{Alg}$ – but has no access to the compression functions contained in $\mathcal{G}$ – cannot distinguish it from a random oracle. For hash function constructions, indistinguishability makes little sense as, for any concrete hash function, the compression functions in $\mathcal{G}$ are public and hence accessible to the adversary. As opposed to block cipher constructions, there is no secret key or any other information the attacker has not. For them, indistinguishable from a random permutation seems to suffice (at least in the ideal cipher model).

Therefore, we will focus in this work on 'indifferentiability from a random oracle' since this seems to be the only security model known that is applicable in all contexts of cryptographic hash functions. The (open) challenge is to find an ideal world security model that is strong enough to defeat all known attacks but it should not be so strong that it leads to real world ambiguities. As we will show in this paper, the notion of indifferentiability has such ambiguities, namely we can prove one and the same real world hash function secure and insecure at the same time.

Security definitions that are based on a random oracle. Note that by assuming *ideal primitives* even in the ALGORITHM case, this definition is inherently based on the random oracle model. In the standard model we cannot assume ideal primitives (at least not without allowing an exponentially-sized memory to store a description of the function), so this notion of security only makes sense in the random oracle model.

Nevertheless, as we understand [8], a part of their motivation was to introduce a formalism for aiding the design of practical hash functions. Showing the above kind of "security" in the random oracle model ought to indicate the absence of structural flaws in the hash function.

On the other hand, if one can efficiently differentiate a hash function (using ideal primitives) from a random oracle, this appears to indicate a weakness in the hash function structure. With this reasoning, we again follow the example of Coron et al., who debunk certain hash function structures as insecure by pointing out efficient differentiation attacks [8, Sections 3.1 and 3.2].

4   Concatenation of Random Oracles: $\mathcal{RO} - \mathcal{RO}$

We start by investigating a fairly simple construction on what to conclude for the security of a hash function where a pre/post-processing function is available.

Definition 2. *Let*

$$F^{(*\to n)}, G^{(*\to n)} : \{0,1\}^* \to \{0,1\}^n \ and$$
$$F^{(n\to n)}, G^{(n\to n)} : \{0,1\}^n \to \{0,1\}^n$$

be random oracles. A Subindex 'i' denotes an injective random oracle, a subindex 'x' denotes a random oracle where we explicitly don't care whether it is injective or not. Let

$$P^{(*\to n)}, Q^{(*\to n)} : \{0,1\}^* \to \{0,1\}^n \ and$$
$$P^{(n\to n)}, Q^{(n\to n)} : \{0,1\}^n \to \{0,1\}^n$$

be collision resistant one way functions. Let

$$W^{(n\to n)} : \{0,1\}^n \to \{0,1\}^n$$

be a function that is easily invertible.

   (i) The hash function $H_{\mathcal{RO}\circ\mathcal{RO}} : \{0,1\}^* \to \{0,1\}^n$ for a message $M \in \{0,1\}^*$ is defined by

$$H_{\mathcal{RO}\circ\mathcal{RO}}(M) := G^{(n\to n)}(F^{(*\to n)}(M)).$$

(ii) *Modification/Partial instantiation I:* The hash function $H_{\mathcal{RO} \circ X} : \{0,1\}^* \to \{0,1\}^n$ for a message $M \in \{0,1\}^*$ is defined by

$$H_{\mathcal{RO} \circ X}(M) := F^{(n \to n)}(P^{(* \to n)}(M)).$$

(iii) *Modification/Partial instantiation II:* The hash function $H_{X \circ \mathcal{RO}} : \{0,1\}^* \to \{0,1\}^n$ for a message $M \in \{0,1\}^*$ is defined by

$$H_{X \circ \mathcal{RO}}(M) := P^{(n \to n)}(F^{(* \to n)}(M)).$$

(iv) *Extension I:* The hash function $H_{\mathcal{RO} \circ \mathcal{RO} \circ X} : \{0,1\}^* \to \{0,1\}^n$ for a message $M \in \{0,1\}^*$ is defined by

$$H_{\mathcal{RO} \circ \mathcal{RO} \circ X}(M) := F^{(n \to n)}(G^{(n \to n)}(P^{(* \to n)}(M))).$$

(v) *Extension II:* The hash function $H_{X \circ \mathcal{RO} \circ \mathcal{RO}} : \{0,1\}^* \to \{0,1\}^n$ for a message $M \in \{0,1\}^*$ is defined by

$$H_{X \circ \mathcal{RO} \circ \mathcal{RO}}(M) := P^{(n \to n)}(F^{(n \to n)}(G^{(* \to n)}(M))).$$

Theorem 1. *In the indifferentiability framework the following statements must hold:*

(i) $H_{\mathcal{RO} \circ \mathcal{RO}}$ *is secure* (i.e., *indifferentiable from a random oracle*),
(ii) $H_{\mathcal{RO} \circ X}$ *is insecure* (i.e., *differentiable from a random oracle*),
(iii) $H_{X \circ \mathcal{RO}}$ *is insecure*,
(iv) $H_{\mathcal{RO} \circ \mathcal{RO} \circ X}$ *is insecure*,
(v) $H_{X \circ \mathcal{RO} \circ \mathcal{RO}}$ *is insecure*.

Recall that for proving a hash function insecure we have to describe an efficient distinguisher which can decide with non-negligible probability if the hash function is an algorithm utilizing random oracles (the ALGORITHM case) or is a random oracle by itself (the RANDOM case).

Proof. Let H denote the hash oracle.

(i) The proof is easy and will be skipped here. It can be found in Appendix A.
Remark: The proof can be easily generalized to all functions $H_{\mathcal{RO} - \dots - \mathcal{RO}}(M)$.
(ii) This result is essentially equivalent to the Coron et al. insecurity result regarding the composition of a CROWF with a random oracle [8], but we state a version of it here for completeness. We describe a distinguisher D to win this game, regardless of the simulator S:
  1. Choose a random message $M \in \{0,1\}^*$.
  2. Compute $u = P(M)$.
  3. Ask the F-oracle for $v = F(u)$.
  4. Ask the hash-oracle for $z = H(M)$.
  5. If $z = v$ output ALGORITHM, else output RANDOM.

Analysis: Clearly, D is efficient. In the ALGORITHM world we always have

$$z = H(M) = F(P(M)) = F(u) = v.$$

so it always outputs ALGORITHM if it interacts with the algorithm and the ideal primitive F.

A simulator trying to fool the distinguisher D does not know M as he receives the $F(u)$-oracle call. In order to answer correctly he has to come up with $M = P^{-1}(u)$ to ask the hash oracle for $H(M)$. As P is a CROWF this is not possible. Any such simulator can be used to invert P. Furthermore it is information-theoretically impossible to recover $M \in \{0,1\}^*$ from u.

(iii) Again, we describe a distinguisher D to win this game, regardless of the simulator S:
1. Choose a random message $M \in \{0,1\}^*$.
2. Ask the hash-oracle for $z = H(M)$.
3. Ask the F-oracle for $u = F(M)$.
4. Compute $v = P(u)$.
5. If $z = v$ output ALGORITHM, else output RANDOM.

Analysis: Again, D is obviously efficient and always outputs ALGORITHM in the algorithm world as we have

$$z = H(M) = P(F(M)) = P(u) = v.$$

In the random world, D learns a random target z and needs to find u with $z = P(u)$ whereas $u = F(M)$. So any simulator able to answer the F-oracle correctly can be used to invert P which is a CROWF.

(iv) The proof is essentially the same as in (ii).
(v) The proof is essentially the same as in (iii). □

Paradox. As we have proven in (ii) and (iii) the hash functions $H_{\mathcal{RO} \circ X}$ and $H_{X \circ \mathcal{RO}}$ are in fact insecure, if X is a collision resistant one way function. What happens if we substitute that CROWF with an easily invertible function? It turns out that both hash functions get secure again. Taking preimage resistance as an example, we are not able to append an 'additional line of defense' (namely the function X) for preimage attacks. without losing the property of being 'indifferentiabity from a random oracle'. Note that we do *not* point out any paradox in the indifferentiability framework itself. But, indeed, we do show several ambiguities we inevitably have to face if we try to apply this framework as a guide for designing secure and practical hash functions. This situations occur if we try to decrease the size of the gap between a practical hash function and its ideal world mapping.

Theorem 2. *Using the same notations as in Definition 2 it must hold in the indifferentiability framework:*

(i) The hash function

$$H_{\mathcal{RO} \circ W}(M) := F^{(n \to n)}(W^{(* \to n)}(M)).$$

is secure if W is an invertible function.

(ii) The hash function

$$H_{W \circ \mathcal{RO}}(M) := W^{(n \to n)}(F^{(* \to n)}(M)).$$

is secure if W is an invertible function.

Proof. Let H denote the hash oracle.

(i) Note, that the function $W : \{0,1\}^* \to \{0,1\}^n$ is unlikely to be uniquely invertible in practice (normally, it will be information-theoretically impossible). But if we assume W's invertibility we can easily give a simulator S thus proving the hash function secure.
  Description of Simulator S:
    - F-oracle queries (parameter A): As we can invert W we can easily calculate $M = W^{-1}(A)$ and ask the hash oracle $v = H(M)$ and return v to the caller.

(ii) Now we have the function $W : \{0,1\}^n \to \{0,1\}^n$. Again, we can give a simple simulator S. *Description of Simulator S:*
    - F-oracle queries (parameter M): Ask the hash oracle $z = H(M)$ and return $W^{-1}(z)$ to the caller.

$\square$

Recall the example of Section 2. Let us start with the hash function $H_{\mathcal{RO} \circ X}$. For X being a CROWF we have shown it to be insecure. If we *strengthen* X and let it be a random oracle our hash function gets secure. If we weaken X, *i.e.*, let X be an invertible function, our hash function gets secure again. The same is true if we begin our discussion with $H_{X \circ \mathcal{RO}}$.

Furthermore, all the theoretical hash functions $H_{\mathcal{RO} \circ \mathcal{RO}}$, $H_{X \circ \mathcal{RO}}$ and $H_{\mathcal{RO} \circ X}$ are a valid model for the same practical hash function, employing two functions F and G and defined by $H(x) = F(G(x))$. Should we conclude that this construction is secure, since we can prove their security if we model both F and G as random oracles? Or should we conclude that this construction is insecure, as we can disprove security in two other cases?

Insecurity by (partial) Instantiation. Another point of view of the problem is given here. We start with the proven-to-be-secure hash function $H_{\mathcal{RO} \circ \mathcal{RO}}$. So one should assume that there are no structural weaknesses found in our construction. In order to get a practical hash function we have to instantiate the random oracle by efficient collision resistant one way functions. Instead of instantiating both random oracles at the same time we choose to instantiate them one after the other. Our intermediate result is either $H_{X \circ \mathcal{RO}}$ or $H_{\mathcal{RO} \circ X}$. Both of them were proved the be insecure in Theorem 1 – but formally we still have not left the random oracle world. Informally, we start now with an insecure hash function and instantiate the other random oracle. Thus, regarding the structural soundness of our construction we get entirely contradicting messages again.

5   NMAC: $MD_{\mathcal{RO}} - \mathcal{RO}$

Definition 3 (NMAC-Hash). *Let*

$$C^{(n+m \to n)} : \{0,1\}^n \times \{0,1\}^m \to \{0,1\}^n \ and$$

$$D^{(n \to n)} : \{0,1\}^n \to \{0,1\}^n$$

be oracles, $M = (M_1, M_2 \ldots, M_L) \in (\{0,1\}^m)^L$ *be a padded message and* $H_0 \in \{0,1\}^n$ *an arbitrary initial value. The Merkle-Damgård hash function* $MD_C : \{0,1\}^n \times (\{0,1\}^m)^+ \to \{0,1\}^n$ *is defined by*

$$H_1 = C(H_0, M_1), \ \ldots, \ H_L = C(H_{L-1}, M_L),$$

$$MD_C(H_0, M_1, \ldots, M_L) = H_L$$

The hash function $NMAC_{C,D} : \{0,1\}^n \times (\{0,1\}^m)^+ \to \{0,1\}^n$ *is defined for fixed* $H_0 \in \{0,1\}^n$ *as follows:*

$$NMAC_{C,D}(M_1, \ldots, M_L) = D(MD_C(H_0, M_1, \ldots, M_L)).$$

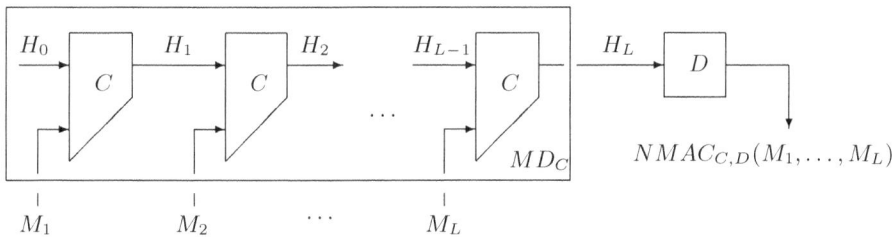

Fig. 2. The NMAC-Hash $NMAC_{C,D}$ – an extension to the plain Merkle-Damgård hash function MD_C

Definition 4. *Let* F, G, P, Q *be as in Definition 2. Additionally, let*

$$F^{(n+m \to n)} : \{0,1\}^{n+m} \to \{0,1\}^n$$

be a random oracle.

(i) The hash function $H_{\mathcal{RO} \circ MD_{\mathcal{RO}}} : \{0,1\}^{m \cdot L} \to \{0,1\}^n$ *for a padded message* $M \in \{0,1\}^{m \cdot L}$ *is defined by*

$$H_{\mathcal{RO} \circ MD_{\mathcal{RO}}}(M) = NMAC_{F^{(n+m \to n)}, G^{(n \to n)}}(M)$$
$$= G^{(n \to n)}(MD_{F^{(n+m \to n)}}(M)).$$

(ii) *Modification/Partial instantiation I: The hash function $H_{\mathcal{RO} \circ MD_X} : \{0,1\}^{m \cdot L} \to \{0,1\}^n$ for a padded message $M \in \{0,1\}^{m \cdot L}$ is defined by*

$$H_{\mathcal{RO} \circ MD_X}(M) = NMAC_{P^{(n+m \to n)}, F^{(n \to n)}}(M)$$
$$= F^{(n \to n)}(MD_{P^{(n+m \to n)}}(M)).$$

(iii) *Modification/Partial instantiation II: The hash function $H_{X \circ MD_{\mathcal{RO}}} : \{0,1\}^{m \cdot L} \to \{0,1\}^n$ for a padded message $M \in \{0,1\}^{m \cdot L}$ is defined by*

$$H_{X \circ MD_{\mathcal{RO}}}(M) = NMAC_{F^{(n+m \to n)}, P^{(n \to n)}}(M)$$
$$= P^{(n \to n)}(MD_{F^{(n+m \to n)}}(M)).$$

(iv) *Extension I: Let*

$$R^{(* \to m \cdot L)} : \{0,1\}^* \to \{0,1\}^{m \cdot L}$$

be a padding function. The hash function $H_{\mathcal{RO} \circ MD_{\mathcal{RO}} \circ R} : \{0,1\}^* \to \{0,1\}^n$ for an (unpadded) message $M \in \{0,1\}^*$ is defined by

$$H_{\mathcal{RO} \circ MD_{\mathcal{RO}} \circ X}(M) = NMAC_{F^{(n+m \to n)}, G^{(n \to n)}}(R^{(* \to m \cdot L)}(M)).$$

(v) *Extension II: The hash function $H_{X \circ \mathcal{RO} \circ MD_{\mathcal{RO}}} : \{0,1\}^{m \cdot L} \to \{0,1\}^n$ for a padded message $M \in \{0,1\}^{m \cdot L}$ is defined by*

$$H_{X \circ \mathcal{RO} \circ MD_{\mathcal{RO}}}(M) = P^{(n \to n)}(NMAC_{F^{(n+m \to n)}, G^{(n \to n)}}(M)).$$

Theorem 3. *Using the indifferentiability framework it must hold:*

(i) *$H_{\mathcal{RO} \circ MD_{\mathcal{RO}}}$ is secure,*
(ii) *$H_{\mathcal{RO} \circ MD_X}$ is insecure,*
(iii) *$H_{X \circ MD_{\mathcal{RO}}}$ is insecure,*
(iv) *$H_{\mathcal{RO} \circ MD_{\mathcal{RO}} \circ X}$ is insecure.*
(v) *$H_{X \circ \mathcal{RO} \circ MD_{\mathcal{RO}}}$ is insecure,*

In [8], the plain Merkle-Damgård hash function with ideal compression functions $MD_{\mathcal{RO}}$ was shown to be insecure.

Proof. (i) Although this result was stated in [8], the proof seems to be missing in the published version of the paper. A proof is provided in the full version of the paper [12] in Appendix C.

(ii) This proof is essentially the same as the proof of Theorem 1 (ii). We only have to take care for the Merkle-Damgård construction. Here, we only give the distinguisher D:

1. Choose a random message $M \in \{0,1\}^*$.
2. Compute $u = MD_P(M)$.
3. Ask the F-oracle for $v = F(u)$.
4. Ask the hash-oracle for $z = H(M)$.
5. If $z = v$ output ALGORITHM, else output RANDOM.

(iii) This proof is essentially the same as the proof of Theorem 1 (iii). We only have to take care for the Merkle-Damgård construction. Here, we only give the distinguisher D:

1. Choose a random message $M \in \{0, 1\}^*$.
2. Ask the hash-oracle for $z = H(M)$.
3. Use the F-oracle to calculate $u = MD_F(M)$.
4. Compute $v = P(u)$.
5. If $z = v$ output ALGORITHM, else output RANDOM.

(iv) The proof is essentially the same as in 1 (iv). So if this hash function is secure, we could use our simulator to efficiently invert the padding function R.

(v) The proof is essentially the same as in 1 (iv). So if this hash function is secure, we could use our simulator to efficiently invert the CROWF G.   □

Paradox. Again, as discussed in Section 4, we get a secure hash function if we substitute the CROWF by an invertible function. For the hash functions (ii)-(v) similar results as were given in theorem 2 can easily be stated.

Part (iv) of the theorem might be (at least in this form) somewhat surprising. Most of the padding functions have the property of being easily invertible. But in the indifferentiability world this is a must-have feature for secure hash functions. If the padding function R is not efficiently invertible, $NMAC$ would be insecure.

In [8] the authors don't care with the padding function. But this turns out to be somewhat shortsighted in the case of indifferentiability secure hash functions. Even such a simple and (in the analysis phase) easily to be forgotten function can drive a hash function insecure if it is added.

We have also analyzed the Mix-Compress-Mix (MCM) construction given by Ristenpart et al. [25]. A short summary is given in the following theorem.

Theorem 4. *(i)* $H_{\mathcal{RO}_i \circ X \circ \mathcal{RO}_i}$ *is secure if X is a $\Delta-$regular function (i.e., every image of H has approximately the same number of preimages, for details see [25]).*

(ii) $H_{\mathcal{RO}_x \circ X \circ Y}$ *is insecure.*

(iii) $H_{Y \circ X \circ \mathcal{RO}_x}$ *is insecure.*

(iv) $H_{X \circ \mathcal{RO} \circ Y}$ *is insecure.*

(v) $H_{\mathcal{RO}_i \circ X \circ \mathcal{RO}_i \circ Y}$ *is insecure.*

(vi) $H_{Y \circ \mathcal{RO}_i \circ X \circ \mathcal{RO}_i}$ *is insecure.*

A proof and formal definitions are given in the full version of the paper [12] in Appendix C.

6 Design Principles for Secure Hash Functions

Definition 5. *Let $k \in \mathbb{N}$. Let $S_1 : \{0, 1\}^* \to \{0, 1\}^m$, $S_i : \{0, 1\}^m \to \{0, 1\}^m$, $(1 < i < k)$, $S_k : \{0, 1\}^m \to \{0, 1\}^n$, be functions. The hash function H for a message M is defined by*

$$H(M) = (S_k \circ S_{k-1} \circ \ldots \circ S_1)(M)$$

Here, we generally don't care whether the functions S_i, $(1 \leq i \leq k)$ are ideal or not. Using the technique applied above it is easy to show

Theorem 5. *(Design Principles for Secure Hash Functions) Let $\mathcal{H}$ be defined as in Definition 5. Let H be a secure (indifferentiable) hash function. Then it must hold:*

(i) S_1 is not a one way function.
(ii) S_k is not a one way function.

Proof. (i) If S_1 is a one way function we could use the simulator (as H is secure) to invert S_1. The proof is essentially the same as the proof of Theorem 1 (ii).
(ii) If S_k is a one way function we could use the simulator (as H is secure) to invert S_k. The proof is essentially the same as the proof of Theorem 1 (iii).   □

This simple design principle is mandatory to all (indifferentiable) secure hash functions. Note that in the MCM construction, the one way function is in the middle of two random oracles. Applying Theorem 5 we can conclude: It is possible to prove a structure secure only if the 'first' and the 'last' functions are random oracles or easily invertible functions, but we might be able – under some circumstances – to choose some functions different to a random oracle. We do not see how this design principle for indifferentiable secure hash function will account for more security if constructing a practical hash function.

7   Discussion and Conclusion

The random oracle model. All ideal world notions and their definitions are inherently based on the random oracle model. Before going into details on indifferentiability itself in Section 7.1 let us recall some results from the literature on the random oracle model. As discussed in Section 1, there had been quite a few uninstantiability results, defining cryptosystems provably secure in the random oracle model, but insecure when instantiated by any efficient function. On can argue that all of these constructions are malicious. They are designed to be insecure. But either one relies on heuristics and intuitions, or one relies on proofs. If one puts proofs above all other aspects, then counter-examples do invalidate the proofs.

7.1   The Ambiguity of Indifferentiability in the Design of Practical Hash Functions

The ideas from Coron et al. [8] have been very influential and inspiring for a lot of researchers. Namely, there have been quite a few proposals for hash function structures provably "indifferentiable from a random oracle", often in addition to other security requirements as, *e.g.*, in [3,5,18].

But the current paper reveals a contradiction in the reasoning from [8]: The *same* formalism can be used to indicate the structural soundness of an implementation, and the presence of structural weaknesses.

The contradiction is not on the formal level – we do not claim any flaw in the theorems or proofs of [8]. If all components (*e.g.*, compression functions) of a hash function are ideal (*i.e.*, random oracles) we don't get ambiguous results. If all components are non-ideal we cannot use the indifferentiability framework to prove anything. But if some of the primitives are ideal and some are not (as for example in [3]) we can get ambiguous results for security proofs. Our research seems to indicate that the indifferentiability model is of limited use for proving the security of

- mixed-model hash functions (using complexity theoretic and ideal components at the same time) *and*
- practical hash functions (*e.g.*, as the SHA-3 candidates).

One might conclude that if any possible description of a structure is insecure (as *e.g.*, is the case for Merkle-Damgård) in the indifferentiability framework then the hash structure is flawed. But it is not clear what we shall conclude for a concrete instantiation if one modeling is secure but another is not.

Taking a secure function (using only ideal components) we have shown in Section 4 and 5 how slight modifications (*i.e.*, adding a pre- or post-processing function) or partial instantiations (*i.e.*, starting our way towards an instantiated hash function) might possibly drive them insecure.

But, in addition to an inherent theoretical motivation, the notion of security in [8] has also been motivated by the need to decide if the structure of a hash function is sound or flawed. A criterion for good hash function structures is very valuable for hash function designers, indeed. On the strictly theoretical side, there is nothing wrong, and studying this kind of security remains an interesting topic for theoretical cryptography.

7.2 Conclusions

The random oracle model makes it possible to design cryptographic functions secure only in the ideal world. As discussed in Section 3, the notions of indistinguishability, preimage awareness and indifferentiability from a public-use random oracle seem to be too weak for designing a secure, practical and general purpose hash structure.

THE RIGHT LEVEL OF ABSTRACTION. If we state the discussion of Section 7.1 somewhat different we can come up with the following: For designing a hash function one might come up with a model/structure that describes the hash function on an abstract level. Then one might try to find a indifferentiability proof for this structure – given that some of the components are ideal. This process usually involves some sort of tweaking of the structure in order to 'find' the proof. Therefore we state that this structure is secure. But if we start with an implementation (*i.e.*, a practical hash function) and want to assess its security in terms of indifferentiability, we are faced with the problem of the right level of abstraction/kind of modeling. If we abstract all the details and come up with a structure only consisting of a random oracle, *all* hash functions are trivially secure (again in terms of indifferentiability). If we abstract nothing, the indistinguishability framework does not have an answer to our question since we have no ideal components. But if we start abstracting some of the components we might be faced with the problem of finding some abstractions that are secure, and some that are not. And we might not know what to conclude for the security of the implementation.

OPEN PROBLEMS. It remains an open problem to derive an ideal world criterion to support the design of general purpose practical hash functions – telling us if the internal structure of a hash function is flawless or not. Certainly, a security proof (*i.e.*, a proof of a hash function being indifferentiable from a random oracle, when modeling some or all the internal functions as random oracles) is comforting. But pursuing this kind of security property requires great care since authors of a new hash function could be tempted to change, *e.g.*, some one-way final transform of their hash function into an easily invertible transformation. This could enable a theoretical security proof in the first place, while, at the same time, practically weaken the hash function.

Designers of practical hash functions, who accept the indifferentiability framework at face value, may be tempted to make poor design decisions. The indifferentiability framework suggests corrections to structures which sometimes make only sense in the ideal world but that have no real-word mapping. Even worse, the danger is that these very corrections drive the corresponding real-world hash function less secure.

Authors of new hash functions are well advised to prove other security properties, such as the established collision-, preimage-, and second-preimage-resistance under some reasonable standard-model assumptions, perhaps in addition to proving theoretical security properties, such as the indifferentiability from a random oracle.

Acknowledgments

The authors would like to thank the anonymous reviewers for helpful comments and suggestions.

References

1. Andreeva, E., Bouillaguet, C., Fouque, P.-A., Hoch, J.J., Kelsey, J., Shamir, A., Zimmer, S.: Second preimage attacks on dithered hash functions. In: Smart [28], pp. 270–288
2. Bellare, M., Boldyreva, A., Palacio, A.: An uninstantiable random-oracle-model scheme for a hybrid-encryption problem. In: Cachin, C., Camenisch, J.L. (eds.) EUROCRYPT 2004. LNCS, vol. 3027, pp. 171–188. Springer, Heidelberg (2004)
3. Bellare, M., Ristenpart, T.: Multi-property-preserving hash domain extension and the emd transform. In: Lai, X., Chen, K. (eds.) ASIACRYPT 2006. LNCS, vol. 4284, pp. 299–314. Springer, Heidelberg (2006)

4. Bellare, M., Rogaway, P.: Random oracles are practical: A paradigm for designing efficient protocols. In: ACM Conference on Computer and Communications Security, pp. 62–73 (1993)
5. Bertoni, G., Daemen, J., Peeters, M., Van Assche, G.: On the indifferentiability of the sponge construction. In: Smart [28], pp. 181–197 (2008)
6. Brassard, G. (ed.): CRYPTO 1989. LNCS, vol. 435. Springer, Heidelberg (1990)
7. Canetti, R., Goldreich, O., Halevi, S.: The random oracle methodology, revisited. J. ACM 51(4), 557–594 (2004)
8. Coron, J.-S., Dodis, Y., Malinaud, C., Puniya, P.: Merkle-damgård revisited: How to construct a hash function. In: Shoup [27], pp. 430–448
9. Damgård, I.: A design principle for hash functions. In: Brassard [6], pp. 416–427
10. Dodis, Y., Oliveira, R., Pietrzak, K.: On the generic insecurity of the full domain hash. In: Shoup [27], pp. 449–466
11. Dodis, Y., Ristenpart, T., Shrimpton, T.: Salvaging merkle-damgård for practical applications. In: Joux, A. (ed.) EUROCRYPT 2009. LNCS, vol. 5479, pp. 371–388. Springer, Heidelberg (2010)
12. Fleischmann, E., Gorski, M., Lucks, S.: Some Observations on Indifferentiability. Cryptology ePrint Archive, Report 2010/264 (2010)
13. Goldwasser, S., Taumann, Y.: On the (in)security of the fiat-shamir paradigm. In: FOCS 2003, pp. 102–115. IEEE Computer Society Press, Los Alamitos (2003)
14. Joux, A.: Multicollisions in iterated hash functions. application to cascaded constructions. In: Franklin, M. K. (ed.) CRYPTO 2004. LNCS, vol. 3152, pp. 306–316. Springer, Heidelberg (2004)
15. Kelsey, J., Kohno, T.: Herding hash functions and the nostradamus attack. In: Vaudenay, S. (ed.) EUROCRYPT 2006. LNCS, vol. 4004, pp. 183–200. Springer, Heidelberg (2006)
16. Kelsey, J., Schneier, B.: Second preimages on n-bit hash functions for much less than 2^n work. In: Cramer, R. (ed.) EUROCRYPT 2005. LNCS, vol. 3494, pp. 474–490. Springer, Heidelberg (2005)
17. Knudsen, L.R.: Hash Functions and SHA-3. In: FSE 2008 (2008) (invited talk), http://fse2008.epfl.ch/docs/slides/day_1_sess_2/Knudsen-FSE2008.pdf
18. Liskov, M.: Constructing an ideal hash function from weak ideal compression functions. In: Biham, E., Youssef, A.M. (eds.) SAC 2006. LNCS, vol. 4356, pp. 358–375. Springer, Heidelberg (2007)
19. Maurer, U.M., Renner, R., Holenstein, C.: Indifferentiability, impossibility results on reductions, and applications to the random oracle methodology. In: Naor, M. (ed.) TCC 2004. LNCS, vol. 2951, pp. 21–39. Springer, Heidelberg (2004)
20. Merkle, R.C.: One way hash functions and des. In: Brassard [6], pp. 428–446
21. Nielsen, J.B.: Separating random oracle proofs from complexity theoretic proofs: The non-committing encryption case. In: Yung, M. (ed.) CRYPTO 2002. LNCS, vol. 2442, pp. 111–126. Springer, Heidelberg (2002)
22. NIST National Institute of Standards and Technology. FIPS 180-1: Secure Hash Standard (April 1995), http://csrc.nist.gov
23. NIST National Institute of Standards and Technology. FIPS 180-2: Secure Hash Standard (April 1995), http://csrc.nist.gov
24. NIST National Institute of Standards and Technology. Tentative Timeline of the Development of New Hash Functions, http://csrc.nist.gov/pki/HashWorkshop/timeline.html
25. Ristenpart, T., Shrimpton, T.: How to build a hash function from any collision-resistant function. In: Kurosawa, K. (ed.) ASIACRYPT 2007. LNCS, vol. 4833, pp. 147–163. Springer, Heidelberg (2007)

26. Rivest, R.L.: RFC 1321: The MD5 Message-Digest Algorithm. Internet Activities Board (April 1992)
27. Shoup, V. (ed.): CRYPTO 2005. LNCS, vol. 3621. Springer, Heidelberg (2005)
28. Smart, N.P. (ed.): EUROCRYPT 2008. LNCS, vol. 4965. Springer, Heidelberg (2008)
29. Stinson, D.R.: Cryptography: Theory and Practice. CRC Press, Boca Raton (1995)

A   Proof of Security for $H_{\mathcal{RO}-\mathcal{RO}}$

Proof of Theorem 1

Proof. Let $H := (\mathcal{RO}_2 \circ \mathcal{RO}_1)(M)$ be the definition of the hash function. We have to describe an efficient simulator S who is able to emulate the random oracles $\mathcal{RO}_1$ and $\mathcal{RO}_2$. The simulator has access to the hash oracle $H_{\mathcal{RO}}$. *Description of the Simulator S:*

1. $\mathcal{RO}_1$ oracle queries: For all queries we perform record keeping. If we have answered the same query before we return the same value again. Else we choose a random value and add it to our database $DB \xleftarrow{add} [query, random]$.
2. $\mathcal{RO}_2$ oracle queries: If $[?, query] \in DB$, then use the first entry to ask the hash oracle $H_{\mathcal{RO}}$ and return the answer. Else choose a random value and add it to our database $DB \xleftarrow{add} [random, query]$. Use the new chosen random value to ask the hash oracle $H_{\mathcal{RO}}$ and return the answer.

Clearly, S is efficient and any distinguisher D cannot differentiate it from a random oracle.

Adaptive and Composable Non-committing Encryptions

Huafei Zhu[1], Tadashi Araragi[2], Takashi Nishide[3], and Kouichi Sakurai[3]

[1] Institute for Infocomm Research, A-STAR, Singapore
huafei@i2r.a-star.edu.sg
[2] NTT Communication Science Laboratories, Kyoto, Japan
araragi@cslab.kecl.ntt.co.jp
[3] Department of Computer Science and Communication Engineering,
Kyushu University, Fukuoka, Japan
nishide@inf.kyushu-u.ac.jp, sakurai@csce.kyushu-u.ac.jp

Abstract. In this paper, a new non-committing encryption protocol without failure during the course of a channel setup procedure is constructed and analyzed in the universally composable (UC) framework. We show that the proposed non-committing scheme realizes the UC-security in the presence of adaptive adversary assuming that the decisional Diffie-Hellman problem is hard.

Keywords: Adaptive security, non-committing encryptions, universal composability.

1 Introduction

Designing protocols securely computing any function dates back to the papers by Yao [15], Goldreich, Micali and Wigderson [11] and Goldreich, Micali and Wigderson [12]. These pioneer works have presented general methods to realize non-adaptively secure multi-party computations in the computational setting. Ben-Or, Goldwasser and Wigderson [3] and independently Chaum, Crépeau and Damgård [8] have proposed general methods to realize adaptively secure multi-party computations in the secure channel setting. Interestingly, Canetti, Feige, Goldreich and Naor [7] have shown that if messages on the secure channel are replaced by ciphertexts of a non-committing encryption (NCE) on open networks, one obtains adaptively secure multi-party computations in the computational setting.

1.1 The State-of-the-Art

The research of non-committing encryptions dates back to the papers by Beaver and Haber [2], Canetti, Feige, Goldreich and Naor [7] and Beaver [1]. Informally, a non-committing encryption is a semantically secure encryption scheme with additional property that a simulator can generate special ciphertexts that can be opened to both a 0 and a 1. However, Nielsen [13] has shown that no non-interactive communication protocol can be adaptively secure in the asynchronous model.

R. Steinfeld and P. Hawkes (Eds.): ACISP 2010, LNCS 6168, pp. 135–144, 2010.

Canetti, Feige, Goldreich and Naor [7] have proposed the first non-committing encryptions based on so called common-domain permutations in the stand-alone, simulation-based framework. To encrypt 1 bit, $\Theta(k^2)$ public key bits are communicated. Subsequently, Damgård and Nielsen [10] have proposed generic constructions of non-committing encryption schemes based on so called simulatable public-key encryption schemes. The Damgård and Nielsen's construction is general which may in turn be realized from the decisional Diffie-Hellman, RSA and worst-cast lattice assumptions. However, the probability that a failure occurs during the course of one bit communication in their scheme is $1/2$.

Very recently, Choi, Soled, Malkin and Wee [9] have presented a new implementation of non-committing encryptions based on a weaker notion called trapdoor simulatable cryptosystems in the stand-alone, simulation-based framework in the presence of adaptive adversaries. The idea behind their construction is simple − a receiver first generates total $4k$ public keys where the first k public keys are generated by a key generation algorithm of the underlying trapdoor simulatable encryption scheme while the rest $3k$ public keys are generated by an oblivious sampling algorithm, where k is a security parameter. To encrypt a bit b, the sender sends $4k$ ciphertexts of which k are encrypted b and the remaining $3k$ ones are obliviously sampled. Although the non-committing encryption scheme in [9] is at the expense of higher computation and communication than the Damgård and Nielsen's protocol [10], such an implementation is definitely interesting since the failure model in [10] is eliminated within their framework.

1.2   This Work

This paper studies non-committing encryption schemes without failure during the course of a channel-step procedure in the presence of adaptive adversaries in the universally composable framework.

Overview of the protocol. The non-committing encryption scheme presented in this paper is constructed from the Decisional Diffie-Hellman assumption. Our protocol is 4-round. In the first round, a sender S randomly selects a bit α and then generates a random Diffie-Hellman quadruple S_α by means of a key generation algorithm and a random quadruple $S_{1-\alpha}$ by an oblivious sampling algorithm (say, the Canetti-Fischlin's oblivious sampling algorithm). In the second round, a receiver R randomly selects a bit β and generates a random Diffie-Hellman quadruple R_β by means of the key generation algorithm and a random quadruple $R_{1-\beta}$ by the oblivious sampling algorithm. R then computes the Naor-Pinkas randomizer $(w_{S,0}, w_{S,1})$ for the given quadruples (S_0, S_1) (notions such as the Naor-Pinkas randomizer, Canetti-Fischlin's oblivious sampling algorithm are described in Section 3). In the third round, the sender S extracts the bit β selected by the receiver R. This is possible since the sender S holds the auxiliary string sk_S of the given Diffie-Hellman quadruple S_α. In the fourth round, the sender S sends back the extracted bit β to the receiver R so that both the sender S and the receiver R share a bit β that will be used to encrypt a bit m through the open networks.

The result. We claim that the proposed non-committing encryption scheme (without failure during the course of a channel setup phase) realizes the UC-security in the presence of adaptive adversary assuming that the decisional Diffie-Hellman problem is hard.

The proof of security. According to the functionality of a non-committing encryption scheme (see Section 2.2 for more details), we know that if either a party is corrupted, then the security definition does not require anything of the protocol as a simulator then knows messages and can thus simulate by running honestly. Therefore *all parts of the protocol should be useful only for the interesting case where both parties are honest and where we hence are interested in hiding the messages.*

To prove the security, a simulator $\mathcal{S}$ first generates two random Diffie-Hellman quadruples S_0 and S_1 on behalf of the honest sender S and two random Diffie-Hellman quadruples R_0 and R_1 for the honest receiver R. When a party $P \in \{S, R\}$ gets corrupted (say, in case that $\alpha = \beta$, where α randomly selected by S while β randomly and independently selected by R), $\mathcal{S}$ will interpret $S_{1-\alpha}$ and $R_{1-\beta}$ as random quadruples by means of the oblivious faking algorithm (intuitively, this captures the idea that a simulator can generate special ciphertexts that can be opened to both a 0 and a 1). This idea applies to the other cases where the first corruption occurs after a ciphertext c has been received successfully; or the first corruption occurs after a secure channel has been set up but before a ciphertext c is generated; or the first corruption occurs during the course of channel setup phase. As a result, we are able to show that $\text{REAL}_{\pi,\mathcal{A},\mathcal{Z}}$ and $\text{IDEAL}_{\mathcal{F},\mathcal{S},\mathcal{Z}}$ are computationally indistinguishable assuming that the decisional Diffie-Hellman problem is hard.

Efficiency. Essentially, the proposed non-committing encryption scheme is 4-round. The total communication for encrypting one bit message requires to generate two Diffie-Hellman quadruples and two random quadruples. Thus, our universally composably secure non-committing encryption protocol is as efficient as the stand-alone, simulation-based non-committing encryption of Beaver [1] (in [1], the probability that a failure occurs during the course of one bit communication is $1/2$. This stand-alone, non-committing encryption scheme is possibly the most efficient implementation of non-committing encryptions so far).

Road-map. The rest of this paper is organized as follows: The functionality and security definition of non-committing encryption protocols are presented in Section 2. The building blocks are sketched in Section 3. In Section 4, a new non-committing encryption scheme without failure during the course of a channel setup phase is proposed and analyzed. We then show that the proposed scheme realizes the UC-security in the presence of adaptive adversaries. We conclude our work in Section 5.

2   Non-committing Encryptions: Functionality and Security Definition

The notion of non-committing encryption schemes introduced in [7] is a cryptographic primitive used to realize secure channels in the presence of adaptive adversaries. In particular, if a non-committing encryption scheme realizes the UC-security in the presence of adaptive adversaries, then a simulator can build a fake transcript to the environment $\mathcal{Z}$ in such a way that the simulator can open this transcript to the actual inputs that the simulator receives from the functionality when the parties get corrupted.

2.1   The Universally Composable Framework

We briefly review the framework and notations for analyzing non-committing encryptions proposed by Canetti [4] and [5]. In this framework one first defines an ideal functionality of a protocol and then proves that a particular implementation of this protocol operating in a given environment securely realizes this functionality. The basic entities involved are n players, an adversary $\mathcal{A}$ and an environment $\mathcal{Z}$. The environment has access only to the inputs and outputs of the parties of π. It does not have direct access to the communication among the parties, nor to the inputs and outputs of the subroutines of π. The task of $\mathcal{Z}$ is to distinguish between two executions sketched below.

In the real world execution, the environment $\mathcal{Z}$ is activated first, generating particular inputs to the other players. Then the protocol π proceeds by having $\mathcal{A}$ exchange messages with the players and the environment. At the end of the protocol execution, the environment $\mathcal{Z}$ outputs a bit.

In the ideal world, the players are replaced by dummy parties, who do not communicate with each other. All dummy parties interact with an ideal functionality $\mathcal{F}$. When a dummy party is activated, it forwards its input to $\mathcal{F}$ and receives the output from the functionality $\mathcal{F}$. In addition, $\mathcal{F}$ may receives messages directly from the ideal world adversary $\mathcal{S}$ and may contain instructions to send message to $\mathcal{S}$.

At the end of the ideal world execution, the environment $\mathcal{Z}$ outputs a bit. Let $\mathrm{REAL}_{\pi,\mathcal{A},\mathcal{Z}}$ be $\mathcal{Z}$'s output after interacting with adversary $\mathcal{A}$ and players running protocol π; Let $\mathrm{IDEAL}_{\mathcal{F},\mathcal{S},\mathcal{Z}}$ be $\mathcal{Z}$'s output after interacting with $\mathcal{S}$ and $\mathcal{F}$ in the ideal execution. A protocol π securely realizes an ideal functionality $\mathcal{F}$ if $\mathrm{REAL}_{\pi,\mathcal{A},\mathcal{Z}}$ and $\mathrm{IDEAL}_{\mathcal{F},\mathcal{S},\mathcal{Z}}$ are computationally indistinguishable.

2.2   Functionality of Non-commitment Encryptions

The functionality of a non-committing encryption scheme depicted in **Fig. 1** (in terms of secure message transmission) is due to Canetti [4].

Definition 1. *We call the functionality $\mathcal{F}_{\mathrm{NCE}}^{l}$ a secure message transmission channel. A real-world protocol π which realizes $\mathcal{F}_{\mathrm{NCE}}^{l}$ is called a secure message transmission protocol.*

Functionality $\mathcal{F}_{\mathrm{NCE}}^l$

$\mathcal{F}_{\mathrm{NCE}}^l$ proceeds as follows, when parameterized by leakage function $l: \{0,1\}^* \to \{0,1\}^*$

1. Upon receiving an input (send, sid, m), do: If $sid = (S, R, sid')$ for some R then send $(\mathsf{send}, sid, l(m))$ to the adversary, generate a private delayed output (send, sid, m) to R and halt. Else, ignore the input.
2. Upon receiving $(\mathsf{corrupt}, sid, P)$ from the adversary, where $P \in \{S, R\}$, disclose m to the adversary. Next, if the adversary provides a value m', and $P = S$, and no output has been yet written to R, then output (send, sid, m') to R and halt.

Fig. 1. The non-committing encryption functionality parameterized by leakage function l

3 Preliminaries

In this section, we sketch the building blocks for constructing non-committing encryptions. The construction of our non-committing encryption is based on the Diffie-Hellman randomizer and the Canetti-Fischlin's oblivious sampling algorithm. The security proof of the given non-committing encryption is relied on the Canetti-Fischlin's oblivious faking algorithm.

3.1 The Naor-Pinkas Randomizer

Let $p = 2q + 1$ and p, q be large prime numbers. Let $G \subseteq Z_p^*$ be a cyclic group of order q. Let g be a random generator of G. For any $0 \neq x \in Z_q$, we define $\mathsf{DLog}_G(x) = \{(g, g^x) : g \in G\}$. On input $(g_1, h_1) \in \mathsf{DLog}_G(x_1)$, and $(g_2, h_2) \in \mathsf{DLog}_G(x_2)$, a mapping ϕ called Naor-Pinkas randomizer is defined below:

$$\phi((g_1, g_2, h_1, h_2) \times (s, t)) = (g_1^s g_2^t \bmod p, \ h_1^s h_2^t \bmod p)$$

where $s, t \in_U Z_q$

Denote $u = g_1^s g_2^t \bmod p$ and $v = h_1^s h_2^t \bmod p$. Naor and Pinkas [14] have shown that

- if $x_1 = x_2 \ (=x)$, then (u, v) is uniformly random in $\mathsf{DLog}_G(x)$;
- if $x_1 \neq x_2$, then (u, v) is uniformly random in G^2.

3.2 The Oblivious Sampling and Faking Algorithms

The oblivious sampling and faking algorithms described below are due to Canetti and Fischlin [6]. The two algorithms combined together allow a simulator to construct a fake transcript to the environment $\mathcal{Z}$ in such a way that the simulator can open this transcript to the actual inputs that the simulator receives from the functionality when the parties get corrupted, a core task to prove the security of protocols against adaptive adversaries in the universally composable security model.

Oblivious sampling algorithm: Let $p = wq + 1$ for some w not divisible by q, and G is a cyclic group of order q in Z_p^*. The Canetti-Fischlin oblivious sampling algorithm **sample** takes $r \in \{0,1\}^{2|p|}$ as input and outputs an element $r_G \in G$ via the following computations

- the sampling algorithm **sample** chooses a string $r \in \{0,1\}^{2|p|}$ uniformly at random, where $|p|$ be the bit length of the prime number p.
- Let $r_p = r \bmod p$ and $r_G = r_p^w \bmod p$.

Lemma 1. *(due to [6]) Let $X = [X = x : x \in_U G]$, and $Y = [Y = y : y \leftarrow$ **sample**$(r), r \in_U \{0,1\}^{2|p|}]$, then the distributions between two random variables X and Y are statistically indistinguishable.*

Oblivious faking algorithm: Let $p = wq + 1$ for some w not divisible by q, and G is a cyclic group of order q in Z_p^*. The Canetti-Fischlin oblivious faking algorithm **fake** takes a random element $h \in G$ as input and outputs $r_h \in \{0,1\}^{2|p|}$ via the following computations

- On input $h \in G$, the faking algorithm **fake** picks a random integer $i \in Z_w$. Let $h_p = h^x g^{iq} \bmod p$, where $xw \equiv 1 \bmod q$;
- **fake** randomly selects $j \in Z_p$ and let $r_h = \mathrm{Len}(jp + h_p)$, where $\mathrm{Len}(x)$ denotes the bit length of an integer x.

Lemma 2. *(due to [6]) Let $X = [X = x : x \in_U \{0,1\}^{2|p|}]$, and $Y = [Y = y : y \leftarrow$ **fake**$(g), g \in_U G]$, then the distributions between two random variables X and Y are statistically indistinguishable.*

4   Universally Composable Non-committing Encryptions

In this section, a new non-committing encryption is described and analyzed. We show that the proposed scheme realizes the UC-security in the presence of adaptive adversaries.

4.1   The Description of Non-committing Encryptions

The non-committing encryption proposed in this paper comprises three phases: an initialization phases, a channel setup phase and a communication phase. In the initialization phase, the global public-key is generated and described for all participants. To set up a secure channel, a sender S randomly selects (S_0, S_1) such that either S_0 or S_1 is a Diffie-Hellman quadruple. The receiver R randomly selects (R_0, R_1) such that either R_0 or R_1 is a Diffie-Hellman quadruple. The protocol is designed so that the sender S is able to extract the bit b selected by R. At this point a secure channel for communicating one bit message is set up between the two parties; The details of protocol are depicted below.

Initialization. The protocol initialization procedure takes security parameter k as input and outputs (p, q, G), where p is a large safe prime number (i.e.,

$p=2q + 1$, q is a prime number) and G is a cyclic group with order q. Let pk $=(p, q, G)$. The protocol initialization procedure then provides a description **des** of algorithm **sample** defined over G. Let $gpk =(pk, \mathbf{des})$ (the global key for all participants).

Channel setup. The channel setup phase comprises the following four steps

Step 1: On input 1^k, the sender S performs the following computations

- S selects a bit $\alpha \in \{0, 1\}$ uniformly at random;
- S randomly generates a Diffie-Hellman quadruple $(S_{\alpha,1}, S_{\alpha,2}, S_{\alpha,3}, S_{\alpha,4})$ such that $\log_{S_{\alpha,1}}(S_{\alpha,3}) = \log_{S_{\alpha,2}}(S_{\alpha,4})$. Let $S_\alpha = (S_{\alpha,1}, S_{\alpha,2}, S_{\alpha,3}, S_{\alpha,4})$ and $sk_S = \log_{S_{\alpha,1}}(S_{\alpha,3}) = (\log_{S_{\alpha,2}}(S_{\alpha,4}))$;
- S invokes **sample** to obliviously generate a random quadruple $(S_{1-\alpha,1}, S_{1-\alpha,2}, S_{1-\alpha,3}, S_{1-\alpha,4})$. Let $S_{1-\alpha} = (S_{1-\alpha,1}, S_{1-\alpha,2}, S_{1-\alpha,3}, S_{1-\alpha,4})$;
- S keeps sk_S secret and sends (S_0, S_1) to R;

Step 2: Upon receiving (S_0, S_1), the receiver R performs the following computations

- R selects a bit $\beta \in \{0, 1\}$ uniformly at random;
- R randomly generates a Diffie-Hellman quadruple $(R_{\beta,1}, R_{\beta,2}, R_{\beta,3}, R_{\beta,4})$ such that $\log_{R_{\beta,1}}(R_{\beta,3}) = \log_{R_{\beta,2}}(R_{\beta,4})$. Let $R_\beta = (R_{\beta,1}, R_{\beta,2}, R_{\beta,3}, R_{\beta,4})$ and $sk_R = \log_{R_{\beta,1}}(R_{\beta,3}) = (\log_{R_{\beta,2}}(R_{\beta,4}))$;
- R invokes **sample** to obliviously generate a random quadruple $(R_{1-\beta,1}, R_{1-\beta,2}, R_{1-\beta,3}, R_{1-\beta,4})$; Let $R_{1-\beta} = (R_{1-\beta,1}, R_{1-\beta,2}, R_{1-\beta,3}, R_{1-\beta,4})$;
- R then selects $x_R \in Z_q$ and $y_R \in Z_q$ uniformly at random, and computes the Naor-Pinkas randomizer $u_{S,\beta} = S_{\beta,1}^{x_R} S_{\beta,2}^{y_R}$ and $v_{S,\beta} = S_{\beta,3}^{x_R} S_{\beta,4}^{y_R}$ for the selected quadruple S_β. Let $w_{S,\beta} =(u_{S,\beta}, v_{S,\beta})$.
- R invokes **sample** to output two random strings $u_{S,1-\beta} \in Z_p^*$ and $v_{S,1-\beta} \in Z_p^*$ for the given quadruple $S_{1-\beta}$. Let $w_{S,1-\beta} =(u_{S,1-\beta}, v_{S,1-\beta})$.
- R keeps sk_R secret and sends (R_0, R_1) and $(w_{S,0}, w_{S,1})$ to S;

Step 3: Upon receiving (R_0, R_1) and $(w_{S,0}, w_{S,1})$, the sender S performs the following computations

- parsing $w_{S,\alpha}$ as $(u_{S,\alpha}, v_{S,\alpha})$, S checks $v_{S,\alpha} \overset{?}{=} u_{S,\alpha}^{sk_S}$:
  - if the check is valid, S selects $x_S \in Z_q$ and $y_S \in Z_q$ uniformly at random, and computes $u_{R,\alpha} = R_{\alpha,1}^{x_S} R_{\alpha,2}^{y_S}$ and $v_{R,\alpha} = R_{\alpha,3}^{x_S} R_{\alpha,4}^{y_S}$. S then invokes **sample** to output random elements $(u_{R,1-\alpha}, v_{R,1-\alpha}) \in G^2$; Let $\gamma =\alpha$ and let $w_{R,\alpha} =(u_{R,\alpha}, v_{R,\alpha})$ and $w_{R,1-\alpha} =(u_{R,1-\alpha}, v_{R,1-\alpha})$;
  - otherwise, S selects $x_S \in Z_q$ and $y_S \in Z_q$ uniformly at random, and computes $u_{R,1-\alpha} = R_{1-\alpha,1}^{x_S} R_{1-\alpha,2}^{y_S}$ and $v_{R,1-\alpha} = R_{1-\alpha,3}^{x_S} R_{1-\alpha,4}^{y_S}$; S then invokes **sample** to output random elements $(u_{R,\alpha}, v_{R,\alpha}) \in G^2$; Let $\gamma =1 - \alpha$ and let $w_{R,\alpha} =(u_{R,\alpha}, v_{R,\alpha})$ and $w_{R,1-\alpha} =(u_{R,1-\alpha}, v_{R,1-\alpha})$;
- S then sends $w_{R,\alpha}$ and $w_{R,1-\alpha}$ to R and outputs γ;

Step 4: Upon receiving $w_{R,0}$ and $w_{R,1}$, the receiver R performs the following computations

- parsing $w_{R,\beta}$ as $(u_{R,\beta}, v_{R,\beta})$, R checks $v_{R,\beta} \overset{?}{=} u_{R,\beta}^{sk_R}$; If the check is valid, let $\gamma = \beta$ and outputs γ; otherwise, output $\perp$ (notice that the probability that the honest party R outputs $\perp$ is negligible in case that the sender is honest).

Message Transfer. On input $m \in \{0,1\}$ and $\gamma \in \{0,1\}$, S computes $m \oplus \gamma$. Let $c = m \oplus \gamma$. S then sends c to R. Upon receiving a ciphertext c, R obtains m by computing $c \oplus \gamma$.

This ends the description of the protocol π.

4.2 The Proof of Security

Theorem 1. *The protocol π realizes the UC-security in the presence of adaptive adversary in the authenticated channel assuming that the decisional Diffie-Hellman problem is hard.*

Proof. We assume the channel between the sender S and the receiver R is authenticated and consider the following cases

- the first corruption occurs after a ciphertext c has been received successfully;
- the first corruption occurs after a secure channel has been setup phase but before a ciphtertext c is generated;
- the first corruption occurs during the course of a channel setup phase.

Case 1: Upon receiving (corrupt, sid, P), where $P \in \{S, R\}$, the simulator $\mathcal{S}$ generates two random Diffie-Hellman quadruples $S_i = (S_{i,1}, S_{i,2}, S_{i,3}, S_{i,4})$ such that $\log_{S_{i,1}}(S_{i,3}) = \log_{S_{i,2}}(S_{i,4})$ $(=: sk_{S,i}, i = 0, 1)$. The simulator $\mathcal{S}$ keeps the trapdoor string $(sk_{S,0}, sk_{S,1})$ secret. Let $sk_S = (sk_{S,0}, sk_{S,1})$. $\mathcal{S}$ then selects $x_{R,i} \in Z_q$ and $y_{R,i} \in Z_q$ uniformly at random, and computes $u_{S,i} = S_{i,1}^{x_{R,i}} S_{i,2}^{y_{R,i}}$ and $v_{S,i} = S_{i,3}^{x_{R,i}} S_{i,4}^{y_{R,i}}$; Similarly, the simulator $\mathcal{S}$ generates two random Diffie-Hellman quadruples $R_i = (R_{i,1}, R_{i,2}, R_{i,3}, R_{i,4})$ such that $\log_{R_{i,1}}(R_{i,3}) = \log_{R_{i,2}}(R_{i,4})$ $(=: sk_{R,i}, i = 0, 1)$. The simulator $\mathcal{S}$ keeps the trapdoor string $(sk_{R,0}, sk_{R,1})$ secret. Let $sk_R = (sk_{R,0}, sk_{R,1})$. $\mathcal{S}$ then selects $x_{S,i} \in Z_q$ and $y_{S,i} \in Z_q$ uniformly at random, and computes $u_{R,i} = R_{i,1}^{x_{S,i}} R_{i,2}^{y_{S,i}}$ and $v_{R,i} = R_{i,3}^{x_{S,i}} R_{i,4}^{y_{S,i}}$.

Let $P \in \{S, R\}$ be the first corrupted party after a ciphertext c has been received by R. The simulator $\mathcal{S}$ corrupts the corresponding dummy party $\widetilde{P} \in \{\widetilde{S}, \widetilde{R}\}$ in the ideal world and learns m from the functionality $\mathcal{F}_{NCE}^l$. Let $\gamma \leftarrow c \oplus m$ and $\beta \leftarrow \gamma$. The simulator $\mathcal{S}$ now invokes **fake** to interpret $R_{1-\gamma}$ as a random quadruple and R_γ as a random Diffie-Hellman quadruple. The randomness $r_{R_{1-\gamma}}$ used to generate $R_{1-\gamma}$ is denoted by $(r_{R_{1-\gamma},1}, r_{R_{1-\gamma},2}, r_{R_{1-\gamma},3}, r_{R_{1-\gamma},4})$. The randomness r_{R_γ} used to generate R_γ is denoted by $(r_{R_\gamma,1}, r_{R_\gamma,2}, r_{R_\gamma,3}, r_{R_\gamma,4})$. $\mathcal{S}$ then interprets $(u_{S,\gamma}, v_{S,\gamma})$ as random strings generated by the Naor-Pinkas randomizer with the auxiliary string $(x_{R,\gamma}, y_{R,\gamma})$. $\mathcal{S}$ further interprets $(u_{S,1-\gamma}, v_{S,1-\gamma})$ as random strings generated by **fake**. The simulator reveals the modified randomness $(r_{R_\gamma}, r_{R_{1-\gamma}})$, together with $(x_{R,\gamma}, y_{R,\gamma})$ and $sk_{R,\gamma}$ to the adversary.

Meanwhile, $\mathcal{S}$ interprets $(u_{R,1-\gamma}, v_{R,1-\gamma})$ as random strings generated by **fake** and interprets $(u_{R,\gamma}, v_{R,\gamma})$ as random strings generated by the Naor-Pinkas randomizer with the auxiliary string $(x_{S,\gamma}, y_{S,\gamma})$. Finally, $\mathcal{S}$ randomly selects a bit $b \in \{0,1\}$ and interprets S_b as a random quadruple by the faking algorithm **fake** and interprets S_{1-b} as a random Diffie-Hellman quadruple and reveals all these random strings together with $sk_{S,b}$ to the adversary $\mathcal{A}$.

Case 2: The first corruption occurs after a secure channel has been set up but before a ciphtertext c is generated; Upon receiving (corrupt, sid, P) from the environment $\mathcal{Z}$, where $P \in \{S, R\}$, the simulator randomly selects a bit $\gamma \in \{0,1\}$ uniformly at random as a random bit selected by the receiver R. The specified bit γ is then shared between the sender S and the receiver R. The rest work of the simulator is same as that described in **case 1** and the details are thus omitted.

Case 3: The first corruption occurs during the course of the channel setup phase. Upon receiving (corrupt, sid, P) from the environment $\mathcal{Z}$, where $P \in \{S, R\}$, the ideal world adversary simulates the following two cases:

- if the receiver R gets corrupted at first, then the simulator $\mathcal{S}$ learns m from the functionality $\mathcal{F}_{NCE}^l$. This means that given a ciphertext c, the simulator can open the ciphertext c to the message m if the shared secret key γ is set to $c \oplus m$. The rest of the simulation is the same as that described in **case 1** and the details are thus omitted.
- if the sender S gets corrupted at first, the simulator $\mathcal{S}$ learns m from the functionality $\mathcal{F}_{NCE}^l$. This means that given a ciphertext c, the simulator can open the ciphertext c to the message m if the shared secret key γ is set to $c \oplus m$. If the adversary provides m' and no output has been yet written to R, then the functionality $\mathcal{F}_{NCE}^l$ outputs m' to R. The simulation of each case is the same as that described in **case 1** and the details are thus omitted.

By the DDH assumption, we know that $REAL_{\pi,\mathcal{A},\mathcal{Z}}$ and $IDEAL_{\mathcal{F},\mathcal{S},\mathcal{Z}}$ are computationally indistinguishable in all cases above. As a result, the real-world protocol π realizes $\mathcal{F}_{NCE}^l$. $\square$

5   Conclusion

In this paper, a new non-committing encryption scheme without failure during the course of the channel setup has been presented and analyzed. We have shown that the proposed non-committing scheme realizes the UC-security in the presence of adaptive adversary assuming that the decisional Diffie-Hellman problem is hard.

References

1. Beaver, D.: Plug and Play Encryption. In: Kaliski Jr., B.S. (ed.) CRYPTO 1997. LNCS, vol. 1294, pp. 75–89. Springer, Heidelberg (1997)
2. Beaver, D., Haber, S.: Cryptographic Protocols Provably Secure Against Dynamic Adversaries. In: Rueppel, R.A. (ed.) EUROCRYPT 1992. LNCS, vol. 658, pp. 307–323. Springer, Heidelberg (1993)
3. Ben-Or, M., Goldwasser, S., Wigderson, A.: Completeness Theorems for Non-Cryptographic Fault-Tolerant Distributed Computation (Extended Abstract). In: STOC 1988, pp. 1–10 (1988)
4. Canetti, R.: Universally Composable Security: A New Paradigm for Cryptographic Protocols, Cryptology ePrint Archive: Report 2000/067 (2000)
5. Canetti, R.: A new paradigm for cryptographic protocols. In: FOCS 2001, pp. 136–145 (2001)
6. Canetti, R., Fischlin, M.: Universally Composable Commitments. In: Kilian, J. (ed.) CRYPTO 2001. LNCS, vol. 2139, pp. 19–40. Springer, Heidelberg (2001)
7. Canetti, R., Feige, U., Goldreich, O., Naor, M.: Adaptively Secure Multi-Party Computation. In: STOC 1996, pp. 639–648 (1996)
8. Chaum, D., Crépeau, C., Damgård, I.: Multiparty Unconditionally Secure Protocols (Extended Abstract). In: STOC 1988, pp. 11–19 (1988)
9. Choi, S.G., Dachman-Soled, D., Malkin, T., Wee, H.: Improved Non-Committing Encryption with Applications to Adaptively Secure Protocols. In: Asiacrypt 2009 (2009)
10. Damgård, I., Nielsen, J.B.: Improved Non-committing Encryption Schemes Based on a General Complexity Assumption. In: Bellare, M. (ed.) CRYPTO 2000. LNCS, vol. 1880, pp. 432–450. Springer, Heidelberg (2000)
11. Goldreich, O., Micali, S., Wigderson, A.: Proofs that Yield Nothing But their Validity and a Methodology of Cryptographic Protocol Design (Extended Abstract). In: FOCS 1986, pp. 174–187 (1986)
12. Goldreich, O., Micali, S., Wigderson, A.: How to Play any Mental Game or A Completeness Theorem for Protocols with Honest Majority. In: STOC 1987, pp. 218–229 (1987)
13. Nielsen, J.B.: Separating Random Oracle Proofs from Complexity Theoretic Proofs: The Non-committing Encryption Case. In: Yung, M. (ed.) CRYPTO 2002. LNCS, vol. 2442, pp. 111–126. Springer, Heidelberg (2002)
14. Naor, M., Pinkas, B.: Efficient oblivious transfer protocols. In: SODA 2001, pp. 448–457 (2001)
15. Yao, A.C.-C.: Protocols for Secure Computations (Extended Abstract). In: FOCS 1982, pp. 160–164 (1982)

Relations among Notions of Complete Non-malleability: Indistinguishability Characterisation and Efficient Construction without Random Oracles

Manuel Barbosa[1] and Pooya Farshim[2]

[1] CCTC/Departamento de Informática, Universidade do Minho,
Campus de Gualtar, 4710-057 Braga, Portugal
mbb@di.uminho.pt
[2] Information Security Group, Royal Holloway, University of London,
Egham, Surrey, TW20 0EX, United Kingdom
Pooya.Farshim@rhul.ac.uk

Abstract. We study relations among various notions of complete non-malleability, where an adversary can tamper with both ciphertexts and public-keys, and ciphertext indistinguishability. We follow the pattern of relations previously established for standard non-malleability. To this end, we propose a more convenient and conceptually simpler indistinguishability-based security model to analyse completely non-malleable schemes. Our model is based on strong decryption oracles, which provide decryptions under arbitrarily chosen public keys. We give the first precise definition of a strong decryption oracle, pointing out the subtleties in different approaches that can be taken. We construct the first *efficient* scheme, which is fully secure against strong chosen-ciphertext attacks, and therefore completely non-malleable, without random oracles.

Keywords: Complete Non-Malleability. Strong Chosen-Ciphertext Attacks. Public-Key Encryption. Provable Security.

1 Introduction

BACKGROUND. The security of public-key encryption schemes has been formalised according to various goals and attack models. Extensive work has been done in establishing relations between these security notions, and converging towards a core set of standard security definitions. Well-studied goals include semantic security, indistinguishability, and non-malleability; whereas chosen-plaintext and (adaptive) chosen-ciphertext are the most common attack scenarios considered in literature.

An important criterion for selecting security models is the guarantee of *necessary security* for a class of applications with practical relevance. Conversely, it is also expected that one can select a security model that is only as strict as required by a specific application. Otherwise, one might rule out valid solutions without justification, possibly sacrificing other important factors such as

R. Steinfeld and P. Hawkes (Eds.): ACISP 2010, LNCS 6168, pp. 145–163, 2010.

set-up assumptions, computational cost or communications bandwidth. Another important criterion is the conceptual simplicity and ease of use of a model.

Indistinguishability of ciphertexts is the most widely used notion of security for public-key encryption schemes. This notion was proposed by Goldwasser and Micali [15] as a convenient formalisation of the more intuitive notion of semantic security. Other notions of security have been proposed in different contexts. Of particular interest to this work is non-malleability, initially proposed by Dolev, Dwork, and Naor [12]. Roughly speaking, an encryption scheme is non-malleable if giving an encryption of a message to an adversary does not increase its chances of producing an encryption of a related message (under a given public key). This is formalised by requiring the existence of a simulator that performs as well as the adversary but without seeing the original encryption.

The relations between different notions of security for public-key encryption schemes were examined in a systematic way by Bellare et al. [4]. There, the authors compare indistinguishability of ciphertexts and non-malleability under chosen-plaintext and chosen-ciphertext attacks. In doing so, they formalise a comparison-based definition of non-malleability and establish important results based on this: non-malleability implies indistinguishability for an equivalent attack model, there is an equivalence between these notions for CCA2 model, and there are separations between the two notions for intermediate attack models.

Bellare and Sahai [8] established a cycle of equivalence between three definitions of non-malleability: a simulation-based definition similar to that of Dolev, Dwork and Naor, a comparison-based definition as introduced in [4], and a new definition called indistinguishability of ciphertexts under parallel chosen-ciphertext attacks. These equivalence relations essentially establish that the three definitions are alternative formulations of the same notion. Pass, Shelat, and Vaikuntanathan [18] revisit this equivalence result, and clarify several technical aspects in the known equivalence proofs. They consider the important question of composability of definitions, and establish a separation between the simulation-based and comparison-based non-malleability definitions, showing that the former is strictly stronger for *general* schemes.

Besides being theoretically interesting, the above results are also relevant in practice. They permit designers of encryption schemes to base their analysis on the simpler and better understood IND-CCA2 security model. This facilitates the presentation of conceptually simpler proofs, which are less prone to errors, as well as the direct application of a well-known set of proof techniques.

COMPLETE NON-MALLEABILITY. Fischlin [13] introduces a stronger notion of non-malleability, known as *complete*, which requires attackers to have negligible advantage, even if they are allowed to transform the public key under which the related message is encrypted. Put differently, the goal of an adversary is to construct a related ciphertext under a new public key pair, for which the attacker might not even know a valid secret key.

Fischlin shows that well-known encryption schemes such as Cramer-Shoup [10] and RSA-OAEP [14] do *not* achieve even the weakest form of complete non-malleability. Furthermore, he proves a negative results with respect to the existence of completely non-malleable schemes for general relations: there is a large class of relations for which completely non-malleable schemes do not exist with respect to black-box simulators. On the other hand, Fischlin establishes a positive result for a modified version of RSA-OAEP, with respect to a restricted class of adversaries, in the random oracle model.

Ventre and Visconti [19] later propose a comparison-based definition of this security notion, which is more in line with the well-studied definitions proposed by Bellare et al. [4,8]. For chosen-plaintext attacks the authors prove that (a restricted version of) their definition is equivalent to that of Fischlin. They also establish equivalence for chosen-ciphertext attacks, for a well-defined class of relations that do not depend on the challenge public key (known as lacking relations). The authors also provide additional feasibility results by proposing two constructions of completely non-malleable schemes, one in the common reference string model using non-interactive zero-knowledge proofs, and another using interactive encryption schemes. Therefore, the only previously known completely non-malleable (and non-interactive) scheme in the standard model, is quite inefficient as it relies on generic zero-knowledge techniques.

MOTIVATION. The initial motivation for complete non-malleability resided on constructing *non-malleable commitment* schemes. A commitment scheme can be constructed from an encryption scheme in the following way. To commit to a message, one generates a key pair and encrypts the message under the generated public key. The resulting public key/ciphertext pair forms the commitment. To de-commit, one reveals a valid secret key or the message/randomness pair used in encryption. In this setting, it is clearly desirable that the encryption scheme should be completely non-malleable in order to guarantee non-malleability of the associated commitment scheme.

Furthermore, new notions of security of high practical relevance have been emerging in the literature that closely relate to different flavours of complete non-malleability. The pattern connecting these notions is that adversaries are allowed to tamper with the keys, under which they are challenged, in order to gain extra advantage. *Robust encryption* [1] is one such notion, and it is pitched at applications where ciphertext anonymity is relevant. This notion requires it to be infeasible to construct a ciphertext which is valid under two distinct public keys. Another such notion is security under *related-key attacks* [5], where cipher operations can be executed over perturbed versions of the challenge secret key. This model is of particular relevance in the symmetric encryption setting. Also worth mentioning are concrete attacks on key-agreement protocols and public-key signature schemes, where attackers are able to introduce public keys of their choice in the protocol execution [13].

The relations between these new notions of security are understudied and constitute a novel challenge in theoretical cryptography. A deeper understanding of the relations between these notions of security should permit identifying a core set of security models that facilitate the design and analysis of strongly secure schemes with practical relevance. The main motivation of this work is, therefore, to take an important step in this direction. We aim to expand the current understanding of complete non-malleability, by establishing relations among notions of complete non-malleability and ciphertext indistinguishability that are akin to those already known for standard non-malleability. To this end, we introduce a new indistinguishability based notion, and demonstrate its applicability by constructing an efficient and completely non-malleable scheme.

STRONG CHOSEN-CIPHERTEXT ATTACKS. Our search for a suitable indistingui-shability-based definition of complete non-malleability resulted in a natural extension of the standard IND-CCA2 security model, in which the adversary can get decryptions of ciphertexts under arbitrary public keys of its choice. We call this a *strong chosen-ciphertext attack* scenario, and say that the adversary is given access to a *strong decryption oracle*. This, in turn, brings together two fields which previously remained unrelated in provable security, namely complete non-malleability and certificateless cryptography [2,11]. Indeed, strong CCA attacks model multi-user scenarios where public keys might not be authenticated, and were initially proposed as a natural attack model for certificateless schemes that aimed to do away with public-key certificates.

The question of whether the weakness captured by such a strong model should be seen as a real vulnerability of public-key encryption schemes has caused some discussion [11]. Arguments against this approach are based on the fact that such an attack model is not realistic, since it is highly unlikely that the adversary is able to get such assistance in a practical scenario. Another way to put this objection is that security models should be defined through experiments that are guaranteed to execute in polynomial time: providing decryptions under unknown secret keys assists the adversary through a super-polynomial time oracle.

The results we present in this paper show that the strength of the complete non-malleability notion is comparable to that of the strong chosen-ciphertext attack scenario. This connection allows us to take a more constructive view of strong decryption oracles, and argue that they can indeed be useful to analyse the security of practical schemes. To support this view, we show that *indistinguishability under strong CCA attacks is a convenient formalisation to establish that a scheme is completely non-malleable*. Furthermore, by proposing a concrete scheme, we also show that both notions are realisable without random oracles.

Finally, we note that strong decryption oracles are closely related to the recently proposed paradigm of adaptive one-way functions [17], which can be used to construct a number of cryptographic protocols that previously remained open in the literature. Indeed, the assumptions that underlie the proposed constructions of adaptive one-way functions rely on similar "magic" oracles. It would be interesting to investigate whether the techniques that we use can be useful in constructing adaptive one-way functions based on standard assumptions. Conversely, the public-key encryption scheme given in [17] seems to achieve strong chosen-ciphertext security. The relationship between adaptive one-way functions and strong security models are left for future work.

CONTRIBUTIONS. The first contribution of our paper is a general definition of a strong decryption oracle, which unifies previous definitional approaches. Our definition is flexible and expressive in the sense that it allows identifying the exact power of the decryption oracle that is provided to an adversary in security analysis. We also show that variants of the strong decryption oracle definition map to interesting properties of encryption schemes. We establish a connection with the validity checks that an encryption scheme performs (message validity, ciphertext validity, public key validity, etc.). More precisely, *we identify a simple and very convenient definition of the strong decryption oracle, which can be used to analyse schemes that incorporate a well-defined and natural set of validity checks.* For schemes that fail to perform these checks, care must be taken to identify the exact strength of the strong decryption oracle under which the scheme can be proven secure.

We then extend the standard indistinguishability and non-malleability models using strong decryption oracles, and examine the relations between the resulting notions. Our approach is consistent with that proposed by Bellare et al. [8,4], which allows us to naturally describe the relation between these stronger models and the more established ones. We also identify the relation between the strong chosen-ciphertext models we propose and the existing notions of complete non-malleability. To the best of our knowledge, this relation was not previously known. It permits fully characterising how these independently proposed models relate to the more standard definitions of non-malleability. The relation we establish between strong decryption oracles and complete non-malleability *provides the first convincing argument that the strong CCA models are useful in analysing the security of practical encryption schemes.*

Finally, we propose a concrete scheme that *efficiently achieves strong chosen-ciphertext security* based on the decisional bilinear Diffie-Hellman assumption. The scheme is secure under a very general definition of the strong decryption oracle, which is made possible by the insights regarding validity checks we described above. The scheme is derived from Waters' identity-based encryption scheme [20] using techniques previously employed in constructing certificateless public-key encryption schemes [11]. Our equivalence result also establishes our scheme as the first efficient completely non-malleable scheme without random oracles. We stress that our scheme is based on a standard and well-known problem and does *not* rely on interactive assumptions or "magic" oracles.

ORGANISATION. In the next section we fix notation by defining public-key encryption schemes and various algorithms associated to them. In Section 3 we discuss different approaches in defining strong decryption oracles and propose a new generic definition. In Section 4 we look at indistinguishability and non-malleability security models for encryption schemes where adversaries have access to strong decryption oracles. We establish relations between these models and also to models existing literature. We present our scheme in the final section.

2   Preliminaries

NOTATION. We write $x \leftarrow y$ for assigning value y to variable x, and $x \leftarrow_\$ X$ for sampling x from set X uniformly at random. If X is empty, we set $x \leftarrow \perp$, where

$\perp \notin \{0,1\}^\star$ is a special failure symbol. If A is a probabilistic algorithm, we write $x \leftarrow_\$ A(I_1, I_2, \ldots)$ for the action of running A on inputs $I_1, I_2, \ldots$ with random coin chosen uniformly at random, and assigning the result to x. Sometimes we run A on specific coins r and write $x \leftarrow A(I_1, I_2, \ldots; r)$. We denote boolean values, namely the output of checking whether a relation holds, by T (true) and F (false). For a space $\mathsf{Sp} \subseteq \{0,1\}^\star$, we identify Sp with its characteristic function. In other words, $\mathsf{Sp}(s) = \mathsf{T}$ if and only if $s \in \mathsf{Sp}$. We say s is valid with respect to Sp if and only if $\mathsf{Sp}(s) = \mathsf{T}$. When this is clear from the context, we also use Sp for sampling uniformly from Sp. Unless stated otherwise, the range of a variable s is assumed to be $\{0,1\}^\star$. The symbol : is used for appending an element to a list. We indicate vectors using bold font.

GAMES. We will be using the code-based game-playing language [7]. Each game has an **Initialize** and a **Finalize** procedure. It also has specifications of procedures to respond to an adversary's various oracle queries. A game Game is run with an adversary $\mathcal{A}$ as follows. First **Initialize** runs and its outputs are passed to $\mathcal{A}$. Then $\mathcal{A}$ runs and its oracle queries are answered by the procedures of Game. These procedures return $\perp$ if queried on $\perp$. When $\mathcal{A}$ terminates, its output is passed to **Finalize** which returns the outcome of the game y. This interaction is written as $\mathsf{Game}^{\mathcal{A}} \Rightarrow y$. In each game, we restrict attention to *legitimate* adversaries. Legitimacy is defined specifically for each game.

PUBLIC-KEY ENCRYPTION. We adopt the standard multi-user syntax with the extra Setup algorithm [3], which we believe is the most natural one for security models involving multiple public keys. A public-key encryption scheme $\Pi = (\mathsf{Setup}, \mathsf{Gen}, \mathsf{MsgSp}, \mathsf{Enc}, \mathsf{Dec})$ is specified by five polynomial-time algorithms (in the length of their inputs) as follows. Setup is the probabilistic setup algorithm which takes as input the security parameter and returns the common parameters I (we fix the security parameter implicitly, as we will be dealing with concrete security). Although all algorithms are parameterised by I, we often omit I as an explicit input for readability. $\mathsf{Gen}(\mathsf{I})$ is the probabilistic key-generation algorithm. On input common parameters I, this algorithm returns a secret key SK and a matching public key PK. Algorithm $\mathsf{MsgSp}(\mathsf{m}, \mathsf{PK})$ is a deterministic message space recognition algorithm. On input m and PK this algorithm returns T or F. $\mathsf{Enc}(\mathsf{m}, \mathsf{PK}; r)$ is the probabilistic encryption algorithm. On input a message m, a public key PK, and possibly some random coins r, this algorithm outputs a ciphertext c or a special failure symbol $\perp$. Finally, $\mathsf{Dec}(\mathsf{c}, \mathsf{SK}, \mathsf{PK})$ is the deterministic decryption algorithm. On input of a ciphertext c and keys SK and PK, it outputs a message m or a special failure symbol $\perp$. The correctness of a public-key encryption scheme requires that for any $\mathsf{I} \leftarrow_\$ \mathsf{Setup}()$, any $(\mathsf{SK}, \mathsf{PK}) \leftarrow_\$ \mathsf{Gen}()$, all $\mathsf{m} \in \mathsf{MsgSp}(\mathsf{PK})$, and any random coins r we have $\mathsf{Dec}(\mathsf{Enc}(\mathsf{m}, \mathsf{PK}; r), \mathsf{SK}, \mathsf{PK}) = \mathsf{m}$.

REMARK. We note that the multi-user syntax permits capturing in a single framework schemes that execute in the plain model, in which case the global parameters are empty, as well as those which execute in the CRS model. The relations that we establish between different models hold in both cases.

VALIDITY CHECKING ALGORITHMS. The following spaces (and associated functions) will be used throughout the paper. All of these spaces are parameterised by I and are subsets of $\{0, 1\}^\star$.

$$\mathsf{MsgSp}(\mathsf{PK}) := \{m : \mathsf{MsgSp}(m, \mathsf{PK})\}$$
$$\mathsf{KeySp} := \{(\mathsf{SK}, \mathsf{PK}) : \exists r \; (\mathsf{SK}, \mathsf{PK}) = \mathsf{Gen}(r)\}$$
$$\mathsf{PKSp} := \{\mathsf{PK} : \exists r, \mathsf{SK} \; (\mathsf{SK}, \mathsf{PK}) = \mathsf{Gen}(r)\}$$
$$\mathsf{SKSp} := \{\mathsf{SK} : \exists r, \mathsf{PK} \; (\mathsf{SK}, \mathsf{PK}) = \mathsf{Gen}(r)\}$$

VALIDITY ASSUMPTIONS. We assume throughout the paper that the encryption and decryption algorithms check if $m \in \mathsf{MsgSp}(\mathsf{PK})$ and return $\perp$ if it does not hold. Often the algorithm MsgSp does not depend on PK in the sense that for any $\mathsf{PK}, \mathsf{PK}' \in \mathsf{PKSp}$ and any $m \in \{0, 1\}^\star$ we have $\mathsf{MsgSp}(m, \mathsf{PK}) = \mathsf{MsgSp}(m, \mathsf{PK}')$. For general schemes, case one can consider the infinite message space $\mathsf{MsgSp}(\mathsf{PK}) = \{0, 1\}^\star$. However, given that in this paper we will often consider the set of all valid messages and sample from it, we restrict our attention to schemes with finite message spaces. As pointed out by Pass et al. [18], this means that to avoid degenerate cases we must also restrict our attention to schemes for which all the elements in the range of decryption can be efficiently encrypted, including the special failure symbol $\perp$. A distribution M on messages is *valid* with respect to a public key PK if it is computable in polynomial time and its support contains strings of equal length which lie in $\mathsf{MsgSp}(\mathsf{PK})$. We also assume that key-pair validity KeySp is efficiently implementable and require that decryption returns $\perp$ if this check fails on the keys passed to it (note that this can easily be achieved for general public key encryption schemes, by including the input randomness to Gen in SK). We also assume various algorithms check for structural properties such as correct encoding, membership in a group, etc.

3 Defining Strong Decryption Oracles

The idea behind a strong chosen-ciphertext attack is to give the adversary access to an oracle that decrypts ciphertexts of the adversary's choice with respect to arbitrary public keys. There are a number technicalities involved in defining such an oracle precisely, which we now discuss.

$$\boxed{\begin{array}{l} \textbf{proc. } \mathbf{SDec}_{\mathsf{U},\mathsf{V}}(\mathsf{c}, \mathsf{PK}, \mathsf{R})\text{:} \\ \hline \mathsf{WitSp} \leftarrow \{(m, r) : \mathsf{V}(\mathsf{c}, \mathsf{PK}, m, r, \mathsf{st}[\mathsf{V}])\} \\ (m, r) \leftarrow_\$ \{(m, r) \in \mathsf{WitSp} : \mathsf{R}(m)\} \\ \mathsf{st}[\mathsf{V}] \leftarrow \mathsf{U}(\mathsf{c}, \mathsf{PK}, \mathsf{R}, m, r, \mathsf{st}[\mathsf{V}]) \\ \text{Return } m \end{array}}$$

Fig. 1. Generic definition of a strong decryption oracle. In the first step the search is performed over sufficiently long bit strings and, for messages, it also includes the special symbol $\perp$. The state $\mathsf{st}[\mathsf{V}]$ is initialised to some value st_0.

We will base our presentation on the generic definition of a strong decryption oracle presented in Figure 1, which we thoroughly explain and justify in the discussion that follows. The oracle proceeds in three steps. The first step models the general procedure of constructing a set of candidate (valid) decryption results. The second step consists of choosing one of these candidate solutions to return to the adversary. The final step updates the state of the oracle, if it keeps one.

More precisely, in the first step, the oracle constructs a set of possible decryption results WitSp using a polynomial-time validity relation V^1. Note that the search for messages includes the special failure symbol $\perp$. This permits making the subtle distinction of returning $\perp$ when a candidate decryption result has not been found[2], or when it has been established that the oracle may return $\perp$ when queried on a given (c, PK) pair. In the second step, it selects the message to return from WitSp. To make sure the security model is not restricting the adversary by choosing the decryption result in a particular way, we allow the adversary to provide a polynomial-time relation R to characterise a set of messages of interest to her. The oracle then samples a message at random from this set and returns it to the adversary. In the third and final step, the oracle updates any state it may have stored from previous queries. We require that the update procedure to be polynomial in the size of its inputs, excluding the state[3].

Although we have constrained the algorithms in our definition (i.e. V, R and U) to be polynomial-time, the calculations carried out in the first two steps may not be computable in polynomial time and may require an exponential number of executions of these algorithms. Nevertheless, we emphasise that the search space must be finite. This is guaranteed by the assumption that the message space of the encryption scheme is finite, and by the fact that the algorithms associated with the scheme run in polynomial time in their inputs.

The motivation for having such a general definition is that the notion of *the message encapsulated by the ciphertext* can be defined in various ways. For concreteness, let us fix U so that $st[V]$ is empty throughout the game execution, and look at two alternative definitions of V. These derive from two interpretations as to which message(s) might be encapsulated in a public key/ciphertext pair: they can be seen as alternative witnesses to the validity of the public key/ciphertext pair. Concretely one can define validity via the encryption operation, in which case a message/randomness pair is the witness or via the decryption algorithm, in which case the natural witness is a message/secret key pair[4]:

$$V(c, PK, m, r) := c \stackrel{?}{=} Enc(m, PK; r) \tag{1}$$

---

[1] This constitutes an NP-relation for the language of valid decryption results.

[2] Recall that we assume that sampling from an empty set returns $\perp$.

[3] Discarding the state size ensures that the run-time of this procedure does not increase exponentially with queries.

[4] Note that we have assumed Dec always performs the key-pair validity check, and so this is redundant in V'. We include it for the sake of clarity: for schemes which do not perform the key-pair validity check, this issue must be considered.

$$\mathsf{V}'(\mathsf{c}, \mathsf{PK}, \mathsf{m}, r) := (\mathsf{SK}, \mathsf{PK}) \overset{?}{=} \mathsf{Gen}(r) \wedge \mathsf{m} \overset{?}{=} \mathsf{Dec}(\mathsf{c}, \mathsf{SK}, \mathsf{PK}). \qquad (2)$$

The first observation to make on these validity criteria is that neither of them guarantees that if a message is found to be a valid decryption result, it will be unique. This is because the correctness restriction only guarantees unique decryptability for correctly constructed $(\mathsf{c}, \mathsf{PK})$ pairs: it says nothing about the result of decryption when an invalid public key and/or an invalid ciphertext are provided as inputs. In particular, the validity criterion in Equation 1 could accept multiple messages as valid, when run on an invalid public key. Ambiguity can also occur for the validity criterion in Equation 2, when multiple valid secret keys correspond to the queried public key, and decrypt an invalid ciphertext inconsistently. This discussion justifies the need for the second step in the definition we propose: there could be many valid decryption results to choose from, and it is left to the adversary to control how this is done. In the simplest scenario, where there is only one candidate decryption result, one can assume without loss of generality that the adversary will choose to retrieve that result by passing in the trivial relation T.

The need for the first step of the definition is justified by observing that the two witness sets associated with the above validity algorithms do not always coincide. To see this, consider an encryption scheme where decryption does not necessarily fail when run on a ciphertext that falls outside the range of the encryption algorithm. Then the first witness set will be empty whereas the second may not be. A concrete example is the Cramer-Shoup [10] encryption scheme. For other schemes, such as RSA-OAEP [14], it may happen that the encryption algorithm produces apparently valid ciphertexts for invalid public keys. When this is the case, the first witness set may not be empty, whereas the second one will surely contain no messages, given that no valid secret key exists.

We note that the above issues do not arise in the standard definition of a decryption oracle, in which decryption is always carried out with a fixed secret key. In other words, the decryption oracle is stateful. To allow capturing this sort of behaviour in strong decryption oracles, we add the last step to the oracle definition. This manages the decryption oracle state, and ensures that the validity checking algorithm can access it in each query.

Specific definitions. Previous attempts to define strong decryption oracles have been introduced for certificateless public-key encryption, where public keys are not authenticated [2,11]. These definitions implicitly adopt validity criteria which are adequate only for the concrete schemes discussed in the referred works.

In the definition proposed in [2] the authors simply describe the oracle as providing "correct decryptions" even though the secret key could be unknown. A close analysis of the presentation in this work indicates that "correct decryption" is defined through a search for a message/randomness pair in the domain of the encryption, similarly to the first validity criterion presented above. However, the unique decryptability issue is implicit in the definition, since the concrete scheme the authors consider ensures that the encryption algorithm fails when queried with an invalid public key. Extending this definition to encryption schemes in general results in the following validity criterion:

$$\mathsf{V_{PK}}(\mathsf{c}, \mathsf{PK}, \mathsf{m}, r||r') := \mathsf{c} \stackrel{?}{=} \mathsf{Enc}(\mathsf{m}, \mathsf{PK}; r) \wedge (\star, \mathsf{PK}) \stackrel{?}{=} \mathsf{Gen}(r').$$

Note that this is equivalent to the validity relation in Equation 1 for schemes which check for public key validity in the encryption algorithm. Alternatively, a solution adopted in literature [13] is to restrict the class of adversaries to those which query only valid public keys. In our view, such a restriction on the adversary's behaviour is unjustified, and we will look for alternatives which guarantee stronger security.

In a more recent work [11], the strong decryption oracle is described as constructing a private key that corresponds to the queried valid public key, and then using that key to decrypt the ciphertext. The oracle then stores the extracted secret key to be reused in subsequent queries under the same public key. This definition is more in line with the intuition that a decryption oracle should reflect the behaviour of the decryption algorithm, and it is also consistent with the stateful operation of the standard decryption oracle. We can capture this definition through the algorithms presented in Figure 2. Note that, for those schemes in which there is a unique valid private key per public key or for those schemes where all valid secret keys behave consistently for all possible, even invalid, ciphertexts, the oracle resulting from these algorithms will be identical to the one using the criterion in Equation 2.

The previous discussion indicates that different definitions of a strong decryption oracle can be seen as natural for particular classes of schemes. However, we can also consider other approaches, which are not so easy to characterise. For example, a straightforward fix to the ambiguity problem described above is to have the oracle simply return $\perp$ when it arises. Agreeably, this approach addresses the problem of ambiguity directly, but it is hardly intuitive with respect to the operation of public-key encryption schemes. In particular, this definition is best suited for the class of encryption schemes for which the ambiguity never occurs. However, there is no natural characterisation of this class of schemes.

As a final motivation for a general definition of a strong decryption oracle, let us look at RSA-OAEP [14]. The non-malleability properties of (a modified version of) this scheme are analysed by Fischlin [13] using a model related to the

proc. $\mathsf{V}(\mathsf{c}, \mathsf{PK}, \mathsf{m}, r, \mathsf{st}[\mathsf{V}])$:	proc. $\mathsf{U}(\mathsf{c}, \mathsf{PK}, \mathsf{R}, \mathsf{m}, r, \mathsf{st}[\mathsf{V}])$:
$(\mathsf{SK}', \mathsf{PK}') \leftarrow \mathsf{Gen}(r)$	$(\mathsf{SK}', \mathsf{PK}') \leftarrow \mathsf{Gen}(r)$
If $((\mathsf{SK}, \mathsf{PK}) \in \mathsf{st}[\mathsf{V}] \wedge \mathsf{SK}' \neq \mathsf{SK})$	If $\mathsf{PK}' \neq \mathsf{PK} \vee (\mathsf{SK}, \mathsf{PK}) \in \mathsf{st}[\mathsf{V}]$
$\quad\quad \mathsf{PK}' \neq \mathsf{PK}$ Return F	$\quad\quad$ Return $\mathsf{st}[\mathsf{V}]$
If $\mathsf{m} = \mathsf{Dec}(\mathsf{c}, \mathsf{SK}', \mathsf{PK}')$ Return T	$\mathsf{st}[\mathsf{V}] \leftarrow (\mathsf{SK}', \mathsf{PK}') : \mathsf{st}[\mathsf{V}]$
Return F	Return $\mathsf{st}[\mathsf{V}]$

Fig. 2. Update and validity algorithms for a stateful strong decryption oracle with initial state $\mathsf{st}_0 = (\mathsf{SK}^\star, \mathsf{PK}^\star)$

decryption oracle associated with Equation 1. However, the analysis is restricted to adversaries that only query valid public keys. For such adversaries, the resulting oracle is identical to that resulting from Equation 2, as the decryption algorithm of the scheme checks for key-pair validity and recovers the random coins used in encryption. However, once this restriction is dropped, the oracles are no longer equivalent. Security with respect to Equation 2 is still implied by Fischlin's analysis but, with respect to Equation 1 it remains an open issue.

SIMPLIFICATION. We now characterise a class of schemes for which the above variants of strong decryption oracle collapse into a simpler definition. This class consists of encryption schemes which perform checks both at encryption and decryption stages. They check for public key validity upon encryption, returning a failure symbol if the key is invalid. Furthermore, in decryption, they check both key-pair validity and that the input ciphertext lies in the range of the encryption algorithm. Note that for such schemes, whenever encryption and decryption do not fail, then correctness ensures that the set of messages which can be obtained using any of the validity criteria above coincide, and have cardinality 1. The simplified version of the strong decryption oracle that we arrive at is shown in Figure 3. The scheme that we present in Section 5 has been designed so that it belongs to this class of encryption schemes, and could therefore be analysed using this simpler oracle. Indeed, this observation is central to our argument that *we propose a simpler and more convenient security model in which to analyse schemes that aim to achieve complete non-malleability.*

---

proc. SDec(c, PK):
$m \leftarrow_\$ \{m : \exists SK, m = Dec(c, SK, PK)\}$
Return m

Fig. 3. Simplified definition of strong decryption for schemes which perform all checks. The search over m excludes $\perp$.

4   Security under Strong Chosen-Ciphertext Attacks

In this section, we use the general definition of a strong decryption oracle in Figure 1 to extend different security models for encryption schemes. This allows for a uniform treatment of strong security models, some of which have been independently proposed in literature. Then, we investigate the relations among the resulting security notions, as well as those in [13,19].

4.1   Indistinguishability of Ciphertexts

We now introduce ciphertext indistinguishability under strong chosen-ciphertext attacks as the natural extension of the standard notions of security for public-key encryption schemes. The IND-SCCAx advantage of an adversary $\mathcal{A}$ for $x = 0, 1, 2$ against a public-key encryption scheme Π is defined by

$$\mathbf{Adv}_{\Pi}^{\mathsf{ind\text{-}sccax}}(\mathcal{A}) := 2 \cdot \Pr\left[\mathsf{IND\text{-}SCCAx}_{\Pi}^{\mathcal{A}} \Rightarrow \mathsf{T}\right] - 1,$$

where game IND-SCCAx is shown in Figure 4. Implicit in this definition are the descriptions of the U and V algorithms, which are fixed when analysing a scheme in the resulting IND-SCCAx model. As seen in the previous section, *one can make general claims of security and still use a simple definition for the strong decryption oracle (Figure 3) by showing that the scheme satisfies a well-defined set of natural properties.*

---

proc. Initialize():
$b \leftarrow_\$ \{0,1\}; \mathsf{I} \leftarrow_\$ \mathsf{Setup}()$
$(\mathsf{SK}^\star, \mathsf{PK}^\star) \leftarrow_\$ \mathsf{Gen}()$
$\mathsf{List} \leftarrow []; \mathsf{st}[\mathsf{V}] \leftarrow \mathsf{st}_0$
Return $(\mathsf{I}, \mathsf{PK}^\star)$

proc. LoR($\mathsf{m}_0, \mathsf{m}_1$):   Game IND-SCCAx$_{\Pi}$
$\mathsf{c} \leftarrow_\$ \mathsf{Enc}(\mathsf{m}_b, \mathsf{PK}^\star)$
$\mathsf{List} \leftarrow (\mathsf{c}, \mathsf{PK}^\star) : \mathsf{List}$
Return c

proc. SDec($\mathsf{c}, \mathsf{PK}, \mathsf{R}$):
Return $\mathbf{SDec}_{\mathsf{U},\mathsf{V}}(\mathsf{c}, \mathsf{PK}, \mathsf{R})$

proc. Finalize(b'):
Return $(b' = b)$

Fig. 4. Game defining indistinguishability under strong chosen-ciphertext attacks. An adversary $\mathcal{A}$ is legitimate if: 1) It calls **LoR** only once with $\mathsf{m}_0, \mathsf{m}_1 \in \mathsf{MsgSp}(\mathsf{PK})$ such that $|\mathsf{m}_0| = |\mathsf{m}_1|$; and 2) R is polynomial-time and, if $\mathsf{x} = 0$ it does not call **SDec**, if $\mathsf{x} = 1$ it does not call **SDec** after calling **LoR**, and if $\mathsf{x} = 2$ it does not call **SDec** with a tuple $(\mathsf{c}, \mathsf{PK})$ in List.

STRONG PARALLEL ATTACKS. Bellare and Sahai [8] define a security notion known as indistinguishability under parallel chosen-ciphertext attacks. Here the adversary can query a vector of ciphertexts to a parallel decryption oracle exactly once and after its left-or-right query, receiving the corresponding component-wise decryptions. It is proved in [8] that parallel security maps well to non-malleability of encryption schemes. We extend this model to incorporate strong attacks by defining the IND-SPCAx advantage of an adversary $\mathcal{A}$ against an encryption scheme Π similarly to above, where game IND-SPCAx is shown in Figure 5. Note that under this definition, and consistently with previous results, IND-SPCA2 is equivalent to IND-SCCA2: the parallel oracle is subsumed by the strong decryption oracle that the adversary is allowed to call adaptively after the challenge phase. We remark that a stronger definition can be adopted, whereby the adversary is allowed to query the parallel oracle with a relation that takes all the ciphertexts simultaneously. We will return to this issue in the next section.

KEM/DEM COMPOSITION. The standard proof technique [10] to establish the security of hybrid encryption schemes consisting of a secure keys encapsulation mechanism (KEM) and a secure data encryption mechanism (DEM), fails to extend to the strong chosen-ciphertext models (strong security for KEMs can be defined in the natural way). This failure is due to the non-polynomial nature of the decryption oracle, which cannot be simulated even if one generates the

proc. Initialize():	**proc. LoR**(m_0, m_1):	Game IND-SPCAx$_\Pi$
$b \leftarrow_\$ \{0,1\}$; $I \leftarrow_\$ \mathsf{Setup}()$	$c \leftarrow_\$ \mathsf{Enc}(m_b, \mathsf{PK}^\star)$	
$(\mathsf{SK}^\star, \mathsf{PK}^\star) \leftarrow_\$ \mathsf{Gen}()$	$\mathsf{List} \leftarrow (c, \mathsf{PK}^\star) : \mathsf{List}$	**proc. PSDec**$(\mathbf{c}, \mathbf{PK}, \mathbf{R})$:
$\mathsf{List} \leftarrow []$; $\mathsf{st}[\mathsf{V}] \leftarrow \mathsf{st}_0$	Return c	For i from 1 to $\#\mathbf{c}$ do
Return $(I, \mathsf{PK}^\star)$		$\quad \mathbf{m}[i] \leftarrow_\$ \mathbf{SDec}_{\mathsf{U},\mathsf{V}}(\mathbf{c}[i], \mathbf{PK}[i], \mathbf{R}[i])$
		Return $\mathbf{m}$
proc. SDec$(c, \mathsf{PK}, \mathsf{R})$:		**proc. Finalize**(b'):
Return $\mathbf{SDec}_{\mathsf{U},\mathsf{V}}(c, \mathsf{PK}, \mathsf{R})$		Return $(b' = b)$

Fig. 5. Game defining indistinguishability under strong parallel chosen-ciphertext attacks. An adversary $\mathcal{A}$ is legitimate if: 1) It calls **LoR** only once with $m_0, m_1 \in \mathsf{MsgSp}(\mathsf{PK})$ such that $|m_0| = |m_1|$; 2) It calls **PSDec** exactly once and after calling **LoR**, on a tuple $(\mathbf{c}, \mathbf{PK}, \mathbf{R})$ such that for $i = 1, \ldots, \#\mathbf{c}$, the tuples $(\mathbf{c}[i], \mathbf{PK}[i])$ do not appear in List and $\mathbf{R}[i]$ are polynomial-time; and 3) R is polynomial-time and, if $\mathsf{x} = 0$ it does not call **SDec**, or if $\mathsf{x} = 1$ it does not call **SDec** after calling **LoR**, or if $\mathsf{x} = 2$ it does not call **SDec** with a tuple (c, PK) in List.

challenge public key. One way to go around this obstacle is to build schemes which permit embedding an *escrow* trapdoor in the common parameters, enabling decryption over *all* public keys.

4.2   Complete Non-malleability

Turning our attention to strong notions of non-malleability, or so-called complete non-malleability, we shall see in this section how strong decryption oracles can be used to bring coherence to existing definitional approaches. In particular, we introduce new definitions using strong decryption oracles that can be used to establish clear relations with the strong indistinguishability notion introduced above. We also clarify how the definitions we propose relate to those previously described in literature.

SIMULATION-BASED DEFINITION. The first definition of complete non-malleability was introduced by Fischlin in [13]. We propose an alternative definition. We define the SNM-SCCAx advantage of an adversary $\mathcal{A}$ with respect to a polynomial-time relation R and a polynomial-time simulator $\mathcal{S}$ against a public-key encryption scheme Π by

$$\mathbf{Adv}^{\mathsf{snm\text{-}sccax}}_{\Pi,\mathsf{R},\mathcal{S}}(\mathcal{A}) := \Pr\left[\mathsf{Real\text{-}SNM\text{-}SCCAx}^{\mathcal{A}}_{\Pi,\mathsf{R}} \Rightarrow \mathsf{T}\right] - \Pr\left[\mathsf{Ideal\text{-}SNM\text{-}SCCAx}^{\mathcal{S}}_{\Pi,\mathsf{R}} \Rightarrow \mathsf{T}\right]$$

where games Real-SNM-SCCAx and Ideal-SNM-SCCAx are as shown in Figure 6. The syntax of public-key encryption that we use includes a Setup procedure and hence we explicitly include the common parameters I as an input to the malleability relation. This approach is consistent with the explicit inclusion of the challenge public key, which is shown in [13] to strictly strengthen the definition. Additionally, for backward compatibility with [8], our relations also include the state information st_R. For strong decryption oracles that behave consistently

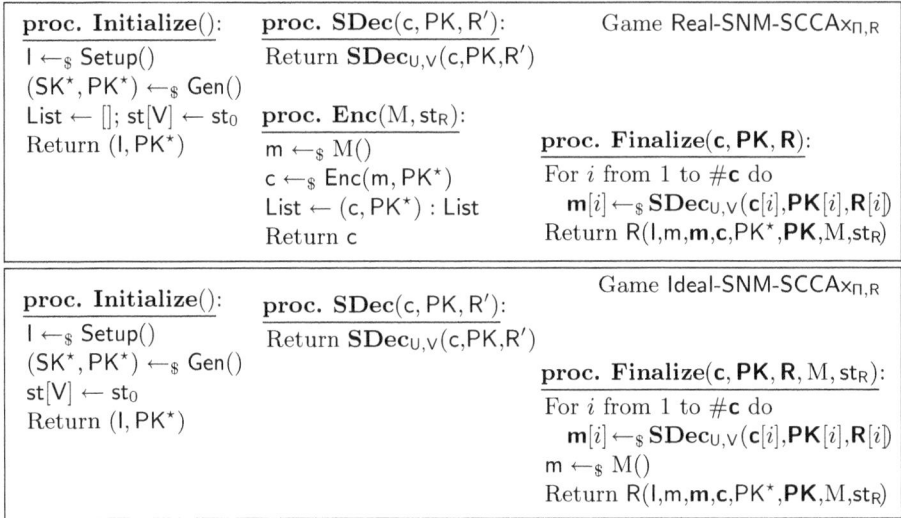

Fig. 6. Games defining simulation-based complete non-malleability under strong chosen-ciphertext attacks. An adversary $\mathcal{A}$, playing the real game, is legitimate if: 1) It calls **Enc** once with a valid M; 2) R′ queried to **SDec** is computable in polynomial time; if x = 0 it does not call **SDec**; if x = 1 it does not call **SDec** after calling **LoR**; and if x = 2 it does not call **SDec** with a tuple in List; and 3) It calls **Finalize** with a tuple such that all relations in **R** are computable in polynomial time and, for $i = 1, \ldots, \#\mathbf{c}$, the tuples $(\mathbf{c}[i], \mathbf{PK}[i])$ do not appear in List. A *non-assisted* simulator, playing the ideal game, $\mathcal{S}$ is legitimate if: 1) It calls **Finalize** with a valid M; and 2) It does not call **SDec**. An *assisted* simulator, playing the ideal game, is legitimate if: 1) It calls **Finalize** with a valid M; 2) R′ queried to **SDec** is computable in polynomial time; and 3) If x = 0 it does not call **SDec**.

with the standard one for $\mathsf{PK}^\star$, and for a class of relations that matches those in the original definition, our definition implies standard assisted and non-assisted simulation-based non-malleability as defined in [8].

A similar line of reasoning does not permit concluding that our definition also implies Fischlin's complete non-malleability. A legitimate adversary under Fischlin's definition is also a legitimate adversary under the definition in 6. However, we cannot identify a concrete version of the strong decryption oracle that captures the environment under which such an adversary should run. This is because Fischlin's model implicitly uses two definitions of decryption oracle: one during the interactive stages of the game, where the adversary has access to a standard decryption oracle that decrypts using the challenge secret key, and a second one in the **Finalize** stage, where the ciphertext produced by the adversary is decrypted by searching through the message/randomness space. We justify our modelling choice with two arguments. Firstly, the construction of **Finalize** in Fischlin's definition makes it impossible to prove that this security model is stronger than the apparently weaker definition of non-malleability proposed in [8], which uses the standard decryption oracle to recover messages from the ciphertexts output by the adversary (recall the particular case of invalid ciphertexts under a valid public key, for which the two interpretations of valid decryption results do not coincide). This suggests that using a consistent definition of a (strong) decryption oracle in all stages of the game is a better approach. Secondly, if this change were introduced in Fischlin's definition, then this would simply be a special case of our more general definition.

COMPARISON-BASED DEFINITION. The simulation-based definition due to Fischlin was later reformulated by Ventre and Visconti [19] as a comparison-based notion. We introduce an alternative definition based on the CNM-SCCAx game shown in Figure 7 and define CNM-SCCAx advantage of an adversary $\mathcal{A}$ against an encryption scheme Π as

$$\mathbf{Adv}_{\Pi}^{\mathrm{cnm\text{-}sccax}}(\mathcal{A}) := \Pr\left[\mathsf{CNM\text{-}SCCAx}_{\Pi}^{\mathcal{A}} \Rightarrow \mathsf{T} \mid b = 1\right] - \Pr\left[\mathsf{CNM\text{-}SCCAx}_{\Pi}^{\mathcal{A}} \Rightarrow \mathsf{T} \mid b = 0\right]$$

Our definition differs from that given in [19] in the following aspects. We provide the adversary with strong decryption oracles in various stages of the attack. In both models the adversary is allowed to return a vector of ciphertexts, although in [19] it is restricted to returning a single public key. Also, procedure **Finalize** does not automatically return F if any of the ciphertexts is invalid. The definition in [19] would therefore be weaker than ours, were it not for our modelling choice in the **Finalize** procedure. In Ventre and Visconti's definition, the relation R is evaluated by a complete search over $(\mathbf{m}[1], r_1) \times \ldots \times (\mathbf{m}[\#\mathbf{c}], r_{\#\mathbf{c}})$. In our definition we have constrained the adversary to performing the search using the strong decryption oracle independently for *each component* in $\mathbf{c}$, before evaluating R. This option is, not only consistent with the standard notions of non-malleability for encryption schemes [8], but is also essential to proving equivalence among the different notions we propose.

proc. **Initialize**():	proc. **Enc**(M):	Game CNM-SCCAx$_\Pi$
$b \leftarrow_\$ \{0,1\}$; $\mathsf{I} \leftarrow_\$ \mathsf{Setup}()$	$m_0, m_1 \leftarrow_\$ M()$	
$(\mathsf{SK}^\star, \mathsf{PK}^\star) \leftarrow_\$ \mathsf{Gen}()$	$\mathsf{c} \leftarrow_\$ \mathsf{Enc}(m_1, \mathsf{PK}^\star)$	
$\mathsf{List} \leftarrow []$; $\mathsf{st}[\mathsf{V}] \leftarrow \mathsf{st}_0$	$\mathsf{List} \leftarrow \mathsf{List} : (\mathsf{c}, \mathsf{PK}^\star)$	proc. **Finalize**($\mathbf{c}, \mathbf{PK}, \mathsf{R}, \mathbf{R}$):
Return $(\mathsf{I}, \mathsf{PK}^\star)$	Return c	For i from 1 to $\#\mathbf{c}$ do
proc. **SDec**($\mathsf{c}, \mathsf{PK}, \mathsf{R}'$):		$\quad \mathbf{m}[i] \leftarrow_\$ \mathbf{SDec}(\mathbf{c}[i], \mathbf{PK}[i], \mathbf{R}[i])$
Return $\mathbf{SDec}_{\mathsf{U},\mathsf{V}}(\mathsf{c}, \mathsf{PK}, \mathsf{R}')$		Return $\mathsf{R}(\mathsf{I}, m_b, \mathbf{m}, \mathbf{c}, \mathsf{PK}^\star, \mathbf{PK})$

Fig. 7. Game defining comparison-based complete non-malleability under strong chosen-ciphertext attacks. An adversary $\mathcal{A}$ is legitimate if: 1) It calls **Enc** once with a valid M; 2) It always queries **SDec** with R' computable in polynomial time; if $\mathsf{x} = 0$ it does not call **SDec**; if $\mathsf{x} = 1$ it does not call **SDec** after calling **LoR**; and if $\mathsf{x} = 2$ it does not call **SDec** with a tuple $(\mathsf{c}, \mathsf{PK})$ in List; 3) It calls **Finalize** with a tuple $(\mathbf{c}, \mathbf{PK}, \mathsf{R}, \mathbf{R})$ such that R and all the elements of $\mathbf{R}$ are computable in polynomial time and, for $i = 1, \ldots, \#\mathbf{c}$, the tuples $(\mathbf{c}[i], \mathbf{PK}[i])$ do not appear in List.

REMARK. Recall that Ventre and Visconti's proof [19] of equivalence between comparison and (non-assisted) simulation-based complete non-malleability holds (for x $\neq$ 0) for a restricted class of relations, called *lacking* relations, which do not depend on the challenge public key given to the adversary. We note that our equivalence proof for assisted simulators does not restrict the class of relations under which equivalence holds. Furthermore, such a restriction would be pointless in our definitions for non-assisted simulators, since the proof technique of generating a new key-pair is no longer sufficient to guarantee that the simulator can answer *strong* decryption queries under arbitrary public keys.

4.3 Relations among Notions of Security

We now present our main theorem that establishes equivalence between the security notions we have proposed above. The proof, which can be found in the full version of the paper, follows the strategy used by Bellare and Sahai [8]. We note that our result holds for *any* instantiation of the strong decryption oracle as given in Figure 1, providing further evidence that the security models we are relating are, in fact, the same notion presented using different formalisms.

Theorem 1 (Equivalence). *The* IND-SPCAx, CNM-SCCAx *and* SNM-SCCAx *notions of security are equivalent, for any* x $\in \{0, 1, 2\}$.

Using a standard hybrid argument one can show that IND-SPCAx self-composes. Together with our equivalence result, we conclude that our notions of complete non-malleability also self-compose [18].

5 An Efficient Completely Non-Malleable Scheme

The only completely non-malleable scheme (without random oracles) known prior to this work, was that of Ventre and Visconti [19], which relied on generic (and hence inefficient) zero-knowledge techniques. In this section, we will present an efficient and strongly secure scheme based on standard assumptions.

Our scheme, which is shown in Figure 8, uses a computational bilinear group scheme Γ and a family of collision resistant hash functions Σ mapping $\mathbb{G}_T \times \mathbb{G} \times \mathbb{G}^2$ to bit strings of size n. Our scheme relies on the decisional bilinear Diffie-Hellman assumption which requires the distributions $(g, g^a, g^b, g^c, \mathbf{e}(g, g)^{abc})$ and $(g, g^a, g^b, g^c, \mathbf{e}(g, g)^d)$, for random a, b, c, and d, to be computationally indistinguishable. The scheme's design is based on the certificateless encryption scheme of [11], which in turn is based on Water's identity-based encryption scheme [20]. The construction also uses Waters' hash [20], defined by $\mathsf{WH}(w) := u_0 \prod_{i=1}^{n} u_i^{[w]_i}$.

VALIDITY ALGORITHMS. We examine which of the validity algorithms exists for this scheme. We assume that Γ specifies algorithms to check for group membership, which are used implicitly throughout the scheme. The MsgSp algorithm is the same as checking membership in $\mathbb{G}_T$. The SKSp algorithm checks membership in $\mathbb{Z}_p$. The KeySp algorithm checks if $g^{\mathsf{SK}} = X$ and $\alpha^{\mathsf{SK}} = Y$ where

proc. $\mathsf{Setup}_{\Gamma,\Sigma,n}()$:	proc. $\mathsf{Enc}(\mathsf{m},\mathsf{PK})$:	proc. $\mathsf{Dec}(\mathsf{c},\mathsf{SK},\mathsf{PK})$:
$k \leftarrow_\$ \mathsf{Key}()$;	$t \leftarrow_\$ \mathbb{Z}_p$; $(X,Y) \leftarrow \mathsf{PK}$	$(X,Y) \leftarrow \mathsf{PK}$
$(\alpha,\beta,u_0,\ldots,u_n) \leftarrow_\$ \mathbb{G}^\star \times \mathbb{G}^{n+2}$	If $\mathbf{e}(X,\alpha) \neq \mathbf{e}(g,Y)$	If $g^{\mathsf{SK}} \neq X \vee \alpha^{\mathsf{SK}} \neq Y$
$\mathsf{I} \leftarrow (\Gamma, \mathsf{H_k}, \alpha, \beta, u_0, \ldots, u_n)$	Return $\bot$	Return $\bot$
Return I	$C_1 \leftarrow \mathsf{m} \cdot \mathbf{e}(Y, \beta^t)$;	$(C_1, C_2, C_3) \leftarrow \mathsf{c}$
	$C_2 \leftarrow \alpha^t$	$w \leftarrow \mathsf{H_k}(C_1, C_2, \mathsf{PK})$
proc. $\mathsf{Gen}()$:	$w \leftarrow \mathsf{H_k}(C_1, C_2, \mathsf{PK})$	If $\mathbf{e}(C_2, \mathsf{WH}(w)) \neq \mathbf{e}(\alpha, C_3)$
$x \leftarrow_\$ \mathbb{Z}_p$; $X \leftarrow g^x$; $Y \leftarrow \alpha^x$	$C_3 \leftarrow \mathsf{WH}(w)^t$	Return $\bot$
$\mathsf{PK} \leftarrow (X,Y)$; $\mathsf{SK} \leftarrow x$	$\mathsf{c} \leftarrow (C_1, C_2, C_3)$	$\mathsf{m} \leftarrow C_1 / \mathbf{e}(C_2, \beta^x)$
Return $(\mathsf{SK}, \mathsf{PK})$	Return c	Return m

Fig. 8. A strongly secure public-key encryption scheme without random oracles

$(X,Y) = \mathsf{PK}$. The PKSp algorithm checks if $\mathbf{e}(X,\alpha) = \mathbf{e}(g,Y)$. Finally, we show that decryption rejects all ciphertexts outside the range of encryption. Let (C_1, C_2, C_3) be a ciphertext. Then, there exists a message m and a t such that this ciphertext can be written as $(\mathsf{m} \cdot \mathbf{e}(Y, \beta)^t, \alpha^t, C_3)$. If this ciphertext is outside the range of encryption, then $C_3 = \mathsf{WH}(w)^{t'}$ for some $t' \neq t$. But then $\mathbf{e}(C_2, \mathsf{WH}(w)) = \mathbf{e}(\alpha, \mathsf{WH}(w))^t \neq \mathbf{e}(\alpha, \mathsf{WH}(w))^{t'} = \mathbf{e}(\alpha, C_3)$ and the equality check in decryption fails.

The next theorem states the security properties of our scheme. Its proof uses technique recently proposed by Bellare and Ristenpart [6] and is given in the full version of the paper.

Theorem 2 (Informal). *Under the decisional bilinear Diffie-Hellman assumption in Γ and the collision resistance of the hash function family Σ, the above scheme is* $\mathsf{IND\text{-}SCCA2}$ *secure (with respect to* $\mathbf{SDec}$ *oracle defined in Figure 3).*

Although our equivalence theorems imply that this scheme admits a black-box assisted simulator, it does not contradict Fischlin's impossibility results on black-box simulation [13]. First note that Fischlin's impossibility result is in the plain model whereas our scheme has a setup procedure. Furthermore, our definitions do not require the opening of message/randomness pairs, whereas Fischlin requires this to derive his impossibility result for *assisted* simulators. We can indeed construct a non-assisted simulator for our scheme through a direct proof, but this requires modifying the common parameters in an essential way to simulate the strong decryption oracle. Hence this result does not hold for general relations, but only for those which ignore the I presented at their inputs (consistently with [19] we call these I-lacking relations). Furthermore, using a similar technique, we are also able to show (through a direct proof) that the zero-knowledge-based construction in [19] is completely non-malleable with respect to black-box simulators for a class of relations that are I-lacking (I in this case comprises the common reference string). We note that this is a better result than that obtained in [19], since there the class of relations must be both I-lacking and PK-lacking (i.e. they must also ignore the PK at their inputs).

Acknowledgments. The authors were funded in part by eCrypt II (EU FP7 - ICT-2007-216646) and FCT project PTDC/EIA/71362/2006. The second author was also funded by FCT grant BPD-47924-2008.

References

1. Abdalla, M., Bellare, M., Neven, G.: Robust encryption. In: Micciancio, D. (ed.) TCC 2010. LNCS, vol. 5978, pp. 480–497. Springer, Heidelberg (2010)
2. Al-Riyami, S.S., Paterson, K.G.: Certificateless public key cryptography. In: Laih, C.-S. (ed.) ASIACRYPT 2003. LNCS, vol. 2894, pp. 452–473. Springer, Heidelberg (2003)
3. Bellare, M., Boldyreva, A., Micali, S.: Public-key encryption in a multi-user setting: Security proofs and improvements. In: Preneel, B. (ed.) EUROCRYPT 2000. LNCS, vol. 1807, pp. 259–274. Springer, Heidelberg (2000)
4. Bellare, M., Desai, A., Pointcheval, D., Rogaway, P.: Relations among notions of security for public-key encryption schemes. In: Krawczyk [16], pp. 26–45
5. Bellare, M., Kohno, T.: A theoretical treatment of related-key attacks: Rka-prps, rka-prfs, and applications. In: Biham, E. (ed.) EUROCRYPT 2003. LNCS, vol. 2656, pp. 491–506. Springer, Heidelberg (2003)
6. Bellare, M., Ristenpart, T.: Simulation without the artificial abort: Simplified proof and improved concrete security for waters' ibe scheme. In: Joux, A. (ed.) EURO-CRYPT 2009. LNCS, vol. 5479, pp. 407–424. Springer, Heidelberg (2010)
7. Bellare, M., Rogaway, P.: The security of triple encryption and a framework for code-based game-playing proofs. In: Vaudenay, S. (ed.) EUROCRYPT 2006. LNCS, vol. 4004, pp. 409–426. Springer, Heidelberg (2006)
8. Bellare, M., Sahai, A.: Non-malleable encryption: Equivalence between two notions, and an indistinguishability-based characterization. Cryptology ePrint Archive, Report 2006/228 (2006), http://eprint.iacr.org/2006/228
9. Cramer, R. (ed.): PKC 2008. LNCS, vol. 4939. Springer, Heidelberg (2008)
10. Cramer, R., Shoup, V.: A practical public key cryptosystem provably secure against adaptive chosen ciphertext attack. In: Krawczyk [16], pp. 13–25
11. Dent, A.W., Libert, B., Paterson, K.G.: Certificateless encryption schemes strongly secure in the standard model. In: Cramer [9], pp. 344–359
12. Dolev, D., Dwork, C., Naor, M.: Nonmalleable cryptography. SIAM J. Comput. 30(2), 391–437 (2000)
13. Fischlin, M.: Completely non-malleable schemes. In: Caires, L., Italiano, G.F., Monteiro, L., Palamidessi, C., Yung, M. (eds.) ICALP 2005. LNCS, vol. 3580, pp. 779–790. Springer, Heidelberg (2005)
14. Fujisaki, E., Okamoto, T., Pointcheval, D., Stern, J.: Rsa-oaep is secure under the rsa assumption. J. Cryptology 17(2), 81–104 (2004)
15. Goldwasser, S., Micali, S.: Probabilistic encryption. J. Comput. Syst. Sci. 28(2), 270–299 (1984)
16. Krawczyk, H. (ed.): CRYPTO 1998. LNCS, vol. 1462. Springer, Heidelberg (1998)
17. Pandey, O., Pass, R., Vaikuntanathan, V.: Adaptive one-way functions and applications. In: Wagner, D. (ed.) CRYPTO 2008. LNCS, vol. 5157, pp. 57–74. Springer, Heidelberg (2008)

18. Pass, R., Shelat, A., Vaikuntanathan, V.: Relations among notions of non-malleability for encryption. In: Kurosawa, K. (ed.) ASIACRYPT 2007. LNCS, vol. 4833, pp. 519–535. Springer, Heidelberg (2007)
19. Ventre, C., Visconti, I.: Completely non-malleable encryption revisited. In: Cramer [9], pp. 65–84
20. Waters, B.: Efficient identity-based encryption without random oracles. In: Cramer, R. (ed.) EUROCRYPT 2005. LNCS, vol. 3494, pp. 114–127. Springer, Heidelberg (2005)

Strong Knowledge Extractors for Public-Key Encryption Schemes

Manuel Barbosa[1] and Pooya Farshim[2]

[1] CCTC/Departamento de Informática, Universidade do Minho,
Campus de Gualtar, 4710-057 Braga, Portugal
mbb@di.uminho.pt
[2] Information Security Group, Royal Holloway, University of London,
Egham, Surrey, TW20 0EX, United Kingdom
Pooya.Farshim@rhul.ac.uk

Abstract. Completely non-malleable encryption schemes resist attacks which allow an adversary to tamper with both ciphertexts *and* public keys. In this paper we introduce two extractor-based properties that allow us to gain insight into the design of such schemes and to go beyond known feasibility results in this area. We formalise *strong plaintext awareness* and *secret key awareness* and prove their suitability in realising these goals. Strong plaintext awareness imposes that it is infeasible to construct a ciphertext under *any* public key without knowing the underlying message. Secret key awareness requires it to be infeasible to produce a new public key without knowing a corresponding secret key.

Keywords: Secret Key Awareness. Strong Plaintext Awareness. Complete Non-Malleability. Strong Chosen-Ciphertext Attacks.

1 Introduction

BACKGROUND. Indistinguishability of ciphertexts under chosen-ciphertext attacks (IND-CCA2) is a convenient reformulation of a more intuitive security notion known as *non-malleability*. Roughly speaking, an encryption scheme is non-malleable if, given a challenge ciphertext, it is infeasible to output a new ciphertext encrypting a plaintext related in a "meaningful" or "interesting" way to that enclosed in the challenge. The advantages of the indistinguishability formulation become apparent when one considers various subtleties which arise when defining what a meaningful relation is [22,10]. Recently, Fischlin [18] has considered the problem of using public key encryption schemes to build non-malleable commitment schemes. It has been shown that the standard definition of non-malleability is not sufficient for this application and that a stronger variant, referred to as *complete non-malleability*, is required. This security definition allows the adversary to maul the challenge public key, as well as the ciphertext. Put differently, the adversary can output a related ciphertext under a new public key of its choice. Unlike standard non-malleability, it has been shown in [18] that completely non-malleable schemes are hard to construct. In particular, such

R. Steinfeld and P. Hawkes (Eds.): ACISP 2010, LNCS 6168, pp. 164–181, 2010.
© Springer-Verlag Berlin Heidelberg 2010

schemes do not exist for general relations with respect to black-box simulators that cannot access a decryption oracle (i.e. non-assisted simulators).

Complete non-malleability has recently been shown to be equivalent to *indistinguishability under strong chosen-ciphertext attacks* [2,1]. This model enhances the adversary's capabilities to forge public keys and ask the decryption oracle to provide decryptions under the corresponding (possibly unknown) secret keys. It was also shown that it is possible to construct efficient completely non-malleable schemes using the strong chosen-ciphertext attack model, which is more convenient than performing the proof in the original simulation-based definition. Unfortunately, the equivalence result connecting strong chosen-ciphertext security to complete non-malleability holds only for simulators *assisted* by a strong decryption oracle. It therefore remains an open problem to construct efficient schemes that achieve complete non-malleability in the strongest sense.

The impossibility result from [18] dictates that to construct a scheme that achieves complete non-malleability with respect to *non-assisted* simulators, one must resort to a non-black-box simulator. In this paper we consider extractor-based properties that allow us to gain insight into the design of completely non-malleable schemes and provide a technique to go beyond known feasibility results in this area. We formalise *strong plaintext awareness* and *secret key awareness* and prove their suitability in realising these goals. We show that if such properties are realisable, and one considers non-black-box simulators, then the impossibility result established for non-assisted simulators no longer holds. We also look at how such notions can be realised with and without random oracles.

STRONG PLAINTEXT AWARENESS. Plaintext awareness formalises the intuition that one can achieve security under chosen-ciphertext attacks by making it infeasible to construct a valid ciphertext without *knowing*, a priori, the message hidden inside it. In fact, it has been shown that the combination of plaintext awareness and semantic security is enough to achieve chosen-ciphertext security [5]. We formulate a natural strengthening of plaintext awareness that requires the existence of a strong plaintext extractor that decrypts ciphertexts, even if they are encrypted under adversarially generated public keys. We prove a fundamental theorem according to which a *strongly plaintext-aware* (SPA) and IND-CPA secure scheme also withstands strong chosen-ciphertext attacks[1]. This implies, through the results in [2], that such a scheme is also completely non-malleable with respect to assisted simulators. We extend this result by showing that strong plaintext awareness allows us to directly build *non-assisted* simulators. The resulting simulators depend on the adversary and hence they are not black-box. This permits going around the impossibility result established by Fischlin [18]. Furthermore, strong plaintext awareness generalises a proof technique used by Fischlin to demonstrate that (a slightly modified version of) RSA-OAEP is completely non-malleable for non-assisted simulators[2].

---

[1] This result also has applications in certificateless encryption [1].

[2] A corollary of this result is that we obtain a new perspective on the *standard* notion of plaintext awareness. Indeed, a similar proof strategy can be used to construct non-assisted simulators for standard non-malleability.

SECRET KEY AWARENESS. We also propose a new extractor-based security definition that takes a different perspective on how to achieve strong plaintext awareness and complete non-malleability. Roughly speaking, this notion that we call *secret key awareness* (SKA), requires it to be infeasible to generate new valid public keys without knowing their corresponding secret keys. It therefore looks at enhancing the security of key-generation mechanisms. We show that an encryption scheme that is secret key aware and IND-CCA2 is also secure under strong chosen-ciphertext attacks, and therefore completely non-malleable. We derive this result via a stronger indistinguishability security notion, where the adversary has access to a public key inversion oracle[3]. Furthermore, we prove that secret key awareness, together with standard plaintext awareness, implies strong plaintext awareness. Hence, secret key awareness provides all of the benefits of strong plaintext awareness. Additionally, secret key awareness permits the construction of a complete non-malleability simulator that *opens* the secret key associated with the public created by the adversary. This is particularly relevant when the scheme is used in commitment schemes, where to de-commit one reveals a secret key rather a message/randomness pair. Strong plaintext awareness in not sufficient to open a ciphertext in the sense of de-commitment, as it does not guarantee knowledge of the *randomness* used in encryption.

SCHEMES. We propose a generic transformation that permits transforming any IND-CCA2 scheme into a secret key aware (and still IND-CCA2) scheme in the random oracle model. The resulting schemes are therefore completely non-malleable for non-assisted simulators. We also take first steps towards building fully secret-key-aware schemes without random oracles. We propose a *generic* construction inspired in escrow public-key encryption [12], relying on schemes whose key-generation routines themselves operate as an encryption scheme. We are, however, unable to instantiate this scheme and leave it as an interesting open problem. Next, we move to specific constructions based on knowledge assumptions. A natural candidate for building a secret key aware scheme is the Diffie-Hellman Knowledge assumption [5]. This approach, however, fails once we notice that secret key awareness allows adaptive attacks on the public keys, whereas Diffie-Hellman tuples are malleable. We therefore introduce a new knowledge assumption stating, roughly speaking, that it is impossible to compute integers of the form P^2Q, where P and Q are prime, without knowing the factors and even if provided with another integer of this form. This assumption can be used to demonstrate that variants of RSA satisfy weak forms of secret key awareness.

ORGANISATION. We first review related work. Then, in Section 2 we settle notation and recall the syntax for public-key encryption schemes. We also recall the definition of strong decryption oracles and IND-SCCA2 security. In the same section we also introduce invert and chosen-ciphertext attacks, which we will use later on in the paper. In Section 3 we introduce our extractor-based notions. In Section 4 we discuss constructions of secret key aware schemes.

---

[3] This type of oracle has been shown to have numerous applications in the context of adaptive one-way functions [21].

1.1 Related Work

Plaintext awareness was originally formulated by Bellare and Rogaway [7] in the random oracle model. Later, Bellare and Palacio [5] gave the first definition of PA in the standard model. It is well known [5,4] that plaintext awareness together with IND-CPA imply a level of security that is strictly stronger than IND-CCA2. The authors in [24] showed that plaintext awareness is an "all-or-nothing property" in the sense that *one-wayness* (or even a weaker condition called non-triviality) together with PA2 plaintext awareness is enough to guarantee IND-CCA2 security. Birkett and Dent [11] settled the relations between notions of plaintext awareness from [16], and showed that schemes with infinite message spaces that are plaintext-aware and one-way do not exist using techniques from [24].

Non-malleability (as a general notion) was originally introduced in the seminal work of Dolev, Dwork, and Naor [17]. In order to establish relations with other notions of security, non-malleability for public-key encryption was reformulated by Bellare et al. [4] as a comparison-based security model. Bellare and Sahai [10,9] later fully established the relation between this comparison-based definition and the original simulation-based definition of Dolev et al. Pass, Shelat, and Vaikuntanathan [22] provide a full characterisation of non-malleability, identifying some shortcomings in previous results and considering their robustness under a form of composition where the adversary is provided with a polynomial number of challenge ciphertexts.

Complete non-malleability, was proposed by Fischlin [18]. Here the adversary is allowed to choose the public key under which the target ciphertext is produced. The same author presented impossibility results as to the construction of completely non-malleable schemes with respect to black-box simulators and general relations, and showed that a modified version of RSA-OAEP is completely non-malleable in the random oracle model. Visconti and Ventre [26] proposed a comparison-based definition of complete non-malleability, studied its relation with the simulation-based definition of Fischlin, and also gave a generic construction of completely non-malleable schemes based on NIZK-PoK. The authors in [2] define strong decryption oracles, use this to introduce indistinguishability under strong chosen-ciphertext attacks and establish relations with assisted simulation-based and comparison-based complete non-malleability. A practical and strongly secure scheme (without random oracles) based on the decisional bilinear Diffie-Hellman problem is also given.

Adaptive one-way functions [21], where an adversary has access to an inversion oracle, and extractable one-way functions [13,14], where one requires knowledge of pre-image, have been recently proposed. Secret key awareness can be seen as a refinement of these notions for public-key encryption.

2 Preliminaries

NOTATION. We write $x \leftarrow y$ for assigning value y to variable x, and $x \leftarrow_\$ X$ for sampling x from set X uniformly at random. If X is empty, we set $x \leftarrow \perp$, where $\perp \notin \{0,1\}^\star$ is a special failure symbol. If A is a probabilistic algorithm we write $x \leftarrow_\$ A(I_1, I_2, \ldots)$ for the action of running A on inputs $I_1, I_2, \ldots$ with random coins chosen uniformly at random, and assigning the result to x. When A is run on specific coins r, we write $x \leftarrow A(I_1, I_2, \ldots; r)$. We denote boolean values, namely the output of checking whether a relation holds, by T (true) and F (false). For a space $\mathsf{Sp} \subseteq \{0,1\}^\star$, we identify Sp with its characteristic function. In other words, $\mathsf{Sp}(s) = \mathsf{T}$ if and only if $s \in \mathsf{Sp}$. The function $\mathsf{Sp}(\cdot)$ always exists, although it may not be computable in polynomial time. We say s is valid with respect to Sp if and only if $\mathsf{Sp}(s) = \mathsf{T}$. When this is clear from the context, we also use Sp for sampling uniformly from Sp. Unless stated otherwise, the range of a variable s is assumed to be $\{0,1\}^\star$. The symbol : is used for appending an element to a list, and we indicate vectors using bold-faced font. We say $f(\lambda)$ is negligible if $f(\lambda) \in \cap_{c \in \mathbb{N}} O(\lambda^{-c})$.

GAMES. In this paper we will be using code-based game-playing [8]. Each game has an **Initialize** and a **Finalize** procedure. It also has specifications of procedures to respond to an adversary's various oracle queries. A game Game is run with an adversary $\mathcal{A}$ as follows. First **Initialize** runs and its outputs are passed to $\mathcal{A}$. Then $\mathcal{A}$ runs and its oracle queries are answered by the procedures of Game. These procedures return $\perp$ if queried on $\perp$. When $\mathcal{A}$ terminates, its output is passed to **Finalize** which returns the outcome of the game y. This interaction is written as $\mathsf{Game}^{\mathcal{A}} \Rightarrow y$. In each game, we restrict attention to *legitimate* adversaries. Legitimacy is defined specifically for each game. All algorithms (adversaries, extractors and plaintext/public-key creators) are assumed to run in probabilistic polynomial time (PPT).

PUBLIC-KEY ENCRYPTION. We adopt the standard multi-user syntax with the extra Setup algorithm [3], which we believe is the most natural one for security models involving multiple public keys. A public-key encryption scheme $\Pi = (\mathsf{Setup}, \mathsf{Gen}, \mathsf{MsgSp}, \mathsf{Enc}, \mathsf{Dec})$ is specified by five polynomial-time algorithms (in the length of their inputs) as follows. Setup is the probabilistic setup algorithm which takes as input the security parameter 1^λ and returns the common domain parameter[4] I. $\mathsf{Gen}(\mathsf{I})$ is the probabilistic key-generation algorithm. On input global parameters I, this algorithm returns a secret key SK and a matching public key PK. Algorithm $\mathsf{MsgSp}(\mathsf{m}, \mathsf{PK})$ is a deterministic message space recognition algorithm. On input m and PK this algorithm returns T or F. $\mathsf{Enc}(\mathsf{m}, \mathsf{PK}; r)$ is the probabilistic encryption algorithm. On input a message m, a public key PK, and possibly some random coins r, this algorithm outputs a ciphertext c or a special failure symbol $\perp$. Finally, $\mathsf{Dec}(\mathsf{c}, \mathsf{SK}, \mathsf{PK})$ is the deterministic decryption algorithm. On input of a ciphertext c and keys SK and PK, this algorithm outputs a message m or a special failure symbol $\perp$. The correctness of a public-key encryption scheme requires that for any $\mathsf{I} \leftarrow_\$ \mathsf{Setup}(1^\lambda)$, any $(\mathsf{SK}, \mathsf{PK}) \leftarrow_\$ \mathsf{Gen}()$ and all $\mathsf{m} \in \mathsf{MsgSp}(\mathsf{PK})$ we have $\Pr[\mathsf{Dec}(\mathsf{Enc}(\mathsf{m}, \mathsf{PK}), \mathsf{SK}, \mathsf{PK}) = \mathsf{m}] = 1$.

---

[4] Although all algorithms are parameterised by I, we often omit I as an explicit input for readability. Furthermore, we assume that the security parameter is included in I.

REMARK. We note that the multi-user syntax permits capturing in the same framework schemes that execute in the standard model, in which case the global parameters are empty; and also schemes which execute in the Common Reference String (CRS) model. All the relations that we establish between the different models apply to both cases.

VALIDITY CHECKING ALGORITHMS. The following spaces (and associated functions) will be used throughout the paper. All of these spaces are parameterised by l and are subsets of $\{0, 1\}^*$.

$$\mathsf{MsgSp(PK)} := \{\mathsf{m} : \mathsf{MsgSp(m, PK)}\}$$
$$\mathsf{KeySp} := \{(\mathsf{SK, PK}) : \exists r \; (\mathsf{SK, PK}) = \mathsf{Gen}(r)\}$$

We assume throughout the paper that the encryption and decryption algorithms check if $\mathsf{m} \in \mathsf{MsgSp(PK)}$ and return $\perp$ if it does not hold. Often the algorithm MsgSp does not depend on PK in the sense that for any two valid public keys PK and PK' and any $\mathsf{m} \in \{0, 1\}^*$ we have $\mathsf{MsgSp(m, PK)} = \mathsf{MsgSp(m, PK')}$. For general schemes, one can consider the infinite message space $\mathsf{MsgSp(PK)} = \{0, 1\}^*$ case. However, given that in this paper we will often consider the set of all valid messages and sample from it, we restrict our attention to schemes with finite message spaces. As pointed out by Pass et al. [22], this means that to avoid degenerate cases we must also restrict our attention to schemes for which all the elements in the range of decryption can be efficiently encrypted[5], including the special failure symbol $\perp$. We also assume that the key-pair validity algorithm KeySp is polynomial-time and require that decryption returns $\perp$ if this check fails on the key-pair passed to it. We also assume various algorithms check for structural properties such as correct encoding, membership in a group, etc.

2.1   Strong Chosen-Ciphertext Security

The idea behind a strong chosen-ciphertext attack is to give the adversary access to an oracle that decrypts ciphertexts of the adversary's choice with respect to arbitrary public keys.

We follow [2] and adopt a generic definition of strong decryption as shown in Figure 1. The oracle proceeds in three steps. The first step models the general

---

> **proc. SDec$_{\mathsf{U,V}}$(c, PK, R):**
> $\mathsf{WitSp} \leftarrow \{(\mathsf{m}, r) : \mathsf{V(c, PK, m}, r, \mathsf{st[V]})\}$
> $(\mathsf{m}, r) \leftarrow_\$ \{(\mathsf{m}, r) \in \mathsf{WitSp} : \mathsf{R(m)}\}$
> $\mathsf{st[V]} \leftarrow \mathsf{U(c, PK, R, m}, r, \mathsf{st[V]})$
> Return m

Fig. 1. Generic definition of a strong decryption oracle

---

[5] This can be easily achieved for schemes used in practice.

procedure of constructing a set of candidate (valid) decryption results[6]. The second step consists of choosing one of these candidate solutions to return to the adversary. The final step updates the state of the oracle, if it keeps one[7]. As discussed in [2] the motivation for having such a general definition is that the notion of *the message encapsulated by the ciphertext* can be defined in a number of ways, depending on the witnesses that are taken to assess the validity of the public key/ciphertext pair. For example, one can define validity via the encryption operation, in which case a message/randomness pair is the witness

$$V(c, PK, m, r) := c \overset{?}{=} Enc(m, PK; r), \tag{1}$$

or via the decryption algorithm, where a message/secret key pair is the witness

$$V'(c, PK, m, r) := (SK, PK) \overset{?}{=} Gen(r) \wedge m \overset{?}{=} Dec(c, SK, PK). \tag{2}$$

Note that neither criterion guarantees that, if a message is found to be a valid decryption result, then it will be unique. This justifies the need for the second step in the definition we propose: there could be many valid decryption results to choose from, and it is left to the adversary to control how this is done by providing a relation R on messages as input to the oracle. For a well-defined and broad class of schemes [2], this general definition collapses into a much simpler one. However, we follow this approach for the sake of generality.

We now present the definition of ciphertext indistinguishability under strong chosen-ciphertext attacks, introduced in [2] as the natural extension of the standard notion of security for public-key encryption schemes. The IND-SCCAx advantage of an adversary $\mathcal{A}$ for $x = 0, 1, 2$ against Π is defined by

$$\mathbf{Adv}_{\Pi,\mathcal{A}}^{\text{ind-sccax}}(\lambda) := 2 \cdot \Pr\left[\text{IND-SCCAx}_{\Pi}^{\mathcal{A}}(\lambda) \Rightarrow T\right] - 1,$$

where game IND-SCCAx is shown in Figure 2. Implicit in this definition are the descriptions of the U and V algorithms, which are fixed when analysing a scheme in the resulting IND-SCCAx model. We say a scheme is IND-SCCAx secure if the advantage of any PPT adversary is negligible.

2.2 Security under Invert and Chosen-Ciphertext Attacks

We introduce a new security model for encryption that helps us clarify the relations among notions we establish later on. In this model, the adversary has access to an oracle that, given a public key generated by the adversary, provides it with the corresponding secret key. Figure 4 presents the general form of the **Inv** procedure, which is analogous to the **SDec** procedure presented in the previous section. When many secret keys satisfy the validity criterion, the adversary is

---

[6] Search for messages is over sufficiently long bit strings together with the special symbol $\bot$. Search for random coins is over sufficiently long bit strings.

[7] The state is initialized to some value st_0. A natural use of the state is to make sure that decryption is consistent in different calls.

proc. Initialize(λ):	**proc. LoR**(m_0, m_1):	Game IND-SCCAx$_\Pi$(λ)
$b \leftarrow_\$ \{0,1\}$; $I \leftarrow_\$ \mathsf{Setup}(1^\lambda)$	$c \leftarrow_\$ \mathsf{Enc}(m_b, \mathsf{PK}^\star)$	
$(\mathsf{SK}^\star, \mathsf{PK}^\star) \leftarrow_\$ \mathsf{Gen}()$	$\mathsf{List} \leftarrow (c, \mathsf{PK}^\star) : \mathsf{List}$	
$\mathsf{List} \leftarrow []$; $\mathsf{st}[\mathsf{V}] \leftarrow \mathsf{st}_0$	Return c	
Return $(I, \mathsf{PK}^\star)$		
	proc. SDec($c, \mathsf{PK}, \mathsf{R}$):	**proc. Finalize**(b'):
	Return $\mathsf{SDec}_{\mathsf{U,V}}(c, \mathsf{PK}, \mathsf{R})$	Return $(b' = b)$

Fig. 2. Game defining indistinguishability under strong chosen-ciphertext attacks. An adversary $\mathcal{A}$ is legitimate if: 1) It calls **LoR** only once with $m_0, m_1 \in \mathsf{MsgSp}(\mathsf{PK})$ such that $|m_0| = |m_1|$; and 2) R is polynomial-time and, if $x = 0$ it does not call **SDec**, if $x = 1$ it does not call **SDec** after calling **LoR**, and if $x = 2$ it does not call **SDec** with a tuple (c, PK) in List.

proc. Initialize(λ):	**proc. LoR**(m_0, m_1):	Game IND-ICAx$_\Pi$(λ)
$b \leftarrow_\$ \{0,1\}$; $I \leftarrow_\$ \mathsf{Setup}(1^\lambda)$	$c \leftarrow_\$ \mathsf{Enc}(m_b, \mathsf{PK}^\star)$	
$(\mathsf{SK}^\star, \mathsf{PK}^\star) \leftarrow_\$ \mathsf{Gen}()$	$\mathsf{List} \leftarrow (c, \mathsf{PK}^\star) : \mathsf{List}$	**proc. Inv**(PK, R):
$\mathsf{List} \leftarrow []$; $\mathsf{st}[\mathsf{V}] \leftarrow \mathsf{st}_0$	Return c	Return $\mathsf{Inv}_{\mathsf{U,V}}(\mathsf{PK}, \mathsf{R})$
Return $(I, \mathsf{PK}^\star)$		
	proc. Dec(c):	**proc. Finalize**(b'):
	Return $\mathsf{Dec}(c, \mathsf{SK}^\star, \mathsf{PK}^\star)$	Return $(b' = b)$

Fig. 3. Game defining indistinguishability under invert and chosen-ciphertext attacks. An adversary $\mathcal{A}$ is legitimate if: 1) It calls **LoR** only once with $m_0, m_1 \in \mathsf{MsgSp}(\mathsf{PK})$ such that $|m_0| = |m_1|$; 2) R is polynomial-time and, if $x = 0$ it does not call **Dec** or **Inv**, if $x = 1$ it does not call **Dec** or **Inv** after calling **LoR**, and if $x = 2$ it does not call **Dec** with a c in List; and 3) It does not call **Inv** on $\mathsf{PK}^\star$.

proc. Inv$_{\mathsf{U,V}}$(PK, R):
$\mathsf{WitSp} \leftarrow \{(\mathsf{SK}, r) : \mathsf{V}(\mathsf{PK}, \mathsf{SK}, r, \mathsf{st}[\mathsf{V}])\}$
$(\mathsf{SK}, r) \leftarrow_\$ \{(\mathsf{SK}, r) \in \mathsf{WitSp} : \mathsf{R}(\mathsf{SK})\}$
$\mathsf{st}[\mathsf{V}] \leftarrow \mathsf{U}(\mathsf{PK}, \mathsf{R}, \mathsf{SK}, r, \mathsf{st}[\mathsf{V}])$
Return SK

Fig. 4. Generic definition of an invert oracle

also allowed to restrict the set of "interesting" secret keys from which the answer is sampled by providing a relation R on secret keys as input to the oracle. A natural validity criteria for this oracle is

$$\mathsf{V}(\mathsf{PK}, \mathsf{SK}, r) := (\mathsf{SK}, \mathsf{PK}) \overset{?}{=} \mathsf{Gen}(r)$$

accepting all key-pairs that may be output by the key-generation algorithm[8].

---

[8] Another validity criteria, corresponding to a natural stateful invert oracle, will ensure that repeat queries will be answered consistently.

algorithm $V'(c, PK, m, r, st[V'])$:	algorithm $U'(c, PK, R, m, r, st[V'])$:
$(st[V], (SK^*, PK^*)) \leftarrow st[V']$	$(st[V], (SK^*, PK^*)) \leftarrow st[V']$
If $PK = PK^*$	$(SK, r') \leftarrow r$
If $m = Dec(c, SK^*, PK^*)$	$R'(SK) := R(Dec(c, SK, PK))$
Return T Else Return F	$st[V] \leftarrow U(PK, R', SK, r', st[V])$
$(SK, r') \leftarrow r$	Return $(st[V], (SK^*, PK^*))$
If $V(PK, SK, r', st[V]) \wedge m = Dec(c, SK, PK)$	
Return T Else Return F	

Fig. 5. U' and V' for $\mathbf{SDec}_{U',V'}$ corresponding to $\mathbf{Inv}_{U,V}$ with $st_0 = (SK^*, PK^*)$

We define the IND-ICAx advantage of an adversary $\mathcal{A}$ for $x = 0, 1, 2$ against encryption scheme Π is defined by

$$\mathbf{Adv}_{\Pi,\mathcal{A}}^{\text{ind-icax}}(\lambda) := 2 \cdot \Pr\left[\text{IND-ICAx}_{\Pi}^{\mathcal{A}}(\lambda) \Rightarrow T\right] - 1,$$

where Game IND-ICAx is shown in Figure 3. We say a scheme is IND-ICAx secure if the advantage of any PPT adversary is negligible. This new model can be related to IND-SCCAx as follows. For a given $\mathbf{Inv}_{U,V}$ procedure, define the associated $\mathbf{SDec}_{U',V'}$ procedure through the algorithms shown in Figure 5. The following theorem shows that security under each possible definition of an invert oracle implies IND-SCCAx security under a well-defined version of the strong decryption oracle.

Theorem 1 (IND-ICAx $\Rightarrow$ IND-SCCAx). *Let $\mathcal{A}$ be an IND-SCCAx adversary against encryption scheme Π with respect to $\mathbf{SDec}_{U',V'}$ associated to $\mathbf{Inv}_{U,V}$ as defined in Figure 5. Then there exists an IND-ICAx adversary $\mathcal{A}_1$ against Π with at least the same advantage as that of $\mathcal{A}$.*

The reduction is constructed by simulating the strong decryption oracle using both the standard decryption oracle (for queries under the challenge public key) and the invert oracle (for adversarially chosen-keys) available in the IND-ICAx game. The details are given in the full version of the paper. The interesting part of the proof is an argument showing that this simulation fits into the generic structure of $\mathbf{SDec}$ given in Figure 1 and, in particular, that the effect of the relation R passed to the strong decryption oracle can be emulated through a relation R' passed to the invert oracle. The intuition here is that the strong decryption oracle associated with a particular invert oracle maps the relation R that allows the adversary to restrict the set of interesting messages onto a relation R' that selects the set of secret keys that decrypt the queried ciphertext into the same set of interesting messages. Technically, the relation $R'_{c,PK}(SK) := R(Dec(c, SK, PK))$ allows us to simulate the oracle in Figure 5 with the correct distribution.

3   Extractor-Based Properties

The strong chosen-ciphertext security model that was recalled in Section 2 suggests that any secure scheme under this definition must ensure, by construction, that strong decryption queries are of no help to the adversary even when the associated public keys are chosen adaptively. Plaintext awareness [5] formalises this intuition when a standard decryption oracle is used (and a public key is fixed). We therefore propose strong plaintext awareness as a natural extension for strong security models. This notion, however, is not the only way to render strong decryption oracles ineffective. An alternative approach is to require that any adversary which outputs a new valid public key must know a valid secret key for it. We refer to this property as secret key awareness. In the next two subsections we formalise these extractor-based notions precisely and demonstrate their adequacy for the security analysis of completely non-malleable schemes.

3.1 Strong Plaintext Awareness

We follow the approach adapted in [5] to define strong plaintext awareness in the standard model. We run an adversary in two possible environments and require that its behaviour does not change in any significant way. In the first world, the adversary has access to a real strong decryption oracle while in the second the oracle executes a polynomial-time extractor. Furthermore, in these environments, the adversary may obtain ciphertexts on "unknown-but-controlled" plaintexts though an encryption oracle, fed with messages produced by a plaintext creator. More formally, the SPAx advantage of an adversary, for $x = 1, 2$, against encryption scheme Π with respect to plaintext creator $\mathcal{P}$ (mapping bit strings to messages), strong plaintext extractor $\mathcal{K}$, and distinguisher $\mathcal{D}$, is defined by

$$\mathbf{Adv}^{\text{spax}}_{\Pi,\mathcal{P},\mathcal{D},\mathcal{K},\mathcal{A}}(\lambda) := \Pr\left[\text{Dec-SPAx}^{\mathcal{A}}_{\Pi,\mathcal{P},\mathcal{D}}(\lambda) \Rightarrow \mathsf{T}\right] - \Pr\left[\text{Ext-SPAx}^{\mathcal{A}}_{\Pi,\mathcal{P},\mathcal{D},\mathcal{K}}(\lambda) \Rightarrow \mathsf{T}\right]$$

where games Dec-SPAx and Ext-SPAx are shown in Figure 6. We say a scheme is SPAx if, for every PPT adversary $\mathcal{A}$, there exists an efficient strong plaintext extractor $\mathcal{K}$ such that, for all distinguishers[9] and plaintext creators, advantage is negligible.

The next theorem, which is proved in the full version of the paper, shows that the above formulation of strong plaintext-awareness, together with semantic security is enough to achieve strong chosen-ciphertext security.

Theorem 2 (SPAx ∧ IND-CPA ⇒ IND-SCCAx). *Fix a definition of* $\mathbf{SDec}_{\mathsf{U,V}}$ *and let* $\mathcal{A}$ *be an* IND-SCCAx *adversary against* Π *in the resulting model. Then there exist an* SPAx *ciphertext creator* $\mathcal{A}_1$, *an* IND-CPA *adversary* $\mathcal{A}_2$, *plaintext creators* $\mathcal{P}_0$, $\mathcal{P}_1$, *and distinguishers* $\mathcal{D}_0$, $\mathcal{D}_1$ *such that*

$$\mathbf{Adv}^{\text{ind-sccax}}_{\Pi,\mathcal{A}}(\lambda) \leq \mathbf{Adv}^{\text{spax}}_{\Pi,\mathcal{P}_0,\mathcal{D}_0,\mathcal{K},\mathcal{A}_1}(\lambda) + \mathbf{Adv}^{\text{spax}}_{\Pi,\mathcal{P}_1,\mathcal{D}_1,\mathcal{K},\mathcal{A}_1}(\lambda) + \mathbf{Adv}^{\text{ind-cpa}}_{\Pi,\mathcal{A}_2}(\lambda),$$

where $\mathcal{K}$ *is the plaintext extractor for* $\mathcal{A}_1$ *implied by the* SPAx *property of* Π.

As we mentioned in the introduction, the equivalence between indistinguishability under strong chosen-ciphertext security and simulation-based complete non-malleability is established for *assisted* simulators [2]. The next theorem shows

---

[9] If unbounded distinguishers are allowed, we get statistical strong plaintext awareness.

---

proc. **Initialize**(λ):   Game Dec-SPAx$_{\Pi,\mathcal{P},\mathcal{D}}(\lambda)$
I $\leftarrow_\$$ Setup(1^λ)
(SK*, PK*) $\leftarrow_\$$ Gen()   proc. **Enc**(Q):
Choose coins Rnd[$\mathcal{A}$] for $\mathcal{A}$ (m, st[$\mathcal{P}$]) $\leftarrow_\$$ $\mathcal{P}(Q,$ st[$\mathcal{P}$])
st[$\mathcal{P}$] $\leftarrow \epsilon$; List $\leftarrow$ []; st[V] $\leftarrow$ st$_0$ c $\leftarrow_\$$ Enc(m, PK*)
Return (I, PK*, Rnd[$\mathcal{A}$])   List $\leftarrow$ (c, PK*) : List
                                                                          Return c
proc. **SDec**(c, PK, R):
Return **SDec**$_{U,V}$(c, PK, R)   proc. **Finalize**(x):
                                                                          Return $\mathcal{D}(x)$

---

proc. **Initialize**(λ):   Game Ext-SPAx$_{\Pi,\mathcal{P},\mathcal{D},\mathcal{K}}(\lambda)$
I $\leftarrow_\$$ Setup(1^λ)
(SK*, PK*) $\leftarrow_\$$ Gen()   proc. **Enc**(Q):
Choose coins Rnd[$\mathcal{A}$] for $\mathcal{A}$; List $\leftarrow$ []   (m, st[$\mathcal{P}$]) $\leftarrow_\$$ $\mathcal{P}(Q,$ st[$\mathcal{P}$])
st[$\mathcal{P}$] $\leftarrow \epsilon$; st[$\mathcal{K}$] $\leftarrow$ (I, PK*, Rnd[$\mathcal{A}$])   c $\leftarrow_\$$ Enc(m, PK*)
Return (I, PK*, Rnd[$\mathcal{A}$])   List $\leftarrow$ (c, PK*) : List
                                                                          Return c
proc. **SDec**(c, PK, R):
(m, st[$\mathcal{K}$]) $\leftarrow_\$$ $\mathcal{K}$(c, PK, R, List, st[$\mathcal{K}$])   proc. **Finalize**(x):
Return m   Return $\mathcal{D}(x)$

Fig. 6. The Dec-SPAx and Ext-SPAx games for defining the strong plaintext-awareness of encryption scheme Π. An adversary $\mathcal{A}$ is legitimate if: 1) R is polynomial-time and if x = 1 it never calls **Enc**; and 2) It never calls **SDec** with a tuple (c, PK) in List.

that using strong plaintext awareness one can strengthen this result to non-assisted simulators[10].

Theorem 3 (SPAx $\wedge$ SNM-CPA $\Rightarrow$ **Non-Assisted** SNM-SCCAx). *Fix a definition of* **SDec**$_{U,V}$ *and let* $\mathcal{A}$ *be a* Real-SNM-SCCAx *adversary against* Π*. Then there exist an* SPAx *ciphertext creator* $\mathcal{A}_1$*, a* Real-SNM-CPA *adversary* $\mathcal{A}_2$*, a plaintext creator* $\mathcal{P}$*, a distinguisher* $\mathcal{D}$*, and a (non-assisted) simulator* $\mathcal{S}$ *such that for all* R

$$\mathbf{Adv}^{\mathsf{snm\text{-}sccax}}_{\Pi,R,\mathcal{S},\mathcal{A}}(\lambda) \leq \mathbf{Adv}^{\mathsf{spax}}_{\Pi,\mathcal{P},\mathcal{D},\mathcal{K},\mathcal{A}_1}(\lambda) + \mathbf{Adv}^{\mathsf{snm\text{-}cpa}}_{\Pi,R,\mathcal{S}_2,\mathcal{A}_2}(\lambda),$$

where $\mathcal{K}$ *is the strong plaintext extractor for* $\mathcal{A}_1$ *implied by the* SPAx *property of* Π *and* $\mathcal{S}_2 = \mathcal{S}$ *is the simulator for* $\mathcal{A}_2$ *implied by the* SNM-CPA *security of* Π*.*

The proof of this theorem, included in the full version of the paper, proceeds in a different way than that in [10] for standard security models. There, a new key-pair is generated to enable the simulator to answer decryption queries, whereas in our proof this is not necessary. As pointed out by Pass et al. [22], the proof in [10] relies on the existence of an algorithm for efficiently encrypting all possible outputs of decryption, including special symbol $\perp$. Plaintext awareness in general does not imply that this must be the case, and so our results extend the results

---

[10] Due to space constraints, we refer the interested reader to [2] for the SNM definitions.

in [22]: schemes that do not have the property identified in [22] may still be plaintext aware, and therefore achieve simulation-based non-malleability for non-assisted simulators. However, as shown in [11, Theorem 2], if an encryption scheme is PA2 and has an infinite message space, then it is not OW-CPA. This also applies to strong plaintext awareness, and hence no scheme with an infinite message space will be captured by the above theorem.

REMARK. In the above theorem we do not need to restrict the class of relations, in particular to those which are independent of the challenge public key (called lacking relations in [26]). This means that through strong plaintext awareness one can improve on the results in [26], where this security level can only be achieved at the cost of relation being independent of the common parameters.

REMARK. Using the techniques introduced by Dent [16], the scheme in [2] might satisfy strong plaintext awareness under an appropriate (bilinear) Diffie-Hellman knowledge assumptions. We leave this and constructing a strongly plaintext-aware scheme in the standard model as an open problem.

3.2   Secret Key Awareness

We now formalise secret key awareness as an alternative route to achieve strong security guarantees. We take a similar approach to plaintext awareness and give an adversary access to an oracle which is either a real *inversion* oracle (as defined in Section 2) or one which uses a polynomial-time secret key extractor. Once again, our requirement is that the behaviour of the adversary is computationally indistinguishable in the two environments. We also provide the adversary with a decryption and a controlled encryption oracle which model the extra auxiliary information that might be useful in producing a new public key. Formally, the SKAx advantage of an adversary $\mathcal{A}$ against encryption scheme Π with respect to secret key extractor $\mathcal{K}$, plaintext creator $\mathcal{P}$, and distinguisher $\mathcal{D}$ is defined by

$$\mathbf{Adv}^{\mathsf{skax}}_{\Pi,\mathcal{P},\mathcal{D},\mathcal{K},\mathcal{A}}(\lambda) := \Pr\left[\mathsf{Inv\text{-}SKAx}^{\mathcal{A}}_{\Pi,\mathcal{P},\mathcal{D}}(\lambda) \Rightarrow \mathsf{T}\right] - \Pr\left[\mathsf{Ext\text{-}SKAx}^{\mathcal{A}}_{\Pi,\mathcal{P},\mathcal{D},\mathcal{K}}(\lambda) \Rightarrow \mathsf{T}\right]$$

where games Inv-SKA and Ext-SKA are shown in Figure 7. We say a scheme is SKAx secure if, for every PPT adversary $\mathcal{A}$, there exists an efficient secret key extractor $\mathcal{K}$ such that, for all distinguishers[11] and plaintext creators, advantage is negligible.

We are now ready to state the main theorem of this section, which permits concluding that secret key awareness combined with IND-CCA2 is strong enough to guarantee IND-SCCA2 security. The proof is analogous to that of Theorem 2 and is included in the full version of the paper.

Theorem 4 (SKAx $\wedge$ IND-CCAx $\Rightarrow$ IND-ICAx). *Fix a definition of* $\mathbf{Inv}_{\mathsf{U},\mathsf{V}}$ *and let* $\mathcal{A}$ *be an* IND-ICAx *adversary against* Π. *Then, there exist an* SKAx *public key creator* $\mathcal{A}_1$, *an* IND-CCAx *adversary* $\mathcal{A}_2$, *plaintext creators* $\mathcal{P}_0$, $\mathcal{P}_1$, *and distinguishers* $\mathcal{D}_0$, $\mathcal{D}_1$ *such that*

---

[11] If unbounded distinguishers are allowed, we get statistical secret key awareness.

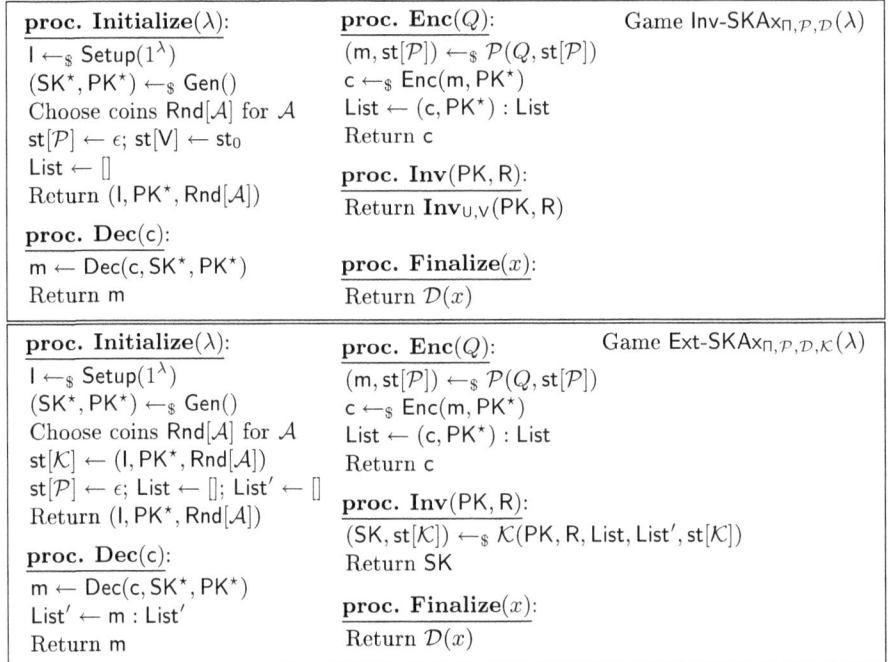

Fig. 7. The Inv-SKAx and Ext-SKAx games for defining secret key awareness. An adversary $\mathcal{A}$ is legitimate if: 1) R is polynomial-time and, if $x = 0$ it never calls **Dec** or **Enc** and if $x = 1$ it never calls **Enc**; 2) It never queries PK^* to **Inv**; and 3) It never calls **Dec** with a ciphertext c such that $(c, PK^*) \in$ List.

$$\mathbf{Adv}^{\mathsf{ind\text{-}icax}}_{\Pi,\mathcal{A}}(\lambda) \leq \mathbf{Adv}^{\mathsf{skax}}_{\Pi,\mathcal{P}_0,\mathcal{D}_0,\mathcal{K},\mathcal{A}_1}(\lambda) + \mathbf{Adv}^{\mathsf{skax}}_{\Pi,\mathcal{P}_1,\mathcal{D}_1,\mathcal{K},\mathcal{A}_1}(\lambda) + \mathbf{Adv}^{\mathsf{ind\text{-}ccax}}_{\Pi,\mathcal{A}_2}(\lambda),$$

where $\mathcal{K}$ is the secret key extractor for $\mathcal{A}_1$ implied by the SKAx property of Π.

To further justify the definition of secret key awareness, we show that it can be used to achieve strong plaintext awareness. The next theorem, proved in in the full version of the paper, states that secret key awareness combined with standard plaintext awareness gives rise to strong plaintext awareness.

Theorem 5 (SKAx $\wedge$ PAx $\Rightarrow$ SPAx). *Fix a definition of* $\mathbf{Inv}_{U,V}$ *and let* $\mathcal{A}$ *be an SPAx ciphertext creator against* Π, *with respect to the* $\mathbf{SDec}_{U',V'}$ *procedure associated to* $\mathbf{Inv}_{U,V}$ *as defined in Figure 5. Then there exists an SKAx public key creator* $\mathcal{A}_1$, *a PAx ciphertext creator* $\mathcal{A}_2$, *and an SPAx plaintext extractor* $\mathcal{K}$ *such that for any plaintext creator* $\mathcal{P}$, *and any distinguisher* $\mathcal{D}$ *we have*

$$\mathbf{Adv}^{\mathsf{spax}}_{\Pi,\mathcal{P},\mathcal{D},\mathcal{K},\mathcal{A}}(\lambda) \leq \mathbf{Adv}^{\mathsf{skax}}_{\Pi,\mathcal{P},\mathcal{D},\mathcal{K}_1\mathcal{A}_1}(\lambda) + \mathbf{Adv}^{\mathsf{pax}}_{\Pi,\mathcal{P},\mathcal{D},\mathcal{K}_2,\mathcal{A}_2}(\lambda),$$

where $\mathcal{K}_1$ is the a secret key extractor for $\mathcal{A}_1$ implied by the SKAx property of Π, and $\mathcal{K}_2$ is the plaintext extractor for $\mathcal{A}_2$ implied by the PAx property of Π.

The intuition behind this theorem is the following. Secret key awareness ensures that a strong plaintext awareness adversary cannot come up with a ciphertext under a *new* public key, for which it does not know the underlying message (as it must know the decryption key). However, no such guarantee is provided for the *challenge* public key, and this justifies the plaintext awareness requirement.

REMARK. An extra feature that comes with secret key awareness is the ability to open ciphertexts via the secret key[12]. In other words, one can convert a non-malleability simulator that only returns $(\mathsf{PK}^\star, \mathsf{c}^\star)$ to another one[13] which also outputs the corresponding opening $(\mathsf{SK}^\star, \mathsf{m}^\star)$. This means that the output of the simulator can indeed be seen as a de-commitment. The same observation does not apply to strong plaintext awareness, as this notion does not guarantee the knowledge of the *randomness* used in encryption.

4   Secret Key Aware Schemes

4.1   Generic Construction with a Random Oracle

We have defined strong plaintext and secret key awareness in the standard model, but the definitions can be adapted to the random oracle model [6] in the natural way[14]. Interestingly, in the random oracle model, there is a simple transformation that turns any encryption scheme into one which is secret key aware without any loss in security: one just changes the key-generation algorithm by attaching the hash of the key-pair to the public key. More formally, the transformed set-up algorithm Setup′ is identical to Setup except that it also specifies a new independent hash function H (i.e. one which is not used by the scheme), which will be modelled as a random oracle in the security analysis. The remaining algorithms are shown in Figure 8.

proc. Gen′(I):	proc. Enc′(m, PK′):	proc. Dec′(c, SK′, PK′):
$(\mathsf{SK}, \mathsf{PK}) \leftarrow_\$ \mathsf{Gen}()$	$(\mathsf{PK}, h) \leftarrow \mathsf{PK}'$	$(\mathsf{PK}, h) \leftarrow \mathsf{PK}'; \mathsf{SK} \leftarrow \mathsf{SK}'$
$\mathsf{PK}' \leftarrow (\mathsf{PK}, \mathsf{H}(\mathsf{SK}, \mathsf{PK}))$	$\mathsf{c} \leftarrow_\$ \mathsf{Enc}(\mathsf{m}, \mathsf{PK})$	If $h \neq \mathsf{H}(\mathsf{SK}, \mathsf{PK})$ Return $\perp$
Return $(\mathsf{SK}, \mathsf{PK}')$	Return c	Return $\mathsf{Dec}(\mathsf{c}, \mathsf{SK}, \mathsf{PK})$

Fig. 8. Generic transformation to a secret-key-aware scheme Π' in the ROM

It can be easily shown that the scheme obtained through the above transformation is secret key aware, as long as the range of H is large enough to ensure

---

[12] Although not considered in this paper, the secret key awareness property can also be used in composition theorems where direct access to secret keys is required. This could be useful, for example, in signcryption.

[13] This new simulator must be given the secret key for the challenge public key, which is consistent with the notion of a non-malleable commitment.

[14] Furthermore, the standard stronger definition requiring the existence of a universal extractor may also be easily formulated [4].

that the transformed scheme has only one valid secret key for each valid public key with overwhelming probability. The idea behind the proof is that adversarially created public keys will be invalid with high probability, unless the random oracle has already been queried on the corresponding secret key. In this case the (unique) secret key can be recovered using the extractability property of the random oracle. Note that simply attaching $H(SK)$ is not enough, as extraction will fail in case the challenge public key is malleable. See the details in the full version of the paper. From this result, together with Theorems 1 and 4, we can deduce that if Π is IND-CCAx-secure then Π' is IND-SCCAx-secure.

REMARK. The previous generic construction can be applied to RSA-OAEP. In the random oracle model, this new version of RSA-OAEP is SKA2 because of the modified keys, and it preserves the original PA2 and IND-CPA security of RSA-OAEP. It follows that this scheme is completely non-malleable with respect to non-assisted simulators. Note that we do not need to restrict the adversary to querying only valid public keys, as is the case in [18], since the secret key extractor implied by the SKA2 property will permit detecting such invalid queries.

4.2   Towards Secret-Key-Aware Schemes without Random Oracles

We present two approaches to constructing secret key aware schemes without random oracles. Both are intended as stepping stones towards achieving the strongest forms of secret key awareness. We first introduce a new knowledge-based assumption and use it to construct a concrete scheme, which is "weakly" secret key aware. Then we propose a generic construction inspired by techniques used in encryption schemes with key escrow [12]. We leave it as an interesting open problem to instantiate this generic construction or show its unrealisability.

THE KNOWLEDGE OF FACTOR ASSUMPTION. We take advantage of the fact that k-bit integers of the form $N = P^2Q$ have a negligible density in the set of all k-bit integers (note that this is *not* the case for the integers of the form PQ), and we postulate that the only way to generate such integers is to start with the two prime factors and calculate N. This assumption is similar to Diffie-Hellman knowledge type assumptions [5] where one exploits the sparse image of the $r \mapsto (g^r, (g^a)^r)$ map. Our assumption, however, has the extra property of being "non-malleable" in the sense that there does not seem to be any way to use the knowledge of one (or in fact many) integers of this form to find an alternative way to construct new ones. Diffie-Hellman tuples on the other hand are malleable. For concreteness, we now present a formal definition of our *knowledge of factorisation assumptions*.

Take $\mathcal{G}_e$ to be the algorithm that, for a given value of the security parameter, generates numbers of the form P^2Q, with P and Q random primes of the appropriate size such that[15] $\gcd(e, \varphi(P^{\star 2}Q^\star)) = 1$. Figure 9 depicts the KFAx game for x = 0, 1, 2, where an adversary is required to construct a new integer of the same form *without knowing* the factorization. We define the KFAx advantage of an adversary against $\mathcal{G}_e$, with respect to knowledge extractor $\mathcal{K}$ as:

---

[15] We use φ to denote Euler's totient function.

```
proc. Initialize(λ):                                          Game KFAx_{𝒢_c,𝒦,ℓ}(λ)
(P★, Q★) ←$ 𝒢_e(1^λ); N★ ← P★²Q★
d ← 1/e (mod φ(N★)); List ← []              proc. Fact(N):
Choose Rnd[𝒜] for 𝒜; Flag ← F               ((P,Q), st[𝒦]) ←$ 𝒦(N, List, st[𝒦])
st[𝒦] ← (N★, Rnd[𝒜])                         If P²Q = N ∧ P ≠ 1 Return (P, Q)
Return (N★, Rnd[𝒜])                          If ∃P', Q' s.t. P'²Q' = N
                                               Set Flag ← T
proc. Root(y):                                Return (⊥, ⊥)
t ← y^d (mod N★); (x, x') ← t
List ← (x, y) : List                         proc. Finalize():
Return x                                      Return Flag
```

Fig. 9. Game defining the knowledge of factor assumption. An adversary $\mathcal{A}$ is legitimate if: 1) If $x = 0$ it queries **Fact** once, and if $x = 0, 1$ it does not query **Root**; and 2) It never queries **Fact** on $N^\star$. **Root** returns the first ℓ bits of t, i.e. $|x| = \ell$.

$$\mathbf{Adv}^{\mathsf{kfax}}_{\mathcal{G}_e,\mathcal{K},\ell,\mathcal{A}}(\lambda) := \Pr[\mathsf{KFAx}^{\mathcal{A}}_{\mathcal{G}_e,\mathcal{K},\ell}(\lambda) \Rightarrow \mathsf{T}].$$

The KFAx assumption states that, for every PPT adversary, there exists an efficient knowledge extractor such that advantage is negligible.

RSA-BASED SECRET-KEY-AWARE SCHEMES. The KFA1 assumption immediately implies that an RSA-based scheme with P^2Q modulus [23] is SKA1. Random padding before encryption allows one to construct an IND-CPA secure scheme. In order to extend this to IND-CCA1 security, a non-adaptive **Root** oracle is added to the RSA problem. As a result, we arrive at an IND-SCCA1 secure encryption scheme without random oracles and with no setup assumptions. We refer the reader to the full version of the paper for the details on this concrete scheme and a proof of the secret key awareness property.

The only factorisation/RSA-based IND-CCA2 secure encryption scheme in the standard model is a recent scheme of Hofheinz and Kiltz [19]. This scheme, with appropriately modified public keys is a candidate for achieving IND-SCCA2 security under the KFAx assumptions through secret key awareness. Such a construction would also admit a (non-black-box) non-assisted complete non-malleability simulator *with no set-up assumptions*. This would solve the open problem [18] of constructing an encryption scheme that is suitable for the implementation of non-malleable commitment schemes in the plain model.

REMARK. The KFAx assumptions lead to a construction of extractable one-way functions analogous to that obtained using the knowledge-of-exponent assumptions [13,14]. However, we can use even the weakest form KFA0 to go beyond. Indeed, this assumption states that one cannot come up with an N of the correct form, even if *given another* integer of this form as auxiliary information. Under this assumption the function $f(P, Q) = P^2Q$, where P and Q are k-bit primes, is an extractable one-way function with *dependent* auxiliary information.

REMARK. We note that knowledge assumptions seem necessary to establish plaintext and secret key awareness. It remains an open problem to construct plaintext-aware schemes without relying on extractor-based assumptions such as Diffie-Hellman Knowledge. NIZK techniques do not provide an answer to this problem, as extractors should work with the *provided* common reference string.

A GENERIC TECHNIQUE BASED ON SCHEMES WITH KEY ESCROW. Consider a public-key encryption scheme Π where the key-generation procedure first generates a secret key in the appropriate range, and then encrypts it under an auxiliary encryption scheme[16] Π_{PK}. Then, the plaintext awareness property of Π_{PK} naturally maps to (a weak form of) secret key awareness for Π. The caveat to this design technique is that plaintext awareness is an all-or-nothing notion [24], which could render this construction unrealisable. Indeed, full plaintext awareness in key-generation would imply a form of indistinguishability for secret keys that is contradicted by the correctness of the scheme[17]. However, by restricting the plaintext awareness property of Π_{PK} to the class of plaintext creators that return a random message from the message space, we can show that Π achieves SKA0 if the auxiliary scheme Π_{PK} is PA2.

Acknowledgments. The authors were funded in part by eCrypt II (EU FP7 - ICT-2007-216646) and FCT project PTDC/EIA/71362/2006. The second author was also funded by FCT grant BPD-47924-2008.

References

1. Al-Riyami, S.S., Paterson, K.G.: Certificateless public key cryptography. In: Laih, C.-S. (ed.) ASIACRYPT 2003. LNCS, vol. 2894, pp. 452–473. Springer, Heidelberg (2003)
2. Barbosa, M., Farshim, P.: Relations among notions of complete non-malleability: Indistinguishability characterisation and efficient construction without random oracles. Accepted for ACISP 2010 (Preprint, 2010)
3. Bellare, M., Boldyreva, A., Micali, S.: Public-key encryption in a multi-user setting: Security proofs and improvements. In: Preneel, B. (ed.) EUROCRYPT 2000. LNCS, vol. 1807, pp. 259–274. Springer, Heidelberg (2000)
4. Bellare, M., Desai, A., Pointcheval, D., Rogaway, P.: Relations among notions of security for public-key encryption schemes. In: Krawczyk [20]
5. Bellare, M., Palacio, A.: Towards plaintext-aware public-key encryption without random oracles. In: Lee, P.J. (ed.) ASIACRYPT 2004. LNCS, vol. 3329, pp. 48–62. Springer, Heidelberg (2004)
6. Bellare, M., Rogaway, P.: Random oracles are practical: A paradigm for designing efficient protocols. In: ACM Conference on Computer and Communications Security, pp. 62–73 (1993)
7. Bellare, M., Rogaway, P.: Optimal asymmetric encryption. In: De Santis, A. (ed.) EUROCRYPT 1994. LNCS, vol. 950, pp. 92–111. Springer, Heidelberg (1995)

---

[16] This means, of course, that a public key for the auxiliary scheme Π_{PK} must be fixed in the global parameters for Π.

[17] Given a public key and two secret keys one can check which secret key is valid through the encryption and decryption algorithms.

8. Bellare, M., Rogaway, P.: The security of triple encryption and a framework for code-based game-playing proofs. In: Vaudenay [25], pp. 409–426
9. Bellare, M., Sahai, A.: Non-malleable encryption: Equivalence between two notions, and an indistinguishability-based characterization. In: Wiener, M.J. (ed.) CRYPTO 1999. LNCS, vol. 1666, pp. 519–536. Springer, Heidelberg (1999)
10. Bellare, M., Sahai, A.: Non-malleable encryption: Equivalence between two notions, and an indistinguishability-based characterization. Cryptology ePrint Archive, Report 2006/228 (2006), http://eprint.iacr.org/2006/228
11. Birkett, J., Dent, A.W.: Relations among notions of plaintext awareness. In: Cramer [15], pp. 47–64
12. Brown, J., Gonzalez Nieto, J.M., Boyd, C.: Efficient and secure self-escrowed public-key infrastructures. In: ASIACCS 2007: Proceedings of the 2nd ACM symposium on Information, computer and communications security, pp. 284–294. ACM, New York (2007)
13. Canetti, R., Dakdouk, R.R.: Extractable perfectly one-way functions. In: Aceto, L., Damgård, I., Goldberg, L.A., Halldórsson, M.M., Ingólfsdóttir, A., Walukiewicz, I. (eds.) ICALP 2008, Part II. LNCS, vol. 5126, pp. 449–460. Springer, Heidelberg (2008)
14. Canetti, R., Dakdouk, R.R.: Towards a theory of extractable functions. In: Reingold, O. (ed.) TCC 2009. LNCS, vol. 5444, pp. 595–613. Springer, Heidelberg (2009)
15. Cramer, R. (ed.): PKC 2008. LNCS, vol. 4939. Springer, Heidelberg (2008)
16. Dent, A.W.: The cramer-shoup encryption scheme is plaintext aware in the standard model. In: Vaudenay [25], pp. 289–307.
17. Dolev, D., Dwork, C., Naor, M.: Non-malleable cryptography (extended abstract). In: STOC, pp. 542–552. ACM, New York (1991)
18. Fischlin, M.: Completely non-malleable schemes. In: Caires, L., Italiano, G.F., Monteiro, L., Palamidessi, C., Yung, M. (eds.) ICALP 2005. LNCS, vol. 3580, pp. 779–790. Springer, Heidelberg (2005)
19. Hofheinz, D., Kiltz, E.: Practical chosen ciphertext secure encryption from factoring. In: Joux, A. (ed.) EUROCRYPT 2009. LNCS, vol. 5479, pp. 313–332. Springer, Heidelberg (2010)
20. Krawczyk, H. (ed.): CRYPTO 1998. LNCS, vol. 1462. Springer, Heidelberg (1998)
21. Pandey, O., Pass, R., Vaikuntanathan, V.: Adaptive one-way functions and applications. In: Wagner, D. (ed.) CRYPTO 2008. LNCS, vol. 5157, pp. 57–74. Springer, Heidelberg (2008)
22. Pass, R., Shelat, A., Vaikuntanathan, V.: Relations among notions of non-malleability for encryption. In: Kurosawa, K. (ed.) ASIACRYPT 2007. LNCS, vol. 4833, pp. 519–535. Springer, Heidelberg (2007)
23. Takagi, T.: Fast RSA-type cryptosystem modulo $p^k q$. In: Krawczyk [20]
24. Teranishi, I., Ogata, W.: Relationship between standard model plaintext awareness and message hiding. IEICE Transactions 91-A(1), 244–261 (2008)
25. Vaudenay, S. (ed.): EUROCRYPT 2006. LNCS, vol. 4004. Springer, Heidelberg (2006)
26. Ventre, C., Visconti, I.: Completely non-malleable encryption revisited. In: Cramer [15], pp. 65–84

A Multi-trapdoor Commitment Scheme from the RSA Assumption

Ryo Nishimaki[1,2], Eiichiro Fujisaki[1], and Keisuke Tanaka[2]

[1] NTT Laboratories, 3-9-11 Midori-cho, Musashino-shi, Tokyo, 180-8585, Japan
{nishimaki.ryo,fujisaki.eiichiro}@lab.ntt.co.jp
[2] Department of Mathematical and Computing Sciences, Tokyo Institute of
Technology, W8-55, 2-12-1 Ookayama Meguro-ku, Tokyo 152-8552, Japan
keisuke@is.titech.ac.jp

Abstract. Gennaro introduced the notion of multi-trapdoor commitments which is a stronger form of trapdoor commitment schemes at CRYPTO 2004. Multi-trapdoor commitments have several cryptographic applications. For example, Gennaro proposed a conversion that makes a non-interactive multi-trapdoor commitment scheme into a non-interactive and reusable non-malleable commitment scheme and a compiler that transforms any proof of knowledge into concurrently non-malleable one. Gennaro gave constructions of multi-trapdoor commitments, but they rely on stronger assumptions, such as the *strong RSA* assumption, the *q-strong Diffie-Hellman* assumption.

In this paper, we propose a non-interactive multi-trapdoor commitment scheme from the *standard RSA* assumption. Thus, as a corollary of our result, we obtain a non-interactive and reusable non-malleable commitment scheme from the standard RSA assumption. Our scheme is based on the Hohenberger-Waters signature scheme proposed at CRYPTO 2009. Several non-interactive and reusable non-malleable commitment schemes (in the common reference string model) have been proposed, but all of them rely on stronger assumptions (e.g., strong RSA).

Keywords: non-interactive commitment, multi-trapdoor commitment, non-malleability, reusability, RSA assumption.

1 Introduction

Commitment is the digital analogue of sealed envelopes. It consists of two phases. In the first phase (the commitment phase), a sender provides a receiver with a sealed envelope that contains a message (the sender sends a *commitment*). The receiver cannot learn anything about the message from the commitment. This secrecy property is called *hiding* in commitment schemes. In the second phase (the decommitment, or opening, phase), the sender opens the sealed envelope and the receiver can obtain the message (the sender sends a *decommitment*). However, the sender cannot change the message in the sealed envelope. This property is called *binding* in commitment schemes.

R. Steinfeld and P. Hawkes (Eds.): ACISP 2010, LNCS 6168, pp. 182–199, 2010.

Commitment is one of the most fundamental cryptographic primitives like public-key encryption (PKE) and digital signature since it is an essential building block in many cryptographic protocols, such as zero-knowledge (ZK) protocols [19].

Commitment schemes can be used for bid auction on the Internet as a direct application. In a scenario of bid auction, parties who want to bid for an item make commitments to their bids in the bidding period. After the period, all bids are revealed by sending the decommitments. If a commitment scheme has the hiding and binding properties, we can bid even in a situation that synchronicity is not guaranteed.

1.1   Background

NON-MALLEABLE COMMITMENTS. Hiding and binding are basic security requirements for secure commitment schemes in the *stand-alone setting* [27]. These basic security notions were in the past thought to be sufficient.

However, when a party send a commitment to message m in the bidding scenario, we want to avoid the situation that a malicious party can generate a valid commitment to $m + 1$ after he saw the commitment to m from another party *without knowing* m. He may be able to open his bid to $m+1$ after seeing a decommitment of the original commitment to m. This situation is not considered in the stand-alone setting.

Dolev, Dwork, and Naor proposed a stronger security notion, *non-malleability* [12] to prevent the above situation. Informally speaking, a commitment scheme is *non-malleable* if any adversary cannot make a commitment to a message that has some relation to the original message m in the *man-in-the-middle setting*, where an adversary is allowed to tap into the communication between a sender and a receiver and may change the content of the communication.

There are two definitions of non-malleable commitment. A commitment scheme is called *non-malleable with respect to decommitment* (or with respect to opening) [9], if the adversary cannot make a commitment from an honest sender's commitment, such that the adversary can correctly open his commitment to a related message after he saw the decommitment of the honest sender's commitment. A commitment scheme is called *non-malleable with respect to commitment* [12], if the adversary cannot make a commitment to a related message (he does not need to open it later). Non-malleable commitment with respect to decommitment (we abbreviate this to NMd) schemes are suffice for most cryptographic applications and bid auction since parties are often required to open their commitments. NMd schemes are used as building blocks of many cryptographic protocols, such as universally composable (UC) commitment schemes [8], UCZK protocols [16]. Thus, it is useful to construct efficient NMd schemes and we does not treat non-malleability with respect to commitment in this paper.

MULTI-TRAPDOOR COMMITMENTS. Gennaro introduced the notion of multi-trapdoor commitments [17]. We explain multi-trapdoor commitments very informally. If we can make a fake commitment that will be opened to an arbitrary value later by using trapdoors, a commitment scheme is called equivocal (this is also known as a trapdoor commitment scheme). A multi-trapdoor scheme consists of a family of trapdoor commitments. Each scheme in the family is statistically hiding. There exists a master trapdoor whose knowledge allows the owner to open any commitment in the family in any way. For each commitment scheme in the family, there exists its own specific trapdoor, which allows to equivocate that specific scheme but does not allow to obtain trapdoors of other schemes. It is required that we cannot open commitments in two different ways without knowing the trapdoors.

It is useful to construct efficient multi-trapdoor commitment schemes because there are cryptographic applications of them. The main applications are non-malleable commitments and concurrently non-malleable proofs of knowledge. Gennaro proposed a conversion that makes a non-interactive multi-trapdoor commitment scheme into a non-interactive and reusable non-malleable commitment scheme and a compiler that transforms any proof of knowledge [3] into concurrently non-malleable one. Such a proof of knowledge yields *concurrently* secure identification schemes.

PROOFS OF KNOWLEDGE AND IDENTIFICATION SCHEMES. A proof of knowledge is a cryptographic protocol, which allows a prover to convince a verifier that he knows some secret information without revealing any information about the secret. Proofs of knowledge are used to construct secure identification schemes [13], which enable a prover to identify himself to a verifier. We can consider non-malleable proofs of knowledge as in the case of non-malleable commitments. Loosely speaking, a proof of knowledge is non-malleable if any man-in-the-middle adversary cannot convince an honest verifier that he has some information which relates to the secret information of an honest prover without knowing it. In this setting, the adversary interacts with only one prover. We can consider a stronger security, *concurrent* non-malleability. In this setting, a man-in-the-middle adversary can interact many different (honest) prover clones and honest verifiers concurrently. All clones have the same secret information but use independent random coins. A proof of knowledge is concurrent non-malleable if any adversary cannot convince an honest verifier in the above setting.

SET-UP ASSUMPTIONS. When we design a cryptographic protocol, we sometimes use some set-up assumptions. Many non-interactive NMd schemes are constructed in the common reference string model (CRS), where it is assumed that all parties have access to a public string that is guaranteed to be selected with a prescribed distribution since it is difficult to design non-interactive NM protocols without non-standard assumptions in the plain model. The plain model means that we use no set-up assumptions like the CRS model. In the CRS model, it is desirable that one CRS can be reused polynomially many times. This property is called *reusability*.

We consider only the CRS model in this paper since the definition of multi-trapdoor commitments by Gennaro is considered in the CRS model and there is no non-interactive NMd scheme without non-standard assumptions in the plain model so far [1,2,8,9,10,12,14,17,25,26,28,29].

1.2   Our Contributions and Constructions

We construct a non-interactive multi-trapdoor commitment scheme from the *standard RSA* assumption. As a corollary, we obtain a non-interactive and reusable NMd scheme from the *standard RSA* assumption since Gennaro proposed a conversion that makes a non-interactive multi-trapdoor commitment scheme into a non-interactive reusable NMd scheme by using a one-time signature scheme. Previous non-interactive and reusable NMd schemes are constructed from stronger assumptions, such as the *strong RSA* assumption, the *q-strong Diffie-Hellman (DH)* assumption. This is the first construction of a non-interactive reusable NMd scheme from the RSA assumption (i.e., standard assumption).

OUR TECHNIQUES. Here is an outline of our techniques.

We explain the security notions of multi-trapdoor commitments informally. For any commitment in the family, commitments are statistically hiding and it is impossible to open a commitment in two different ways even if the adversary is given trapdoors of selected schemes. The adversary must select the schemes *before seeing the definition of the whole family. Of course, the trapdoor of the target scheme is not given to the adversary.* We call this binding property *strongly secure binding* in this paper.

We consider a trapdoor commitment scheme whose trapdoors consist of signatures for some public strings. If breaking the binding property of this commitment scheme is equivalent to obtaining a valid signature (i.e., its trapdoor), the security game of strongly secure binding is very similar to the security game of signature.

In the strongly secure binding game of multi-trapdoor commitments, the adversary selects public strings which specify schemes, then he is given a master public key that defines the whole family and trapdoors for the selected public strings. At this point, he outputs a commitment, two distinct messages, and their decommitments. If a master public key, a master trapdoor, public strings, specific trapdoors for public strings, and an oracle that give the adversary trapdoors of a multi-trapdoor commitment scheme correspond to a public verification key, a private signing key, messages, signatures for messages, and a signing oracle of a signature scheme, respectively, then the above game corresponds to the security game of existential unforgeability against *weak* chosen message attacks. Loosely speaking, the adversary must query *before he is given a verification key* in the weak chosen message attack game.

In the strongly secure binding game, the adversary must query public strings *before given a master public key*, so the security against *weak* chosen message attacks is sufficient for our purpose (we do not need the security against adaptive chosen message attacks).

Thus, intuitively, we may be able to construct multi-trapdoor commitment schemes from signature schemes secure against *weak* chosen message attacks. We use the Hohenberger-Waters signature scheme proposed at CRYPTO'09 [21] to construct multi-trapdoor schemes since their schemes are secure against *weak* chosen message attacks under standard assumptions. Our commitment scheme comes from the Guillou-Quisquater (GQ) identification scheme [4,20].

From the above observation, we find that the identity-based trapdoor commitment scheme by Dodis, Shoup, and Walfish [11] can be seen as a multi-trapdoor commitment scheme based on Waters' signature [31], so we obtain a multi-trapdoor commitment scheme from the computational Diffie-Hellman (CDH) assumption (in bilinear groups). It immediately yields a non-interactive reusable NMd scheme from the CDH assumption (in bilinear groups).

1.3   Related Works and Comparisons

Several NMd schemes in the CRS model have been proposed, Di Crescenzo, Ishai, and Ostrovsky, Fischlin and Fischlin, and Di Crescenzo, Katz, Ostrovsky, and Smith constructed NMd schemes from standard assumption, such as the RSA or the discrete logarithm (DL) assumption, but they does not have reusability [9,10,14,15]. Damgård and Groth, MacKenzie and Yang, and Gennaro constructed non-interactive and reusable NMd schemes, but they used stronger assumptions, such as the strong RSA or the q-strong DH assumption [8,17,26].

A comparison of our results and previous results for NMd schemes is shown in Table 1. Reusability indicates whether a CRS can be reused polynomially many times or not. The DSA assumption in the table means that Digital Signature Algorithm (DSA) [23] is secure.

Table 1. Previous and our schemes in the CRS model

Reference	Interaction	Reusability	Assumptions
DIO [9]	non-interactive	No	One-Way Functions
FF [14,15]	interactive	No	DL, RSA, or Factoring
DKOS [10]	non-interactive	No	DL or RSA
DG [8]	non-interactive	Yes	strong RSA
MY [26]	non-interactive	Yes	strong RSA or DSA
Gennaro [17]	non-interactive	Yes	strong RSA or q-strong DH
Ours	*non-interactive*	*Yes*	*RSA*

2   Preliminaries

2.1   Notations and Conventions

For any $n \in \mathbb{N}$, let $[n]$ be the set $\{1, \ldots, n\}$. We describe probabilistic algorithms using standard notations and conventions. For probabilistic polynomial-time (PPT) algorithm A, $A(x_1, x_2, ...; r)$ denotes the random variable of A's output on inputs $x_1, x_2, ...$ and random coins r. We let $y \xleftarrow{\text{R}} A(x_1, x_2, ...)$ denote that y is randomly selected from $A(x_1, x_2, ...; r)$ according to its distribution. If S is a finite set, then $x \xleftarrow{\text{U}} S$ denotes that x is uniformly selected from S. If α is neither an algorithm nor a set, $x := \alpha$ indicates that α is assigned to x. We say that function $f : \mathbb{N} \to \mathbb{R}$ is negligible in security parameter $\lambda \in \mathbb{N}$ if for every constant $c \in \mathbb{N}$ there exists $k_c \in \mathbb{N}$ such that $f(\lambda) < \lambda^{-c}$ for any $\lambda > k_c$. Hereafter, we use $f < \mathsf{negl}(\lambda)$ (or $f < \mathsf{negl}$) to mean that f is negligible in λ.

2.2   Indistinguishability

The statistical distance between two random variables X and Y over a countable set S is defined as $\Delta(X, Y) := \frac{1}{2} \sum_{\alpha \in S} |\Pr[X = \alpha] - \Pr[X = \alpha]|$. Let $\mathcal{X} = \{X_\lambda\}_{\lambda \in I}$ and $\mathcal{Y} = \{Y_\lambda\}_{\lambda \in I}$ denote two ensembles of random variables indexed by λ (where I is a countable set).

Definition 1. *We say that $\mathcal{X}$ and $\mathcal{Y}$ are statistically indistinguishable if the statistical distance between these variables is negligible, that is, $\Delta(X_\lambda, Y_\lambda) <$ negl.*

Definition 2. *We say that $\mathcal{X}$ and $\mathcal{Y}$ are computationally indistinguishable if for every non-uniform PPT algorithm D, there exists negligible function* negl *such that for every $\lambda \in I$,*

$$|\Pr[D(X_\lambda) = 1] - \Pr[D(Y_\lambda) = 1]| < \mathsf{negl}.$$

We write $\mathcal{X} \approx_c \mathcal{Y}$ (resp., $\mathcal{X} \approx_s \mathcal{Y}$) to denote that $\mathcal{X}$ and $\mathcal{Y}$ are computationally (resp., statistically) indistinguishable.

2.3   Complexity Assumptions

We recall standard complexity assumptions and related basic facts.

THE RSA ASSUMPTION. Let λ be the security parameter. Let $\mathsf{GenRSA}(1^\lambda)$ be a PPT algorithm that outputs modulus $N = pq$ where p and q are two λ bit, distinct odd primes, randomly chosen integer $e > 0$ with $\gcd(e, \phi(N)) = 1$ and less than $\phi(N) = (p-1)(q-1)$, and integer d satisfying $ed = 1 \bmod \phi(N)$. The RSA problem is as follows: Given (N, e) and random $y \overset{U}{\leftarrow} \mathbb{Z}_N^*$, computing x such that $x^e = y \bmod N$. The advantage is

$$\mathsf{Adv}_{\mathcal{A}}^{RSA}(\lambda) := \Pr[x^e = y \bmod N | (N, e, d) \overset{R}{\leftarrow} \mathsf{GenRSA}(1^\lambda); y \overset{U}{\leftarrow} \mathbb{Z}_N^*; x \overset{R}{\leftarrow} \mathcal{A}(N, y, e)].$$

Definition 3 (RSA assumption). *We say that the RSA assumption holds if the RSA problem is hard, that is, for any PPT $\mathcal{A}$, $\mathsf{Adv}_{\mathcal{A}}^{RSA}(\lambda)$ is negligible.*

Lemma 1 (Shamir [30]). *Given $x, y \in \mathbb{Z}_N$ and $a, b \in \mathbb{Z}$ such that $x^a = y^b$ and $\gcd(a, b) = 1$, there is an efficient algorithm for computing $z \in \mathbb{Z}_N$ such that $z^a = y$.*

THE COMPUTATIONAL DIFFIE-HELLMAN (CDH) ASSUMPTION. We consider a cyclic group $\mathbb{G}$ and $\mathbb{G}_T$ of prime order p. A bilinear map is an efficient mapping $e : \mathbb{G} \times \mathbb{G} \to \mathbb{G}_T$ satifying the following properties:

bilinearity: for all $g \in \mathbb{G}$ and $a, b \overset{U}{\leftarrow} \mathbb{Z}_p$, $e(g^a, g^b) = e(g, g)^{ab}$.
non-degeneracy: If g generates $\mathbb{G}$, then $e(g, g) \neq 1$.

Let g generate a group $\mathbb{G}$. Let $\mathcal{BG}$ be a standard bilinear group generator that takes as input the security parameter λ and outputs $(\mathbb{G}, \mathbb{G}_T, p, g, e)$. The CDH problem in bilinear groups is as follows: Given $(\mathbb{G}, \mathbb{G}_T, p, g, e, g^a, g^b)$, computing g^{ab}. The advantage is

$$\mathsf{Adv}_{\mathcal{A}}^{\mathrm{CDH}}(\lambda) := \Pr[z = g^{ab} | (\mathbb{G}, \mathbb{G}_T, p, g, e) \xleftarrow{\mathsf{U}} \mathcal{BG}(1^\lambda); a, b \xleftarrow{\mathsf{U}} \mathbb{Z}_p; z \xleftarrow{\mathsf{R}} \mathcal{A}(g, g^a, g^b)].$$

Definition 4 (CDH assumption in bilinear groups). *We say that the CDH assumption holds in bilinear groups if the CDH problem in bilinear groups is hard, that is, for any PPT $\mathcal{A}$, $\mathsf{Adv}_{\mathcal{A}}^{CDH}(\lambda)$ is negligible.*

2.4 Signature Schemes

Signature scheme Sig consists of three PPT algorithms $\mathsf{Sig} = \mathsf{Sig}.\{\mathsf{Gen}, \mathsf{Sign}, \mathsf{Vrfy}\}$ satisfying the following properties.

- Key generation algorithm $\mathsf{Sig.Gen}$ takes as input security parameter 1^λ and outputs a pair of keys, that is, $(vk, sk) \xleftarrow{\mathsf{R}} \mathsf{Sig.Gen}(1^\lambda)$. They are called the (public) verification key and the (private) signing key, respectively.
- Signing algorithm $\mathsf{Sig.Sign}$ takes as input a signing key and a message and outputs signature σ. That is, $\sigma \xleftarrow{\mathsf{R}} \mathsf{Sig.Sign}_{sk}(m)$, where $m \in \mathcal{M}$ and $\mathcal{M}$ is a message space.
- Verification algorithm $\mathsf{Sig.Vrfy}$ is deterministic and takes as input a verification key, a message, and a signature and outputs bit b. If $b = 1$ then the signature is valid. Else, it is invalid. That is, $b = \mathsf{Sig.Vrfy}_{vk}(m, \sigma)$.

It is required that for any λ, $(vk, sk) \xleftarrow{\mathsf{R}} \mathsf{Sig.Gen}(1^\lambda)$, and $m \in \mathcal{M}$, it holds that $\mathsf{Sig.Vrfy}_{vk}(m, \mathsf{Sig.Sign}_{sk}(m)) = 1$.

WEAK CHOSEN MESSAGE ATTACKS. We consider a weaker definition called existential unforgeability against *weak* chosen message attacks. Signature scheme $\mathsf{Sig} = \mathsf{Sig}.\{\mathsf{Gen}, \mathsf{Sign}, \mathsf{Vrfy}\}$ is said to be unforgeable under weak chosen message attacks (EUF-wCMA) if the advantage of the following game is negligible.

Queries: The adversary firstly sends the challenger list $\mathcal{Q}$ of messages $M_1, \dots, M_q \in \mathcal{M}$.

Response: The challenger obtains $(vk, sk) \xleftarrow{\mathsf{R}} \mathsf{Sig.Gen}(1^\lambda)$ and signs each queried message, that is, generates $\sigma_i \xleftarrow{\mathsf{R}} \mathsf{Sig.Sign}_{sk}(M_i)$ for $i = 1$ to n. The challenger sends $(vk, \sigma_1, \dots, \sigma_n)$ to the adversary.

Output: The adversary outputs (M^*, σ^*). If $M^* \notin \mathcal{Q}$ and $\mathsf{Sig.Vrfy}_{vk}(M^*, \sigma^*) = 1$, it is said to win the game.

We define $\mathsf{Adv}_{\mathcal{A}}^{\mathrm{EUF\text{-}weak}}(\lambda)$ to be the probability that adversary $\mathcal{A}$ wins in the game.

Definition 5 (Unforgeability against Weak Chosen Message Attacks). *Signature scheme Sig is existentially unforgeable against weak chosen message attacks if for all PPT $\mathcal{A}$, $\mathsf{Adv}_{\mathcal{A}}^{EUF\text{-}weak}$ is negligible.*

We briefly review two signature schemes by Hohenberger and Waters [21,31].

THE HOHENBERGER-WATERS SIGNATURE SCHEME *[21].*

Key Generation. On input security parameter λ, generates an RSA modulus N, such that $2^\ell < \phi(N) < 2^{\ell+2}$, where ℓ depends on 1^λ and chooses random $h \xleftarrow{U} \mathbb{Z}_N^*$, random key K for pseudo-random function (PRF, see [18,22] for the definition) $F : \{0,1\}^* \to \{0,1\}^\ell$, and random $c \in \{0,1\}^\ell$. It defines function $H : \{0,1\}^* \to \{0,1\}^\ell$ as follows

$$H_{K,c}(z) := F_K(i, z) \oplus c,$$

where value i is the smallest $i \geq 1$ such that $F_K(i, z) \oplus c$ is odd and prime. Value i is called *resolving index.* It outputs verification key $vk = (N, h, c, K)$ and signing key $sk = (p, q, vk)$.

Signing. Let $M^{(i)}$ denote the first i bits of M, that is, the length i of prefix M. For $i = 1$ to n, it computes $e_i = H_{K,c}(M^{(i)})$ and then outputs signature $\sigma = h^{\prod_{i=1}^n e_i^{-1}} \bmod N$.

Verification. On input vk, M, signature σ, $e_i = H_{K,c}(M^{(i)})$, outputs 1 if and only if

$$\sigma^{\prod_{i=1}^n e_i} \equiv h \pmod{N}.$$

Theorem 1 ([21]). *If the RSA assumption holds, the above signature scheme is existentially unforgeable against weak chosen message attacks.*

THE WATERS SIGNATURE SCHEME *[21,31].*

Key Generation. On input security parameter λ, selects a bilinear group $\mathbb{G}$ of prime order $p > 2^\lambda$ and chooses random $a \xleftarrow{U} \mathbb{Z}_p$ and group elements $g, v_0, v_1, \ldots, v_n \in \mathbb{G}$. It outputs verification key $vk = (g, v_0, v_1, \ldots, v_n, e(g,g)^a)$ and signing key $sk = a$.

Signing. Let M_i denote the i-th bit of M. It chooses random $r \xleftarrow{U} \mathbb{Z}_p$ and computes

$$\sigma_1 = g^a \left(v_0 \prod_{i=1}^n v_i^{M_i} \right)^r, \quad \sigma_2 = g^r.$$

Signature is $\sigma = (\sigma_1, \sigma_2)$.

Verification. On input vk, M, signature σ, outputs 1 if and only if

$$e(\sigma_1, g) = e(g, g)^a \cdot e(v_0 \prod_{i=1}^n v_i^{M_i}, \sigma_2).$$

Theorem 2 ([21,31]). *If the CDH assumption holds, the above signature scheme is existentially unforgeable against weak chosen message attacks.*

2.5 Σ-Protocol for Waters Signature [11]

A Σ-protocol for a polynomial-time relation $\mathcal{R} = \{(x, w)\}$ is a 3-move protocol for prover $\mathbf{P}$ and verifier $\mathbf{V}$ [7]. Statement x is the common input and $\mathbf{P}$ has w as private input (called witness). Transcripts in the protocol have form (a, c, z) where c is a random challenge sent by $\mathbf{V}$.

1. $\mathbf{P}$ chooses random coin r_a, computes $a := \Sigma_1(x, w; r_a)$, and sends it to $\mathbf{V}$.
2. $\mathbf{V}$ sends random challenge $c \xleftarrow{\text{R}} \Sigma_2(x, a)$ to $\mathbf{P}$.
3. Finally, $\mathbf{P}$ responds with $z := \Sigma_3(x, w, r_a, c)$. $\mathbf{V}$ computes and returns bit $b = \Sigma_{\text{Vrfy}}(x, a, c, z)$.

Σ-protocols satisfy the following properties:

Completeness. If $(x, w) \in \mathcal{R}$, then $\mathbf{V}$ outputs $b = 1$ with probability 1.
Special Soundness. If one has two valid transcripts (a, c, z) and (a, c', z') where $c \neq c'$, one can efficiently compute witness w such that $(x, w) \in \mathcal{R}$.
Honest Verifier Zero-Knowledge. There exists PPT simulator $\mathcal{S}$ such that for all PPT distinguisher $\mathcal{D}$, and for all challenge c, for any $(x, w) \in \mathcal{R}$,

$$\left| \Pr[\mathcal{D}(\text{trans}) = 1 | r_a \xleftarrow{\text{U}} \mathcal{C}_{\mathbf{P}}; a := \Sigma_1(x, w; r_a); z := \Sigma_3(x, w, r_a, c)] - \right.$$
$$\left. \Pr[\mathcal{D}(\text{trans}) = 1 | (a, z) \xleftarrow{\text{R}} \mathcal{S}(x, c)] \right| < \text{negl} ,$$

where $\text{trans} = (x, w, a, c, z)$ and $\mathcal{C}_{\mathbf{P}}$ is the set of random coins of the honest prover.

Dodis, Shoup, and Walfish proposed a Σ-protocol for proving knowledge of the Waters signature $(\sigma_1, \sigma_2) \in \mathbb{G} \times \mathbb{G}$ on a message $M \in \{0, 1\}^*$. The protocol is as follows:

1. The prover chooses $\bar{r}_1, \bar{r}_2 \xleftarrow{\text{U}} \mathbb{Z}_p$ and computes $\bar{\sigma}_1 := \sigma_1^{1/\bar{r}_1}$, $\bar{\sigma}_2 := \sigma_2^{1/\bar{r}_2}$. Let $\gamma_1 := e(\bar{\sigma}_1, g), \gamma_2 := e(\bar{\sigma}_2, v_M^{-1}), \gamma := e(g, g)^a$. Note that $\gamma_1^{\bar{r}_1} \gamma_2^{\bar{r}_2} = \gamma$ due to the verification condition of Waters signature. Next, chooses $\widehat{r}_1, \widehat{r}_2 \xleftarrow{\text{U}} \mathbb{Z}_p$ and computes $\widehat{\gamma} := \gamma_1^{\widehat{r}_1} \gamma_2^{\widehat{r}_2}$. The prover sends $(\bar{\sigma}_1, \bar{\sigma}_2, \widehat{\gamma})$ to the verifier.
2. The verifier chooses challenge $c \xleftarrow{\text{U}} \mathbb{Z}_p$ and sends it to the prover.
3. The prover computes $\widetilde{r}_1 \leftarrow \widehat{r}_1 - c \cdot \bar{r}_1 \bmod p$, $\widetilde{r}_2 \leftarrow \widehat{r}_2 - c \cdot \bar{r}_2 \bmod p$. and sends $(\widetilde{r}_1, \widetilde{r}_2)$ to the verifier.
4. The verifier checks that $\gamma_1^{\widetilde{r}_1} \gamma_2^{\widetilde{r}_2} \gamma^c = \widehat{\gamma}$.

2.6 Commitment Scheme

There are several types of commitment. We review their definitions.

Definition 6 (Trapdoor Commitment). *A trapdoor commitment scheme consists of the following algorithms:*

Key Generation. *On input security parameter* λ, *outputs a public key and a trapdoor,* $(pk, td) \xleftarrow{\text{R}} \mathsf{KGen}(1^\lambda)$.

Commitment. *On input* pk, *message* m, *and randomness* r, *it computes* $(C, D) = \mathsf{Com}(pk, m; r)$. *Value* C *is the commitment and value* D *is the decommitment.*

Verification. *On input* pk, *message* m *and* C, D, *it output bit* $b = \mathsf{Vrfy}(pk, m, C, D)$.

Equivocation. *On input* pk, td, C, D, *and message* $\widetilde{m} \neq m$, *outputs fake decommitment information* $\widetilde{D} = \mathsf{Equiv}(pk, td, C, D, \widetilde{m})$ *such that* $(C, \widetilde{D}) = \mathsf{Com}(pk, \widetilde{m}; \widetilde{r})$ *for some* $\widetilde{r}$, *with the same distribution as if* $\widetilde{r}$ *has been chosen at random.*

Definition 7 (Multi-Trapdoor Commitment [17]). *A (non-interactive) multi-trapdoor commitment scheme consists of six algorithms* ($\mathsf{KGen}, \mathsf{Sel}, \mathsf{TGen}, \mathsf{Com}, \mathsf{Vrfy}, \mathsf{Equiv}$).

Master Key Generation Algorithm KGen. *On input the security parameter, outputs master public key associated with the family of commitment schemes* PK *and master trapdoor* TK, $(PK, TK) \xleftarrow{\text{R}} \mathsf{KGen}(1^\lambda)$.

Selection Algorithm Sel. Sel *selects a commitment in the family. On input* PK, *outputs specific public key* pk *that identifies one of the schemes.*

Trapdoor Generation Algorithm TGen. *On input* PK, pk, *and* TK, *outputs specific trapdoor* $td = \mathsf{TGen}(PK, pk, TK)$ *relative to* pk.

Commitment Algorithm Com. *On input* PK, pk, *and message* m, *computes* $(C, D) = \mathsf{Com}(PK, pk, m; r)$ *where* r *are random coins. Output* C *is the commitment string, while value* D *is the decommitment of* C *and secret information.*

Verification Algorithm Vrfy. *On input* PK, pk, *message* m, *and two values* C, D, *outputs bit* $b = \mathsf{Vrfy}(PK, pk, m, C, D)$.

Equivocation Algorithm Equiv. *We can open a commitment to any value given the trapdoor.* Equiv *takes as input keys* PK, pk, *commitment* C, *its decommitment* D, *message* $\widetilde{m} \neq m$ *and string* T. *If* $T = TK$ *or* $T = td$ *(for* pk) *then outputs* $\widetilde{D}$ *such that* $(C, \widetilde{D}) = \mathsf{Com}(PK, pk, \widetilde{m}; \widetilde{r})$ *for some* $\widetilde{r}$, *with the same distribution as if* $\widetilde{r}$ *has been chosen at random.*

A multi-trapdoor commitment schemes satisfy the following three properties.

Correctness. *For all message* m, *if* $(C, D) = \mathsf{Com}(PK, pk, m; r)$, *then it holds that* $\mathsf{Vrfy}(PK, pk, m, C, D) = 1$.

Information Theoretic Secrecy. *For every* m, m', *it holds that* $C \approx_s C'$, *where* $(C, D) = \mathsf{Com}(PK, pk, m; r)$ *and* $(C', D') = \mathsf{Com}(PK, pk, m'; r')$. *That is, the commitment is statistically hiding.*

Secure Binding. *We define secure binding game* $\mathsf{SBnd}_{\mathcal{A}}(\lambda)$ *as follows. First, adversary* $\mathcal{A}$ *selects* k *strings* $(pk_1, \dots, pk_k)$. *It is then given master public key* PK *for a multi-trapdoor commitment family generated with the same distribution as the ones generated by* $\mathsf{KGen}(1^\lambda)$. *Adversary* $\mathcal{A}$ *has access to equivocation oracle* $\mathcal{EQ}$. *When it queries* (C, D, m, r, pk, m') *where* $(C, D) =$

$\mathsf{Com}(PK, pk, m, r)$ and $m' \neq m$, if $pk = pk_i$ for some i, and is a valid public key, then $\mathcal{EQ}$ answers with decommitment $D' = \mathsf{Equiv}(PK, pk, C, D, td)$ $(td = \mathsf{TGen}(PK, pk, TK))$ such that $(C, D') = \mathsf{Com}(PK, pk, m', r')$, otherwise it outputs $\perp$. The game output 1 if and only if $\mathcal{A}$ outputs $(C, m, D, \widetilde{m}, \widetilde{D}, pk)$ such that $\mathsf{Vrfy}(PK, pk, m, C, D) = \mathsf{Vrfy}(PK, pk, \widetilde{m}, C, \widetilde{D}) = 1$ such that $m \neq \widetilde{m}$ and $pk \neq pk_i$ for all i. We require that for all PPT $\mathcal{A}$,

$$\Pr[\mathsf{SBnd}_{\mathcal{A}}(\lambda) = 1] < \mathsf{negl}(\lambda).$$

STRONGLY SECURE BINDING. Gennaro considered a stronger version of the Secure Binding property by requiring that adversary $\mathcal{A}$ receives trapdoors td_i's matching public keys pk_i's, instead of accessing equivocation oracle $\mathcal{EQ}$. In this case, we call *strongly* secure binding. It is easily seen that the secure binding property is implied by strongly secure binding property. We consider only *strongly* secure binding in this paper.

Theorem 3 ([17]). *If there exist multi-trapdoor commitment schemes and one-time signature schemes, we can construct reusable non-malleable trapdoor commitment schemes.*

2.7 Description of a Commitment Scheme from the CDH Assumption

The construction below is essentially same as the identity-based trapdoor commitment scheme based on the Waters signature introduced by Dodis, Shoup, and Walfish [11]. We find that their identity-based trapdoor commitment scheme can be seen as a multi-trapdoor commitment scheme. The proof is in the appendix.

Let WSig denote the Waters signature scheme from the CDH assumption. The master public key is the verification key of WSig, the master trapdoor is the signing key of WSig, and trapdoors are signatures of WSig. We describe a multi-trapdoor commitment scheme from the CDH assumption, named CDHCom.

Master Key Generation. On input security parameter λ, selects a bilinear group $\mathbb{G}$ of prime order $p > 2^{\lambda}$, chooses $a \xleftarrow{\mathsf{U}} \mathbb{Z}_p$ and random group elements $g, v_0, v_1, \ldots, v_n \in \mathbb{G}$, and outputs master trapdoor a and master public key $PK = (g, v_0, v_1, \ldots, v_n, e(g, g)^a)$.

Selection. Outputs public key pk. This is an arbitrary n-bit string $ID \in \{0, 1\}^n$.

Trapdoor Generation. Chooses random $\omega \xleftarrow{\mathsf{U}} \mathbb{Z}_p$ and outputs $td = (\sigma_1, \sigma_2)$, where

$$\sigma_1 = g^a \left(v_0 \prod_{i=1}^n v_i^{ID_i} \right)^{\omega}, \quad \sigma_2 = g^{\omega}.$$

Let ID_i denote the i-th bit of ID. Let $v_{ID} := v_0 \prod_{i=1}^n v_i^{ID_i}$ for notational convention.

Commitment. On input PK, $pk = ID$, chooses $\bar{\sigma}_1, \bar{\sigma}_2 \overset{\cup}{\leftarrow} \mathbb{G}$ and let $\gamma_1 := e(\bar{\sigma}_1, g), \gamma_2 := e(\bar{\sigma}_2, v_{ID}^{-1}), \gamma := e(g, g)^a$. In order to commit $m \in \mathbb{Z}_p$, chooses $r_1, r_2 \overset{\cup}{\leftarrow} \mathbb{Z}_p$ and computes

$$(C, D) = \mathsf{Com}(PK, ID, m; r) = ((\bar{\sigma}_1, \bar{\sigma}_2, \gamma_1^{r_1} \gamma_2^{r_2} \gamma^m), (m, r_1, r_2)).$$

Verification. On input $PK, pk = ID$, commitment C, decommitment $D = (r_1, r_2, m)$, outputs 1 if and only if

$$C = (\bar{\sigma}_1, \bar{\sigma}_2, \gamma_1^{r_1} \gamma_2^{r_2} \gamma^m).$$

Equivocation. In order to make a fake commitment, the same procedure as ordinary commitment is executed except that using trapdoor (σ_1, σ_2). On input $PK, pk = ID, C = (\bar{\sigma}_1, \bar{\sigma}_2, \gamma_1^{\hat{r}_1} \gamma_2^{\hat{r}_2} \gamma^m)$, $D = (\hat{r}_1, \hat{r}_2)$, and $td = (\sigma_1, \sigma_2)$, to open to $\tilde{m} \neq m$, let $\tilde{r}_1 := \hat{r}_1 - (\tilde{m} - m) \cdot \bar{r}_1 \bmod p$, $\tilde{r}_2 := \hat{r}_2 - (\tilde{m} - m) \cdot \bar{r}_2 \bmod p$ and output $\tilde{D} = \mathsf{Equiv}(PK, pk, C, D, td) = (\tilde{m}, \tilde{r}_1, \tilde{r}_2)$.

3 The Construction from the RSA Assumption

We present our multi-trapdoor commitment scheme. Let HWSig denote the Hohenberger-Waters signature scheme from the RSA assumption. The master public key is the verification key of HWSig, the master trapdoor is the signing key of HWSig, public keys are arbitrary strings (corresponding to messages of signature), and trapdoors are signatures of HWSig [21].

3.1 Description of a Commitment Scheme from the RSA Assumption

We describe our multi-trapdoor commitment scheme from the RSA assumption, named RSACom.

Master Key Generation. On input security parameter λ, outputs an RSA modulus N, such that $2^\ell < \phi(N) < 2^{\ell+2}$, where ℓ depends on 1^λ, and chooses random $h \overset{\cup}{\leftarrow} \mathbb{Z}_N^*$, random key K for PRF function $F : \{0,1\}^* \to \{0,1\}^\ell$ and random $c \in \{0,1\}^\ell$. Function $H : \{0,1\}^* \to \{0,1\}^\ell$ is defined as follows:

$$H_{K,c}(z) := F_K(i, z) \oplus c,$$

where value i is the smallest $i \geq 1$ such that $F_K(i, z) \oplus c$ is odd and prime. Value i is called *resolving index*. Outputs master public key $PK = (N, h, c, K)$ and master trapdoor $TK = (p, q, PK)$.

Selection. Outputs public key pk. It is an arbitrary n-bit string $ID \in \{0,1\}^n$.

Trapdoor Generation. Let $ID^{(i)}$ denote the first i bits of ID, that is, the length i of prefix ID. For $i = 1$ to n, computes $e_i = H_{K,c}(ID^{(i)})$ and outputs trapdoor $td = h^{\prod_{i=1}^{n} e_i^{-1}} \bmod N$.

Commitment. On input PK, $pk = ID$, for $i = 1$ to n, computes $e_i = H_{K,c}(ID^{(i)})$. Let $e_{\min}$ denote the smallest value among $(e_1, \ldots, e_n)$. The *message space is* $\mathbb{Z}_{e_{\min}}$. *This restriction is important for the security proof.* In order to commit $m \in \mathbb{Z}_{e_{\min}}$, computes

$$(C, D) = \mathsf{Com}(PK, ID, m; r) = (h^m r^{\prod_{i=1}^n e_i} \bmod N, (m, r)).$$

Verification. On input PK, $pk = ID$, commitment C, decommitment $D = (m, r)$, $e_i = H_{K,c}(ID^{(i)})$, outputs 1 if and only if

$$C = h^m r^{\prod_{i=1}^n e_i}.$$

Equivocation. In order to make a fake commitment, the same procedure as ordinary commitment is executed. In order to open to $\tilde{m} \neq m$, using trapdoor $td = h^{\prod_{i=1}^n e_i^{-1}}$ for $pk = ID$ where $e_i = H_{K,c}(ID^{(i)})$, set $\tilde{r} = td^{m-\tilde{m}} \cdot r$. That is, $\tilde{D} = \mathsf{Equiv}(PK, pk, C, D, td) = (\tilde{m}, \tilde{r})$. Then, it holds that

$$\begin{aligned}
h^{\tilde{m}} \cdot \tilde{r}^{\prod_{i=1}^n e_i} &= h^{\tilde{m}} \cdot td^{(m-\tilde{m}) \prod_{i=1}^n e_i} \cdot r^{\prod_{i=1}^n e_i} \\
&= h^{\tilde{m}} \cdot (h^{\prod_{i=1}^n e_i^{-1}})^{\prod_{i=1}^n e_i (m-\tilde{m})} \cdot r^{\prod_{i=1}^n e_i} \\
&= h^{\tilde{m}} \cdot h^{m-\tilde{m}} \cdot r^{\prod_{i=1}^n e_i} \\
&= h^m \cdot r^{\prod_{i=1}^n e_i}.
\end{aligned}$$

REMARK. We restrict the message space to $\mathbb{Z}_{e_{\min}}$ because if the message space is $\mathbb{Z}_{\prod_{i=1}^n e_i}$ then an adversary can achieve equivocation by receiving trapdoors in the strongly secure binding game. If the adversary query ID and obtain its trapdoor $td = h^{\prod_{i=1}^n e_i^{-1}}$ where $e_i = H_{K,c}(ID^{(i)})$ for $i = 1$ to n, he makes commitment $C = h^m r^{\prod_{i=1}^n e_i^*}$ to m under public key ID^* and later open it to $\tilde{m} = m - k \cdot E^*/E$ by setting $\tilde{r} = r \cdot td^k = r \cdot h^{k/E}$ where $E = \prod_{i=1}^n e_i$, $E^* = \prod_{i=1}^n e_i^*$, $e_i^* = H_{K,c}(ID^{*(i)})$, and k is some integer. It holds that $h^{\tilde{m}} \tilde{r}^{E^*} = h^{m-k \cdot E^*/E} \cdot r^{E^*} \cdot h^{k \cdot E^*/E} = h^m r^{E^*} = C$.

ON THE MESSAGE SPACE. If we restrict the message space to $\mathbb{Z}_{e_{\min}}$, the message space is determined *after* a specific public key is determined. If we want to determine the message space in advance, we can use $\ell - 1$ bits integer as a message. In this case, value m is smaller than $e_{\min}$ since e_i is ℓ bits prime.

3.2   Security Proof of **RSACom**

In this section, we show that RSACom is a multi-trapdoor commitment scheme.

Theorem 4. *If the RSA assumption holds, then* RSACom *is a multi-trapdoor commitment scheme.*

Proof. We prove the theorem by proving that RSACom is statistically hiding, computationally binding, and strongly secure binding.

Lemma 2. RSACom *is statistically hiding.*

Proof of lemma: Given value $C = h^m \cdot r^{\prod_{i=1}^{n} e_i} \mod N$, for each value $m' \neq m$ there exists unique value r' such that $C = h^{m'} \cdot (r')^{\prod_{i=1}^{n} e_i}$, because e_i is relatively prime to $\phi(N)$ for all i and taking elements to the $\prod_{i=1}^{n} e_i$-th power is a permutation on $\mathbb{Z}_N^*$. Thus, C is a uniformly distributed element in $\mathbb{Z}_N^*$ and reveal nothing about the message.   ■

Lemma 3. *If the RSA assumption holds,* RSACom *is strongly secure binding.*

Proof of lemma: We show that if HWSig is existentially unforgeable against weak chosen message attacks, RSACom is strongly secure binding. It implies that RSACom is strongly secure binding if the RSA assumption holds since Hohenberger and Waters showed that HWSig is existentially unforgeable against weak chosen message attacks under the RSA assumption [21].

Assume that RSACom is not strongly secure binding, that is, there exists adversary $\mathcal{A}$ such that outputs $(C{=}h^m \cdot r^{\prod_{i=1}^{n} e_i}, m, D{=}(m,r), \widetilde{m}, \widetilde{D}, pk)$ where $\mathsf{Vrfy}(PK, pk, m, C, D) = \mathsf{Vrfy}(PK, pk, \widetilde{m}, C, \widetilde{D}) = 1$, $m \neq \widetilde{m}$, and $pk \neq pk_i$ for all i. We construct forger $\mathcal{F}$ for scheme HWSig in the weak chosen message attack game by using $\mathcal{A}$ in a black-box manner.

Adversary $\mathcal{A}$ in the strongly secure binding game firstly sends k strings $(pk_1, \ldots, pk_k)$. Forger $\mathcal{F}$ sends the challenger list Q of messages $(M_1, \ldots, M_k) := (pk_1, \ldots, pk_k)$ as queries. Then, the challenger runs $\mathsf{Sig.Gen}(1^\lambda)$ to generate verification key $vk = (N, h, c, K)$ and signing key $sk = (p, q, vk)$ and signs each queried message as $\sigma_i \xleftarrow{\mathrm{R}} \mathsf{Sig.Sign}_{sk}(M_i)$ for $i = 1$ to n. The challenger sends $(vk, \sigma_1, \ldots, \sigma_n)$ to $\mathcal{F}$.

$\mathcal{F}$ sets $PK = vk$ and $td_i = \sigma_i$ and sends PK and td_i for $i = 1$ to n (these are trapdoors matching public keys pk_i selected first) to adversary $\mathcal{A}$ in the strongly secure binding game. This is a perfect simulation. At this point, $\mathcal{A}$ outputs $(C = h^m \cdot r^{\prod_{i=1}^{n} e_i}, m, D = (m,r), \widetilde{m}, \widetilde{D} = (\widetilde{m}, \widetilde{r}), pk)$ such that $\mathsf{Vrfy}(PK, pk, m, C, D) = \mathsf{Vrfy}(PK, pk, \widetilde{m}, \widetilde{D}) = 1$, $m \neq \widetilde{m}$, and $pk \neq pk_i$ for all i. By the definition, it holds that $e_i = H_{K,c}(pk^{(i)})$ and

$$C = h^m \cdot r^{\prod_{i=1}^{n} e_i} = h^{\widetilde{m}} \cdot \widetilde{r}^{\prod_{i=1}^{n} e_i}.$$

Let $E = \prod_{i=1}^{n} e_i$ and $\delta = m - \widetilde{m}$ for notational convention. From the above equation, we can rewrite $h^\delta = (\widetilde{r}/r)^E$. It holds that $\gcd(\delta, E) = 1$ since $m - \widetilde{m} < e_{\min}$ (recall that the message space is restricted to $\mathbb{Z}_{e_{\min}}$). By using Lemma 1, we can compute $h^{E^{-1}}$ as follows: We can find integers α and β such that $\alpha \cdot \delta + \beta \cdot E = 1$, so $h = h^{\alpha\delta + \beta E} = (h^\delta)^\alpha \cdot h^{\beta E} = (\widetilde{r}/r)^{\alpha E} \cdot h^{\beta E}$, that is,

$$h^{E^{-1}} = h^{\prod_{i=1}^{n} e_i^{-1}} = \left(\frac{\widetilde{r}}{r}\right)^\alpha \cdot h^\beta.$$

This is a valid forgery of HWSig for message $pk \neq pk_i$ for all i. Thus, we succeed constructing forger $\mathcal{F}$ by using adversary $\mathcal{A}$ in the strongly secure binding game of multi-trapdoor commitment. This contradicts the security of HWSig. ∎

Lemma 4. *If the RSA assumption holds,* RSACom *is computationally binding.*

Proof of lemma: This lemma is implied by Lemma 3, so we omit the proof. ∎

From the three lemmas, the theorem follows. □

Corollary 1. *We can construct a non-interactive and reusable non-malleable commitment with respect to decommitment scheme from the RSA assumption in the CRS model.*

4   Conclusion

We presented two non-interactive multi-trapdoor commitment schemes. One is constructed from the RSA assumption, and the other is obtained from the identity-based trapdoor commitment scheme (from the CDH assumption) by Dodis, Shoup, and Walfish. Thus, we can obtain non-interactive and reusable non-malleable commitment with respect to decommitment schemes from above two standard assumptions by applying the Gennaro's conversion to the multi-trapdoor commitment schemes. One-time signature schemes can be constructed efficiently from one-way functions [5,6,24].

Acknowledgements

The authors would like to thank the anonymous referees of PKC 2010 and ACISP 2010 for their useful comments.

References

1. Barak, B.: Constant-round coin-tossing with a man in the middle or realizing the shared random string model. In: FOCS, pp. 345–355. IEEE Computer Society, Los Alamitos (2002)
2. Barak, B.: Non-Black-Box Techniques in Cryptography. PhD thesis, Department of Computer Science and Applied Mathematics, Weizmann Institute of Science (2004)
3. Bellare, M., Goldreich, O.: On defining proofs of knowledge. In: Brickell, E.F. (ed.) CRYPTO 1992. LNCS, vol. 740, pp. 390–420. Springer, Heidelberg (1993)
4. Bellare, M., Ristov, T.: Hash Functions from Sigma Protocols and Improvements to VSH. In: Pieprzyk, J. (ed.) ASIACRYPT 2008. LNCS, vol. 5350, pp. 125–142. Springer, Heidelberg (2008)
5. Bleichenbacher, D., Maurer, U.: On the Efficiency of One-Time Digital Signatures. In: Kim, K.-c., Matsumoto, T. (eds.) ASIACRYPT 1996. LNCS, vol. 1163, pp. 145–158. Springer, Heidelberg (1996)

6. Bleichenbacher, D., Maurer, U.: Optimal Tree-Based One-Time Digital Signature Schemes. In: Puech, C., Reischuk, R. (eds.) STACS 1996. LNCS, vol. 1046, pp. 363–374. Springer, Heidelberg (1996)

7. Cramer, R., Damgård, I., Schoenmakers, B.: Proofs of Partial Knowledge and Simplified Design of Witness Hiding Protocols. In: Desmedt, Y.G. (ed.) CRYPTO 1994. LNCS, vol. 839, pp. 174–187. Springer, Heidelberg (1994)

8. Damgård, I., Groth, J.: Non-interactive and reusable non-malleable commitment schemes. In: STOC, pp. 426–437. ACM, New York (2003)

9. Di Crescenzo, G., Ishai, Y., Ostrovsky, R.: Non-interactive and non-malleable commitment. In: STOC, pp. 141–150 (1998)

10. Di Crescenzo, G., Katz, J., Ostrovsky, R., Smith, A.: Efficient and Non-interactive Non-malleable Commitment. In: Pfitzmann, B. (ed.) EUROCRYPT 2001. LNCS, vol. 2045, pp. 40–59. Springer, Heidelberg (2001)

11. Dodis, Y., Shoup, V., Walfish, S.: Efficient Constructions of Composable Commitments and Zero-Knowledge Proofs. In: Wagner, D. (ed.) CRYPTO 2008. LNCS, vol. 5157, pp. 515–535. Springer, Heidelberg (2008)

12. Dolev, D., Dwork, C., Naor, M.: Non-malleable cryptography. SIAM J. Computing 30, 391–437 (2000)

13. Feige, U., Fiat, A., Shamir, A.: Zero-knowledge proofs of identity. Journal of Cryptology 1, 77–94 (1988)

14. Fischlin, M., Fischlin, R.: Efficient Non-Malleable Commitment Schemes. In: Bellare, M. (ed.) CRYPTO 2000. LNCS, vol. 1880, pp. 413–431. Springer, Heidelberg (2000)

15. Fischlin, M., Fischlin, R.: Efficient Non-Malleable Commitment Schemes. Journal of Cryptology 22(4), 530–571 (2009)

16. Garay, J.A., MacKenzie, P.D., Yang, K.: Strengthening zero-knowledge protocols using signatures. J. Cryptology 19(2), 169–209 (2006); Preliminary version appeared in Eurocrypt 2003

17. Gennaro, R.: Multi-trapdoor Commitments and Their Applications to Proofs of Knowledge Secure Under Concurrent Man-in-the-Middle Attacks. In: Franklin, M. (ed.) CRYPTO 2004. LNCS, vol. 3152, pp. 220–236. Springer, Heidelberg (2004)

18. Goldreich, O.: Foundations of Cryptography: Basic Tools, vol. 1. Cambridge Press, New York (2001)

19. Goldreich, O., Micali, S., Wigderson, A.: Proofs that Yield Nothing But Their Validity, or All Languages in NP Have Zero-Knowledge Proof Systems. Journal of the ACM 38(3), 691–729 (1991); preliminary version appeared in FOCS 1986

20. Guillou, L.C., Quisquater, J.-J.: A "paradoxical" Indentity-Based Signature Scheme Resulting From Zero-Knowledge. In: Goldwasser, S. (ed.) CRYPTO 1988. LNCS, vol. 403, pp. 216–231. Springer, Heidelberg (1990)

21. Hohenberger, S., Waters, B.: Short and Stateless Signatures from the RSA Assumption. In: Halevi, S. (ed.) Advances in Cryptology - CRYPTO 2009. LNCS, vol. 5677, pp. 654–670. Springer, Heidelberg (2009)

22. Katz, J., Lindell, Y.: Introduction to Modern Cryptography. Chapman & Hall/CRC Press (2007)

23. Kravitz, D.W.: Digital signature algorithm. U.S. Patent 5,231,668 (July 27, 1993)

24. Lamport, L.: Constructing digital signatures from a one-way function. Technical report, SRI Intl. (1979) CSL 98

25. Lin, H., Pass, R.: Non-malleability amplification. In: STOC, pp. 189–198. ACM, New York (2009)

26. MacKenzie, P.D., Yang, K.: On Simulation-Sound Trapdoor Commitments. In: Cachin, C., Camenisch, J.L. (eds.) EUROCRYPT 2004. LNCS, vol. 3027, pp. 382–400. Springer, Heidelberg (2004)

27. Naor, M.: Bit commitment using pseudorandomness. J. Cryptology 4(2), 151–158 (1991)

28. Pandey, O., Pass, R., Vaikuntanathan, V.: Adaptive One-Way Functions and Applications. In: Wagner, D. (ed.) CRYPTO 2008. LNCS, vol. 5157, pp. 57–74. Springer, Heidelberg (2008)

29. Pass, R., Rosen, A.: New and Improved Constructions of Nonmalleable Cryptographic Protocols. SIAM J. Comput. 38(2), 702–752 (2008)

30. Shamir, A.: On the Generation of Cryptographically Strong Pseudorandom Sequences. ACM Trans. Comput. Syst. 1(1), 38–44 (1983)

31. Waters, B.: Efficient identity-based encryption without random oracles. In: Cramer, R. (ed.) EUROCRYPT 2005. LNCS, vol. 3494, pp. 114–127. Springer, Heidelberg (2005)

A Multi-Trapdoor Commitment Scheme from the CDH Assumption

A.1 Security Proof of CDHCom

In this section, we show that CDHCom is a multi-trapdoor commitment scheme.

First, we show that we can achieve equivocation if we use trapdoor $td = (\sigma_1, \sigma_2)$, where

$$\sigma_1 = g^a \left(v_0 \prod_{i=1}^n v_i^{ID_i} \right)^\omega, \quad \sigma_2 = g^\omega.$$

Let commitment value $\gamma_1^{\hat{r}_1} \gamma_2^{\hat{r}_2} \gamma^m$ where $\bar{r}_1, \bar{r}_2 \overset{\cup}{\leftarrow} \mathbb{Z}_p$, $\bar{\sigma}_1 := \sigma_1^{1/\bar{r}_1}$, $\bar{\sigma}_2 := \sigma_2^{1/\bar{r}_2}$, $\gamma_1 := e(\bar{\sigma}_1, g), \gamma_2 := e(\bar{\sigma}_2, v_{ID}^{-1}), \gamma = e(g, g)^a$.

In order to open arbitrary $\tilde{m} \neq m$, let $\tilde{r}_1 := \hat{r}_1 - (\tilde{m} - m) \cdot \bar{r}_1 \bmod p$, $\tilde{r}_2 := \hat{r}_2 - (\tilde{m} - m) \cdot \bar{r}_2 \bmod p$. Then

$$
\begin{aligned}
\gamma_1^{\tilde{r}_1} \gamma_2^{\tilde{r}_2} \gamma^{\tilde{m}} &= \gamma_1^{\hat{r}_1 - (\tilde{m}-m)\cdot\bar{r}_1} \gamma_2^{\hat{r}_2 - (\tilde{m}-m)\cdot\bar{r}_2} \gamma^{\tilde{m}} \\
&= \gamma_1^{\hat{r}_1} \gamma_2^{\hat{r}_2} \gamma^{\tilde{m}} \cdot \gamma_1^{(m-\tilde{m})\bar{r}_1} \cdot \gamma_2^{(m-\tilde{m})\bar{r}_2} \\
&= \gamma_1^{\hat{r}_1} \gamma_2^{\hat{r}_2} \gamma^{\tilde{m}} \cdot e(\bar{\sigma}_1, g)^{(m-\tilde{m})\bar{r}_1} \cdot e(\bar{\sigma}_2, v_{ID}^{-1})^{(m-\tilde{m})\bar{r}_2} \\
&= \gamma_1^{\hat{r}_1} \gamma_2^{\hat{r}_2} \gamma^{\tilde{m}} \cdot e(\sigma_1, g)^{m-\tilde{m}} \cdot e(\sigma_2, v_{ID}^{-1})^{m-\tilde{m}} \\
&= \gamma_1^{\hat{r}_1} \gamma_2^{\hat{r}_2} (\gamma \cdot e(\sigma_1, g)^{-1} \cdot e(\sigma_2, v_{ID}^{-1})^{-1})^{\tilde{m}} \cdot (e(\sigma_1, g) \cdot e(\sigma_2, v_{ID}^{-1}))^m \\
&= \gamma_1^{\hat{r}_1} \gamma_2^{\hat{r}_2} \gamma^m
\end{aligned}
$$

The last equation holds since equation $\gamma \cdot e(\sigma_1, g)^{-1} \cdot e(\sigma_2, v_{ID}^{-1})^{-1} = 1$ holds by the verification condition of Waters signature.

Theorem 5. *If the CDH assumption holds, then* CDHCom *is a multi-trapdoor commitment scheme.*

Proof. We prove the theorem by proving that CDHCom is statistically hiding, computationally binding, and strongly secure binding.

Lemma 5. CDHCom *is statistically hiding.*

Proof of lemma: For generator $g_T \in \mathbb{G}_T$, we can rewrite $\gamma_1^{r_1}\gamma_2^{r_2}\gamma^m = g_T^{r_1 x + r_2 y + mz}$ for some $x, y, z \in \mathbb{Z}_p$, so it is easily seen that the commitment does not reveal any information about m. ∎

Lemma 6. *If the CDH assumption holds,* CDHCom *is strongly secure binding.*

Proof of lemma: We show that if WSig is existentially unforgeable against weak chosen message attacks, CDHCom is strongly secure binding. It implies that CDHCom is strongly secure binding under CDH assumption since Hohenberger and Waters showed that WSig is existentially unforgeable against weak chosen message attacks under RSA assumption [21].

Assume that CDHCom is not strongly secure binding, that is, there exists adversary $\mathcal{A}$ such that outputs $(C = \gamma_1^{r_1}\gamma_2^{r_2}\gamma^m, m, D = (m, r), \widetilde{m}, \widetilde{D}, pk)$ where $\mathsf{Vrfy}(PK, pk, m, C, D) = \mathsf{Vrfy}(PK, pk, \widetilde{m}, \widetilde{D}) = 1$ such that $m \neq \widetilde{m}$ and $pk \neq pk_i$ for all i. We construct forger $\mathcal{F}$ for scheme WSig in the weak chosen message attack game that outputs a valid forgery by using $\mathcal{A}$ in a black-box manner.

Adversary $\mathcal{A}$ in the strongly secure binding game firstly sends k strings $(pk_1, \ldots, pk_k)$. $\mathcal{F}$ sends the challenger list Q of messages $(M_1, \ldots, M_k) = (pk_1, \ldots, pk_k)$ as queries. Then, the challenger runs $\mathsf{Sig.Gen}(1^\lambda)$ to generate verification key $vk = (g, v_0, v_1, \ldots, v_n, e(g, g)^a)$ and signing key $sk = a$ and signs each queried message as $\sigma_i \xleftarrow{\mathsf{R}} \mathsf{Sig.Sign}_{sk}(M_i)$ for $i = 1$ to n. The challenger sends $(vk, \sigma_1, \ldots, \sigma_n)$ to $\mathcal{F}$.

$\mathcal{F}$ sets $PK = vk$ and $tk_i = \sigma_i$ and sends PK and td_i for $i = 1$ to n (these are trapdoors matching public keys pk_i selected first) to adversary $\mathcal{A}$ in the strongly secure binding game. At this point, $\mathcal{A}$ outputs $(C = \bar{\gamma} = \gamma_1^{r_1}\gamma_2^{r_2}\gamma^m, m, D = (m, r_1, r_2), \widetilde{m}, C, \widetilde{D}, pk)$ where $\mathsf{Vrfy}(PK, pk, m, C, D) = \mathsf{Vrfy}(PK, pk, \widetilde{m}, \widetilde{D}) = 1$ such that $m \neq \widetilde{m}$ and $pk \neq pk_i$ for all i. By the definition, these can be seen as two accepted conversation with the same first message of Σ-protocol for proving knowledge of Waters signature (Here, m and $\widetilde{m}$ can be seen as distinct challenges).

Therefore, by the special soundness property of Σ-protocols, we can obtain witness $w_1, w_2 \in \mathbb{Z}_p$ such that $e(\bar{\sigma}_1, 4g)^{w_1} \cdot e(\bar{\sigma}_2, v_{pk}^{-1})^{w_2} = e(g, g)^a$, that is, we can obtain Waters signature $(\sigma_1, \sigma_2) = (\bar{\sigma}_1^{w_1}, \bar{\sigma}_2^{w_2})$. This is a valid forgery of WSig for message $pk \neq pk_i$ for all i. Thus, we succeed constructing forger $\mathcal{F}$ by using adversary $\mathcal{A}$ in the strongly secure binding game of multi-trapdoor commitment. This contradicts the security of WSig. ∎

Lemma 7. *If the CDH assumption holds,* CDHCom *is computationally binding.*

Proof of lemma: This is implied by the above lemma. ∎

From the three lemmas, the theorem follows. □

Identity-Based Chameleon Hash Scheme without Key Exposure

Xiaofeng Chen[1], Fangguo Zhang[2], Willy Susilo[3], Haibo Tian[2],
Jin Li[4], and Kwangjo Kim[5]

[1] Key Laboratory of Computer Networks and Information Security,
Ministry of Education, Xidian University, Xi'an 710071, P.R.China
xfchen@xidian.edu.cn
[2] School of Information Science and Technology,
Sun Yat-sen University, Guangzhou 510275, P.R. China
{isszhfg,tianhb}@mail.sysu.edu.cn
[3] School of Computer Science and Software Engineering,
University of Wollongong, New South Wales 2522, Australia
wsusilo@uow.edu.au
[4] School of Computer Science and Educational Software,
Guangzhou University, Guangzhou 510006, P.R. China
jinli71@gmail.com
[5] Department of Computer Science, KAIST, Daejeon 305-714, Korea
kkj@kaist.ac.kr

Abstract. The notion of chameleon hash function without key exposure plays an important role in designing chameleon signatures. However, all of the existing key-exposure free chameleon hash schemes are presented in the setting of certificate-based systems. In 2004, Ateniese and de Medeiros questioned whether there is an efficient construction for identity-based chameleon hashing without key exposure.

In this paper, we propose the first identity-based chameleon hash scheme without key exposure based on the three-trapdoor mechanism, which provides an affirmative answer to the open problem.

Keywords: Chameleon hashing, Identity-based system, Key exposure.

1 Introduction

Chameleon signatures, introduced by Krawczyk and Rabin [28], are based on well established hash-and-sign paradigm, where a chameleon hash function is used to compute the cryptographic message digest. A chameleon hash function is a trapdoor one-way hash function, which prevents everyone except the holder of the trapdoor information from computing the collisions for a randomly given input. Chameleon signatures simultaneously provide the properties of non-repudiation and non-transferability for the signed message as undeniable signatures [3,10,11,12,14,16,21,22,23,25,26,27,33] do, but the former allows for simpler and more efficient realization than the latter. In particular, chameleon signatures are non-interactive and less complicated. More precisely, the signer can generate

R. Steinfeld and P. Hawkes (Eds.): ACISP 2010, LNCS 6168, pp. 200–215, 2010.
© Springer-Verlag Berlin Heidelberg 2010

the chameleon signature without interacting with the designated recipient, and the recipient will be able to verify the signature without the collaboration of the signer. On the other hand, if presented with a forged signature, the signer can deny its validity by only revealing certain values. That is, the forged-signature denial protocol is also non-interactive. Besides, since the chameleon signatures are based on well established hash-and-sign paradigm, it provides more generic and flexible constructions.

One limitation of the original chameleon signature scheme is that signature forgery (*i.e.*, collision computation) results in the signer recovering the recipient's trapdoor information, *i.e.*, the private key. This is named as the key exposure problem of chameleon hashing, firstly addressed by Ateniese and de Medeiros [1] in 2004. If the signer knows the recipient's trapdoor information, he then can use it to deny *other* signatures given to the recipient. In the worst case, the signer could collaborate with other individuals to invalidate any signatures which were designated to be verified by the same public key. This will create a strong disincentive for the recipient to compute the hash collisions and thus weakens the property of non-transferability.

The original two constructions of chameleon hashing [28] both suffer from the key exposure problem. Ateniese and de Medeiros [1] first introduced the idea of identity-based chameleon hashing to solve this problem. Due to the distinguishing property of identity-based system [37], the signer can sign a message to an intended recipient, without having to first retrieve the recipient's certificate. Moreover, the signer uses a different public key (corresponding to a different private key) for each transaction with a recipient, so that signature forgery only results in the signer recovering the trapdoor information associated to a single transaction. Therefore, the signer will not be capable of denying signatures on any message in other transactions. However, this kind of transaction-specific chameleon hash scheme still suffers from the key exposure problem unless an identity is never reused in the different chameleon signatures, which requires that the public/secret key pair of the recipient must be changed for each transaction. We argue that this idea only provides a partial solution for the key exposure problem of chameleon hashing.[1]

Chen et al. [17] proposed the first full construction of a key-exposure free chameleon hash function in the gap Diffie-Hellman (GDH) groups with bilinear pairings. Ateniese and de Medeiros [2] then presented three key-exposure free chameleon hash functions, two based on the RSA assumption, as well as a new construction based on bilinear pairings. Gao et al. [19] proposed a factoring-based chameleon hash scheme without key exposure. Recently, Gao et al. [20] claimed to present a key-exposure free chameleon hash scheme based on the Schnorr signature. However, it requires an interactive protocol between the signer and the recipient and thus violates the basic definition of chameleon hashing and signatures. Chen et al. [18] propose the first discrete logarithm based

---

[1] A trivial solution for the key exposure problem is that the signer changes his key pair frequently in the chameleon signature scheme. However, it is only meaningful in theoretical sense because the key distribution problem arises simultaneously.

key-exposure free chameleon hash scheme without using the GDH groups. However, all of the above constructions are presented in the setting of certificate-based systems where the public key infrastructure (PKI) is required.Zhang et al. [38] presented two identity-based chameleon hash schemes from bilinear pairings, but neither of them is key-exposure free. As pointed out by Ateniese and de Medeiros, the single-trapdoor commitment schemes are not sufficient for the construction of key-exposure free chameleon hashing and the double-trapdoor mechanism [24] can be used to construct either an identity-based chameleon hash scheme or a key-exposure free one, but not both. Therefore, an interesting open problem is whether there is an efficient construction for identity-based chameleon hashing without key exposure [2].

Our Contribution. In this paper, we propose the first identity-based chameleon hash scheme without key exposure, which provides an affirmative answer to the open problem introduced by Ateniese and de Medeiros in 2004. Moreover, the proposed chameleon hash scheme is proved to achieve all the desired security notions in the random oracle model.

Organization. The rest of the paper is organized as follows: Some preliminaries are given in Section 2. The definitions associated with identity-based chameleon hashing are introduced in Section 3. The proposed identity-based key-exposure free chameleon hash scheme and its security analysis are given in Section 4. Finally, conclusions will be made in Section 5.

2   Preliminaries

In this section, we first introduce the basic definition and properties of bilinear pairings and some well-known number-theoretic problems in the gap Diffie-Hellman groups. We then present some proof systems for knowledge of discrete logarithms.

2.1   Bilinear Pairings and Number-Theoretic Problems

Let $\mathbb{G}_1$ be a cyclic additive group generated by P, whose order is a prime q, and $\mathbb{G}_2$ be a cyclic multiplicative group of the same order q. Let a and b be elements of $\mathbb{Z}_q^*$. A bilinear pairing is a map $e : \mathbb{G}_1 \times \mathbb{G}_1 \to \mathbb{G}_2$ with the following properties:

1. Bilinear: $e(aR, bQ) = e(R, Q)^{ab}$ for all $R, Q \in \mathbb{G}_1$ and $a, b \in \mathbb{Z}_q^*$.
2. Non-degenerate: There exists R and $Q \in \mathbb{G}_1$ such that $e(R, Q) \neq 1$.
3. Computable: There is an efficient algorithm to compute $e(R, Q)$ for all $R, Q \in \mathbb{G}_1$.

In the following we introduce some problems in $\mathbb{G}_1$.

- Discrete Logarithm Problem (DLP): Given two elements P and Q, to find an integer $n \in \mathbb{Z}_q^*$, such that $Q = nP$ whenever such an integer exists.
- Computation Diffie-Hellman Problem (CDHP): Given P, aP, bP for $a, b \in \mathbb{Z}_q^*$, to compute abP.

- Decision Diffie-Hellman Problem (DDHP): Given P, aP, bP, cP for $a, b, c \in \mathbb{Z}_q^*$, to decide whether $c \equiv ab \bmod q$.

It is proved that the CDHP and DDHP are not equivalent in the group $\mathbb{G}_1$ and thus called a gap Diffie-Hellman (GDH) group. More precisely, we call $\mathbb{G}$ a GDH group if the DDHP can be solved in polynomial time but there is no polynomial time algorithm to solve the CDHP with non-negligible probability. The examples of such a group can be found in supersingular elliptic curves or hyperelliptic curves over finite fields. For more details, see [4,5,6,9,29,30,32,35]. Moreover, we call $< P, aP, bP, cP >$ a valid Diffie-Hellman tuple if $c \equiv ab \bmod q$.

Since the DDHP in the group $\mathbb{G}_1$ is easy, it cannot be used to design cryptosystems in $\mathbb{G}_1$. Boneh and Franklin [6] introduced a new problem in $(\mathbb{G}_1, \mathbb{G}_2, e)$ named Bilinear Diffie-Hellman Problem:

- Bilinear Diffie-Hellman Problem (BDHP): Given P, aP, bP, cP for $a, b, c \in \mathbb{Z}_q^*$, to compute $e(P, P)^{abc} \in \mathbb{G}_2$.

Trivially, the BDHP in $(\mathbb{G}_1, \mathbb{G}_2, e)$ is no harder than the CDHP in $\mathbb{G}_1$ or $\mathbb{G}_2$. However, the converse is still an open problem. On the other hand, currently it seems that there is no polynomial time algorithm to solve the BDHP in $(\mathbb{G}_1, \mathbb{G}_2, e)$ with non-negligible probability. The security of our proposed identity-based chameleon hash scheme without key exposure is also based on the hardness of the BDHP in $(\mathbb{G}_1, \mathbb{G}_2, e)$.

2.2 Proofs of Knowledge

A prover with possession a secret number $x \in \mathbb{Z}_q$ wants to show a verifier that $x = \log_g y$ without exposing x. This is named the proof of knowledge of a discrete logarithm.

This proof of knowledge is basically a Schnorr signature [36] on message (g, y): The prover chooses a random number $r \in_R \mathbb{Z}_q$, and then computes $c = H(g, y, g^r)$, and $s = r - cx \bmod q$, where $H : \{0,1\}^* \rightarrow \{0,1\}^k$ is a collision-resistant hash function. The verifier accepts the proof if and only if $c = H(g, y, g^s y^c)$.

Definition 1. *A pair* $(c, s) \in \{0,1\}^k \times \mathbb{Z}_q$ *satisfying the equation*

$$c = H(g, y, g^s y^c)$$

is a proof of knowledge of a discrete logarithm of the element y to the base g.

Similarly, we can define the proof of knowledge for the equality of two discrete logarithms: A prover with possession a secret number $x \in \mathbb{Z}_q$ wants to show that $x = \log_g u = \log_h v$ without exposing x.

Chaum and Pedersen [15] firstly proposed the proof as follows: The prover chooses a random number $r \in_R \mathbb{Z}_q$, and then computes $c = H(g, h, u, v, g^r, h^r)$, and $s = r - cx \bmod q$, where $H : \{0,1\}^* \rightarrow \{0,1\}^k$ is a collision-resistant hash function. The verifier accepts the proof if and only if $c = H(g, h, u, v, g^s u^c, h^s v^c)$. Trivially, the verifier can efficiently decide whether $< g, u, h, v >$ is a valid Diffie-Hellman tuple with the pair (c, s).

Definition 2. *A pair $(c, s) \in \{0, 1\}^k \times \mathbb{Z}_q$ satisfying the equation*

$$c = H(g, h, u, v, g^s u^c, h^s v^c)$$

is a proof of knowledge for the equality of two discrete logarithms of elements u, v with respect to the base g, h.

The identity-based proof of knowledge for the equality of two discrete logarithms, first introduced by Baek and Zheng [8] from bilinear pairings. Define $g = e(P, P)$, $u = e(P, S_{ID})$, $h = e(Q, P)$ and $v = e(Q, S_{ID})$, where P and Q are independent elements of $\mathbb{G}_1$. The following non-interactive protocol presents a proof of knowledge that $\log_g u = \log_h v$: The prover chooses a random number $r \in_R \mathbb{Z}_q$, and then computes $c = H(g, h, u, v, g^r, h^r)$, and $S = rP - cS_{ID}$, where $H : \{0, 1\}^* \to \{0, 1\}^k$ is a collision-resistant hash function. The verifier accepts the proof if and only if $c = H(g, h, u, v, e(P, S)u^c, e(Q, S)v^c)$.

Definition 3. *A pair $(c, S) \in \{0, 1\}^k \times \mathbb{G}_1$ satisfying the equation*

$$c = H(g, h, u, v, e(P, S)u^c, e(Q, S)v^c)$$

is an identity-based proof of knowledge for the equality of two discrete logarithms of elements u, v with respect to the base g, h.

3   Definitions

In this section, we introduce the formal definitions and security requirements of identity-based chameleon hashing [1,2].

3.1   Identity-Based Chameleon Hashing

A chameleon hash function is a trapdoor collision-resistant hash function, which is associated with a trapdoor/hash key pair (TK, HK). Anyone who knows the public key HK can efficiently compute the hash value for each input. However, there exists no efficient algorithm for anyone except the holder of the secret key TK, to find collisions for every given input. In the identity-based chameleon hash scheme, the hash key HK is just the identity information ID of the user. A trusted third party called Private Key Generator (PKG) computes the trapdoor key TK associated with HK for the user.

Definition 4. *An identity-based chameleon hash scheme consists of four efficiently computable algorithms:*

- **Setup:** *PKG runs this probabilistic polynomial-time algorithm to generate a pair of keys (SK, PK) defining the scheme. PKG publishes the system parameters SP including PK, and keeps the master key SK secret. The input to this algorithm is a security parameter k.*

- **Extract:** *A deterministic polynomial-time algorithm that, on input the master key SK and an identity string ID, outputs the trapdoor key TK associated to the hash key ID.*
- **Hash:** *A probabilistic polynomial-time algorithm that, on input the master public key PK, an identity string ID, a customized identity L,[2] a message m, and a random string r,[3] outputs the hash value $h = \mathsf{Hash}(PK, ID, L, m, r)$. Note that h does not depend on TK and we denote $h = \mathsf{Hash}(ID, L, m, r)$ for simplicity throughout this paper.*
- **Forge:** *A deterministic polynomial-time algorithm $\mathcal{F}$ that, on input the trapdoor key TK associated to the identity string ID, a customized identity L, a hash value h of a message m, a random string r, and another message $m' \neq m$, outputs a string r' that satisfies*

$$h = \mathsf{Hash}(ID, L, m, r) = \mathsf{Hash}(ID, L, m', r').$$

More precisely,
$$r' = \mathcal{F}(TK, ID, L, h, m, r, m').$$

Moreover, if r is uniformly distributed in a finite space $\mathcal{R}$, then the distribution of r' is computationally indistinguishable from uniform in $\mathcal{R}$.

3.2   Security Requirements

The most dangerous attack on the identity-based chameleon hashing is the recovery of either the master key SK or the trapdoor key TK. In this case, the chameleon hash scheme would be totally broken. A weaker attack is that an active adversary computes a collision of the chameleon hashing without the knowledge of the trapdoor TK. In this security model, the adversary is allowed to compromise various users and obtain their secrets, and makes queries to the algorithm **Extract** on the adaptively chosen identity strings except the target one. Therefore, the first essential requirement for identity-based chameleon hashing is the collision resistance against active attackers.

Definition 5. (Collision resistance against active attackers): *Let ID be a target identity string and m be a target message. Let k be the security parameter. The chameleon hash scheme is collision resistance against active attackers if, for all non-constant polynomials $f_1()$ and $f_2()$, there exists no efficient algorithm $\mathcal{A}$ that, on input a customized identity L, outputs a message $m' \neq m$, and two random strings r and r' such that $\mathsf{Hash}(ID, L, m', r') = \mathsf{Hash}(ID, L, m, r)$, with non-negligible probability. Suppose that $\mathcal{A}$ runs in time less than $f_1(k)$, and makes at most $f_2(k)$ queries to the **Extract** oracle on the adaptively chosen identity strings other than ID.*

---

[2] A customized identity is actually a label for each transaction. For example, we can let $L = ID_S || ID_R || ID_T$, where ID_S, ID_R, and ID_T denote the identity of the signer, recipient, and transaction, respectively [1].

[3] Note that r can be either a randomly chosen element in a finite space $\mathcal{R}$, or a bijective function of a random variant which is uniformly distributed in a domain $\mathcal{D}$.

The second requirement for identity-based chameleon hashing is the semantic security, *i.e.*, the chameleon hash value does not reveal anything about the possible message that was hashed.

Definition 6. (Semantic security): *Let $H[X]$ denote the entropy of a random variable X, and $H[X|Y]$ the entropy of the variable X given the value of a random function Y of X. Semantic security is the statement that the conditional entropy $H[m|h]$ of the message given its chameleon hash value h equals the total entropy $H[m]$ of the message space.*

The identity-based chameleon hashing must also be key-exposure free. It was pointed out that all key-exposure free chameleon hash schemes must have (at least) double trapdoors: a master trapdoor, and an ephemeral trapdoor associated with a customized identity [2]. Loosely speaking, key exposure freeness means that even if the adversary $\mathcal{A}$ has obtained polynomially many ephemeral trapdoors associated with the corresponding customized identities, there is no efficient algorithm for $\mathcal{A}$ to compute a new ephemeral trapdoor. Formally, we have the following definition.

Definition 7. (Key exposure freeness): *If a recipient with identity ID has never computed a collision under a customized identity L, then there is no efficient algorithm for an adversary $\mathcal{A}$ to find a collision for a given chameleon hash value $\mathsf{Hash}(ID, L, m, r)$. This must remain true even if the adversary $\mathcal{A}$ has oracle access to $\mathcal{F}$ and is allowed polynomially many queries on triples (L_j, m_j, r_j) of his choice, except that L_j is not allowed to equal the challenge L.*

4   Identity-Based Key-Exposure Free Chameleon Hashing

All of the existing identity-based chameleon hash schemes [1,38] are based on the double-trapdoor mechanism and suffer from the key exposure problem. In more detail, there are two trapdoors in these chameleon hash schemes: One is the master key x of PKG, and the other is the secret key S_{ID} of the user with identity information ID (In identity-based systems, S_{ID} is actually a signature of PKG on message ID with the secret key x). Given a collision of the chameleon hash function, the trapdoor key S_{ID} will be revealed. Ateniese and de Medeiros [2] thus concluded that the double-trapdoor mechanism cannot be used to construct an efficient chameleon hash scheme that is simultaneously identity-based and key-exposure free, but the multiple-trapdoor (more than two, and consecutive trapdoors) mechanism *perhaps* could provide such a construction.

In this section, we first propose an identity-based key-exposure free chameleon hash scheme based on bilinear pairings. There are three consecutive trapdoors in our chameleon hash scheme: The first one is the master key x of PKG, the second one is the secret key $S_{ID} = xH(ID)$ of the user with identity information ID, and the third one is the ephemeral trapdoor $e(H(L), S_{ID})$ for each transaction with the customized identity L. Given a collision of the chameleon hash function, only the ephemeral trapdoor $e(H(L), S_{ID})$ is revealed, but the permanent trapdoors x and S_{ID} still remain secret. Actually, even given polynomially many ephemeral trapdoors $e(H(L_i), S_{ID})$ associated with the label L_i, it is infeasible to compute a new ephemeral trapdoor $e(H(L), S_{ID})$ associated with the label $L \neq L_i$. Trivially, it is more difficult to compute the trapdoor x or S_{ID}. Therefore, the identity information ID and the corresponding secret key S_{ID} can be used repeatedly for different transactions.

4.1   The Proposed Identity-Based Chameleon Hash Scheme

- **Setup:** Let k be a security parameter. Let $\mathbb{G}_1$ be a GDH group generated by P, whose order is a prime q, and $\mathbb{G}_2$ be a cyclic multiplicative group of the same order q. A bilinear pairing is a map $e : \mathbb{G}_1 \times \mathbb{G}_1 \to \mathbb{G}_2$. Let $H : \{0,1\}^* \to \mathbb{G}_1$ be a full-domain collision-resistant hash function [7,13,34]. PKG picks a random integer $x \in_R \mathbb{Z}_q^*$ and computes $P_{pub} = xP$. The system parameters are $SP = \{\mathbb{G}_1, \mathbb{G}_2, q, e, P, P_{pub}, H, k\}$.
- **Extract:** Given an identity string ID, computes the trapdoor key $S_{ID} = xH(ID) = xQ_{ID}$.
- **Hash:** On input the hash key ID, a customized identity L, a message m, chooses a random integer $a \in_R \mathbb{Z}_q^*$, and computes $r = (aP, e(aP_{pub}, Q_{ID}))$. Our proposed chameleon hash function is defined as

$$\mathcal{H} = \mathsf{Hash}(ID, L, m, r) = aP + mH(L).$$

Note that $\mathcal{H}$ does not depend on the trapdoor key S_{ID}. Besides, if a is a uniformly random integer in $\mathbb{Z}_q^*$, then the string $r = (aP, e(aP_{pub}, Q_{ID}))$ can be viewed as a random input of the chameleon hash function $\mathcal{H}$. We argue that a is not an input of $\mathcal{H}$.

- **Forge:** For any valid hash value $\mathcal{H}$, the algorithm $\mathcal{F}$ can be used to compute a string r' with the trapdoor key S_{ID} as follows:

$$r' = \mathcal{F}(S_{ID}, ID, L, \mathcal{H}, m, aP, e(aP_{pub}, Q_{ID}), m') = (a'P, e(a'P_{pub}, Q_{ID})),$$

where

$$a'P = aP + (m - m')H(L),$$

$$e(a'P_{pub}, Q_{ID}) = e(aP_{pub}, Q_{ID})e(H(L), S_{ID})^{m-m'}.$$

Note that

$$\mathsf{Hash}(ID, L, m', a'P, e(a'P_{pub}, Q_{ID})) = \mathsf{Hash}(ID, L, m, aP, e(aP_{pub}, Q_{ID}))$$

and

$$
\begin{aligned}
e(a'P_{pub}, Q_{ID}) &= e(a'P, S_{ID}) \\
&= e(aP + (m - m')H(L), S_{ID}) \\
&= e(aP, S_{ID})e(H(L), S_{ID})^{m-m'} \\
&= e(aP_{pub}, Q_{ID})e(H(L), S_{ID})^{m-m'}
\end{aligned}
$$

Therefore, the forgery is successful. Moreover, if $(aP, e(aP_{pub}, Q_{ID}))$ is uniformly distributed, then the distribution of $(a'P, e(a'P_{pub}, Q_{ID}))$ is computationally indistinguishable from uniform.

Remark 1. Given a string $r = (aP, e(aP_{pub}, Q_{ID}))$, a necessary condition is the equality of two discrete logarithms of elements aP and $e(aP_{pub}, Q_{ID})$ with respect to the base P and $e(P_{pub}, Q_{ID})$, i.e., $\log_P aP = \log_{e(P_{pub}, Q_{ID})} e(aP_{pub}, Q_{ID})$. Obviously, the holder R of the trapdoor key S_{ID} can be convinced of the fact if the equation $e(aP, S_{ID}) = e(aP_{pub}, Q_{ID})$ holds: If $e(aP, S_{ID}) = e(aP_{pub}, Q_{ID})$ holds, then we have $\log_P aP = \log_{e(P, S_{ID})} e(aP, S_{ID}) = \log_{e(P, S_{ID})} e(aP_{pub}, Q_{ID})$ $= \log_{e(P_{pub}, Q_{ID})} e(aP_{pub}, Q_{ID})$.

In the chameleon signatures, it is also essential for any third party without knowing S_{ID} (e.g., a Judge) to verify the validity of r. Due to the identity-based proof of knowledge for the equality of two discrete logarithms in section 2.2, R can prove that $< e(P, P), e(P_{pub}, Q_{ID}), e(aP, P), e(aP_{pub}, Q_{ID}) >$ is a valid Diffie-Hellman tuple. If $< e(P, P), e(P_{pub}, Q_{ID}), e(aP, P), e(aP_{pub}, Q_{ID}) >$ is a valid Diffie-Hellman tuple, then $< e(P, P), e(aP, P), e(P_{pub}, Q_{ID}), e(aP_{pub}, Q_{ID}) >$ is also a valid Diffie-Hellman tuple. So, we have $\log_P aP = \log_{e(P, P)} e(aP, P) = \log_{e(P_{pub}, Q_{ID})} e(aP_{pub}, Q_{ID})$. Moreover, it also holds for any other string $r' = (a'P, e(a'P_{pub}, Q_{ID}))$. That is to say, for any given string r', R can prove that $< e(P, P), e(P_{pub}, Q_{ID}), e(a'P, P), e(a'P_{pub}, Q_{ID}) >$ is a valid Diffie-Hellman tuple in a computationally indistinguishable way. For more details, please refer to Appendix A.

4.2   Security Analysis

Theorem 1. *In the random oracle model, the proposed identity-based chameleon hash scheme is collision resistance against active attackers under the assumption that the BDHP in $(\mathbb{G}_1, \mathbb{G}_2, e)$ is intractable.*

Proof. Given a random instance $< P, xP, yP, zP >$ of BDHP, the aim of algorithm $\mathcal{B}$ is to compute $e(P, P)^{xyz}$. $\mathcal{B}$ runs the **Setup** algorithm of the proposed identity-based chameleon hash scheme and sets $P_{pub} = xP$. The resulting system parameters $\{\mathbb{G}_1, \mathbb{G}_2, q, e, P, H, k, P_{pub}\}$ are given to the adversary $\mathcal{A}$. The security analysis will view H as a random oracle.

Let ID be the target identity string and m be the target message. Suppose that $\mathcal{A}$ makes at most $f_1(k)$ queries to the **Extract** oracle, where $f_1(k)$ is a non-constant polynomial. $\mathcal{B}$ randomly chooses $b_i \in \mathbb{Z}_q^*$ for $i \in \{1, 2, \cdots, f_1(k)\}$, and responds to the H query and **Extract** query of $\mathcal{A}$ as follows:

$$H(L) = yP$$

$$H(ID_i) = \begin{cases} b_iP, & \text{if } ID_i \neq ID \\ zP, & \text{Otherwise} \end{cases}$$

$$S_{ID_i} = \begin{cases} b_iP_{pub}, & \text{if } ID_i \neq ID \\ \text{``Fail''}, & \text{Otherwise} \end{cases}$$

if $\mathcal{A}$ can output a message $m' \neq m$, and two strings $r = (aP, e(aP_{pub}, Q_{ID}))$ and $r' = (a'P, e(a'P_{pub}, Q_{ID}))$ such that $\mathsf{Hash}(ID, L, m', r') = \mathsf{Hash}(ID, L, m, r)$ in time T with a non-negligible probability ϵ, then $\mathcal{B}$ can compute

$$e(H(L), S_{ID}) = (e(a'P_{pub}, Q_{ID})/e(aP_{pub}, Q_{ID}))^{(m-m')^{-1}}$$

in time T as the solution of the BDHP. The success of probability of $\mathcal{B}$ is also ϵ.

Theorem 2. *The proposed identity-based chameleon hash scheme is semantically secure.*

Proof. Given an identity ID and a customized identity L, there is a one-to-one correspondence between the hash value $\mathcal{H} = \mathsf{Hash}(ID, L, m, r)$ and the string $r = (aP, e(aP_{pub}, Q_{ID}))$ for each message m. Therefore, the conditional probability $\mu(m|\mathcal{H}) = \mu(m|r)$. Note that m and r are independent variables, the equation $\mu(m|\mathcal{H}) = \mu(m)$ holds. Then, we can prove that the conditional entropy $H[m|\mathcal{H}]$ equals the entropy $H[m]$ as follows:

$$H[m|\mathcal{H}] = -\sum_m \sum_{\mathcal{H}} \mu(m, \mathcal{H}) \log(\mu(m|\mathcal{H})) = -\sum_m \sum_{\mathcal{H}} \mu(m, \mathcal{H}) \log(\mu(m))$$
$$= -\sum_m \mu(m) \log(\mu(m)) = H[m].$$

Theorem 3. *In the random oracle model, the proposed identity-based chameleon hash scheme is key-exposure free under the assumption that the BDHP in $(\mathbb{G}_1, \mathbb{G}_2, e)$ is intractable.*

Proof. Loosely speaking, the ephemeral trapdoor $e(H(L), S_{ID})$ can be viewed as the partial signature on message L in the Libert and Quisquater's identity-based undeniable signature scheme [31]. Also, in the random oracle model, their undeniable signature scheme is proved secure against existential forgery on adaptively chosen message and ID attacks under the assumption that the BDHP in $(\mathbb{G}_1, \mathbb{G}_2, e)$ is intractable. That is, even if the adversary has obtained polynomially many signatures $e(H(L_j), S_{ID})$ on message L_j, he cannot forge a signature $e(H(L), S_{ID})$ on message $L \neq L_j$. So, our chameleon hash scheme satisfies the property of key exposure freeness.

Now we give the formal proof of our chameleon hash scheme in details. Given a random instance $< P, xP, yP, zP >$ of BDHP, the aim of algorithm $\mathcal{B}$ is to compute $e(P, P)^{xyz}$ using the adversary $\mathcal{A}$. $\mathcal{B}$ firstly provides $\mathcal{A}$ the system parameters $\{\mathbb{G}_1, \mathbb{G}_2, q, e, P, H, k, P_{pub}\}$ such that $P_{pub} = xP$. The security analysis will view H as a random oracle.

Note that in our chameleon hash scheme, the ephemeral trapdoor $e(H(L), S_{ID})$ can be used to compute a collision (m', r') of the given chameleon hash value $\mathcal{H}$ in any desired way. On the other hand, any collision (m', r') will result in the recovery of the ephemeral trapdoor $e(H(L), S_{ID})$. For the ease of explanation, in the following we let the output of the algorithm $\mathcal{F}$ be the ephemeral trapdoor $e(H(L), S_{ID})$ instead of a collision (m', r'), i.e., $\mathcal{F}(\cdot) = e(H(L), S_{ID})$.

Let ID_t and L_t be the target identity and customized identity, respectively. We stress that L_t is a label only related to the target identity ID_t. That is, (ID_i, L_t) cannot be the input of the query to oracle $\mathcal{F}$ for any other identity $ID_i \neq ID_t$. Suppose that $\mathcal{A}$ makes at most $f(k)$ queries to the **Extract** oracle, where $f(k)$ is a non-constant polynomial. For each $i \in \{1, 2, \cdots, f(k)\}$, assume that $\mathcal{A}$ makes at most $g_i(k)$ queries to the $\mathcal{F}$ oracle on four-triples $(L_{i_j}, m_{i_j}, a_{i_j}P, e(a_{i_j}P_{pub}, Q_{ID_i}))$ of his choice, where $g_i(k)$ are non-constant polynomials and $j \in \{1, 2, \cdots, g_i(k)\}$. That is, $\mathcal{A}$ could obtain $g_i(k)$ ephemeral trapdoors $e(H(L_{i_j}), S_{ID_i})$ for each $i \in \{1, 2, \cdots, f(k)\}$. At the end of the game, the output of $\mathcal{A}$ is a collision of the hash value $\mathcal{H} = \mathsf{Hash}(ID_t, L_t, m, aP, e(aP_{pub}, Q_{ID_t}))$ where $L_t \neq L_{t_j}$ and $j \in \{1, 2, \cdots, g_t(k)\}$, i.e., a new ephemeral trapdoor $e(H(L_t), S_{ID_t})$ for $H(L_t) \neq H(L_{t_j})$.

$\mathcal{B}$ randomly chooses $b_i \in \mathbb{Z}_q^*$ and $c_{i_j} \in \mathbb{Z}_q^*$ for $i \in \{1, 2, \cdots, f(k)\}$, $j \in \{1, 2, \cdots, g_i(k)\}$, and then responds to the H query, **Extract** query, and $\mathcal{F}$ query of $\mathcal{A}$ as follows:

$$H(L_{i_j}) = \begin{cases} c_{i_j}P, & \text{if } L_{i_j} \neq L_t \\ yP, & \text{Otherwise} \end{cases}$$

$$H(ID_i) = \begin{cases} b_iP, & \text{if } ID_i \neq ID_t \\ zP, & \text{Otherwise} \end{cases}$$

$$S_{ID_i} = \begin{cases} b_iP_{pub}, & \text{if } ID_i \neq ID_t \\ \text{``Fail''}, & \text{Otherwise} \end{cases}$$

$$\mathcal{F}(\cdot) = \begin{cases} e(c_{i_j}P, b_iP_{pub}), & \text{if } ID_i \neq ID_t \\ e(c_{t_j}P_{pub}, zP), & \text{if } ID_i = ID_t \text{ and } L_{i_j} \neq L_t \\ \text{``Fail''}, & \text{if } ID_i = ID_t \text{ and } L_{i_j} = L_t \end{cases}$$

We say $\mathcal{A}$ wins the game if $\mathcal{A}$ outputs a new valid trapdoor $e(H(L_t), S_{ID_t})$ in time T with a non-negligible probability ϵ. Note that $e(H(L_t), S_{ID_t}) = e(P, P)^{xyz}$, so $\mathcal{B}$ can solve the BDHP in time T with the same probability ϵ.

5   Conclusions

Chameleon signatures simultaneously provide the properties of non-repudiation and non-transferability for the signed message, thus can be used to solve the conflict between authenticity and privacy in the digital signatures. However, the original constructions suffer from the so-called key exposure problem of chameleon hashing. Recently, some constructions of key-exposure free chameleon hash schemes [2,17] are presented using the idea of "Customized Identities" while in the setting of certificate-based systems. Besides, all of the existing identity-based chameleon hash schemes suffer from the key exposure problem. To the best of our knowledge, there seems no research work on the identity-based chameleon hash scheme without key exposure.

In this paper, we propose the first identity-based chameleon hash scheme without key exposure, which gives an affirmative answer for the open problem introduced by Ateniese and de Medeiros in 2004.

Acknowledgement

We are grateful to the anonymous referees of ACISP 2010 for their invaluable suggestions. This work is supported by the National Natural Science Foundation of China (No. 60970144 and 60773202), Doctoral Fund of Ministry of Education of China for New Teachers (No. 20090171120006), Guangdong Natural Science Foundation (No. 8451027501001508), Program of the Science and Technology of Guangzhou, China (No. 2008J1-C231-2) and ARC Future Fellowship (FT0991397).

References

1. Ateniese, G., de Medeiros, B.: Identity-based chameleon hash and applications. In: Juels, A. (ed.) FC 2004. LNCS, vol. 3110, pp. 164–180. Springer, Heidelberg (2004)
2. Ateniese, G., de Medeiros, B.: On the key exposure problem in chameleon hashes. In: Blundo, C., Cimato, S. (eds.) SCN 2004. LNCS, vol. 3352, pp. 165–179. Springer, Heidelberg (2005)
3. Boyar, D., Chaum, D., Damgå, I., Pedersen, T.: Convertible undeniable signatures. In: Menezes, A., Vanstone, S.A. (eds.) CRYPTO 1990. LNCS, vol. 537, pp. 183–195. Springer, Heidelberg (1991)
4. Barreto, P., Kim, H., Lynn, B., Scott, M.: Efficient algorithms for Pairing-based cryptosystems. In: Yung, M. (ed.) CRYPTO 2002. LNCS, vol. 2442, pp. 354–368. Springer, Heidelberg (2002)
5. Boneh, D., Lynn, B., Shacham, H.: Short signatures from the Weil pairings. In: Boyd, C. (ed.) ASIACRYPT 2001. LNCS, vol. 2248, pp. 514–532. Springer, Heidelberg (2001)
6. Boneh, D., Franklin, M.: Identity-based encryption from the Weil pairing. In: Kilian, J. (ed.) CRYPTO 2001. LNCS, vol. 2139, pp. 213–229. Springer, Heidelberg (2001)
7. Bellare, M., Rogaway, P.: The exact security of digital signatures-How to sign with RSA and Rabin. In: Maurer, U.M. (ed.) EUROCRYPT 1996. LNCS, vol. 1070, pp. 399–416. Springer, Heidelberg (1996)
8. Baek, J., Zheng, Y.: Identity-based threshold decryption. In: Bao, F., Deng, R., Zhou, J. (eds.) PKC 2004. LNCS, vol. 2947, pp. 248–261. Springer, Heidelberg (2004)
9. Cha, J., Cheon, J.: An identity-based signature from gap Diffie-Hellman groups. In: Desmedt, Y.G. (ed.) PKC 2003. LNCS, vol. 2567, pp. 18–30. Springer, Heidelberg (2002)
10. Chaum, D.: Zero-knowledge undeniable signatures. In: Damgård, I.B. (ed.) EUROCRYPT 1990. LNCS, vol. 473, pp. 458–464. Springer, Heidelberg (1991)
11. Chaum, D.: Designated confirmer signatures. In: De Santis, A. (ed.) EUROCRYPT 1994. LNCS, vol. 950, pp. 86–91. Springer, Heidelberg (1995)
12. Chaum, D., van Antwerpen, H.: Undeniable signatures. In: Brassard, G. (ed.) CRYPTO 1989. LNCS, vol. 435, pp. 212–216. Springer, Heidelberg (1990)
13. Coron, J.: On the exact security of full domain hash. In: Bellare, M. (ed.) CRYPTO 2000. LNCS, vol. 1880, pp. 229–235. Springer, Heidelberg (2000)
14. Chaum, D., van Heijst, E., Pfitzmann, B.: Cryptographically strong undeniable signatures, unconditionally secure for the signer. In: Feigenbaum, J. (ed.) CRYPTO 1991. LNCS, vol. 576, pp. 470–484. Springer, Heidelberg (1992)

15. Chaum, D., Pedersen, T.: Wallet databases with observers. In: Brickell, E.F. (ed.) CRYPTO 1992. LNCS, vol. 740, pp. 89–105. Springer, Heidelberg (1993)
16. Camenisch, J., Michels, M.: Confirmer signature schemes secure against adaptive adversaries. In: Preneel, B. (ed.) EUROCRYPT 2000. LNCS, vol. 1807, pp. 243–258. Springer, Heidelberg (2000)
17. Chen, X., Zhang, F., Kim, K.: Chameleon hashing without key exposure. In: Zhang, K., Zheng, Y. (eds.) ISC 2004. LNCS, vol. 3225, pp. 87–98. Springer, Heidelberg (2004)
18. Chen, X., Zhang, F., Tian, H., Wei, B., Kim, K.: Key-exposure free chameleon hashing and signatures based on discrete logarithm systems, Cryptology ePrint Archive: Report 2009/035 (2009)
19. Gao, W., Wang, X., Xie, D.: Chameleon hashes without key exposure based on factoring. Journal of Computer Science and Technology 22(1), 109–113 (2007)
20. Gao, W., Li, F., Wang, X.: Chameleon hash without key exposure based on Schnorr signature. Computer Standards and Interfaces 31, 282–285 (2009)
21. Galbraith, S., Mao, W., Paterson, K.G.: RSA-based undeniable signatures for general moduli. In: Preneel, B. (ed.) CT-RSA 2002. LNCS, vol. 2271, pp. 200–217. Springer, Heidelberg (2002)
22. Galbraith, S., Mao, W.: Invisibility and anonymity of undeniable and confirmer signatures. In: Joye, M. (ed.) CT-RSA 2003. LNCS, vol. 2612, pp. 80–97. Springer, Heidelberg (2003)
23. Gennaro, S., Krawczyk, H., Rabin, T.: RSA-based undeniable signatures. In: Kaliski Jr., B.S. (ed.) CRYPTO 1997. LNCS, vol. 1294, pp. 132–149. Springer, Heidelberg (1997)
24. Gennaro, R.: Multi-trapdoor commitments and their applications to proofs of knowledge secure under concurrent man-in-the-middle attacks. In: Franklin, M. (ed.) CRYPTO 2004. LNCS, vol. 3152, pp. 220–236. Springer, Heidelberg (2004)
25. Jakobsson, M., Sako, K., Impagliazzo, R.: Designated verifier proofs and their applications. In: Maurer, U.M. (ed.) EUROCRYPT 1996. LNCS, vol. 1070, pp. 143–154. Springer, Heidelberg (1996)
26. Kurosawa, K., Heng, S.: 3-move undeniable signature scheme. In: Cramer, R. (ed.) EUROCRYPT 2005. LNCS, vol. 3494, pp. 181–197. Springer, Heidelberg (2005)
27. Kurosawa, K., Heng, S.: Relations among security notions for undeniable signature schemes. In: De Prisco, R., Yung, M. (eds.) SCN 2006. LNCS, vol. 4116, pp. 34–48. Springer, Heidelberg (2006)
28. Krawczyk, H., Rabin, T.: Chameleon signatures. In: Proc. of NDSS, pp. 143–154 (2000); A preliminary version can be found at Cryptology ePrint Archive: Report 1998/010
29. Hess, F.: Efficient identity based signature schemes based on pairings. In: Nyberg, K., Heys, H.M. (eds.) SAC 2002. LNCS, vol. 2595, pp. 310–324. Springer, Heidelberg (2003)
30. Joux, A.: The Weil and Tate pairings as building blocks for public key cryptosystems. In: Fieker, C., Kohel, D.R. (eds.) ANTS 2002. LNCS, vol. 2369, pp. 20–32. Springer, Heidelberg (2002)
31. Libert, B., Quisquater, J.: ID-based undeniable signatures. In: Okamoto, T. (ed.) CT-RSA 2004. LNCS, vol. 2964, pp. 112–125. Springer, Heidelberg (2004)
32. Miller, V.: The Weil pairing, and its efficient calculation. Journal of Cryptology 17(4), 235–261 (2004)
33. Monnerat, J., Vaudenay, S.: Generic homomorphic undeniable signatures. In: Lee, P.J. (ed.) ASIACRYPT 2004. LNCS, vol. 3329, pp. 354–371. Springer, Heidelberg (2004)

34. Ogata, W., Kurosawa, K., Heng, S.: The security of the FDH variant of Chaum's undeniable signature scheme. In: Vaudenay, S. (ed.) PKC 2005. LNCS, vol. 3386, pp. 328–345. Springer, Heidelberg (2005)
35. Okamoto, T., Pointcheval, D.: The gap-problems: a new class of problems for the security of cryptographic schemes. In: Kim, K.-c. (ed.) PKC 2001. LNCS, vol. 1992, pp. 104–118. Springer, Heidelberg (2001)
36. Schnorr, C.P.: Efficient signature generation for smart cards. Journal of Cryptology 4(3), 239–252 (1991)
37. Shamir, A.: Identity-based cryptosystems and signature schemes. In: Blakely, G.R., Chaum, D. (eds.) CRYPTO 1984. LNCS, vol. 196, pp. 47–53. Springer, Heidelberg (1985)
38. Zhang, F., Safavi-Naini, R., Susilo, W.: ID-based chameleon hashes from bilinear pairings, Cryptology ePrint Archive: Report 2003/208 (2003)

Appendix A: The Resulting Chameleon Signature Scheme

Since chameleon signatures are based on well established hash-and-sign paradigm, we can construct an identity-based chameleon signature scheme by incorporating the proposed identity-based chameleon hash scheme Hash and any secure identity-based signature scheme SIGN.

There are two users, a signer S and a recipient R, in the proposed identity-based chameleon signature scheme. When dispute occurs, a judge J is involved in the scheme. Our signature scheme consists of four efficient algorithms **Setup**, **Extract**, **Sign**, **Verify**, and a specific protocol **Deny**. The algorithms of **Setup** and **Extract** are the same as in section 4.1. Let (S_{ID_S}, ID_S) be the signing/verification key pair of S, and (S_{ID_R}, ID_R) be the trapdoor/hash key pair of R.

Given a message m and a customized identity L, S randomly chooses an integer $a \in_R \mathbb{Z}_q^*$, and computes $r = (aP, e(aP_{pub}, Q_{ID_R}))$. The signature σ for message m is $\sigma = (m, r, L, \mathsf{SIGN}_{S_{ID_S}}(\mathcal{H}))$, where $\mathcal{H} = \mathsf{Hash}(ID_R, L, m, r)$.

Given a signature σ, R first uses his trapdoor key S_{ID_R} to verify whether the equation $e(aP, S_{ID_R}) = e(aP_{pub}, Q_{ID_R})$ holds. If the verification fails, he rejects the signature; else, he computes the chameleon hash value $\mathcal{H} = \mathsf{Hash}(ID_R, L, m, r)$ and verifies the validity of $\mathsf{SIGN}_{S_{ID_S}}(\mathcal{H})$ with the verification key ID_S.

When dispute occurs, R provides J a signature $\sigma = (m', r', L, \mathsf{SIGN}_{S_{ID_S}}(\mathcal{H}))$ and a non-interactive identity-based proof of knowledge Π' for the equality of two discrete logarithms that $\log_{e(P,P)} e(P_{pub}, Q_{ID_R}) = \log_{e(a'P,P)} e(a'P_{pub}, Q_{ID_R})$. If either $\mathsf{SIGN}_{S_{ID_S}}(\mathcal{H})$ or Π' is invalid, J rejects it. Otherwise, J summons S to accept/deny the claim. If S wants to accept the signature, he just confirms to J this fact. Otherwise, he provides a collision of the chameleon hash function as follows:

- If S wants to achieve the property of "message recovery", *i.e.*, he wants to prove which message was the one originally signed. In this case, S provides J the tuple (m, r, Π) as a collision, where Π is a non-interactive proof of knowledge for the equality of two discrete logarithms that $a =$

$\log_{e(P,P)} e(aP, P) = \log_{e(P_{pub}, Q_{ID_R})} e(aP_{pub}, Q_{ID_R})$. If and only if $m \neq m'$, $\mathcal{H} = \mathsf{Hash}(ID_R, L, m, r)$, and Π is valid, then J can be convinced that R forged the signature on message m' and S only generated a valid signature on message m.

- If S wants to achieve the property of "message hiding", *i.e.*, he wants to protect the confidentiality of the original message even against the judge. In this case, S provides J the tuple (m'', r'') such that $\mathcal{H} = \mathsf{Hash}(ID_R, L, m'', r'')$ as a collision. Note that given two pairs (m, r) and (m', r') such that $\mathcal{H} = \mathsf{Hash}(ID_R, L, m', r') = \mathsf{Hash}(ID_R, L, m, r)$, S can compute the ephemeral trapdoor $e(H(L), S_{ID_R}) = (e(a'P_{pub}, Q_{ID_R})/e(aP_{pub}, Q_{ID_R}))^{(m-m')^{-1}}$. Then, for a randomly chosen message m'', the string $r'' = (a''P, e(a''P_{pub}, Q_{ID_R}))$ can be computed as follows: $a''P = aP + (m - m'')H(L)$, $e(a''P_{pub}, Q_{ID_R}) = e(aP_{pub}, Q_{ID_R})e(H(L), S_{ID_R})^{m-m''}$. If R accepts the collision (m'', r''), J can be convinced that R forged the signature on message m' and the original message m is never revealed. Otherwise, R provides a non-interactive knowledge proof that r'' is not valid: Let $r'' = (U, V)$, R provide a value $W \neq V$ and a non-interactive knowledge proof that $\log_{e(P,P)} e(P_{pub}, Q_{ID_R}) = \log_{e(U,P)} W$, then J can be convinced that S generated a valid signature on message m'.[4]

Remark 2. Note that if (g, g^a, g^b, g^{ab}) is a valid Diffie-Hellman tuple, then (g, g^b, g^a, g^{ab}) is also a valid Diffie-Hellman tuple, vice versa. That is, there are two different ways (based on the knowledge a or b, respectively) to prove that (g, g^a, g^b, g^{ab}) is a valid Diffie-Hellman tuple when using the proof of knowledge for the equality of two discrete logarithms: $\log_g g^a = \log_{g^b} g^{ab}$ or $\log_g g^b = \log_{g^a} g^{ab}$. This is the main trick of the **Deny** protocol in our signature scheme. We explain it in more details.

For any random string $r' = (a'P, e(a'P_{pub}, Q_{ID_R}))$, R cannot provide a proof that $\log_P a'P = \log_{e(P_{pub}, Q_{ID_R})} e(a'P_{pub}, Q_{ID_R})$. However, R (with the knowledge of S_{ID_R}) could provide a proof that

$$\log_{e(P,P)} e(P_{pub}, Q_{ID_R}) = \log_{e(a'P,P)} e(a'P_{pub}, Q_{ID_R}).$$

That is, $\log_{e(P,P)} e(a'P, P) = \log_{e(P_{pub}, Q_{ID_R})} e(a'P_{pub}, Q_{ID_R})$. So, we can easily deduce that $\log_P a'P = \log_{e(P,P)} e(a'P, P) = \log_{e(P_{pub}, Q_{ID_R})} e(a'P_{pub}, Q_{ID_R})$. In particular, it is also holds even when $r' = r$. That is, the original input r is totally indistinguishable with any collision r'. Moreover, we stress that it is **NOT** required for R to know the value a' or a in the knowledge proof that $\log_{e(P,P)} e(P_{pub}, Q_{ID_R}) = \log_{e(a'P,P)} e(a'P_{pub}, Q_{ID_R})$.

---

[4] We must consider the case that R provides the original collision (m', r') (that is, m' is the original message to be signed) while S provides an invalid collision (m'', r'') to cheat J. Note that if $\log_{e(P,P)} e(P_{pub}, Q_{ID_R}) = \log_{e(U,P)} W$, then we have $W = e(U, S_{ID_R}) = e(a''P, S_{ID_R})$. Trivially, $V \neq e(a''P_{pub}, Q_{ID_R})$. This means that the tuple (m'', r'') provide by S is not a valid collision.

On the other hand, note that only S knows the knowledge a and no one knows the knowledge $a' \neq a$. Therefore, only S can provide a proof of knowledge that $a = \log_{e(P,P)} e(aP,P) = \log_{e(P_{pub},Q_{ID_R})} e(aP_{pub},Q_{ID_R})$, and no one can provide a proof of knowledge that $a'=\log_{e(P,P)} e(a'P,P)=\log_{e(P_{pub},Q_{ID_R})} e(a'P_{pub},Q_{ID_R})$ when $a' \neq a$. This ensures that S can efficiently prove which message was the original one if he desires.

Remark 3. We can also give a new solution to achieve the property of "message hiding" in the resulting identity-based chameleon signature scheme. S chooses a random integer $\theta \in_R \mathbb{Z}_q^*$ and computes $m'' = \theta m$ and $a'' = \theta a$. Let $\mathcal{H}'' = a''P + m''H(L)$, S then provides J the tuple (m'',r'',Σ,Π) as a collision, where $r'' = (a''P, e(a''P_{pub},Q_{ID_R}))$, Σ is a non-interactive proof of knowledge of a discrete logarithm that $\theta = \log_{\mathcal{H}} \mathcal{H}''$, and Π is a non-interactive proof of knowledge for the equality of two discrete logarithms that $a'' = \log_P a''P = \log_{e(P_{pub},Q_{ID_R})} e(a''P_{pub},Q_{ID_R})$. If and only if $m'\mathcal{H}'' \neq m''\mathcal{H}$, and Σ and Π are both valid, then J can be convinced that R forged the signature on message m' and the original message m is still confidential. The reason is as follows: if $\mathcal{H}'' = \theta\mathcal{H}$, then the pair $(m,a) = (\theta^{-1}m'',\theta^{-1}a'')$ is the original tuple of S due to the hardness of discrete logarithm assumption. Otherwise, we could compute the discrete logarithm $\log_P H(L)$ while $H(L)$ can be viewed a random element in $\mathbb{G}_1$. Obviously, $(m,r) = (m,(aP,e(aP_{pub},Q_{ID_R})))$ is the original input of chameleon hashing. Besides, $m'\mathcal{H}'' \neq m''\mathcal{H}$ implies $m \neq m'$. This means that S is capable of providing a new collision different from (m',r'). Due to the randomness of θ, the original message m is kept secret in the sense of semantic security. The more detailed proof will be presented in the full version of this paper.

Remark 4. Compared with the confirm protocol of the identity-based undeniable signature scheme [31], the **Verify** algorithm in our proposed identity-based chameleon signature scheme is non-interactive, *i.e.*, the recipient can verify the signature without the collaboration of the signer. The **Deny** protocol is also non-interactive in our signature scheme. Moreover, our signature scheme is based on the well established hash-and-sign paradigm and thus can provide more flexible constructions. Another distinguishing advantage of our scheme is that the property of "message hiding" or "message recovery" can be achieved freely by the signer.

Compared with the existing identity-based chameleon signature schemes [1,38], our proposed scheme is as efficient as them in the **Sign** and **Verify** algorithms. While in the **Deny** protocol, it requires a (very) little more computation and communication cost for the *non-interactive* proofs of knowledge. However, none of the schemes [1,38] is key-exposure free. Currently, it seems that our proposed scheme is the unique choice for the efficient and secure identity-based chameleon signature scheme in the real applications.

The Security Model of Unidirectional Proxy Re-Signature with Private Re-Signature Key

Jun Shao[1,2], Min Feng[3], Bin Zhu[3], Zhenfu Cao[4], and Peng Liu[2]

[1] College of Computer and Information Engineering
Zhejiang Gongshang University
[2] College of Information Sciences and Technology
Pennsylvania State University
[3] Microsoft Research Asia
[4] Department of Computer Science and Engineering
Shanghai Jiao Tong University
chn.junshao@gmail.com, mfeng@microsoft.com, bzhu@microsoft.com,
zfcao@cs.sjtu.edu.cn, pliu@ist.psu.edu

Abstract. In proxy re-signature (PRS), a semi-trusted proxy, with some additional information (a.k.a., re-signature key), can transform Alice's (delegatee) signature into Bob's (delegator) signature on the same message, but cannot produce an arbitrary signature on behalf of either the delegatee or the delegator. In this paper, we investigate the security model of proxy re-signature, and find that the previous security model proposed by Ateniese and Honhenberger at ACM CCS 2005 (referred to as the AH model) is not complete since it does not cover all possible attacks. In particular, the attack on the unidirectional proxy re-signature with private re-signature key. To show this, we artificially design such a proxy re-signature scheme, which is proven secure in the AH model but suffers from a specific attack. Furthermore, we propose a new security model to solve the problem of the AH model. Interestingly, the previous two private re-signature key, unidirectional proxy re-signature schemes (one is proposed by Ateniese and Honhenberger at ACM CCS 2005, and the other is proposed by Libert and Vergnaud at ACM CCS 2008), which are proven secure in the AH model, can still be proven secure in our security model.

Keywords: Security model, Unidirectional PRS, Private re-signature key, AH model.

1 Introduction

1.1 What Is Proxy Re-Signature?

Proxy re-signature (PRS), introduced by Blaze, Bleumer, and Strauss [7], and formalized by Ateniese and Hohenberger [5], allows a semi-trusted proxy to transform a delegatee's (Alice) signature into a delegator's (Bob) signature on the same message by using some additional information (a.k.a., re-signature key).

R. Steinfeld and P. Hawkes (Eds.): ACISP 2010, LNCS 6168, pp. 216–232, 2010.

The proxy, however, cannot generate arbitrary signatures on behalf of either the delegatee or the delegator.

Eight desired properties of proxy re-signatures are given in [5].

1. **Unidirectional:** In a *unidirectional* scheme, a re-signature key allows the proxy to transform A's signature to B's but not vice versa. In a *bidirectional* scheme, on the other hand, the re-signature key allows the proxy to transform A's signature to B's as well as B's signature to A's.
2. **Multi-use:** In a *multi-use* scheme, a transformed signature can be re-transformed again by the proxy. In a *single-use* scheme, the proxy can transform only the signatures that have not been transformed.
3. **Private re-signature key:** The proxy can keep the re-signature key as a secret in a *private re-signature key* scheme, but *anyone* can recompute the re-signature key by observing the proxy passively in a *public re-signature key* scheme.
4. **Transparent:** In a *transparent* scheme, users may not even know that a proxy exists.
5. **Key-optimal:** In a *key-optimal* scheme, a user is required to protect and store only a small constant amount of secrets no matter how many signature delegations the user gives or accepts.
6. **Non-interactive:** The delegatee is not required to participate in delegation process.
7. **Non-transitive:** A re-signing right cannot be re-delegated by the proxy alone.
8. **Temporary:** A re-signing right is temporary. This can be done by either revoking the right as in [5] or expiring the right.

1.2   Applications of PRS

One interesting application of PRS is the interoperable architecture of ditigal rights management (DRM). A DRM system is designed to prevent illegal redistribution of digital content. With DRM systems, the digital content can only be played in a specified device (regime). For example, a song playable in device (regime) A cannot be played in device (regime) B. However, it is reported that 86% consumers prefer paying twice price for a song that runs on any device rather than runs on only one device [2]. Most of current interoperability architectures require to change the existing DRM systems a lot [11]; this modification cannot be adopted due to business reasons. Based on proxy re-signature and proxy re-encryption [3,4], Taban *et al.* [14] proposed a new interoperability architecture, which does not change the existing DRM systems a lot but keep the DRM systems' security. In their architecture, only a new module called Domain Interoperability Manager (DIM) is introduced. PRS allows the DIM to transform licenses (signatures) in regime A into another in regime B, while DIM cannot generate valid licenses (signatures) either in regime A or in regime B.

PRS can also be used in other applications according to its properties: Space-efficient proof that a path was taken (by using the multi-usability and private re-signature key properties) [5], management of group signatures (by using the unidirection and private re-signature key properties) [5]. We refer the reader to [1] for more applications.

1.3 History of Proxy Re-Signatures

Proxy re-signature has a rather short history. This primitive was introduced at Eurocrypt 1998 by Blaze, Bleumer and Strauss [7]. Their scheme, referred to as the BBS scheme, is a *multi-use, public re-signature key* and *bidirectional* scheme. However, from the re-signature key (which is public), the delegator can easily obtain the delegatee's signing key or vice versa. As a result, the BBS scheme is not suitable for many practical applications described in [7].

After the first proxy re-signature scheme appeared in 1998, there was no follow-ups until the work by Ateniese and Hohenberger published at ACM CCS 2005 [5]. One of the reasons for such a long quiet time is, as stated in [5], that the definition of proxy re-signatures given in [7] is informal and can be easily confused with other signature variations. In [5], Ateniese and Hohenberger first formalized the definition of security for a proxy re-signature, referred to as the AH model in this paper, and then proposed three proxy re-signature schemes with proven security. The first one is *multi-use, private re-signature key* and *bidirectional*, the second one is *single-use, public re-signature key* and *unidirectional*, and the third one is *single-use, private re-signature key* and *unidirectional*. Later, Shao *et al.* [13] proposed a *multi-use, private re-signature key* and *bidirectional* proxy re-signature scheme based on Waters' identity based signature [15], and Libert and Vergnaud [12] proposed a *multi-use, private re-signature key* and *unidirectional* scheme based on the ℓ-FlexDH assumption. All the proxy re-signature schemes in [5,13,12] are proven secure in the AH model.

1.4 This Paper's Contributions

The AH model covers two types of forgeries for bidirectional proxy re-signature: (1) an outsider who is neither the proxy nor one of the delegation parties aims to produce signatures on behalf of either delegation party; (2) the proxy aims to produce signatures on behalf of either delegation party. For unidirectional proxy re-signature, in addition to these two types of forgeries, the AH model covers another two types of forgeries: (3) the delegator colludes with the proxy to produce signatures on behalf of the delegatee; (4) the delegatee colludes with the proxy to produce the *first-level* signatures[1].

The AH model covers almost all types of forgeries for unidirectional proxy re-signatures, but it omits one type of forgeries that the delegatee may aim to produce signatures on behalf of the delegator *without* colluding with the proxy. For example, for the transformation path: Alice → Proxy → Bob. Alice may attempt to produce a *second-level* signature on a message on behalf of Bob without the transformation of Proxy. This situation is not allowed in the proxy

---

[1] See Remark 1 in Section 2.

re-signature schemes with private re-signature key, since Bob has delegated his signing rights via Proxy but not to Alice directly. The schemes suffering from this attack cannot be used in most of the applications of PRS listed in [1].

To show the deficiency of the AH model, we artificially design in Section 3.3 a scheme, denoted as S_{us}, which suffers from the above attack but is proven secure in the AH model.

On one hand, one of the possible reasons for the deficiency of the AH model is that the AH model tried to model *all* types of attacks on *all* types of proxy re-signatures. Hence, it is more complex than security models of other types of signatures, which makes the AH model hard to be verified. It would be better if one security model is associated to only one type of proxy re-signatures.

On the other hand, as shown in [1], the unidirectional proxy re-signature with private re-signature key is more useful than other types of proxy re-signatures. Hence, it is desired to propose a security model to instruct people to design the unidirectional proxy re-signature with private re-signature key.

As a result, in this paper, we only focus on the security model of the unidirectional proxy re-signature with private re-signature key. The proposed security model is clearer and simpler than the AH model (one security game vs. four security games). For the simple expression, we refer to this type of proxy re-signatures as UPRS-prk in the rest of this paper. Fortunately, the previous UPRS-prk[2] schemes [5,12], which are proven secure in the AH model, can still be proven secure in our model.

1.5 Paper Organization

The remaining paper is organized as follows. In Section 2, the definitions of UPRS-prk are introduced and the AH model is described. In Section 3, we present a scheme which is proven secure in the AH model but insecure. In Section 4, we present our security model of UPRS-prk and point out that the previous UPRS-prk schemes are still secure in our model. We conclude the paper in Section 5.

2   Definitions

2.1   Unidirectional Proxy Re-Signature with Private Re-Signature Key

The following definitions are from [5].

Definition 1 (UPRS-prk). *A UPRS-prk scheme PRS consists of the following five probabilistic algorithms:* KeyGen, ReKey, Sign, ReSign, *and* Verify *where:*

KeyGen: *It takes as input the security parameter 1^k, and returns a verification key pk and a signing key sk. This algorithm is denoted as $(pk, sk) \leftarrow KeyGen(1^k)$.*

---

[2] All existing UPRS-prk schemes are non-interactive. Hence, we only focus on non-interactive UPRS-prk in this paper.

ReKey: *It takes as input delegatee Alice's verification key pk_A, and delegator Bob's key pair (pk_B, sk_B), and returns a re-signature key $rk_{A \to B}$ for the proxy. This algorithm is denoted as $rk_{A \to B} \leftarrow$ ReKey(pk_A, pk_B, sk_B).*

Sign: *It takes as input a signing key sk, a positive integer ℓ and a message m, and returns a signature σ at level ℓ. This algorithm is denoted as $\sigma \leftarrow$ Sign(sk, m, ℓ).*

ReSign: *It takes as input a re-signature key $rk_{A \to B}$, a signature σ_A on a message m under pk_A at level ℓ, and returns the signature σ_B on the same message m under pk_B at level $\ell + 1$ if Verify$(pk_A, m, \sigma_A, \ell) = 1$, or $\perp$ otherwise. This algorithm is denoted as $\sigma_B \leftarrow$ ReSign$(rk_{A \to B}, pk_A, m, \sigma_A, \ell)$.*

Verify: *It takes as input a verification key pk, a message m, a signature σ and a positive integer ℓ, and returns 1 if σ is a valid signature under pk at level ℓ, or 0 otherwise. This algorithm is denoted as $(1 \text{ or } 0) \leftarrow$ Verify(pk, m, σ, ℓ).*

Correctness. The following property must be satisfied for the correctness of a UPRS-prk scheme: For any message m in the message space and any two key pairs (pk_A, sk_A) and (pk_B, sk_B), let $rk_{A \to B} \leftarrow$ Rekey(pk_A, pk_B, sk_B), the following two equations must hold:

$$\text{Verify}(pk_A, m, \sigma_A, \ell) = 1,$$

where σ_A is a signature on message m under pk_A at level ℓ from Sign. If the UPRS-prk scheme is single-use, then $\ell \in \{1, 2\}$; $\ell \geq 1$ otherwise.

$$\text{Verify}(pk_B, m, \text{ReSign}(rk_{A \to B}, pk_A, m, \sigma'_A, \ell - 1), \ell) = 1.$$

If the UPRS-prk scheme is single-use, σ'_A is a signature on message m under pk_A from Sign, and $\ell = 2$; if the proxy re-signature scheme is multi-use, σ'_A could also be a signature on message m under pk_A from ReSign, and $\ell \geq 2$.

Remark 1 (Two Types of Signatures.). In all existing unidirectional proxy re-signature schemes, a signature manifests in two types: the *owner-type* (i.e., the first-level defined in [5], $\ell = 1$) and the *non-owner-type* (i.e., the second-level signatures in [5], $\ell > 1$). An *owner-type* signature can be computed only by the owner of the signing key, while a *non-owner-type* signature can be computed not only by the owner of the signing key, but also by collaboration between his proxy and delegatee.

2.2   The AH Model

In this subsection, we review the AH model for UPRS-prk. It contains two aspects (four security games): the external security and the internal security. The details are as follows.

External Security: This security deals with adversaries other than the proxy and any delegation parties. A UPRS-prk scheme has external security if and only if for security parameter k, any non-zero $n \in \mathrm{poly}(k)$, and all probabilistic polynomial time (p.p.t.) algorithms $\mathcal{A}$, the following probability is *negligible*:

$$\Pr[\{(pk_i, sk_i) \leftarrow \mathtt{KeyGen}(1^k)\}_{i \in [1,n]}, (t, m^*, \sigma^*, \ell^*) \leftarrow \mathcal{A}^{\mathcal{O}_s(\cdot), \mathcal{O}_{rs}(\cdot)}(\{pk_i\}_{i \in [1,n]}):$$
$$\mathtt{Verify}(pk_t, m^*, \sigma^*, \ell^*) = 1 \wedge (t, m^*) \notin Q],$$

where oracle $\mathcal{O}_s$ takes as input a verification key pk_i and a message $m \in \mathcal{M}$, and produces an output $\mathtt{Sign}(sk_i, m, 1)$; oracle $\mathcal{O}_{rs}$ takes as input two distinct verification keys pk_i and pk_j, a message m, a signature σ and a positive integer ℓ, and produces an output $\mathtt{ReSign}(\mathtt{ReKey}(pk_i, pk_j, sk_j), pk_i, m, \sigma, \ell)$; and Q denotes the set of $(index, message)$ pairs (i, m) that $\mathcal{A}$ obtains a signature on a message m under the verification key pk_i by querying $\mathcal{O}_s$ on (pk_i, m) or querying $\mathcal{O}_{rs}$ on $(\cdot, pk_i, m, \cdot, \cdot)$. Note that if the treated PRS scheme is single-use, then $\ell^* \in \{1, 2\}$ and $\ell = 1$; otherwise, $\ell^* \geq 1$ and $\ell \geq 1$.

Internal Security: This security protects a user from inside adversaries who can be any parties, i.e., the proxy, the delegatee, and the delegator, in a proxy re-signature scheme. It can be classified into the following three types.

    `Limited Proxy`: In this case that only the proxy is a potential adversary $\mathcal{A}$. We must guarantee that the proxy cannot produce signatures on behalf of either the delegator or the delegatee except the signatures produced by the delegatee and delegated to the proxy to re-sign. Internal security in this case is very similar to external security described above except that $\mathcal{A}$ queries a rekey oracle $\mathcal{O}_{rk}$ instead of a re-signature oracle $\mathcal{O}_{rs}$. A `UPRS-prk` scheme is said to have limited proxy security if and only if for security parameter k, any non-zero $n \in \mathrm{poly}(k)$, and all p.p.t. algorithms $\mathcal{A}$, the following probability is *negligible*:

$$\Pr[\{(pk_i, sk_i) \leftarrow \mathtt{KeyGen}(1^k)\}_{i \in [1,n]}, (t, m^*, \sigma^*, \ell^*) \leftarrow \mathcal{A}^{\mathcal{O}_s(\cdot), \mathcal{O}_{rk}(\cdot)}(\{pk_i\}_{i \in [1,n]}):$$
$$\mathtt{Verify}(pk_t, m^*, \sigma^*, \ell^*) = 1 \wedge (t, m^*) \notin Q],$$

where $\mathcal{O}_s$ and ℓ^* are the same as that in **external security**, oracle $\mathcal{O}_{rk}$ takes as input two distinct verification keys pk_i, pk_j, and returns the output of $\mathtt{ReKey}(pk_i, pk_j, sk_j)$; and Q denotes the set of $(index, message)$ tuples (i, m) that $\mathcal{A}$ obtained a signature on m under verification key pk_i or one of its delegatees' keys by querying $\mathcal{O}_s$.

    `Delegatee Security`: In this case that the proxy and delegator may collude with each other. This security guarantees that their collusion cannot produce any signatures on behalf of the delegatee. We associate the index 0 to the delegatee. A `UPRS-prk` scheme is said to have delegatee security if and only if for security parameter k, any non-zero $n \in \mathrm{poly}(k)$, and all p.p.t. algorithms $\mathcal{A}$, the following probability is *negligible*:

$$\Pr[\{(pk_i, sk_i) \leftarrow \mathtt{KeyGen}(1^k)\}_{i \in [0,n]}, (m^*, \sigma^*, \ell^*) \leftarrow \mathcal{A}^{\mathcal{O}_s(\cdot)}(pk_0, \{pk_i, sk_i\}_{i \in [1,n]}):$$
$$\mathtt{Verify}(pk_0, m^*, \sigma^*, \ell^*) = 1 \wedge (0, m^*) \notin Q],$$

where $\mathcal{O}_s$ and ℓ^* are the same as that in `external security`, Q is the set of pairs $(0, m)$ that $\mathcal{A}$ obtains a signature by querying oracle $\mathcal{O}_s$ on (pk_0, m). Note that $\mathcal{O}_{rk}$ is useless in this case, since the adversary is able to compute re-signature keys $rk_{0 \rightarrow i}$ himself by using sk_i.

Delegator Security: In this case that the proxy and delegatee may collude with each other. This security guarantees that their collusion cannot produce any owner-type signatures on behalf of the delegator. We associate the index 0 to the delegator. A UPRS-prk scheme is said to have delegator security if and only if for security parameter k, any non-zero $n \in \text{poly}(k)$, and all p.p.t. algorithms $\mathcal{A}$, the following probability is *negligible*:

$$\Pr[\{(pk_i, sk_i) \leftarrow \texttt{KeyGen}(1^k)\}_{i \in [0,n]}, (m^*, \sigma^*, 1) \leftarrow \mathcal{A}^{\mathcal{O}_s(\cdot), \mathcal{O}_{rk}(\cdot)}(pk_0, \{pk_i, sk_i\}_{i \in [1,n]}) :$$
$$\texttt{Verify}(pk_0, m^*, \sigma^*, 1) = 1 \wedge (0, m^*) \notin Q],$$

where $\mathcal{O}_s$ is the same as that in **external security**, $\mathcal{O}_{rk}$ is the same as that in limited proxy security, Q is the set of pairs $(0, m)$ for which $\mathcal{A}$ obtains an owner-type signature by querying oracle $\mathcal{O}_s$ on (pk_0, m).

Remark 2. According the definition in [5] (page 313), we say a unidirectional proxy re-signature scheme is *public re-signature key* if it does not hold the external security, like scheme S_{uni} in [5]; otherwise, the scheme is private re-signature key, like scheme S^*_{uni} in [5] and the schemes in [12].

3   On the AH Model

Before proposing scheme S_{us}, we first introduce some basic knowledge related to scheme S_{us}.

3.1   Bilinear Groups

Bilinear maps and bilinear map groups are briefly reviewed in this subsection. Details can be found in [8,9].

1. $\mathbb{G}$ and $\mathbb{G}_T$ are two (multiplicative) cyclic groups of prime order q;
2. g is a generator of $\mathbb{G}$;
3. e is a bilinear map, $e : \mathbb{G} \times \mathbb{G} \to \mathbb{G}_T$.

Let $\mathbb{G}$ and $\mathbb{G}_T$ be two groups as above. An *admissible bilinear map* is a map $e : \mathbb{G} \times \mathbb{G} \to \mathbb{G}_T$ with the following properties:

1. *Bilinearity:* For all $P, Q, R \in \mathbb{G}$, $e(P \cdot Q, R) = e(P, R) \cdot e(Q, R)$ and $e(P, Q \cdot R) = e(P, Q) \cdot e(P, R)$.
2. *Non-degeneracy:* If $e(P, Q) = 1$ for all $Q \in \mathbb{G}$, then $P = \mathcal{O}$, where $\mathcal{O}$ is a point at infinity.

We say that $\mathbb{G}$ is a bilinear group if the group action in $\mathbb{G}$ can be computed efficiently and there exist a group $\mathbb{G}_T$ and an efficiently computable bilinear map as above. We use BSetup to denote an algorithm that, on input the security parameter 1^k, outputs the parameters for a bilinear map as $(q, g, \mathbb{G}, \mathbb{G}_T, e)$, where $q \in \Theta(2^k)$.

3.2   Complexity Assumption

The security of scheme S_{us} can be proved based on the extended Computational Diffie-Hellman (eCDH) assumption in the AH model.

Definition 2 (Extended Computational Diffie-Hellman Assumption).
Let $(q, g, \mathbb{G}, \mathbb{G}_T, e) \leftarrow$ BSetup(1^k). The extended computational Diffie-Hellman problem (eCDH) in $(\mathbb{G}, \mathbb{G}_T)$ is defined as follows: given 4-tuple $(g, g^u, g^v, g^{1/v}) \in \mathbb{G}^5$ as input, output g^{uv} or $g^{u/v}$. An algorithm $\mathcal{A}$ has advantage ε in solving the eCDH problem in $(\mathbb{G}, \mathbb{G}_T)$ if

$$Pr[\mathcal{A}(g, g^u, g^v, g^{1/v}) = g^{uv} \text{ or } g^{u/v}] \geq \varepsilon,$$

where the probability is taken over the random choices of $u, v \in Z_q^*$ and the random bits of $\mathcal{A}$.

We say that the (t, ε)-extended computational Diffie-Hellman (eCDH) assumption holds in $(\mathbb{G}, \mathbb{G}_T)$ if no t-time algorithm has advantage ε at least in solving the eCDH problem in $(\mathbb{G}, \mathbb{G}_T)$.

In this paper, we drop the t and ε and refer to the eCDH assumption rather than the (ε, t)-eCDH assumption.

3.3   The Scheme S_{us}

In this subsection, we propose a UPRS-prk scheme, named S_{us}, which is proven secure in the AH model, but it cannot provide all the required security properties. This fact shows that the AH model is not complete. The public parameters of scheme S_{us} are $(q, g, \mathbb{G}, \mathbb{G}_T, e)$, where $(q, g, \mathbb{G}, \mathbb{G}_T, e) \leftarrow$ BSetup(1^k), H is a cryptographic hash function: $\{0, 1\}^* \rightarrow \mathbb{G}$.

KeyGen: On input the security parameter 1^k, it selects a random number $a \in Z_q^*$, and outputs the key pair $pk = g^a$ and $sk = a$.

ReKey: On input the delegatee's verification key $pk_A = g^a$ and the delegator's signing key $sk_B = b$, it outputs the re-signature key

$$rk_{A \rightarrow B} = (rk_{A \rightarrow B}^{(1)}, rk_{A \rightarrow B}^{(2)}, rk_{A \rightarrow B}^{(3)}) = (r', (pk_A)^{r'}, H(g^{a \cdot r'}||2)^{1/b}),$$

where r' is a random number in Z_q^* determined by Bob.

Sign: On input a signing key $sk = a$, a message $m \in \mathcal{M}$ and an integer $\ell \in \{1, 2\}$,

 – if $\ell = 1$, it outputs an *owner-type* signature

$$\sigma = (A, B, C) = (H(m||0)^r, g^r, H(g^r||1)^a),$$

 – if $\ell = 2$, it outputs a *non-owner-type* signature

$$\sigma = (A, B, C, D, E) = (H(m||0)^{r_1}, g^{r_1}, H(g^{r_1}||1)^{r_2}, g^{r_2}, H(g^{r_2}||2)^{1/a}).$$

ReSign: Given an *owner-type* signature σ at level 1, a re-signature key $rk_{A\rightarrow B} = (rk^{(1)}_{A\rightarrow B}, rk^{(2)}_{A\rightarrow B}, rk^{(3)}_{A\rightarrow B})$, a verification key pk_A, and a message m, this algorithm first checks $\text{Verify}(pk_A, m, \sigma, 1) \overset{?}{=} 1$. If it does not hold, outputs $\perp$; otherwise, outputs

$$
\begin{aligned}
\sigma' &= (A', & B', & \quad C', & D', & E') \\
&= (A, & B, & \quad C^{rk^{(1)}_{A\rightarrow B}}, & rk^{(2)}_{A\rightarrow B}, & rk^{(3)}_{A\rightarrow B}) \\
&= (H(m\|0)^r, & g^r, & \quad H(g^r\|1)^{ar'}, & (pk_A)^{r'}, & H((pk_A)^{r'}\|2)^{1/b}) \\
&= (H(m\|0)^{r_1}, & g^{r_1}, & \quad H(g^{r_1}\|1)^{r_2}, & g^{r_2}, & H(g^{r_2}\|2)^{1/b})
\end{aligned}
$$

Note that we set $r_1 = r \bmod q$ and $r_2 = ar' \bmod q$.

Verify: On input a verification key pk, a message m at level $\ell \in \{1, 2\}$, and a signature σ,

– if σ is an *owner-type* signature $\sigma = (A, B, C)$ (i.e., $\ell = 1$), it checks

$$
e(pk, H(B\|1)) \overset{?}{=} e(g, C),
$$
$$
e(B, H(m\|0)) \overset{?}{=} e(g, A).
$$

If the two equations both hold, it outputs 1; otherwise, outputs 0.

– if σ is a *non-owner-type* signature $\sigma = (A, B, C, D, E)$ (i.e., $\ell = 2$), it checks

$$
e(g, H(D\|2)) \overset{?}{=} e(pk, E),
$$
$$
e(D, H(B\|1)) \overset{?}{=} e(g, C),
$$
$$
e(B, H(m\|0)) \overset{?}{=} e(g, A).
$$

If all the equations hold, it outputs 1; otherwise, outputs 0.

3.4 Correctness

Scheme S_{us} has the correctness due to the following equations.

– owner-type signature:

$$
\begin{aligned}
e(pk, H(B\|1)) &= e(g^a, H(B\|1)) = e(g, H(B\|1)^a) = e(g, C), \\
e(B, H(m\|0)) &= e(g^r, H(m\|0)) = e(g, H(m\|0)^r) = e(g, A),
\end{aligned}
$$

– non-owner-type signatures:

$$
\begin{aligned}
e(pk, E) &= e(g^b, H(D\|2)^{1/b}) = e(g, H(D\|2)), \\
e(D, H(B\|1)) &= e(g^{r_2}, H(B\|1)) = e(g, H(B\|1)^{r_2}) = e(g, C), \\
e(B, H(m\|0)) &= e(g^{r_1}, H(m\|0)) = e(g, H(m\|0)^{r_1}) = e(g, A).
\end{aligned}
$$

3.5 Security Analysis

Theorem 1. *Scheme S_{us} is secure in the AH model if the eCDH problem is hard, and hash function H is treated as a random oracle.*

Proof. We prove the security in the AH model in two parts, similar to that in [5].

We show that if adversary $\mathcal{A}$ can break scheme S_{us} in the AH model, then we can build another algorithm $\mathcal{B}$ that can solve the eCDH problem. Given $(q, g, \mathbb{G}, \mathbb{G}_T, e, g^u, g^v, g^{1/v})$, $\mathcal{B}$ aims to output g^{uv} or $g^{u/v}$. The proxy re-signature security game is as follows.

External Security.

- *Random oracle $\mathcal{O}_h$:* On input string R, $\mathcal{B}$ first checks whether $(R, R_h, r_h, *)$ exists in Table T_h. If yes, $\mathcal{B}$ returns R_h and terminates; otherwise, $\mathcal{B}$ chooses a random number $r_h \in Z_q^*$, and the next performance of $\mathcal{B}$ has three situations.

  - The input string R satisfies the format $m\|0$, where $m \in \mathcal{M}$. $\mathcal{B}$ guesses whether m is the target message m^*. If yes, $\mathcal{B}$ outputs $R_h = (g^u)^{r_h}$; otherwise, $\mathcal{B}$ outputs $R_h = g^{r_h}$.
  - The input string R satisfies the format $m\|1$ or $m\|2$, where $m \in \mathbb{G}$. $\mathcal{B}$ outputs $R_h = (g^u)^{r_h}$.
  - The input string R does not satisfy any of the above formats. $\mathcal{B}$ outputs $R_h = g^{r_h}$.

  At last, $\mathcal{B}$ records $(R, R_h, r_h, \perp)$ in Table T_h.

- *Verification keys oracle $\mathcal{O}_{pk}$:* As the adversary requests the creation of system users, $\mathcal{B}$ first chooses a random number $x_i \in Z_q^*$, and guesses whether it is pk_t. For $pk_i \neq pk_t$, it sets $pk_i = g^{x_i}$; for $pk_i = pk_t$, it sets $pk_i = (g^v)^{x_i}$. At last, $\mathcal{B}$ records (pk_i, x_i) in Table T_{pk}.

- *Signature oracle $\mathcal{O}_s$:* On input (pk_i, m_i).

  - If $m_i = m^*$, then $pk_i \neq pk_t$, $\mathcal{B}$ chooses a random number $r \in Z_q^*$, and outputs

  $$\sigma = (A, B, C) = (H(m^*\|0)^r, g^r, H(g^r\|1)^{x_i}).$$

  - If $m_i \neq m^*$, then $\mathcal{B}$ chooses a random number $r \in Z_q^*$, and checks whether $((g^v)^r\|1, *, *, *)$ exists in Table T_h. If it exists, $\mathcal{B}$ reports "failure" and aborts; otherwise $\mathcal{B}$ chooses a random number $r_1 \in Z_q^*$, and records $((g^v)^r\|1, g^{r_1}, r_1, r)$ in Table T_h. And then $\mathcal{B}$ checks whether $(m_i\|0, \star_1, \star_2, \perp)$ exists in Table T_h. If it exists, then $\mathcal{B}$ sets $r_2 = \star_2$; otherwise, $\mathcal{B}$ chooses a random number $r_2 \in Z_q^*$ and records $(m_i\|0, g^{r_2}, r_2, \perp)$ in Table T_h. At last, $\mathcal{B}$ outputs

  $$\sigma = (A, B, C) = ((g^v)^{r r_2}, (g^v)^r, pk_i^{r_1}) = (H(m_i\|0)^{vr}, g^{vr}, H(g^{vr}\|1)^{vx_i}).$$

- *Re-signature oracle $\mathcal{O}_{rs}$:* On input $(pk_i, pk_j, m, \sigma, 1)$, where $\sigma = (A, B, C)$. If $\texttt{Verify}(pk_i, m, \sigma, 1) = 1$, then B does the following performances; otherwise, outputs $\perp$.

- If $pk_j \neq pk_t$, $\mathcal{B}$ uses x_j, associated to pk_j in Table T_{pk}, to run ReKey and ReSign, and gets the required re-signature.
- If $pk_j = pk_t$, then $m \neq m^*$, and $\mathcal{B}$ chooses a random number $r \in Z_q^*$, and checks whether $((g^v)^{x_i r}||2, *, *, *)$ exists in Table T_h. If it exists, $\mathcal{B}$ reports "failure" and aborts; $\mathcal{B}$ chooses a random number $r_1 \in Z_q^*$, and records $((g^v)^{x_i r}||2, g^{r_1}, r_1, x_i r)$ in Table T_h. $\mathcal{B}$ searches $(B||1, \star_1, \star_2, \star_3)$ in Table T_h, and outputs

$$
\begin{aligned}
\sigma' &= (\ A',\ B',\ C', & D', & \quad E') \\
&= (\ A,\ B,\ (g^v)^{x_i \star_2 r}, & (g^v)^{x_i r}, & \quad (g^{1/v})^{r_1/x_t}) \\
&= (\ A,\ B,\ H(B||1)^{x_i v r}, & g^{x_i v r}, & \quad H(g^{x_i v r}||2)^{1/(v x_t)}),
\end{aligned}
$$

- *Forgery:* At some point, the adversary must output a forgery (pk_t, m^*, σ^*).

Now, we show how $\mathcal{B}$ gets the eCDH solution from the forgery.

- If σ^* is an owner-type signature, such as $\sigma^* = (A^*, B^*, C^*)$, then we have the following analysis.
  - If $(*, B^*, C^*)$ did not appear in one owner-type signature of pk_t from $\mathcal{O}_s$, then $\mathcal{B}$ finds $(B^*||1, \star_1', \star_2', \star_3')$ in Table T_h, and gets the solution of the eCDH problem:

  $$
  (C^*)^{1/(x_t \star_2')} = (H(B^*||1)^{v x_t})^{1/(x_t \star_2')} = (g^{u \star_2' v x_t})^{1/(x_t \star_2')} = g^{uv}.
  $$

  Note that $pk_t = (g^v)^{x_t}$.
  - If $(*, B^*, C^*)$ appeared in one owner-type signature of pk_t from $\mathcal{O}_s$, then $\mathcal{B}$ finds $(m^*||0, \star_1, \star_2, \star_3)$ and $(B^*||1, \star_1', \star_2', \star_3')$ in Table T_h, and gets the solution of the eCDH problem:

  $$
  (A^*)^{1/(\star_2 \star_3')} = (H(m^*||0)^{v \star_3'})^{1/(\star_2 \star_3')} = (g^{uv \star_2 \star_3'})^{1/(\star_2 \star_3')} = g^{uv}.
  $$

  Note that $B^* = (g^v)^{\star_3'}$.
- If σ^* is a non-owner-type signature, such as $\sigma^* = (A^*, B^*, C^*, D^*, E^*)$, then we have the following analysis.
  - If $(*, *, *, D^*, E^*)$ did not appear in any one signature of pk_t from $\mathcal{O}_{rs}$, then $\mathcal{B}$ finds $(D^*||2, \star_1'', \star_2'', \star_3'')$ in Table T_h, and gets the solution of the eCDH problem:

  $$
  (E^*)^{x_t/\star_2''} = (H(D^*||2)^{1/(v x_t)})^{x_t/\star_2''} = (g^{u \star_2''/(v x_t)})^{x_t/\star_2''} = g^{u/v}.
  $$

  Note that $pk_t = (g^v)^{x_t}$.
  - If $(*, *, *, D^*, E^*)$ appeared in one signature of pk_t, but $(*, B^*, C^*, D^*, E^*)$ did not appear in any one signature of pk_t from $\mathcal{O}_{rs}$, then $\mathcal{B}$ finds $(D^*||2, \star_1'', \star_2'', \star_3'')$ and $(B^*||1, \star_1', \star_2', \star_3')$ in Table T_h, and gets the solution of the eCDH problem:

  $$
  (C^*)^{1/(\star_3'' \star_2')} = (H(B^*||1)^{v \star_3''})^{1/(\star_3'' \star_2')} = (g^{u \star_2' v \star_3''})^{1/(\star_3'' \star_2')} = g^{uv}.
  $$

  Note that $D^* = (g^v)^{\star_3''}$.

- If $(*, B^*, C^*, D^*, E^*)$ appeared in one owner-type signature of pk_t from $\mathcal{O}_{rs}$, then $\mathcal{B}$ finds $(m^*||0, \star_1, \star_2, \star_3)$ and $(B||1, \star'_1, \star'_2, \star'_3)$ in Table T_h, and gets the solution of the eCDH problem:

$$(A^*)^{1/(\star_2\star'_3)} = (H(m^*||0)^{v\star'_3})^{1/(\star_2\star'_3)} = (g^{uv\star_2\star'_3})^{1/(\star_2\star'_3)} = g^{uv}.$$

  Note that $B^* = (g^v)^{\star'_3}$.

Note that $\mathcal{B}$ guessed the right target verification key with the probability $1/n$ at least, and $\mathcal{B}$ reports "failure" and aborts in $\mathcal{O}_s$ and $\mathcal{O}_{rs}$ with the probabilities $(q_h + q_s)/q$ and $(q_h + q_{rs})/q$ at most, respectively. Here, q_h, q_s, and q_{rs} are the maximum numbers that $\mathcal{A}$ can query to random oracle $\mathcal{O}_h$, signature oracle $\mathcal{O}_s$, re-signature oracle $\mathcal{O}_{rs}$ respectively. As a result, $\mathcal{B}$ solves the eCDH problem with a non-negligible probability.

Internal Security: Internal security includes three parts: Limited Proxy Security, Delegatee Security, Delegator Security.

Limited Proxy Security:
  - *Random oracle $\mathcal{O}_h$:* Identical to that in the external security.
  - *Verification keys oracle $\mathcal{O}_{pk}$:* As the adversary requests the creation of system users, $\mathcal{B}$ first chooses a random number $x_i \in Z_q^*$, and then outputs $pk_i = (g^v)^{x_i}$. At last, $\mathcal{B}$ records (pk_i, x_i) in Table T_{pk}.
  - *Signature oracle $\mathcal{O}_s$:* On input (pk_i, m_i).
    - If $m = m^*$, then $pk_i \neq pk_t$, and $\mathcal{B}$ chooses a random number r, and checks whether $(g^r||1, *, *, *)$ exists in Table T_h. If it exists, then $\mathcal{B}$ reports "failure" and aborts; otherwise, $\mathcal{B}$ chooses a random number r_1, and records $(g^r||1, g^{r_1}, r_1, \perp)$ into Table T_h. At last, $\mathcal{B}$ outputs $(H(m^*||0)^r, g^r, pk_i^{r_1})$.
    - If $m \neq m^*$, then $\mathcal{B}$ performs the same as that in the external security.
  - *Re-signature key generation oracle $\mathcal{O}_{rk}$:* On input (pk_i, pk_j), $\mathcal{B}$ chooses a random number $r \in Z_q^*$, and checks whether $((g^v)^{x_i r}||2, *, *, *)$ exists in Table T_h. If it exists, $\mathcal{B}$ reports "failure" and aborts; otherwise, $\mathcal{B}$ chooses a random number $r_1 \in Z_q^*$, records $((g^v)^{x_i r}||2, g^{r_1}, r_1, x_i r)$ in Table T_h, and outputs $(r, (g^v)^{x_i r}, (g^{1/v})^{r_1/x_j})$.
    Note that $pk_i = (g^v)^{x_i}$ and $pk_j = (g^v)^{x_j}$.

  With the similar analysis in the external security, $\mathcal{B}$ solves the eCDH problem with a non-negligible probability.

Delegatee Security: Compared to the limited proxy, $\mathcal{B}$ needs to change verification keys oracle $\mathcal{O}_{pk}$, signature oracle $\mathcal{O}_s$, and re-signature key generation oracle $\mathcal{O}_{rk}$ as follows.
  - *Verification keys oracle $\mathcal{O}_{pk}$:* For the delegatee, set the verification key as $(g^v)^{x_0}$, and for all other users g^{x_i}, where x_i's $(i = 0, \cdots, n)$ are random numbers from Z_q^*.
  - *Signature oracle $\mathcal{O}_s$:* On input (pk_0, m_i), $\mathcal{B}$ performs as the same as that in $\mathcal{O}_s$ with input in (pk_t, m_i) in the external security, where pk_0 is treated as pk_t.

- *Re-signature key generation oracle* $\mathcal{O}_{rk}$: On input (pk_i, pk_j), where $pk_j \neq pk_0$, $\mathcal{B}$ performs as the same as that in the real execution since it knows x_j such that $pk_j = g^{x_j}$.

With the similar analysis in the external security, $\mathcal{B}$ solves the eCDH problem with a non-negligible probability.

Delegator Security: Compared to the limited proxy, $\mathcal{B}$ needs to change verification keys oracle $\mathcal{O}_{pk}$, signature oracle $\mathcal{O}_s$, re-signature key generation oracle $\mathcal{O}_{reky}$ as follows.

- *Verification keys oracle* $\mathcal{O}_{pk}$: For the delegator, set the verification key as $(g^v)^{x_0}$, and for all other users g^{x_i}, where x_i's ($i = 0, \cdots, n$) are random numbers from Z_q^*.
- *Signature oracle* $\mathcal{O}_s$: On input (pk_0, m_i), $\mathcal{B}$ performs as the same as that in $\mathcal{O}_s$ with input in (pk_t, m_i) in the external security, where pk_0 is treated as pk_t.
- *Re-signature key generation oracle* $\mathcal{O}_{rk}$: On input (pk_i, pk_j),
  - if $pk_j \neq pk_0$, then $\mathcal{B}$ gets x_j from Table T_{pk}, and uses x_j to run Rekey(pk_i, pk_j). At last, $\mathcal{B}$ outputs the result from ReKey.
  - if $pk_j = pk_0$, then $\mathcal{B}$ chooses a random number $r \in Z_q^*$, and checks whether $(pk_i^r || 2, *, *, *)$ exists in Table T_h. If it exists, $\mathcal{B}$ reports "failure" and aborts; otherwise, $\mathcal{B}$ chooses a random number $r_1 \in Z_q^*$, records $(pk_i^r || 2, g^{r_1}, r_1, \perp)$ in Table T_h, and outputs $(r, pk_i^r, (g^{1/v})^{r_1/x_0})$.

With the similar analysis in the external security, $\mathcal{B}$ solves the eCDH problem with a non-negligible probability. Note that in this case, the forgery σ^* is an owner-type signature.

Hence, we get this theorem. $\square$

3.6   An Attack on Scheme S_{us}

Now, we consider the following case: Alice $\rightarrow$ Proxy $\rightarrow$ Bob. First, Alice can produce an owner-type signature on m: $\sigma_a = (H(m||0)^r, g^r, H(g^r||1)^a)$, where she knows the value of r. Then Proxy can transform σ_a into Bob's signature $\sigma_b = (H(m||0)^r, g^r, (H(g^r||1)^a)^{rk_{a\rightarrow b}^{(1)}}, rk_{a\rightarrow b}^{(2)}, rk_{a\rightarrow b}^{(3)})$. In this case, Alice can generate signatures on any message, simply by changing m to m', since she knows the value of r. This shows that scheme S_{us} is insecure. Hence, the AH model is not suitable for UPRS-prk. Note that most of the existing unidirectional proxy re-signature schemes are UPRS-prk schemes; hence, it is desired to propose a new security model to solve this problem.

Remark 3. The scheme S_{us} *cannot* be considered as a unidirectional PRS scheme with public re-signature key, since it holds external security which classifies the unidirectional PRS schemes with private re-signature key or not [5,12].

Remark 4. The main reason why scheme S_{us} is insecure is that the re-sign algorithm does not affect the value containing the message m in the owner-type signature. However, the schemes in [5,12] do not have this flaw.

4   The Proposed Security Model for UPRS-prk

In this section, we propose a new security model for UPRS-prk, which covers the attack in Section 3.6. Due to its simplicity, it is easy to verify its completeness. Before giving our security model, we would first define several terms.

1. If user A delegates his signing rights to user B via a proxy P, then both user A and user B are said to be in a *delegation chain*, denoted as (B,A). User B is called user A's *delegation predecessor*. The combination of the proxy and a user, either the delegatee B or the delegator A, is called a *delegation pair*. Therefore user A and proxy P is a *delegation pair*. So is user B and proxy P.
2. If one of parties in a delegation pair is corrupted, then the delegation pair is *corrupted*; otherwise, it is *uncorrupted*.
3. A user can be treated as the smallest *delegation chain*.
4. If two users A and B are in a delegation chain and B is A's delegation predecessor, then B's signature can be transformed by a proxy or proxies into A's signature.
5. A delegation chain is its own *subchain*.
6. *(Only for multi-use UPRS-prk.)* If user A delegates his signing rights to user B via a proxy P, and user B delegates his signing rights to user C via a proxy P', then user A and user C are said to be in a *delegation chain* too. User C is also called user A's *delegation predecessor*. In this case, users A, B, C are in a delegation chain (C,B,A). The delegation chains (B,A) and (C,B) are *delegation subchains* of the delegation chain (C,B,A). The delegation chain (C,B,A) can be extended if C delegates his signing rights to another user via another proxy.

Existential Unforgeability for UPRS-prk. The existential unforgeability for UPRS-prk is defined by the following adaptively chosen-message attack game played between a challenger $\mathcal{C}$ and an adversary $\mathcal{A}$. Note that we work in a static mode, that is, before the game starts, the adversary should decide which users and proxies are corrupted, and all the verification keys in the security model are generated by the challenger.

Queries: The adversary adaptively makes a number of different queries to the challenger. Each query can be one of the following.

- *Verification Key query* $\mathcal{O}_{pk}$. On input an index $i \in \{1, \cdots, n\}$ by the adversary[3], the challenger responds by running KeyGen(1^k) to get a key pair (pk_i, sk_i), and forwards the verification key pk_i to the adversary. At last, the challenger records (pk_i, sk_i) in the list T_K.
- *Signing Key query* $\mathcal{O}_{sk}$. On input a verification key pk_i by the adversary, the challenger responds sk_i which is the associated value with pk_i in the list T_K, if pk_i is corrupted; otherwise, the challenger responds with $\bot$.

---

[3] We assume that the adversary never inputs number twice. If so, the challenger simply returns the previous value.

- *Re-signature Key query $\mathcal{O}_{rk}$.* On input two verification keys (pk_i, pk_j) $(pk_i \neq pk_j)$ by the adversary, the challenger responds with $\texttt{ReKey}(pk_i, pk_j, sk_j)$, where sk_i is the signing key of pk_i.
- *Signature query $\mathcal{O}_s$.* On input a verification key pk_i, and a message m_i by the adversary, the challenger responds with $\texttt{Sign}(sk_i, m, 1)$, where sk_i is the signing key of pk_i.
- *Re-Signature query $\mathcal{O}_{rs}$.* On input two verification keys pk_i, pk_j $(pk_i \neq pk_j)$, a message m_i, and a signature σ_i at level ℓ by the adversary, the challenger responds with $\texttt{ReSign}(\texttt{ReKey}(pk_i, pk_j, sk_j), pk_i, m_i, \sigma_i, \ell)$, where sk_i is the signing key of pk_i.

Forgery: The adversary output a message m^*, a verification key pk^*, and a signature σ^* at level ℓ^*. The adversary *wins* if the following hold true:

1. $\texttt{Verify}(pk^*, m^*, \sigma^*, \ell^*) = 1$.
2. pk^* is uncorrupted.
3. The adversary has not made a signature query on (pk^*, m^*).
4. The adversary has not made a signature query on (pk', m^*), where pk' is uncorrupted, and there exists such a delegation subchain from pk' to pk^* that does not contain any uncorrupted delegation pair;
5. The adversary has not made a re-signature key query on (pk_i, pk_j), which satisfies all the following conditions:
   - pk_i is corrupted,
   - pk_j is uncorrupted,
   - there exists such a delegation subchain from pk_j to pk^* that does not contain any uncorrupted delegation pair.
6. The adversary has not made a re-signature query on $(pk_i, pk_j, m^*, \sigma_i, *)$, where pk_j is uncorrupted, and there exists such a delegation subchain from pk' to pk^* that does not contain any uncorrupted delegation pair.

We define $\mathbf{Adv}_\mathcal{A}$ to be the probability that adversary $\mathcal{A}$ wins in the above game.

Definition 3. *A UPRS-prk scheme is existentially unforgeable with respect to adaptive chosen message attacks if for all p.p.t. adversaries $\mathcal{A}$, $\mathbf{Adv}_\mathcal{A}$ is negligible in k.*

Remark 5 (Winning Requirements). The first requirement guarantees that $(pk^*, m^*, \sigma^*, \ell^*)$ is a valid signature. The second requirement guarantees that the adversary cannot trivially obtained a valid owner-type signature by obtaining the signing key. The third requirement guarantees that the adversary cannot trivially obtained a valid owner-type signature by the signature oracle. The fourth and fifth requirements guarantee that the adversary cannot trivially obtained a valid non-owner-type signature by the re-signature key oracle and re-signature oracle, respectively.

Remark 6 (Chosen Key Model). Following the spirit in [12], we can easily extend our security model into the chosen key model [6], where the central authority does not need to verify that the owner of one verification key indeed knows the corresponding signing key. In particular, the challenger is no longer responsible for replying the query that needs the signing key of the corrupted verification key to answer.

Remark 7 (Relationship between the AH Model and the Proposed Model). It is clear to see that scheme S_{us} cannot be proven secure in the proposed model due to the attack in Section 3. Hence, the proposed model is not weaker than the AH model if the treated PRS scheme is a `UPRS-prk` scheme. However, these two models are incomparable if the treated PRS scheme is not a `UPRS-prk` scheme, since our proposed model only deals with `UPRS-prk`.

Remark 8 (Existential Unforgeability for Existing UPRS-prk Schemes). It is interesting that the private re-signature key, unidirectional proxy re-signature schemes in [5,12], which are proven secure in the AH model, are still proven secure in our security model. The main reason is that unlike scheme S_{us}, the re-sign algorithms of these schemes affect the value including m in the owner-type signature. Due to the limited space, we will give these proofs in the full version.

5 Conclusions

In this paper, we pointed out that the AH model proposed in [5] cannot guarantee all desired security requirements for all kinds of unidirectional proxy re-signatures. To show this, we artificially constructed a single-use, private re-signature key, unidirectional proxy re-signature scheme S_{us} which is insecure but proven secure in the AH model. We then proposed a new security model to address the deficiency of the AH model. Fortunately, the private re-signature key, unidirectional proxy re-signature schemes in [5,12] proven secure in the AH model are still proven secure in the new model. Hence, we can still use them in the applications where it demands private re-signature key, unidirectional proxy re-signature.

Acknowledgements

The authors thank the anonymous reviewers for helpful comments. Jun Shao and Peng Liu were supported by ARO STTR "Practical Efficient Graphical Models for Cyber-Security Analysis in Enterprise Networks," AFOSR FA9550-07-1-0527 (MURI), ARO MURI: Computer-aided Human Centric Cyber Situation Awareness, and NSF CNS-0905131. Zhenfu Cao was supported by the National Natural Science Foundation of China (NSFC) under Grant Nos. 60972034, 60970110 and 60773086.

References

1. http://tdt.sjtu.edu.cn/~jshao/prcbib.htm
2. The Informed Dialogue about Consumer Acceptability of DRM Solutions in Europe (INDICARE). Consumer Survey on Digital Music and DRM (2005), http://www.indicare.org/survey
3. Ateniese, G., Fu, K., Green, M., Hohenberger, S.: Improved Proxy Re-encryption Schemes with Applications to Secure Distributed Storage. In: Internet Society (ISOC): NDSS 2005, pp. 29–43 (2005)

4. Ateniese, G., Fu, K., Green, M., Hohenberger, S.: Improved Proxy Re-encryption Schemes with Applications to Secure Distributed Storage. ACM Transactions on Information and System Security (TISSEC) 9(1), 1–30 (2006)
5. Ateniese, G., Hohenberger, S.: Proxy re-signatures: new definitions, algorithms, and applications. In: ACM CCS 2005, pp. 310–319 (2005)
6. Bellare, M., Neven, G.: Multi-signatures in the plain public-key model and a general forking lemma. In: ACM CCS 2006, pp. 390–399 (2006)
7. Blaze, M., Bleumer, G., Strauss, M.: Divertible protocols and atomic proxy cryptography. In: Nyberg, K. (ed.) EUROCRYPT 1998. LNCS, vol. 1403, pp. 127–144. Springer, Heidelberg (1998)
8. Boneh, D., Franklin, M.: Identity-based encryption from the weil pairing. In: Kilian, J. (ed.) CRYPTO 2001. LNCS, vol. 2139, pp. 213–229. Springer, Heidelberg (2001)
9. Boneh, D., Franklin, M.: Identity-based encryption from the weil pairing. SIAM Journal of Computing 32(3), 586–615 (2003)
10. Boneh, D., Lynn, B., Shacham, H.: Short signatures from the weil pairing. In: Boyd, C. (ed.) ASIACRYPT 2001. LNCS, vol. 2248, pp. 514–532. Springer, Heidelberg (2001)
11. Koenen, R., Lacy, J., Mackey, M., Mitchell, S.: The long march to interoperable digital rights management. Proceedings of the IEEE 92(6), 883–897 (2004)
12. Libert, B., Vergnaud, D.: Multi-use unidirectional proxy re-signatures. In: ACM CCS 2008, pp. 511–520 (2008), http://arxiv.org/abs/0802.1113v1
13. Shao, J., Cao, Z., Wang, L., Liang, X.: Proxy re-signature schemes without random oracles. In: Srinathan, K., Rangan, C.P., Yung, M. (eds.) INDOCRYPT 2007. LNCS, vol. 4859, pp. 197–209. Springer, Heidelberg (2007)
14. Taban, G., Cárdenas, A.A., Gligor, V.D.: Towards a Secure and Interoperable DRM Architecture. In: ACM DRM 2006, pp. 69–78 (2006)
15. Waters, B.: Efficient identity-based encryption without random oracles. In: Cramer, R. (ed.) EUROCRYPT 2005. LNCS, vol. 3494, pp. 114–127. Springer, Heidelberg (2005)

Security Estimates for Quadratic Field Based Cryptosystems

Jean-François Biasse[1], Michael J. Jacobson Jr.[2,*], and Alan K. Silvester[3]

[1] École Polytechnique, 91128 Palaiseau, France
`biasse@lix.polytechnique.fr`
[2] Department of Computer Science, University of Calgary
2500 University Drive NW, Calgary, Alberta, Canada T2N 1N4
`jacobs@cpsc.ucalgary.ca`
[3] Department of Mathematics and Statistics, University of Calgary
2500 University Drive NW, Calgary, Alberta, Canada T2N 1N4
`aksilves@math.ucalgary.ca`

Abstract. We describe implementations for solving the discrete logarithm problem in the class group of an imaginary quadratic field and in the infrastructure of a real quadratic field. The algorithms used incorporate improvements over previously-used algorithms, and extensive numerical results are presented demonstrating their efficiency. This data is used as the basis for extrapolations, used to provide recommendations for parameter sizes providing approximately the same level of security as block ciphers with 80, 112, 128, 192, and 256-bit symmetric keys.

1   Introduction

Quadratic fields were proposed as a setting for public-key cryptosystems in the late 1980s by Buchmann and Williams [7,8]. There are two types of quadratic fields, imaginary and real. In the imaginary case, cryptosystems are based on arithmetic in the ideal class group (a finite abelian group), and the discrete logarithm problem is the computational problem on which the security is based. In the real case, the so-called infrastructure is used instead, and the security is based on the analogue of the discrete logarithm problem in this structure, namely the principal ideal problem.

Although neither of these problems is resistant to quantum computers, cryptography in quadratic fields is nevertheless an interesting alternative to more widely-used settings. Both discrete logarithm problems can be solved in subexponential time using index calculus algorithms, but with asymptotically slower complexity than the state-of-the art algorithms for integer factorization and computing discrete logarithms in finite fields. In addition, the only known relationship to the quadratic field discrete logarithm problems from other computational problems used in cryptography is that integer factorization reduces to both of the quadratic field problems. Thus, both of these are at least as hard as factoring, and the lack

---

* The second author is supported in part by NSERC of Canada.

R. Steinfeld and P. Hawkes (Eds.): ACISP 2010, LNCS 6168, pp. 233–247, 2010.

of known relationships to other computational problems implies that the breaking of other cryptosystems, such as those based on elliptic or hyperelliptic curves, will not necessarily break those set in quadratic fields. Examining the security of quadratic field based cryptosystems is therefore of interest.

The fastest algorithms for solving discrete logarithm problem in quadratic fields are based on an improved version of Buchmann's index-calculus algorithm due to Jacobson [17]. The algorithms include a number of practical enhancements to the original algorithm of Buchmann [5], including the use of self-initialized sieving to generate relations, a single large prime variant, and practice-oriented algorithms for the required linear algebra. These algorithms enabled the computation of a discrete logarithm in the class group of an imaginary quadratic field with 90 decimal digit discriminant [15], and the solution of the principal ideal problem for a real quadratic field with 65 decimal digit discriminant [18].

Since this work, a number of further improvements have been proposed. Biasse [3] presented practical improvements to the corresponding algorithm for imaginary quadratic fields, including a double large prime variant and improved algorithms for the required linear algebra. The resulting algorithm was indeed faster then the previous state-of-the-art and enabled the computation of the ideal class group of an imaginary quadratic field with 110 decimal digit discriminant. These improvements were adapted to the case of real quadratic fields by Biasse and Jacobson [4], along with the incorporation of a batch smoothness test of Bernstein [2], resulting in similar speed-ups in that case.

In this paper, we adapt the improvements of Biasse and Jacobson to the computation of discrete logarithms in the class group of an imaginary quadratic field and the principal ideal problem in the infrastructure of a real quadratic field. We use versions of the algorithms that rely on easier linear algebra problems than those described in [17]. In the imaginary case, this idea is due to Vollmer [26]; our work represents the first implementation of his method. Our data obtained shows that our algorithms are indeed faster than previous methods. We use our data to estimate parameter sizes for quadratic field cryptosystems that offer security equivalent to NIST's five recommended security levels [25]. In the imaginary case, these recommendations update previous results of Hamdy and Möller [14], and in the real case this is the first time such recommendations have been provided.

The paper is organized as follows. In the next section, we briefly recall the required background of ideal arithmetic in quadratic fields, and give an overview of the index-calculus algorithms for solving the two discrete logarithms in Section 3. Our numerical results are described in Section 4, followed by the security parameter estimates in Section 5.

2   Arithmetic in Quadratic Fields

We begin with a brief overview of arithmetic in quadratic fields. For more details on the theory, algorithms, and cryptographic applications of quadratic fields, see [20].

Let $K = \mathbb{Q}(\sqrt{\Delta})$ be the quadratic field of discriminant Δ, where Δ is a nonzero integer congruent to 0 or 1 modulo 4 with Δ or $\Delta/4$ square-free. The integral closure of $\mathbb{Z}$ in K, called the maximal order, is denoted by $\mathcal{O}_\Delta$. The ideals of $\mathcal{O}_\Delta$ are the main objects of interest in terms of cryptographic applications. An ideal can be represented by the two dimensional $\mathbb{Z}$-module

$$\mathfrak{a} = s \left[a\mathbb{Z} + \frac{b + \sqrt{\Delta}}{2}\mathbb{Z} \right] ,$$

where $a, b, s \in \mathbb{Z}$ and $4a \mid b^2 - \Delta$. The integers a and s are unique, and b is defined modulo $2a$. The ideal $\mathfrak{a}$ is said to be primitive if $s = 1$. The norm of $\mathfrak{a}$ is given by $\mathcal{N}(\mathfrak{a}) = as^2$.

Ideals can be multiplied using Gauss' composition formulas for integral binary quadratic forms. Ideal norm respects this operation. The prime ideals of $\mathcal{O}_\Delta$ have the form $p\mathbb{Z} + (b_p + \sqrt{\Delta})/2\mathbb{Z}$ where p is a prime that is split or ramified in K, i.e., the Kronecker symbol $(\Delta/p) \neq -1$. As $\mathcal{O}_\Delta$ is a Dedekind domain, every ideal can be factored uniquely as a product of prime ideals. To factor $\mathfrak{a}$, it suffices to factor $\mathcal{N}(\mathfrak{a})$ and, for each prime p dividing the norm, determine whether the prime ideal $\mathfrak{p}$ or $\mathfrak{p}^{-1}$ divides $\mathfrak{a}$ according to whether b is congruent to b_p or $-b_p$ modulo $2p$.

Two ideals $\mathfrak{a}, \mathfrak{b}$ are said to be equivalent, denoted by $\mathfrak{a} \sim \mathfrak{b}$, if there exist $\alpha, \beta \in \mathcal{O}_\Delta$ such that $(\alpha)\mathfrak{a} = (\beta)\mathfrak{b}$, where (α) denotes the principal ideal generated by α. This is in fact an equivalence relation, and the set of equivalence classes forms a finite abelian group called the class group, denoted by Cl_Δ. Its order is called the class number, and is denoted by h_Δ.

Arithmetic in the class group is performed on reduced ideal representatives of the equivalence classes. An ideal $\mathfrak{a}$ is reduced if it is primitive and $\mathcal{N}(\mathfrak{a})$ is a minimum in $\mathfrak{a}$. Reduced ideals have the property that $a, b < \sqrt{|\Delta|}$, yielding reasonably small representatives of each group element. The group operation then consists of multiplying two reduced ideals and computing a reduced ideal equivalent to the product. This operation is efficient and can be performed in $O(\log^2 |\Delta|)$ bit operations.

In the case of imaginary quadratic fields, we have $h_\Delta \approx \sqrt{|\Delta|}$, and that every element in Cl_Δ contains exactly one reduced ideal. Thus, the ideal class group can be used as the basis of most public-key cryptosystems that require arithmetic in a finite abelian group. The only wrinkle is that computing the class number h_Δ seems to be as hard as solving the discrete logarithm problem, so only cryptosystems for which the group order is not known can be used.

In real quadratic fields, the class group tends to be small; in fact, a conjecture of Gauss predicts that $h_\Delta = 1$ infinitely often, and the Cohen-Lenstra heuristics [11] predict that this happens about 75% of the time for prime discriminants. Thus, the discrete logarithm problem in the class group is not in general suitable for cryptographic use.

Another consequence of small class groups in the real case is that there are no longer unique reduced ideal representatives in each equivalence class. Instead, we have that $h_\Delta R_\Delta \approx \sqrt{\Delta}$, where the regulator R_Δ roughly approximates how many reduced ideals are in each equivalence class. Thus, since h_Δ is frequently small, there are roughly $\sqrt{\Delta}$ equivalent reduced ideals in each equivalence class.

The infrastructure, namely the set of reduced principal ideals, is used for cryptographic purposes instead of the class group. Although this structure is not a finite abelian group, the analogue of exponentiation (computing a reduced principal ideal (α) with $\log \alpha$ as close to a given number as possible) is efficient and can be used as a one-way problem suitable for public-key cryptography. The inverse of this problem, computing an approximation of the unknown $\log \alpha$ from a reduced principal ideal given in $\mathbb{Z}$-basis representation, is called the principal ideal problem or infrastructure discrete logarithm problem, and is believed to be of similar difficulty to the discrete logarithm problem in the class group of an imaginary quadratic field.

3   Solving the Discrete Logarithm Problems

The fastest algorithms in practice for computing discrete logarithms in the class group and infrastructure use the index-calculus framework. Like other index-calculus algorithms, these algorithms rely on finding certain smooth quantities, those whose prime divisors are all small in some sense. In the case of quadratic fields, one searches for smooth principal ideals for which all prime ideal divisors have norm less than a given bound B. The set of prime ideals $\mathfrak{p}_1, \ldots, \mathfrak{p}_n$ with $\mathcal{N}(\mathfrak{p}_i) \leq B$ is called the factor base, denoted by $\mathcal{B}$.

A principal ideal $(\alpha) = \mathfrak{p}_1^{e_1} \cdots \mathfrak{p}_n^{e_n}$ with $\alpha \in K$ that factors completely over the factor base yields the relation $(e_1, \ldots, e_n, \log |\alpha|)$. In the imaginary case, the $\log |\alpha|$ coefficients are not required and are ignored. The key to the index-calculus approach is the fact, proved by Buchmann [5], that the set of all relations forms a sublattice $\Lambda \subset \mathbb{Z}^n \times \mathbb{R}$ of determinant $h_\Delta R_\Delta$ as long as the prime ideals in the factor base generate Cl_Δ. This follows, in part, due to the fact that L, the integer component of Λ, is the kernel of the homomorphism $\phi : \mathbb{Z}^n \mapsto Cl_\Delta$ given by $\mathfrak{p}_1^{e_1} \cdots \mathfrak{p}_n^{e_n}$ for $(e_1, \ldots, e_n) \in \mathbb{Z}^n$. The homomorphism theorem then implies that $\mathbb{Z}^n / L \cong Cl_\Delta$. In the imaginary case, where the $\log |\alpha|$ terms are omitted, the relation lattice consists only of the integer part, and the corresponding results were proved by Hafner and McCurley [12].

The main idea behind the algorithms described in [17] for solving the class group and infrastructure discrete logarithm problems is to find random relations until they generate the entire relation lattice Λ. Suppose A is a matrix whose rows contain the integer coordinates of the relations, and $\boldsymbol{v}$ is a vector containing the real parts. To check whether the relations generate Λ, we begin by computing the Hermite normal form of A and then calculating its determinant, giving us a multiple h of the class number h_Δ. We also compute a multiple of the regulator R_Δ. Using the analytic class number formula and Bach's $L(1, \chi)$-approximation method [1], we construct bounds such that $h_\Delta R_\Delta$ itself is the only integer multiple of the product of the class number and regulator satisfying $h^* < h_\Delta < 2h^*$; if hR satisfies these bounds, then h and R are the correct class number and regulator and the set of relations given in A generates Λ.

A multiple R of the regulator R_Δ can be computed either from a basis of the kernel of the row-space of A (as in [17]) or by randomly sampling from the kernel as described by Vollmer [27]. Every kernel vector $\boldsymbol{x}$ corresponds to a multiple of the regulator via $\boldsymbol{x} \cdot \boldsymbol{v} = mR_\Delta$. Given $\boldsymbol{v}$ and a set of kernel vectors, an algorithm of Maurer [24, Sec 12.1] is used to compute the "real GCD" of the regulator multiples with guaranteed numerical accuracy, where the real GCD of $m_1 R_\Delta$ and $m_2 R_\Delta$ is defined to be $\gcd(m_1, m_2)R_\Delta$.

To solve the discrete logarithm problem in Cl_Δ, we compute the structure of Cl_Δ, i.e., integers $m_1, \ldots, m_k$ with $m_{i+1} \mid m_i$ for $i = 1, \ldots, k-1$ such that $Cl_\Delta \cong \mathbb{Z}/m_1\mathbb{Z} \times \cdots \times \mathbb{Z}/m_k\mathbb{Z}$, and an explicit isomorphism from $\mathbb{Z}^n$ to $\mathbb{Z}/m_1\mathbb{Z} \times \cdots \times \mathbb{Z}/m_k$. Then, to compute x such that $\mathfrak{g}^x \sim \mathfrak{a}$, we find ideals equivalent to $\mathfrak{g}$ and $\mathfrak{a}$ that factor over the factor base and maps these vectors in $\mathbb{Z}^n$ to $\mathbb{Z}/m_1\mathbb{Z} \times \cdots \times \mathbb{Z}/m_k$, where the discrete logarithm problem can be solved easily.

To solve the infrastructure discrete logarithm problem for $\mathfrak{a}$, we find an ideal equivalent to $\mathfrak{a}$ that factors over the factor base. Suppose the factorization is given by $\boldsymbol{v} \in \mathbb{Z}^n$. Then, since L is the kernel of ϕ, if $\mathfrak{a}$ is principal, $\boldsymbol{v}$ must be a linear combination of the elements of L. This can be determined by solving $\boldsymbol{x}A = \boldsymbol{v}$, where as before the rows of A are the vectors in L. Furthermore, we have $\log \alpha = \boldsymbol{x} \cdot \boldsymbol{v} \pmod{R_\Delta}$ is a solution to the infrastructure discrete logarithm problem. The approximation of $\log \alpha$ is computed to guaranteed numerical accuracy using another algorithm of Maurer [24, Sec 5.5].

If it is necessary to verify the solvability of the problem instance, then one must verify that the relations generate all of Λ, for example, as described above. The best methods for this certification are conditional on the Generalized Riemann Hypothesis, both for their expected running time and their correctness. However, in a cryptographic application, it can safely be assumed that the problem instance does have a solution (for example, if it comes from the Diffie-Hellman key exchange protocol), and simplifications are possible. In particular, the correctness of the computed solution can be determined without certifying that the relations generate Λ, for example, by verifying that $\mathfrak{g}^x = \mathfrak{a}$. As a result, the relatively expensive linear algebra required (computing Hermite normal form and kernel of the row space) can be replaced by linear system solving.

In the imaginary case, if the discrete logarithm is known to exist, one can use an algorithm due to Vollmer [26,28]. Instead of computing the structure of Cl_Δ, one finds ideals equivalent to $\mathfrak{g}$ and $\mathfrak{a}$ that factor over the factor base. Then, combining these factorizations with the rest of the relations and solving a linear system yields a solution of the discrete logarithm problem. If the linear system cannot be solved, then the relations do not generate Λ, and the process is simply repeated after generating some additional relations. The expected asymptotic complexity of this method, under reasonable assumptions about the generation of relations, is $O(L_{|\Delta|}[1/2, 3\sqrt{2}/4 + o(1)])$ [28,6], where

$$L_N[e, c] = \exp\left(c \left(\log N\right)^e \left(\log \log N\right)^{1-e}\right)$$

for e, c constants and $0 \le e \le 1$. In practice, all the improvements to relation generation and simplifying the relation matrix described in [3] can be applied. When using practical versions for generating relations, such as sieving as described in [17], it is conjectured that the algorithm has complexity $O(L_{|\Delta|}[1/2, 1 + o(1)])$.

In the real case, we also do not need to compute the Hermite normal form, as only a multiple of R_Δ suffices. The consequence of not certifying that we have the true regulator is that the solutions obtained for the infrastructure discrete logarithm problem may not be minimal. However, for cryptographic purposes this is sufficient, as these values can still be used to break the corresponding protocols in the same way that a non-minimal solution to the discrete logarithm problem suffices to break group-based protocols. Thus, we use Vollmer's approach [27] based on randomly sampling from the kernel of A. This method computes a multiple that is with high probability equal to the regulator in time $O(L_{|\Delta|}[1/2, 3\sqrt{2}/4 + o(1)])$ by computing the multiple corresponding to random elements in the kernel of the row space of A. These random elements can also be found by linear system solving. The resulting algorithm has the same complexity as that in the imaginary case. In practice, all the improvements described in [4] can be applied. When these are used, including sieving as described in [17], we also conjecture that the algorithm has complexity $O(L_{|\Delta|}[1/2, 1 + o(1)])$.

4   Implementation and Numerical Results

Our implementation takes advantage of the latest practical improvements in ideal class group computation and regulator computation for quadratic number fields, described in detail in [3,4]. In the following, we give a brief outline of the methods we used for the experiments described in this paper.

To speed up the relation collection phase, we combined the double large prime variation with the self-initialized quadratic sieve strategy of [17], as descried in [3]. This results in a considerable speed-up in the time required for finding a relation, at the cost of a growth of the dimensions of the relation matrix. We also used Bernstein's batch smoothness test [2] to enhance the relation collection phase as described in [4], by simultaneously testing residues produced by the sieve for smoothness.

The algorithms involved in the linear algebra phase are highly sensitive to the dimensions of the relation matrix. As the double large prime variation induces significant growth in the dimensions of the relation matrix, one needs to perform Gaussian elimination to reduce the number of columns in order to make the linear algebra phase feasible. We used a graph-based elimination strategy first described by Cavallar [9] for factorization, and then adapted by Biasse [3] to the context of quadratic fields. At the end of the process, we test if the resulting matrix A_{red} has full rank by reducing it modulo a word-sized prime. If not, we collect more relation and repeat the algorithm.

For solving the discrete logarithm problem in the imaginary case, we implemented the algorithm due to Vollmer [26,28] . Given two ideals $\mathfrak{a}$ and $\mathfrak{g}$ such that $\mathfrak{g}^x \sim \mathfrak{a}$ for some integer x, we find two extra relations $(e_1, \ldots, e_n, 1, 0)$ and $(f_1, \ldots, f_n, 0, 1)$ such that $\mathfrak{p}_1^{e_1} \cdots \mathfrak{p}_n^{e_n} \mathfrak{g} \sim (1)$ and $\mathfrak{p}_1^{f_1} \cdots \mathfrak{p}_n^{f_n} \mathfrak{a}^{-1} \sim (1)$ over the extended factor base $\mathcal{B} \cup \{\mathfrak{g}, \mathfrak{a}^{-1}\}$. The extra relations are obtained by multiplying $\mathfrak{a}^{-1}$ and $\mathfrak{g}$ by random power products of primes in $\mathcal{B}$ and sieving with the resulting ideal to find an equivalent ideal that is smooth over $\mathcal{B}$. Once these relations have been found, we construct the matrix

$$A' := \left(\begin{array}{ccccc|cc} & & A & & & \vdots & (0) \\ \hline e_1 & \cdots & e_n & & & 1 & 0 \\ f_1 & \cdots & f_n & & & 0 & 1 \end{array} \right),$$

and solve the system $\boldsymbol{x}A' = (0, \ldots, 0, 1)$. The last coordinate of $\boldsymbol{x}$ necessarily equals the discrete logarithm x. We used `certSolveRedLong` from the IML library [10] to solve these linear systems.

As the impact of Vollmer's and Bernstein's algorithms on the overall time for class group and discrete logarithm computation in the imaginary case had not been studied, we provide numerical data in Table 1 for discriminants of size between 140 and 220 bits. The timings, given in seconds, are averages of three different random prime discriminants, obtained with 2.4 GHz Opterons with 8GB or memory. We denote by "DL" the discrete logarithm computation using Vollmer's method and by "CL" the class group computation. "CL Batch" and "DL Batch" denote the times obtained when also using Bernstein's algorithm. We list the optimal factor base size for each algorithm and discriminant size (obtained via additional numerical experiments), the time for each of the main parts of the algorithm, and the total time. In all cases we allowed two large primes and took enough relations to ensure that A_{red} have full rank. Our results show that enhancing relation generation with Bernstein's algorithm is beneficial in all cases. In addition, using Vollmer's algorithm for computing discrete logarithms is faster than the approach of [17] that also requires the class group.

To solve the infrastructure discrete logarithm problem, we first need to compute an approximation of the regulator. For this purpose, we used an improved version of Vollmer's system solving based algorithm [27] described by Biasse and Jacobson [4]. In order to find elements of the kernel, the algorithm creates extra relations r_i, $0 \le i \le k$ for some small integer k (in our experiments, we always have $k \le 10$). Then, we solve the k linear systems $X_i A = r_i$ using the function `certSolveRedLong` from the IML library [10]. We augment the matrix A by adding the r_i as extra rows, and augment the vectors X_i with $k - 1$ zero coefficients and a -1 coefficient at index $n + i$, yielding

$$A' := \left(\begin{array}{c} A \\ \hline r_i \end{array} \right), \quad X_i' := \left(\begin{array}{ccccccc} X_i & \vdots & 0 \ldots 0 & -1 & 0 \ldots 0 \end{array} \right).$$

The X_i' are kernel vectors of A', which can be used along with the vector $\boldsymbol{v}$ containing the real parts of the relations, to compute a multiple of the regulator with Maurer's algorithm [24, Sec 12.1]. As shown in Vollmer [27], this multiple is equal to the regulator with high probability. In [4], it is shown that this method is faster than the one requiring a kernel basis because it only requires the solution

Table 1. Comparison between class group computation and Vollmer Algorithm

Size	Strategy	$|\mathcal{B}|$	Sieving	Elimination	Linear algebra	Total
140	CL	200	2.66	0.63	1.79	5.08
	CL Batch	200	1.93	0.65	1.78	4.36
	DL	200	2.57	0.44	0.8	3.81
	DL batch	200	1.92	0.41	0.76	3.09
160	CL	300	11.77	1.04	8.20	21.01
	CL Batch	300	9.91	0.87	8.19	18.97
	DL	350	10.17	0.73	2.75	13.65
	DL batch	400	6.80	0.96	3.05	10.81
180	CL	400	17.47	0.98	12.83	31.28
	CL Batch	400	14.56	0.97	12.9	28.43
	DL	500	15.00	1.40	4.93	21.33
	DL batch	500	11.35	1.34	4.46	17.15
200	CL	800	158.27	7.82	81.84	247.93
	CL Batch	800	133.78	7.82	81.58	223.18
	DL	1000	126.61	9.9	21.45	157.96
	DL batch	1100	85.00	11.21	26.85	123.06
220	CL	1500	619.99	20.99	457.45	1098.43
	CL Batch	1500	529.59	19.56	447.29	996.44
	DL	1700	567.56	27.77	86.38	681.71
	DL batch	1600	540.37	24.23	73.76	638.36

to a few linear systems, and it can be adapted in such a way that the linear system involves A_{red}.

Our algorithm to solve the infrastructure discrete logarithm problem also makes use of the system solving algorithm. The input ideal $\mathfrak{a}$ is first decomposed over the factor base, as in the imaginary case, yielding the factorization $\mathfrak{a} = (\gamma)\mathfrak{p}_1{}^{e_1} \cdots \mathfrak{p}_n{}^{e_n}$. Then, we solve the system $xA = (e_1, \ldots, e_n)$ and compute a numerical approximation to guaranteed precision of $\log|\alpha|$ modulo our regulator multiple using Maurer's algorithm [24, Sec 5.5] from γ, the coefficients of x, and the real parts of the relation stored in v.

The results of our experiments for the imaginary case are given in Table 2, and for the real case in Table 3. They were obtained on 2.4 GHz Xeon with 2GB of memory. For each bit length of Δ, denoted by "size(Δ)," we list the average time in seconds required to solve an instance of the appropriate discrete logarithm problem $(\overline{t_\Delta})$ and standard deviation (std). In the imaginary case, for each discriminant size less than 220 bits, 14 instances of the discrete logarithm problem were solved. For size 230 and 256 we solved 10, and for size 280 and 300 we solved 5 examples. In the real case, 10 instances were solved for each size up to 256, 6 for size 280, and 4 for size 300.

Table 2. Average run times for the discrete logarithm problem in Cl_Δ, $\Delta < 0$

size(Δ)	$\overline{t_\Delta}$ (sec)	std	$L_{|\Delta|}[1/2, \sqrt{2}]/\overline{t_\Delta}$	$L_{|\Delta|}[1/2, 1]/\overline{t_\Delta}$
140	7.89	2.33	6.44×10^8	1.79×10^8
142	8.80	1.90	7.01×10^8	1.93×10^8
144	9.91	3.13	7.55×10^8	2.06×10^8
146	10.23	1.69	8.86×10^8	2.39×10^8
148	11.80	3.45	9.29×10^8	2.48×10^8
150	12.88	2.66	10.28×10^8	2.71×10^8
152	14.42	3.38	11.09×10^8	2.89×10^8
154	17.64	5.61	10.93×10^8	2.82×10^8
156	22.06	5.57	10.53×10^8	2.69×10^8
158	28.74	12.11	9.73×10^8	2.46×10^8
160	27.12	8.77	12.39×10^8	3.10×10^8
162	32.72	15.49	12.34×10^8	3.05×10^8
164	31.08	6.85	15.58×10^8	3.82×10^8
166	41.93	14.65	13.85×10^8	3.36×10^8
168	51.92	16.51	13.39×10^8	3.21×10^8
170	59.77	15.42	13.92×10^8	3.30×10^8
172	68.39	17.79	14.54×10^8	3.42×10^8
174	99.20	62.61	11.97×10^8	2.78×10^8
176	124.86	80.29	11.35×10^8	2.61×10^8
178	140.50	55.41	12.03×10^8	2.74×10^8
180	202.42	145.98	9.94×10^8	2.24×10^8
182	166.33	63.91	14.40×10^8	3.22×10^8
184	150.76	58.37	18.90×10^8	4.18×10^8
186	198.72	63.23	17.04×10^8	3.73×10^8
188	225.90	94.94	17.79×10^8	3.86×10^8
190	277.67	234.93	17.17×10^8	3.69×10^8
192	348.88	134.36	16.20×10^8	3.45×10^8
194	395.54	192.26	16.93×10^8	3.57×10^8
196	547.33	272.83	14.48×10^8	3.02×10^8
198	525.94	153.63	17.83×10^8	3.68×10^8
200	565.43	182.75	1.96×10^9	4.01×10^8
202	561.36	202.80	2.33×10^9	4.73×10^8
204	535.29	205.68	2.89×10^9	5.80×10^8
206	776.64	243.35	2.35×10^9	4.67×10^8
208	677.43	200.08	3.17×10^9	6.25×10^8
210	1050.64	501.31	2.41×10^9	4.70×10^8
212	1189.71	410.98	2.50×10^9	4.84×10^8
214	1104.83	308.57	3.17×10^9	6.07×10^8
216	1417.64	352.27	2.90×10^9	5.51×10^8
218	2185.80	798.95	2.21×10^9	4.16×10^8
220	2559.79	1255.94	2.22×10^9	4.13×10^8
230	3424.40	1255.94	3.66×10^9	6.52×10^8
256	22992.70	13062.14	4.00×10^9	6.36×10^8
280	88031.08	34148.54	6.09×10^9	8.76×10^8
300	702142.20	334566.51	3.16×10^9	4.19×10^8

Table 3. Average run times for the infrastructure discrete logarithm problem

size(Δ)	$\overline{t_\Delta}$ (sec)	std	$L_{|\Delta|}[1/2, \sqrt{2}]/\overline{t_\Delta}$	$L_{|\Delta|}[1/2, 1]/\overline{t_\Delta}$
140	11.95	3.13	4.25×10^8	1.18×10^8
142	12.47	2.06	4.95×10^8	1.36×10^8
144	15.95	5.79	4.69×10^8	1.28×10^8
146	14.61	2.94	6.20×10^8	1.67×10^8
148	17.05	3.46	6.43×10^8	1.71×10^8
150	21.65	4.55	6.12×10^8	1.61×10^8
152	25.65	7.15	6.23×10^8	1.63×10^8
154	29.01	6.97	6.65×10^8	1.72×10^8
156	27.52	4.79	8.44×10^8	2.16×10^8
158	33.59	8.80	8.32×10^8	2.10×10^8
160	36.27	12.28	9.27×10^8	2.32×10^8
162	43.55	10.73	9.27×10^8	2.29×10^8
164	49.37	11.76	9.81×10^8	2.40×10^8
166	59.73	17.18	9.72×10^8	2.36×10^8
168	73.66	18.56	9.44×10^8	2.26×10^8
170	75.50	19.80	1.10×10^9	2.62×10^8
172	101.00	20.84	9.85×10^8	2.31×10^8
174	94.80	38.87	1.25×10^9	2.91×10^8
176	106.30	23.77	1.33×10^9	3.07×10^8
178	149.70	44.04	1.13×10^9	2.57×10^8
180	132.70	30.25	1.52×10^9	3.42×10^8
182	178.80	25.67	1.34×10^9	2.99×10^8
184	211.40	52.14	1.35×10^9	2.98×10^8
186	258.20	110.95	1.31×10^9	2.87×10^8
188	352.70	94.50	1.14×10^9	2.47×10^8
190	290.90	46.57	1.64×10^9	3.52×10^8
192	316.80	51.75	1.78×10^9	3.80×10^8
194	412.90	71.90	1.62×10^9	3.42×10^8
196	395.40	94.71	2.00×10^9	4.18×10^8
198	492.30	156.69	1.90×10^9	3.94×10^8
200	598.90	187.19	1.85×10^9	3.79×10^8
202	791.40	285.74	1.65×10^9	3.35×10^8
204	888.10	396.85	1.74×10^9	3.49×10^8
206	928.40	311.37	1.96×10^9	3.90×10^8
208	1036.10	260.82	2.07×10^9	4.08×10^8
210	1262.30	415.32	2.00×10^9	3.91×10^8
212	1582.30	377.22	1.88×10^9	3.64×10^8
214	1545.10	432.42	2.27×10^9	4.34×10^8
216	1450.80	453.85	2.84×10^9	5.39×10^8
218	2105.00	650.64	2.30×10^9	4.32×10^8
220	2435.70	802.57	2.33×10^9	4.34×10^8
230	5680.90	1379.94	2.21×10^9	3.93×10^8
256	29394.01	7824.15	3.13×10^9	4.98×10^8
280	80962.80	27721.01	6.62×10^9	9.52×10^8
300	442409.00	237989.12	5.01×10^9	6.64×10^8

For the extrapolations in the next section, we need to have a good estimate of the asymptotic running time of the algorithm. As described in the previous section, the best proven run time is $O(L_{|\Delta|}[1/2, 3\sqrt{2}/4 + o(1)]$, but as we use sieving to generate relations, this can likely be reduced to $O(L_{|\Delta|}[1/2, 1 + o(1)])$. To test which running time is most likely to hold for the algorithm we implemented, we list $L_{|\Delta|}[1/2, 3\sqrt{2}/4]/\overline{t_\Delta}$ and $L_{|\Delta|}[1/2, 1]/\overline{t_\Delta}$ in Table 2 and Table 3. In both cases, our data supports the hypothesis that the run time of our algorithm is indeed closer to $O(L_{|\Delta|}[1/2, 1 + o(1)])$, with the exception of a few outliers corresponding to instances where only a few instances of the discrete logarithm were computed for that size.

5   Security Estimates

General purpose recommendations for securely choosing discriminants for use in quadratic field cryptography can be found in [14] for the imaginary case and [18] for the real case. In both cases, it usually suffices to use prime discriminants, as this forces the class number h_Δ to be odd. In the imaginary case, one then relies on the Cohen-Lenstra heuristics [11] to guarantee that the class number is not smooth with high probability. In the real case, one uses the Cohen-Lenstra heuristics to guarantee that the class number is very small (and that the infrastructure is therefore large) with high probability.

Our goal is to estimate what bit lengths of appropriately-chosen discriminants, in both the imaginary and real cases, are required to provide approximately the same level of security as the RSA moduli recommended by NIST [25]. The five security levels recommended by NIST correspond to using secure block ciphers with keys of 80, 112, 128, 192, and 256 bits. The estimates used by NIST indicate that RSA moduli of size 1024, 2048, 3072, 7680, and 15360 should be used.

To estimate the required sizes of discriminants, we follow the approach of Hamdy and Möller [14], who provided such estimates for the imaginary case. Our results update these in the sense that our estimates are based on our improved algorithms for solving the discrete logarithms in quadratic fields, as well as the latest data available for factoring large RSA moduli. Our estimates for real quadratic fields are the first such estimates produced.

Following, Hamdy and Möller, suppose that an algorithm with asymptotic running time $L_N[e, c]$ runs in time t_1 on input N_1. Then, the running time t_2 of the algorithm on input N_2 can be estimated using the equation

$$\frac{L_{N_1}[e, c]}{L_{N_2}[e, c]} = \frac{t_1}{t_2} \ . \tag{1}$$

We can also use the equation to estimate an input N_2 that will cause the algorithm to have running time t_2, again given the time t_1 for input N_1.

The first step is to estimate the time required to factor the RSA numbers of the sizes recommended by NIST. The best algorithm for factoring large integers is the generalized number field sieve [22], whose asymptotic running time

Table 4. Security Parameter Estimates

RSA	Δ (imaginary, old)	Δ (imaginary)	Δ (real)	Est. run time (MIPS-years)
768	540	640	634	8.80×10^6
1024	687	798	792	1.07×10^{10}
2048	1208	1348	1341	1.25×10^{19}
3072	1665	1827	1818	4.74×10^{25}
7680	0	3598	3586	1.06×10^{45}
15360	0	5971	5957	1.01×10^{65}

is heuristically $L_N[1/3, \sqrt[3]{64/9} + o(1)]$. To date, the largest RSA number factored is RSA-768, a 768 bit integer [21]. It is estimated in [21] that the total computation required 2000 2.2 GHz AMD Opteron years. As our computations were performed on a different architecture, we follow Hamdy and Möller and use the MIPS-year measurement to provide an architecture-neutral measurement. In this case, assuming that a 2.2 GHz AMD Opteron runs at 4400 MIPS, we estimate that this computation took 8.8×10^6 MIPS-years. Using this estimate in conjunction with (1) yields the estimated running times to factor RSA moduli of the sizes recommended by NIST given in Table 4. When using this method, we use $N_1 = 2^{768}$ and $N_2 = 2^b$, where b is the bit length of the RSA moduli for which we compute a run time estimate.

The second step is to estimate the discriminant sizes for which the discrete logarithm problems require approximately the same running time. The results in Table 2 and Table 3 suggest that $L_N[1/2, 1 + o(1)]$ is a good estimate of the asymptotic running time for both algorithms. Thus, we use $L_N[1/2, 1]$ in (1), as ignoring the $o(1)$ results in a conservative under-estimate of the actual running time. For N_1 and t_1, we take the largest discriminant size in each table for which at least 10 instances of the discrete logarithm problem were run and the corresponding running time (in MIPS-years); thus we used 256 in the imaginary case and 230 in the real case. We take for t_2 the target running time in MIPS-years. To convert the times in seconds from Table 2 and Table 3 to MIPS-years, we assume that the 2.4 GHz Intel Xeon machine runs at 4800 MIPS. To find the corresponding discriminant size, we simply find the smallest integer b for which $L_{2^b}[1/2, 1] > L_{N_1}[1/2, 1]t_2/t_1$.

Our results are listed in Table 4. We list the size in bits of RSA moduli (denoted by "RSA"), discriminants of imaginary quadratic fields (denoted by "Δ (imaginary)"), and real quadratic fields (denoted by "Δ (real)") for which factoring and the quadratic field discrete logarithm problems all have the same estimated running time. For comparison purposes, we also list the discriminant sizes recommended in [14], denoted by "Δ (imaginary, old)." Note that these estimates were based on different equivalent MIPS-years running times, as the largest factoring effort at the time was RSA-512. In addition, they are based on an implementation of the imaginary quadratic field discrete logarithm algorithm from [17], which is slower than the improved version from this paper. Consequently, our security parameter estimates are slightly larger than those from [14].

We note also that the recommended discriminant sizes are slightly smaller in the real case, as the infrastructure discrete logarithm problem requires more time to solve on average than the discrete logarithm in the imaginary case.

6   Conclusions

It is possible to produce more accurate security parameter estimates by taking more factors into account as is done, for example, by Lenstra and Verheul [23], as well as using a more accurate performance measure than MIPS-year. However, our results nevertheless provide a good rough guideline on the required discriminant sizes that is likely sufficiently accurate in the inexact science of predicting security levels.

It would also be of interest to conduct a new comparison of the efficiency of RSA as compared to the cryptosystems based on quadratic fields. Due to the differences in the asymptotic complexities of integer factorization and the discrete logarithm problems in quadratic fields, it is clear that there is a point where the cryptosystems based on quadratic fields will be faster than RSA. However, ideal arithmetic is somewhat more complicated than the simple integer arithmetic required for RSA, and in fact Hamdy's conclusion [13] was that even with smaller parameters, cryptography using quadratic fields was not competitive at the security levels of interest. There have been a number of recent advances in ideal arithmetic in both the imaginary and real cases (see, for example, [16] and [19]) that warrant revisiting this issue.

References

1. Bach, E.: Explicit bounds for primality testing and related problems. Math. Comp. 55(191), 355–380 (1990)
2. Bernstein, D.: How to find smooth parts of integers. Submitted to Mathematics of Computation
3. Biasse, J.-F.: Improvements in the computation of ideal class groups of imaginary quadratic number fields. To appear in Advances in Mathematics of Communications, http://www.lix.polytechnique.fr/~biasse/papers/biasseCHILE.pdf
4. Biasse, J.-F., Jacobson Jr., M.J.: Practical improvements to class group and regulator computation of real quadratic fields. To appear in ANTS 9 (2010)
5. Buchmann, J.: A subexponential algorithm for the determination of class groups and regulators of algebraic number fields. Séminaire de Théorie des Nombres (Paris), pp. 27–41 (1988–1989)
6. Buchmann, J., Vollmer, U.: Binary quadratic forms: An algorithmic approach. In: Algorithms and Computation in Mathematics, vol. 20. Springer, Berlin (2007)
7. Buchmann, J., Williams, H.C.: A key-exchange system based on imaginary quadratic fields. Journal of Cryptology 1, 107–118 (1988)
8. Buchmann, J., Williams, H.C.: A key-exchange system based on real quadratic fields. In: Brassard, G. (ed.) CRYPTO 1989. LNCS, vol. 435, pp. 335–343. Springer, Heidelberg (1990)

9. Cavallar, S.: Strategies in filtering in the number field sieve. In: Bosma, W. (ed.) ANTS 2000. LNCS, vol. 1838, pp. 209–232. Springer, Heidelberg (2000)
10. Chen, Z., Storjohann, A., Fletcher, C.: IML: Integer Matrix Library (2007), http://www.cs.uwaterloo.ca/~z4chen/iml.html
11. Cohen, H., Lenstra Jr., H.W.: Heuristics on class groups of number fields. In: Number Theory. Lecture Notes in Math., vol. 1068, pp. 33–62. Springer, New York (1983)
12. Hafner, J.L., McCurley, K.S.: A rigorous subexponential algorithm for computation of class groups. J. Amer. Math. Soc. 2, 837–850 (1989)
13. Hamdy, S.: Über die Sicherheit und Effizienz kryptografischer Verfahren mit Klassengruppen imaginär-quadratischer Zahlkörper. Ph.D. thesis, Technische Universität Darmstadt, Darmstadt, Germany (2002)
14. Hamdy, S., Möller, B.: Security of cryptosystems based on class groups of imaginary quadratic orders. In: Okamoto, T. (ed.) ASIACRYPT 2000. LNCS, vol. 1976, pp. 234–247. Springer, Heidelberg (2000)
15. Hühnlein, D., Jacobson Jr., M.J., Weber, D.: Towards practical non-interactive public-key cryptosystems using non-maximal imaginary quadratic orders. Designs, Codes and Cryptography 30(3), 281–299 (2003)
16. Imbert, L., Jacobson Jr., M.J., Schmidt, A.: Fast ideal cubing in imaginary quadratic number and function fields. To appear in to Advances in Mathematics of Communication (2010)
17. Jacobson Jr., M.J.: Computing discrete logarithms in quadratic orders. Journal of Cryptology 13, 473–492 (2000)
18. Jacobson Jr., M.J., Scheidler, R., Williams, H.C.: The efficiency and security of a real quadratic field based key exchange protocol. In: Public-Key Cryptography and Computational Number Theory, Warsaw, Poland, pp. 89–112. de Gruyter (2001)
19. Jacobson Jr., M.J., Scheidler, R., Williams, H.C.: An improved real quadratic field based key exchange procedure. Journal of Cryptology 19, 211–239 (2006)
20. Jacobson Jr., M.J., Williams, H.C.: Solving the Pell equation. CMS Books in Mathematics. Springer, Heidelberg (2009) ISBN 978-0-387-84922-5
21. Kleinjung, T., Aoki, K., Franke, J., Lenstra, A.K., Thomé, E., Bos, J.W., Gaudry, P., Kruppa, A., Montgomery, P.L., Osvik, D.A., te Riele, H., Timofeev, A., Zimmerman, P.: Factorization of a 768-bit RSA modulus, Eprint archive no. 2010/006 (2010)
22. Lenstra, A.K., Lenstra Jr., H.W.: The development of the number field sieve. Lecture Notes in Mathematics, vol. 1554. Springer, Berlin (1993)
23. Lenstra, A.K., Verheul, E.: Selecting cryptographic key sizes. In: Imai, H., Zheng, Y. (eds.) PKC 2000. LNCS, vol. 1751, pp. 446–465. Springer, Heidelberg (2000)
24. Maurer, M.: Regulator approximation and fundamental unit computation for real-quadratic orders, Ph.D. thesis, Technische Universität Darmstadt, Darmstadt, Germany (2000)
25. National Institute of Standards and Technology (NIST), Recommendation for Key Management — Part 1: General (Revised), NIST Special Publication 800-57 (March 2007), http://csrc.nist.gov/groups/ST/toolkit/documents/ SP800-57Part1_3-8-07.pdf
26. Vollmer, U.: Asymptotically fast discrete logarithms in quadratic number fields. In: Bosma, W. (ed.) ANTS 2000. LNCS, vol. 1838, pp. 581–594. Springer, Heidelberg (2000)

27. Vollmer, U.: An accelerated Buchmann algorithm for regulator computation in real quadratic fields. In: Fieker, C., Kohel, D.R. (eds.) ANTS 2002. LNCS, vol. 2369, pp. 148–162. Springer, Heidelberg (2002)
28. Vollmer, U.: Rigorously analyzed algorithms for the discrete logarithm problem in quadratic number fields, Ph.D. thesis, Technische Universität Darmstadt (2003)

Solving Generalized Small Inverse Problems

Noboru Kunihiro

The University of Tokyo, Japan
kunihiro@k.u-tokyo.ac.jp

Abstract. We introduce a "generalized small inverse problem (GSIP)" and present an algorithm for solving this problem. GSIP is formulated as finding small solutions of $f(x_0, x_1, \ldots, x_n) = x_0 h(x_1, \ldots, x_n) + C = 0 (\mathrm{mod}\ M)$ for an n-variate polynomial h, non-zero integers C and M. Our algorithm is based on lattice-based Coppersmith technique. We provide a strategy for construction of a lattice basis for solving $f = 0$, which are systematically transformed from a lattice basis for solving $h = 0$. Then, we derive an upper bound such that the target problem can be solved in polynomial time in $\log M$ in an explicit form. Since GSIPs include some RSA-related problems, our algorithm is applicable to them. For example, the small key attacks by Boneh and Durfee are re-found automatically.

Keywords: LLL algorithm, small inverse problem, RSA. lattice-based cryptanalysis.

1 Introduction

Since the seminal work of Coppersmith [3,4,5], many cryptanalysis have been proposed by using his technique which is based on LLL algorithm. The first typical application is a small secret exponent attack on RSA proposed by Boneh and Durfee [2]. The second is a proof of deterministic polynomial time equivalence between computing the RSA secret key and factoring [6,16].

In RSA [18], the small secret exponent d is commonly used to speed up the decryption or signature generation. In 1990, Wiener showed that when $d \le \frac{1}{3}N^{1/4}$, the RSA moduli N can be factored in polynomial time [20]. Then, in 1999, Boneh and Durfee [2] improved the Wiener's bound to $d \le N^{0.284}$. Furthermore, they proved that N can be factored in polynomial time when $d \le N^{0.292}$. In their attack, lattice reduction algorithms such as LLL algorithm [14] play an important role. Let us briefly describe their attack. First, they reduce small secret exponent attack to solving a bivariate modular equation:

$$x(A + y) = 1 \pmod{e},$$

where A is a given integer and the solution $(x, y) = (\bar{x}, \bar{y})$ satisfies $|\bar{x}| < e^{\delta}$ and $|\bar{y}| < e^{1/2}$. They referred this problem as "small inverse problem." Then, they proposed a polynomial time algorithm for solving this problem. They obtained the condition on δ such that the algorithm outputs the solution. This leads to the

R. Steinfeld and P. Hawkes (Eds.): ACISP 2010, LNCS 6168, pp. 248–263, 2010.

weaker bound: $d \leq N^{0.284}$ and the stronger bound: $d \leq N^{0.292}$. By extending
their (weaker) algorithm, Durfee and Nguyen showed cryptanalysis on some
variants of RSA with short secret exponent [7]. They proposed an algorithm for
solving trivariate modular equation $f(x, y, z) = x(A + y + z) + 1 = 0 \pmod{e}$
with constraint $yz = N$ in their analysis. It is crucial in their algorithm how
to handle the constraint $yz = N$. To do so, they introduced so-called "Durfee-
Nguyen technique."

May (and Coron-May) proved that if the RSA secret key d is revealed, the
RSA moduli N can be factored in *deterministic* polynomial time [6,16]. We
will focus on the Coron-May's proof [6] rather than May's original proof [16].
Consider a univariate modular equation: $h(y) \equiv A + y = 0 \pmod{S}$, where S
is an unknown divisor of a known positive integer U and A is a known positive
integer. They showed a deterministic polynomial time algorithm which solves
the equation for $S \leq U^{1/2}$ to prove that (balanced) RSA moduli N can be
factored deterministically when d is revealed. They extended their result to the
unbalanced RSA case [6]. They showed the condition that the bivariate modular
equation: $h(y, z) \equiv A + y + z = 0 \pmod{S}$ with constraint $yz = N$, where S
and U are in the same setting as the balanced RSA.

1.1   Our Contribution

In this paper, we introduce "generalized small inverse problem (GSIP)" for an
$n + 1$-variate equation. Let f be an $n + 1$-variate polynomial by

$$f(x_0, x_1, \ldots, x_n) = x_0 h(x_1, \ldots, x_n) + C$$

for an n-variate polynomial h and a non-zero integer C. Let M be a positive integer whose prime factors are unknown. Suppose that the solution of
$f = 0 \pmod{M}$ satisfies $|\bar{x}_0| < X_0, |\bar{x}_1| < X_1, \ldots, |\bar{x}_n| < X_n$ for fixed positive
integers $X_0, X_1, \ldots, X_n$. Then, one wants to find the solution: $(x_0, x_1, \ldots, x_n) =
(\bar{x}_0, \bar{x}_1, \ldots, \bar{x}_n)$. Some cases may have constraints between variables $x_1, \ldots, x_n$.
When $C = 1$, the problem can be viewed as follows: given a function
$h(x_1, x_2, \ldots, x_n)$, find small elements $(\bar{x}_1, \ldots, \bar{x}_n)$ such that the inverse of
$-h(\bar{x}_1, \bar{x}_2, \ldots, \bar{x}_n)$ modulo M is "small". So, we call this problem as generalized small inverse problem. Classical "small inverse problem" [2] corresponds to
$n = 1$, $h(x_1) = A + x_1$ and $C = 1$, where A is a given integer. GSIP is not
only a natural extension of classical small inverse problem, but also is applicable
to many RSA-related cryptanalysis. In our paper, we are concerned with only
modular equations not integer equations.

Second, we propose a polynomial time algorithm for solving this problem. Our
algorithm is based on Coppersmith's approach [3] and has the following property
in the lattice basis construction:

1. First, construct a lattice basis for solving $h(x_1, \ldots, x_n) = 0 \pmod{p}$, where
   p is an unknown divisor of known integer N.
2. Then, construct a lattice basis for $f(x_0, \ldots, x_n) = 0 \pmod{M}$ by employing
   a lattice basis for h.

We introduce 4 restrictions for a lattice in solving $h = 0$. Since many methods in the literature hold these restrictions, they are not too strong restrictions. Then, we propose a simple but effective compiler which transforms a lattice basis for $h = 0$ to that for $f = x_0 h + C = 0$ (**Compiler**). Our compiler works if a lattice for $h = 0$ holds the 4 restrictions. It gives a good insight in construction of a lattice basis for f.

Our compiler is applicable to many kinds of cryptanalysis. For example, we can re-find Boneh-Durfee's small secret exponent attack on RSA [2] by using our compiler and the lattice employed in the proof for deterministic polynomial time equivalence [6]. That is, our compiler builds a bridge between these two works. It is the first time to point out this kind of connection as far as we know. Our compiler is especially effective when one needs to construct a special type of lattice. Suppose that some variables have constraint, ex. $yz = N$. In this case, it is well known that Durfee-Nguyen technique is effective [7]. If one can construct a good lattice for n-variate equation: $h = 0$ built Durfee-Nguyen technique into, one has also a good lattice for $n + 1$-variate equation: f built Durfee-Nguyen technique into. In general, the more variables are involved, the harder the construction of a good lattice is. If one uses our compiler, one just constructs a lattice basis for h not for f. Hence, one can more easily construct a good lattice basis for f.

Next, we obtain the upper bound of the solution such that the equation: $f(x_0, x_1, \ldots, x_n) = 0 \pmod{M}$ is solvable in polynomial time in $\log M$ (but not in n) (Lemma 5 and Theorem 2). That means, letting the solution be $(\bar{x}_0, \ldots, \bar{x}_n)$ and positive integers $X_0, \ldots, X_n$, when $|\bar{x}_i| < X_i$ for each i, one can solve the problem in polynomial time. In deriving X_i, one needs not tedious computation. In particular, when $X_1, \ldots, X_n$ are fixed, one can easily obtain the upper bound of solution X_0.

In Boneh-Durfee's [2] and Durfee-Nguyen's analyses [7], tedious computations are needed. Furthermore, their computations are not applicable to the other kind of attacks. We generalize this kind of calculation to obtain the evaluation formula, which is easy to use and covers many kind of cryptanalysis including Boneh-Durfee's. Hence, we provide another type of "toolkit" for (especially RSA-related) cryptanalysis from that of Blömer-May [1].

Our Strategies vs. General Strategies for Construction of Lattice Basis. It is well known that the shape of Newton polytope of a polynomial to be solved is important. This is suggested by Coppersmith [4] and fully explained by Blömer and May in the case of bivariate integer equation [1]. For general polynomials, Jochemsz and May proposed general methods for construction of optimal lattice basis [11]. Although their method is general and effective, it cannot handle constrained variables case. Actually, when Durfee-Nguyen technique is involved, their method could not generate a good lattice. Using our compiler, Durfee-Nguyen technique is automatically involved in constructing the lattice for f if it is involved in the lattice for h. Our compiler is especially effective for specific type of equations and is applicable to many kinds of RSA-related cryptanalysis.

1.2 Organization

Section 2 gives preliminaries. In Section 3, we show how to solve the "generalized small inverse problems." First, we introduce 4 restrictions for a lattice in solving $h = 0$. Then, we give a compiler which transforms a lattice basis for $h(x_1, \ldots, x_n) = 0$ into that for $f(x_0, x_1, \ldots, x_n) = x_0 h(x_1, \ldots, x_n) + C = 0$. In Section 4, we evaluate the volume of lattice for f and derive the condition among upper bounds of solutions. In Section 5, we argue application of our compiler to GSIP and give details of an application: the small secret exponent attack to RSA, which shows the effectiveness of our compiler. Section 6 concludes the paper. Some of proofs are given in Appendix A. Some examples are given in full version [13].

2 Preliminaries

2.1 Small Secret Exponent Attack on RSA [2]

Let (N, e) be a public key in RSA cryptosystem, where $N = pq$ is the product of two distinct primes. For simplicity, we assume that $\gcd(p - 1, q - 1) = 2$. A secret key d satisfies that $ed = 1 \mod (p - 1)(q - 1)/2$. Hence, there exists an integer k such that $ed + k((N + 1)/2 - (p + q)/2) = 1$. Writing $s = -(p + q)/2$ and $A = (N + 1)/2$, we have $k(A + s) = 1 \pmod{e}$.

We set $f(x, y) = x(A + y) + 1$. If one can solve a bivariate modular equation: $f(x, y) = x(A + y) + 1 = 0 \pmod{e}$, one has k and s and knows the prime factors p and q of N. Suppose that the secret key satisfies $d \leq N^\delta$. Further assume that $e \approx N$. To summarize, the secret key will be recovered by finding the solution $(x, y) = (\bar{x}, \bar{y})$ of the equation: $x(A + y) = 1 \pmod{e}$, where $x \leq e^\delta$ and $|y| \leq e^{1/2}$. They referred this as the *small inverse problem*.

Boneh and Durfee gave an algorithm for solving this problem and obtained the condition on δ so that the algorithm works in polynomial time. Concretely, they showed that if $d \leq N^{0.284}$, N can be factored in polynomial time. Furthermore, they improved the bound to $d \leq N^{0.292}$.

2.2 LLL Algorithm and Howgrave-Graham's Lemma

For a vector $\boldsymbol{b}$, $||\boldsymbol{b}||$ denotes the Euclidean norm of $\boldsymbol{b}$. For a n-variate polynomial $h(x_1, \ldots, x_n) = \sum h_{j_1, \ldots, j_n} x_1^{j_1} \cdots x_n^{j_n}$, define the norm of a polynomial as $||h(x_1, \ldots, x_n)|| = \sqrt{\sum h_{j_1, \ldots, j_n}^2}$. That is, $||h(x_1, \ldots, x_n)||$ denotes the Euclidean norm of the vector which consists of coefficients of $h(x_1, \ldots, x_n)$.

Let $B = \{a_{ij}\}$ be a $w \times w'$ matrix of integers. The rows of B generate a lattice L, a collection of vectors closed under addition and subtraction; in fact the rows forms a basis of L. The lattice L is also represented as follows. Letting $\boldsymbol{a_i} = (a_{i1}, a_{i2}, \ldots, a_{iw'})$, the lattice L spanned by $\langle \boldsymbol{a_1}, \ldots, \boldsymbol{a_w} \rangle$ consists of all integral linear combinations of $\boldsymbol{a_1}, \ldots, \boldsymbol{a_w}$, that is: $L = \{\sum_{i=1}^{w} n_i \boldsymbol{a_i} | n_i \in \mathbb{Z}\}$. The volume of lattice is defined by $\mathrm{vol}(L) = \sqrt{\det({}^t\!B B)}$, where ${}^t\!B$ is a transposed matrix of B. In particular, $\mathrm{vol}(L) = |\det(B)|$ if B is full-rank.

LLL algorithm outputs short vectors in the lattice L.

Proposition 1 (LLL). *Let $B = \{a_{ij}\}$ be a non-singular $w \times w'$ matrix of integers. The rows of B generates a lattice L. Given B, the LLL algorithm outputs a reduced basis $\{\boldsymbol{b_1}, \ldots, \boldsymbol{b_w}\}$ with*

$$||\boldsymbol{b_i}|| \leq 2^{w(w-1)/(4(w+1-i))}(\text{vol }(L))^{1/(w+1-i)}$$

in time polynomial in $(w, \max \log_2 |a_{ij}|)$.

The following lemma is used when a modular equation is reduced into integer equation.

Lemma 1 (Howgrave-Graham [8]). *Let $\hat{h}(x_1, \ldots, x_n) \in \mathbb{Z}[x_1, \ldots, x_n]$ be a polynomial, which is a sum of at most w' monomials. Let m and ϕ be positive integers and $X_1, \ldots, X_n$ be some positive integers. Suppose that*

1. $\hat{h}(\bar{x}_1, \ldots, \bar{x}_n) = 0 \bmod \phi^m$, where $|\bar{x}_1| < X_1, \ldots |\bar{x}_n| < X_n$ and
2. $||\hat{h}(x_1 X_1, \ldots, x_n X_n)|| < \phi^m/\sqrt{w'}$.

Then $\hat{h}(\bar{x}_1, \ldots, \bar{x}_n) = 0$ holds over integers.

3   How to Solve Generalized Small Inverse Problem

For a polynomial $h(x_1, \ldots, x_n)$, consider the following two problems: (I) Given $N(= pq)$, find a small solution of $h(x_1, \ldots, x_n) = 0 \pmod p$. (II) Given M, find a small solution of $x_0 h(x_1, \ldots, x_n) + C = 0 \pmod M$. Problem (II) corresponds to a generalized small inverse problem. We will show a compiler which transforms a lattice basis for (I) to that for (II).

3.1   Lattice-Based Algorithm for (I)

The problem (I) can be solved by combining the LLL algorithm and Lemma 1 as follows. Let $X_1, \ldots, X_n$ be positive integers of Lemma 1. Define a polynomial as $h_{[j_1, \ldots, j_n, k]}(x_1, \ldots, x_n) := x_1^{j_1} \cdots x_n^{j_n} h(x_1, \ldots, x_n)^k$ for non-negative integers $j_1, \ldots, j_n, k$. Let u be a non-negative integer. Using $h_{[j_1, \ldots, j_n, k]}$, we define a shift-polynomial

$$h_{[j_1, \ldots, j_n, k]}^{(u)}(x_1, \ldots, x_n) := h_{[j_1, \ldots, j_n, k]}(x_1, \ldots, x_n)N^{u-k}. \tag{1}$$

Let a solution of $h = 0 \pmod p$ be $(x_1, \ldots, x_n) = (\bar{x}_1, \ldots, \bar{x}_n)$. It is easy to see that

$$h_{[j_1, \ldots, j_n, k]}^{(u)}(\bar{x}_1, \ldots, \bar{x}_n) = 0 \pmod{p^u}$$

for any $(j_1, \ldots, j_n, k)$.

Fix a set $\mathcal{H}^{(u)}$ of $[j_1, \ldots, j_n, k]$ for each u. We construct a lattice $L_h^{(u)}$ spanned by a set of the coefficient vector of $h_{[j_1, \ldots, j_n, k]}^{(u)}(x_1 X_1, \ldots, x_n X_n)$ for $[j_1, \ldots, j_n, k] \in \mathcal{H}^{(u)}$. Then, we apply the LLL algorithm to this lattice. The LLL algorithm yields small vectors of this lattice. Finally, we can obtain polynomial $\hat{h}$ satisfying the condition of Lemma 1 from this small vector. How to choose $\mathcal{H}^{(u)}$ for each u depends on $h(x_1, \ldots, x_n)$.

First, we define the set $M(h_{[j_1, \ldots, j_n, k]})$ of monomials

$$M(h_{[j_1, \ldots, j_n, k]}) \equiv \{x_1^{i_1} \cdots x_n^{i_n} | x_1^{i_1} \cdots x_n^{i_n} \text{ is a monomial of } h_{[j_1, \ldots, j_n, k]}(x_1, \ldots, x_n)\}.$$

Next, we define the set $M(\mathcal{H}^{(u)})$ of monomials

$$M(\mathcal{H}^{(u)}) \equiv \bigcup_{[j_1, \ldots, j_n, k] \in \mathcal{H}^{(u)}} M(h_{[j_1, \ldots, j_n, k]}).$$

We will introduce 4 restrictions for a lattice in solving $h = 0$ and consider only a set $\mathcal{H}^{(u)}$ of $[j_1, \ldots, j_n, k]$ for each u which holds 4 restrictions.

Restriction 1. For any positive integer u, there exist two sets $\mathcal{A} = \{[j_{1i}, \ldots, j_{ni}]\}_{1 \leq i \leq \#\mathcal{A}}$ and $\mathcal{B} = \{[j_{1i}^*, \ldots, j_{ni}^*]\}_{1 \leq i \leq \#\mathcal{B}}$ such that $\mathcal{A} \subseteq \mathcal{B}$ and $\mathcal{H}^{(u)}$ is given by

$$\mathcal{H}^{(u)} = \bigcup_{k=0}^{u-1} \{[j_{1i}, \ldots, j_{ni}, k]\}_{1 \leq i \leq \#\mathcal{A}} \cup \{[j_{1i}^*, \ldots, j_{ni}^*, u]\}_{1 \leq i \leq \#\mathcal{B}}. \tag{2}$$

We call $(\mathcal{A}, \mathcal{B})$ a *generator*.

Restriction 2. For any u, $L_h^{(u)}$ is full rank.

Restriction 3. A generator $\mathcal{B}$ is parametrized by some optimizing parameters $\boldsymbol{t} = (t_1, \ldots, t_k)$. If needed, we use notation: $\mathcal{B}(\boldsymbol{t})$.

Restriction 4. The volume of $L_h^{(u)}$ does not depend on coefficients of h. That is, it is given by

$$\text{vol } L_h^{(u)} = N^{\gamma_U} X_1^{\gamma_1} X_2^{\gamma_2} \cdots X_n^{\gamma_n}. \tag{3}$$

Let w be the dimension of the lattice. Here, $\gamma_U, \gamma_1, \ldots, \gamma_n$ and w are functions of u and $\boldsymbol{t}$. Moreover, each total degree of $\gamma_U, \gamma_1, \ldots, \gamma_n$ and uw is 2. If needed, we use $\text{vol } L_h^{(u;\boldsymbol{t})}, \gamma_U(u; \boldsymbol{t}), \gamma_i(u; \boldsymbol{t})$ for $1 \leq i \leq n$.

Lattices derived in many previous methods [6,9,12,17] hold Restrictions 1–4 as described in Table 1.

Restriction 1 implies that if $[j_1, \ldots, j_n, k] \in \mathcal{H}^{(u)}$ and $k \geq 1$, then $[j_1, \ldots, j_n, k-1] \in \mathcal{H}^{(u-1)}$, which is crucial for our compiler. For convenience, we use the following notation: for a set $\mathcal{A}$ and $k \in \mathbb{Z}_{\geq 0}$, a set $[\mathcal{A}, k]$ is defined by $\{[j_1, \ldots, j_n, k] | [j_1, \ldots, j_n] \in \mathcal{A}\}$. If this notation is used, we can rewrite Eq. (2) as

$$\mathcal{H}^{(u)} = \bigcup_{k=0}^{u-1} [\mathcal{A}, k] \cup [\mathcal{B}, u].$$

Restriction 2 implies that $\#\mathcal{H}^{(u)} = \#M(\mathcal{H}^{(u)})$. The polynomial order of $\mathcal{H}^{(u)}$ and monomial order of $M(\mathcal{H}^{(u)})$ should be adequately defined so as to be linearly ordered. Let $B_h^{(u)}(\mathcal{A}, \mathcal{B})$ denote a $\#\mathcal{H}^{(u)} \times \#\mathcal{H}^{(u)}$ square matrix, where each row of $B_h^{(u)}(\mathcal{A}, \mathcal{B})$ is the coefficient vector of $h_{[j_1, \ldots, j_n, k]}^{(u)}(x_1 X_1, \ldots, x_n X_n)$ when $(\mathcal{A}, \mathcal{B})$ is used as a generator. If $\mathcal{A}$ and $\mathcal{B}$ are clear from the context, we often omit $\mathcal{A}, \mathcal{B}$ and simply write $B_h^{(u)}$. Since $L_h^{(u)}$ is full-rank, $\text{vol } L_h^{(u)} = |\det B_h^{(u)}|$.

3.2   How to Solve (II)

We show how to solve the problem (II). First, we overview our algorithm and then focus on Step 1-2.

Input: $n+1$-variate equation $f(x_0, x_1, \ldots, x_n) = x_0 h(x_1, \ldots, x_n) + C = 0$ (mod M) with small roots
Output: All small roots $(\bar{x}_0, \ldots, \bar{x}_n)$ of $f(x_0, x_1, \ldots, x_n) = 0$ (mod M)
Step1: Construct a lattice for f.
   Step1-1: Construct a lattice $L_h^{(u)}$ for h or choose a generator $\mathcal{A}$ and $\mathcal{B}$ for h.
   Step1-2: Construct a lattice L_f for f by employing the lattice for $L_h^{(u)}$ or $\mathcal{A}$ and $\mathcal{B}$.
Step2: Run LLL algorithm for input L_f to obtain $n + 1$ polynomials $r_1, r_2, \ldots, r_{n+1} \in \mathbb{Z}[x_0, x_1, \ldots, x_n]$ over the integers, where they are non-zero integer combination of $f_{[i,j_1,\ldots,j_n,k]}(x_0 X_0, x_1 X_1, \ldots, x_n X_n)$ with small coefficients.
Step3: Compute a resultant for r_i to obtain a univariate integer equation. Then, solve the equation by using standard technique.

We point out some remarks. Our algorithm cannot always guarantee to output correct solutions. So, our algorithm is heuristic. We assume the following as same as [11].

Assumption 1. *The resultant computations for polynomials r_i yield non-zero polynomials.*

Experiments are needed for specific cases to justify the assumption.
   We move on to the discussion of Step 1-2. Letting m be a positive integer, we define shift-polynomials for $f(x_0, x_1, \ldots, x_n)$ as

$$f_{[i,j_1,\ldots,j_n,k]}(x_0, x_1, \ldots, x_n) := x_0^i x_1^{j_1} \cdots x_n^{j_n} f(x_0, x_1, \ldots, x_n)^k M^{m-k}.$$

Let a solution of $f = 0$ (mod M) be $(x_0, \ldots, x_n) = (\bar{x}_0, \ldots, \bar{x}_n)$. It is easy to see that

$$f_{[i,j_1,\ldots,j_n,k]}(\bar{x}_0, \ldots, \bar{x}_n) = 0 \ (\text{mod } M^m)$$

for any $(i, j_1, \ldots, j_n, k)$.
   Let $\mathcal{F}$ be a set of indexes $[i, j_1, \ldots, j_n, k]$. We construct the lattice L_f spanned by the coefficient vectors of $f_{[i,j_1,\ldots,j_n,k]}(x_0 X_0, \ldots, x_n X_n)$ with $[i, j_1, \ldots, j_n, k] \in \mathcal{F}$. How does one choose a set of indexes $\mathcal{F}$? This is a difficult problem. The choice of $\mathcal{F}$ determines the performance of the algorithm. Indeed, the volume of the lattice derived by $\mathcal{F}$ should be small. Moreover, one must calculate or estimate the volume of lattice. If $\mathcal{F}$ is badly chosen, it might be difficult to calculate (or even though estimate) its volume. So, one must choose in a clever way the set $\mathcal{F}$. We overcome this problem by employing a lattice basis for solving $h = 0$. We propose the following compiler, which transforms a set of shift-polynomial for $h = 0$ into that for $f = 0$. In explanation, we use a notation: a set $[k_1, \mathcal{A}, k_2]$ is defined by $[k_1, \mathcal{A}, k_2] = \{[k_1, j_1, \ldots, j_n, k_2] | [j_1, \ldots, j_n] \in \mathcal{A}\}$.

Compiler. Fix a positive integer m. By using generators $\mathcal{A}$ and $\mathcal{B}$ for $h = 0$, we construct a set $\mathcal{F}$ of shift-polynomials as follows. First, we set

$$\mathcal{F}^{(u)} \equiv \bigcup_{k=0}^{u-1} [u - k, \mathcal{A}, k] \cup [0, \mathcal{B}, u].$$

Then, we set

$$\mathcal{F} \equiv \bigcup_{u=0}^{m} \mathcal{F}^{(u)} = \bigcup_{u=0}^{m} \left\{ \bigcup_{k=0}^{u-1} [u - k, \mathcal{A}, k] \cup [0, \mathcal{B}, u] \right\}.$$

$\mathcal{F}$ is explicitly given by

$$\mathcal{F} = \bigcup_{u=0}^{m} \left\{ \bigcup_{k=0}^{u-1} \{[u - k, j_{1i}, \ldots, j_{ni}, k]\}_{1 \leq i \leq \#\mathcal{A}} \cup \{[0, j_{1i}^*, \ldots, j_{ni}^*, u]\}_{1 \leq i \leq \#\mathcal{B}} \right\}.$$

Obviously, $\#\mathcal{F}^{(u)} = \#\mathcal{H}^{(u)}$. If we define polynomial and monomial orders as follows, the polynomial set $\mathcal{F}$ and the monomial order are linearly ordered.

monomial order: We define $\prec$ as $x_0^u x_1^{j_1} \cdots x_n^{j_n} \prec x_0^{u'} x_1^{j_1'} \cdots x_n^{j_n'}$

$$\text{if} \begin{cases} u < u' \text{ or} \\ u = u' \text{ and } x_1^{j_1} \cdots x_n^{j_n} \prec x_1^{j_1'} \cdots x_n^{j_n'} \text{ in } M(\mathcal{H}^{(u)}). \end{cases}$$

polynomial order: We define $\prec$ as $[i, j_1, \ldots, j_n, k] \prec [i', j_1', \ldots, j_n', k']$

$$\text{if} \begin{cases} i + k < i' + k' \text{ or} \\ i + k = i' + k' \text{ and } [j_1, \ldots, j_n, k] \prec [j_1', \ldots, j_n', k'] \text{ in } \mathcal{H}^{(i+k)} \end{cases}$$

Informally, letting $f' \in \mathcal{F}^{(u')}$ and $f'' \in \mathcal{F}^{(u'')}$, $f' \prec f''$ if $u' < u''$.

Theorem 1. *Suppose that $\mathcal{F}$ is set by our Compiler and $\mathcal{H}^{(u)}$ holding 4 restrictions. Let B be a matrix, where each row of B is the coefficient vectors of $f_{[u-k, j_1, \ldots, j_n, k]}(x_0 X_0, \ldots, x_n X_n)$ according to the order of $\mathcal{F}$. Then, the matrix B is square and blocked lower triangular.*

For Theorem 1, B is written as

$$B = \begin{pmatrix} B_0 & & & \mathbf{0} \\ & B_1 & & \\ \vdots & & \ddots & \\ * & & \cdots & B_m \end{pmatrix},$$

where each B_u is a $\#\mathcal{H}^{(u)} \times \#\mathcal{H}^{(u)}$ matrix for $0 \leq u \leq m$. Note that B_u corresponds to $\#\mathcal{H}^{(u)}$ polynomials $\{f_{[i,j_1,\ldots,j_n,k]} | [i, j_1, \ldots, j_n, k] \in \mathcal{F}^{(u)}\}$ and $\#\mathcal{H}^{(u)}$ monomials which are divisible by x_0^u. The determinant of B is simply given by $\det B = \det B_0 \det B_1 \cdots \det B_m$.

The application to small secret exponent attack will be given in Section 5. Other examples are given in Section 5 and a full version [13].

3.3   Proof of Theorem 1

We define the set of monomials as $M(f_{[u-k,j_1,\ldots,j_n,k]})$
$\equiv \{x_0^{i_0} x_1^{i_1} \cdots x_n^{i_n} | x_0^{i_0} x_1^{i_1} \cdots x_n^{i_n}$ is a monomial of $f_{[u-k,j_1,\ldots,j_n,k]}\}$ and

$$M(\mathcal{F}^{(u)}) \equiv \bigcup_{J \in \mathcal{F}^{(u)}} M(f_J).$$

We use the notation: $x_0^{i_0} \mathcal{M} \equiv \{x_0^{i_0} x_1^{i_1} \cdots x_n^{i_n} | x_1^{i_1} \cdots x_n^{i_n} \in \mathcal{M}\}$ for $\mathcal{M} = \{x_1^{i_1} \cdots x_n^{i_n}\}$.

First, we show the following two lemmas.

Lemma 2. *If $[u - k, j_1, \ldots, j_n, k] \in \mathcal{F}$ for $k \geq 1$, it holds that*

$$M(f_{[u-k,j_1,\ldots,j_n,k-1]}) \subset M(f_{[u-k,j_1,\ldots,j_n,k]}).$$

Furthermore, it holds that for $k \geq 1$, $M(f_{[u-k,j_1,\ldots,j_n,k]}) \setminus M(f_{[u-k,j_1,\ldots,j_n,k-1]})$ $= x_0^u \{x_1^{i_1} \cdots x_n^{i_n} | x_1^{i_1} \cdots x_n^{i_n}$ is a monomial of $h_{[j_1,\ldots,j_n,k]}\}$.

Lemma 3. *It holds that $M(\mathcal{F}^{(0)}) \subset M(\mathcal{F}^{(1)}) \subset \cdots \subset M(\mathcal{F}^{(m)})$. Furthermore, it holds that $M(\mathcal{F}^{(u)}) \setminus M(\mathcal{F}^{(u-1)}) = x_0^u M(\mathcal{H}^{(u)})$.*

Proof (of Lemma 2). For $k \geq 1$, if $[u - k, j_1, \ldots, j_n, k] \in \mathcal{F}^{(u)}$, then $[u - k, j_1, \ldots, j_n, k - 1] \in \mathcal{F}^{(u-1)}$. The expansion of $f_{[u-k,j_1,\ldots,j_n,k]}(x_0, x_1, \ldots, x_n)$ is given by

$$f_{[u-k,j_1,\ldots,j_n,k]}(x_0, x_1, \ldots, x_n) = x_0^{u-k} x_1^{j_1} \cdots x_n^{j_n} (x_0 h + C)^k M^{m-k}$$

$$= x_0^u h_{[j_1,\ldots,j_n,k]} M^{m-k} + \sum_{i=1}^{k} \binom{k}{i} C^i M^{m-k} x_0^{u-i} h_{[j_1,\ldots,j_n,k-i]}.$$

The expansion of $f_{[u-k,j_1,\ldots,j_n,k-1]}(x_0, x_1, \ldots, x_n)$ is given by

$$f_{[u-k,j_1,\ldots,j_n,k-1]}(x_0, x_1, \ldots, x_n) = x_0^{u-k+1} x_1^{j_1} \cdots x_n^{j_n} (x_0 h + C)^{k-1} M^{m-k+1}$$

$$= \sum_{i=1}^{k} \binom{k-1}{i-1} C^{i-1} M^{m-k+1} x_0^{u-i} h_{[j_1,\ldots,j_n,k-i]}.$$

Then, we have the lemma.   □

For Lemma 3, the number of monomials firstly appearing in $\mathcal{F}^{(u)}$ is $\#(M(\mathcal{F}^{(u)}) \setminus M(\mathcal{F}^{(u-1)})) = \#M(\mathcal{H}^{(u)})$. For the construction of our Compiler, the number of polynomials in $\mathcal{F}^{(u)}$ is $\#\mathcal{F}^{(u)} = \#\mathcal{H}^{(u)}$. Restriction 2 implies that $\#\mathcal{H}^{(u)} = \#M(\mathcal{H}^{(u)})$. Then, $\#(M(\mathcal{F}^{(u)}) \setminus M(\mathcal{F}^{(u-1)})) = \#\mathcal{F}^{(u)}$. This implies that B is blocked lower triangular.   □

3.4   Small Example of Our Compiler

We show a small example which shows how our Compiler works. Let $h(y)$ be a univariate monic polynomial with degree 1: $h(y) = A + y$. In this case, a target equation is $f(x,y) = xh(y) + C = x(A + y) + C = 0 \pmod M$.

Let $h_{[j,k]}^{(u)}(y) \doteq y^j h(y)^k N^{u-k}$. Suppose that we use a generator $\mathcal{A}=\{[0]\}$ and $\mathcal{B}= \{[0],[1],[2]\}$. Then, $\mathcal{H}^{(0)}=\{[0,0],[1,0],[2,0]\}$ and $\mathcal{H}^{(1)}=\{[0,0],[0,1],[1,1],[2,1]\}$. Corresponding matrixes $B_h^{(0)}$ and $B_h^{(1)}$ are given as follows.

$$
B_h^{(0)} =
\begin{array}{c|ccc}
 & 1 & y & y^2 \\\hline
h_{[0,0]}^{(0)}(=1) & 1 & 0 & 0 \\
h_{[1,0]}^{(0)}(=Yy) & 0 & Y & 0 \\
h_{[2,0]}^{(0)}(=Y^2y^2) & 0 & 0 & Y^2
\end{array}
,\quad
B_h^{(1)} =
\begin{array}{c|cccc}
 & 1 & y & y^2 & y^3 \\\hline
h_{[0,0]}^{(1)}(=N) & N & 0 & 0 & 0 \\
h_{[0,1]}^{(1)}(=A+Yy) & A & Y & 0 & 0 \\
h_{[1,1]}^{(1)}(=AYy+Y^2y^2) & 0 & AY & Y^2 & 0 \\
h_{[2,1]}^{(1)}(=AY^2y^2+Y^3y^3) & 0 & 0 & AY^2 & Y^3
\end{array}
$$

For example, $M(h_{[1,1]}) = \{y, y^2\}$ and $M(\mathcal{H}^{(1)}) = \{1, y, y^2, y^3\}$.

For a positive integer m, let $f_{[i,j,k]}(x,y) := x^i y^j f(x,y)^k M^{m-k}$. In the example, we fix $m = 1$. Applying our compiler, we obtain $\mathcal{F}$ of $f_{[i,j,k]}$ for solving $f(x,y) = xh(y) + C = 0 \pmod M$ as follows: $\mathcal{F} = \{[0,0,0],[0,1,0],[0,2,0],[1,0,0],[0,0,1],[0,1,1],[0,2,1]\}$. A matrix B generated by $\mathcal{F}$ is given as follows.

$$
B =
\begin{array}{c|ccc|cccc}
 & 1 & y & y^2 & x & xy & xy^2 & xy^3 \\\hline
f_{[0,0,0]}(=M) & M & 0 & 0 & 0 & 0 & 0 & 0 \\
f_{[0,1,0]}(=YMy) & 0 & YM & 0 & 0 & 0 & 0 & 0 \\
f_{[0,2,0]}(=Y^2My^2) & 0 & 0 & Y^2M & 0 & 0 & 0 & 0 \\
f_{[1,0,0]}(=XMx) & 0 & 0 & 0 & XM & 0 & 0 & 0 \\
f_{[0,0,1]}(=C+AXx+XYxy) & C & 0 & 0 & AX & XY & 0 & 0 \\
f_{[0,1,1]}(=CYy+AXYxy+XY^2xy^2) & 0 & CY & 0 & 0 & AXY & XY^2 & 0 \\
f_{[0,2,1]}(=CY^2y^2+AXY^2xy^2+XY^3xy^3) & 0 & 0 & CY^2 & 0 & 0 & AXY^2 & XY^3
\end{array}
$$

Columns and rows are ordered by polynomial and monomial orders in $\mathcal{F}$. The determinant of B is given by the product of diagonal elements. So, $\det B = M^4 X^4 Y^9$.

4   Deriving a Condition for Solving GSIP

In the previous section, we show how to choose a set $\mathcal{F}$. The next thing to do is evaluation of a volume of the lattice L_f or the determinant of the corresponding matrix B. Then, we will derive the condition for solving the problem by combining the value of determinant and Lemma 1.

First, we derive a determinant of matrix B (or a volume of L_f) obtained by our compiler.

Lemma 4. *Let $B_h^{(u;t)}$ be the corresponding matrix for h and $w(u;t)$ be the dimension of the lattice. Then, the determinant of B derived by our Compiler is given by*

$$\det B = M^{mW} \left(\frac{X_0}{M}\right)^{\sum_{u=0}^{m} uw(u;t)} \prod_{u=0}^{m} \det B_h^{(u;t)}(M), \tag{4}$$

where $W (= \sum_{u=0}^{m} w(u;t))$ is the rank of B.

Next, we derive a condition that we can find all solutions of $f = 0 \pmod{M}$.

Lemma 5. *Suppose that the determinant of $B_h^{(u;t)}$ is given as the same as Lemma 4. Under Assumption 1, we can find all solutions of the equation $f = 0 \pmod{M}$ with $|x_0| < X_0, |x_1| < X_1, \ldots, |x_n| < X_n$ if*

$$\prod_{u=0}^{m} \det B_h^{(u;t)}(M) < \left(\frac{M}{X_0}\right)^{\sum uw(u;t)} = \prod_{u=0}^{m} \left(\frac{M}{X_0}\right)^{uw(u;t)}. \tag{5}$$

The time complexity is polynomial in $\log M$ and 2^n.

In case of Maximizing X_0. In many cryptanalysis, all the task is to maximize X_0 for fixed $X_1, X_2, \ldots, X_n$. Hereafter, we focus on this situation. We introduce an operator: $I : m^k \to \frac{1}{k+1} m^{k+1}$. Obviously, the operator I is homomorphic. Hence, we can write $\sum_{u=0}^{m} uw(u;t) = I(mw(m;t))$ and $\sum_{u=0}^{m} \gamma_i(u;t) = I(\gamma_i(m;t))$. We rewrite Eq. (5) by using the operator I as: $(X_0/M)^{I(mw(m;t))} < M^{-I(\gamma_U(m;t))} X_1^{-I(\gamma_1(m;t))} \cdots X_n^{-I(\gamma_n(m;t))}$. Hence, we have

$$X_0 < M/(M^{I(\gamma_U(m;t))} X_1^{I(\gamma_1(m;t))} \cdots X_n^{I(\gamma_n(m;t))})^{1/I(mw(m;t))}.$$

Let A_i be a fixed positive number such that $X_i = M^{A_i}$ for $1 \le i \le n$. We can simplify the above as $X_0 < M/M^{(I(\gamma_U(m;t)) + \sum_{i=1}^{n} A_i I(\gamma_i(m;t)))/I(mw(m;t))}$. Setting

$$l(m;t) \equiv \frac{I(\gamma_U(m;t)) + \sum_{i=1}^{n} A_i I(\gamma_i(m;t))}{I(mw(m;t))} = \frac{I(\gamma_U(m;t) + \sum_{i=1}^{n} A_i \gamma_i(m;t))}{I(mw(m;t))}, \tag{6}$$

we have $X_0 < M^{1-l(m;t)}$. The next thing to do is to obtain t minimizing $l(m;t)$ for fixed m. The values t minimizing $l(m;t)$ is given by solving simultaneous equations:

$$\frac{\partial l(m;t)}{\partial t_1} = \frac{\partial l(m;t)}{\partial t_2} = \cdots = \frac{\partial l(m;t)}{\partial t_k} = 0.$$

Let t' be the solution of the above equations if it exists. If we ignore small terms[1], each $I(\gamma_U(m;t))$, $I(\gamma_1(m;t)), \ldots I(\gamma_n(m;t)), I(mw(m;t)))$ consists of one term with the same total degree 3. Hence, each element t_i' of t' is represented

---

[1] If we don't ignore the small term, we can obtain the optimal value of m. But, we need tedious computation in general. For small secret exponent attack case, Boneh and Durfee gave the details analysis [2].

by $t'_i = \tau'_i m$ for positive integers τ_i's. Letting $\boldsymbol{\tau'} = (\tau'_1, \ldots, \tau'_n)$, we have the condition for X_0:

$$X_0 \leq \max_{\boldsymbol{t}} M^{1-l(m;t)} = M^{1-\min_t l(m;t)} = M^{1-l(m;m\boldsymbol{\tau'})}, \qquad (7)$$

which does not depend on m.

Next, we will analyze the most simple case, that is, $\mathcal{B}$ is parametrized by one parameter. In this case, we have an explicit formula of the upper bound of X_0.

Theorem 2. *Suppose that a lattice for $h = 0$ holds Restrictions 1–4 holds and $\mathcal{B}$ is parametrized by one parameter t. For given positive integers $A_1, \ldots, A_n$, we set $a_2 m^2 + a_1 mt + a_0 t^2 \equiv \gamma_U(m;t) + \sum_{i=1}^{n} A_i \gamma_i(m;t)$ and $w(m;t) = b_2 m + b_1 t$. Suppose that $a_1 b_2 < a_2 b_1$. Under Assumption 1, we can find all solutions of equation: $f = 0 \pmod{M}$ with $|x_0| < X_0, |x_1| < M^{A_1}, \ldots, |x_n| < M^{A_n}$ if $X_0 < M^{1 - \frac{4a_0}{b_1} c' - \frac{a_1}{b_1}}$, where $c' = (\sqrt{4a_0^2 b_2^2 - 3a_0 a_1 b_1 b_2 + 3a_0 a_2 b_1^2} - 2a_0 b_2)/(3a_0 b_1)$. In particular, if $b_1 = b_2$, we simply have the condition as*

$$X_0 < M^{1 - \frac{4\sqrt{4a_0^2 - 3a_0 a_1 + 3a_0 a_2} - 8a_0 + 3a_1}{3b_1}}. \qquad (8)$$

Time complexity is in polynomial in $\log M$ and 2^n.

5   Application of Our Compiler to RSA-Related Cryptanalysis

We show several examples of GSIP and argue applications of our compiler to them. Table 1 summarizes some example of GSIP in the literature. "Constraint" shows what kind of constraint variables have in both of solving $f = 0$ and $h = 0$. "$\mathcal{A}$" and "$\mathcal{B}$" show what kind of generators we use in both of solving $f = 0$ and $h = 0$. We give more explanation for each cases and give details of Case 1. For other cases, see the full version [13].

Table 1. Examples of GSIP

	Case 1	Case 1'	Case 2	Case 3
	Boneh-Durfee [2]	May [15]	Durfee-Nguyen [7]	Itoh et al. [10]
$f = xh + C$	$x(A+y)+1$	$x(y-N)+N$	$x(A+y+z)+1$	$x(y-1)(z-1)+1$
Constraint	-	-	$yz = N$	$y^r z = N$
	Howgrave-Graham [9]		Coron-May [6]	Kunihiro-Kurosawa [12]
h	$y+A$	$y-N$	$A+y+z$	$(y-1)(z-1)$
$\mathcal{A}$	$\{[0]\}$		$\{[0,0],[1,0]\}$	$\{[0,0],[1,0]\} \cup \bigcup_{i=1}^{r-1}\{[i,1]\}$
$\mathcal{B}$	$\bigcup_{i=0}^{t}\{[i]\}$		$\bigcup_{i=0}^{t_1}\{[i,0]\} \cup \bigcup_{j=1}^{t_2}\{[0,j]\}$	$\bigcup_{i=0}^{t_1}\{[i,0]\} \cup \bigcup_{k=0}^{r-1}\bigcup_{j=1}^{t_2}\{[k,j]\}$

Case 1: Consider the small secret exponent attack on RSA by Boneh and Dur-fee [2]. In their attack, they handled $f(x,y) = x(y+A)+1 = 0 \pmod{e}$. Hence, this problem corresponds to $h(y) = y + A$ and $C = 1$. By using our compiler, the lattice basis for $f(x) = 0$ is automatically obtained. Then, one can easily obtain the bound: $d \leq N^{0.284}$. We'll discuss the details later.

Case 1': Consider the small CRT exponent attack on unbalanced RSA by May [15]. In his attack, he handled $f(x, y) = x(y - N) + N = 0 \pmod{e}$. Hence, this problem corresponds to $h(y) = y - N$ and $C = N$. By using our compiler, the lattice basis for $f(x) = 0$ is automatically obtained. Furthermore, one can easily obtain the bound $d_p \leq e^{1-2(\sqrt{\beta^2+3\beta}+\beta)/3}$, where $q < e^\beta$.

Case 2: Consider cryptanalysis on some variants of RSA with small secret exponent by Durfee-Nguyen [7]. In their attack, they handled the trivariate modular equation: $f(x, y, z) = x(A + y + z) + 1 = 0 \pmod{e}$ with constraint $yz = N$. Hence, this problem corresponds to $h(y, z) = A + y + z$ and $C = 1$. By using our compiler, the lattice basis for Durfee-Nguyen's attack is automatically obtained.

Case 3: Consider the small secret exponent attacks on Takagi's variant of RSA [19] by Itoh et al. [10]. This attack can be obtained by our compiler and a lattice basis used in proving a deterministic polynomial equivalence between factoring and computing the secret exponent in that scheme [12]. Note that since Durfee-Nguyen technique is adequately involved in a lattice basis for h, we can easily obtain that for f. One can easily obtain the bound: $d \leq N^{(7-2\sqrt{7})/3(r+1)}$.

Case 1: Transforming Howgrave-Graham's Lattice Basis to Boneh-Durfee's Lattice Basis. Next, we move on to an actual cryptanalysis. We show that our compiler builds a bridge between a lattice basis in [9] and that in [2]. We simply write x, y, X, Y instead of x_0, x_1, X_0 and X_1.

Howgrave-Graham [9] provided an algorithm[2] for solving $h(y) = A + y = 0 \pmod{S}$ for integers A and S, which is an unknown divisor of an known integer U. Set shift-polynomials as $h_{[j,k]}^{(u)}(y) := h_{[j,k]} N^{u-k} = y^j h(y)^k N^{u-k}$. In his paper, he chose the set of the indexes of shift-polynomials as $\mathcal{H}^{(u)} = \bigcup_{k=0}^{u-1}\{[0,k]\} \cup \bigcup_{i=0}^{t}\{[i,u]\}$. We set a polynomial order by this. Note that a generator is given by $\mathcal{A} = \{[0]\}$ and $\mathcal{B} = \cup_{i=0}^{t}\{[i]\}$. Hence, $\mathcal{H}^{(u)}$ holds Restrictions 1–3.

Let $f(x, y) = x(A + y) + 1$. We argue a lattice basis construction for f. Since $f(x, y) = xh(y) + 1$, we can employ our Compiler to construct a lattice basis for f. For a positive integer m, we define shift-polynomials for f as $f_{[i,j,k]}(x, y) := x^i y^j f(x, y)^k M^{m-k}$. By our Compiler and Howgrave-Graham's lattice basis, we have a set $\mathcal{F}$ as

$$\mathcal{F} = \bigcup_{u=0}^{m} \left\{ \bigcup_{k=0}^{u-1} \{[u-k, 0, k]\} \cup \bigcup_{i=0}^{t} \{[0, i, u]\} \right\}$$

for fixed t. We have explicitly

$$\mathcal{F} = \{[0,0,0], [0,1,0], \ldots, [0,t,0], [1,0,0], [0,0,1], [0,1,1], \ldots, [0,t,1],$$
$$[m,0,0], [m-1,0,1], \ldots, [0,0,m], [0,1,m], [0,2,m], \ldots, [0,t,m]\}.$$

---

[2] By employing his algorithm, Coron and May gave the deterministic polynomial time algorithm for factoring the RSA modulus when the secret key d is given [6].

As you can easily verify, Boneh-Durfee's set of shift-polynomials [2] and ours are completely the same as a set (but, a polynomial order is different). Then, they are the same as a lattice basis. So, we obtain the *same* lattice with Boneh-Durfee's by using our compiler and Howgrave-Graham's lattice basis [9].

Next, according to the discussion in Section 4, we will re-derive the bound of the secret key d. In [6], γ_U and γ_Y are given as $\gamma_U(u;t) = u(u + 1)/2$ and $\gamma_Y(u;t) = (u + t)(u + t + 1)/2$. And, the dimension is given by $w(u;t) = u + t + 1$ and $A_Y = \log_e Y = 1/2$. In this case, we can obtain the same bound very easily. Since $\deg(\gamma_U(u;t)) = \deg(\gamma_Y(u;t)) = 2$ and $\deg(w(u;t)) = 1$, Restriction 4 holds. Then, we can use Theorem 2. By ignoring small terms, we have $a_0 = 1/4, a_1 = 1/2, a_2 = 3/4, b_1 = b_2 = 1$. By plugging these values into Eq. (8), one can easily obtain the bound

$$ X < e^{1 - \frac{4\sqrt{4 \cdot 1 - 3 \cdot 1 \cdot 2 + 3 \cdot 1 \cdot 3} - 8 \cdot 1 + 3 \cdot 2}{3 \cdot 4}} = e^{(7 - 2\sqrt{7})/6} \approx N^{0.284}, $$

which is exactly same as the Boneh-Durfee's weaker bound.

Remark 1. By using the same lattice basis as Case1, we re-derive the small CRT-exponent attack [15]. By just replacing $A_Y = \beta$, we can derive the condition: $d_p < e^{1 - 2(\sqrt{\beta^2 + 3\beta} + \beta)/3}$, where $q < e^\beta$.

6   Concluding Remarks and Open Problems

We note that our conversion is not enough. As shown in Sec. 5, our approach just achieves the Boneh-Durfee's *weaker bound*. We need more analysis to achieve the stronger bound: $d \leq N^{0.292}$. Actually, Boneh and Durfee [2] deleted some *bad* lattice bases and introduced the concept *Geometrically Progressive Matrix* to evaluate the upper bound of the determinant of the lattice. By these efforts, they achieved the stronger bound $d \leq N^{0.292}$. We need to develop a general theory including such an improvement.

Acknowledgement

The author thanks Kaoru Kurosawa for helpful discussions.

References

1. Blömer, J., May, A.: A Tool Kit for Finding Small Roots of Bivariate Polynomials over the Integers. In: Cramer, R. (ed.) EUROCRYPT 2005. LNCS, vol. 3494, pp. 251–267. Springer, Heidelberg (2005)
2. Boneh, D., Durfee, G.: Cryptanalysis of RSA with private key d less than $N^{0.292}$. IEEE Transactions on Information Theory 46(4), 1339 (2000) (Firstly appeared in Eurocrypt 1999)
3. Coppersmith, D.: Finding a Small Root of a Univariate Modular Equation. In: Maurer, U.M. (ed.) EUROCRYPT 1996. LNCS, vol. 1070, pp. 155–165. Springer, Heidelberg (1996)

4. Coppersmith, D.: Finding a Small Root of a Bivariate Integer Equation; Factoring with High Bits Known. In: Maurer, U.M. (ed.) EUROCRYPT 1996. LNCS, vol. 1070, pp. 178–189. Springer, Heidelberg (1996)
5. Coppersmith, D.: Small Solutions to Polynomial Equations, and Low Exponent RSA Vulnerabilities. J. Cryptology 10(4), 233–260 (1997)
6. Coron, J.S., May, A.: Deterministic Polynomial Time Equivalence of Computing the RSA Secret Key and Factoring. Journal of Cryptology 20(1), 39–50 (2007); (IACR ePrint Archive: Report 2004/208, 2004)
7. Durfee, G., Nguyen, P.: Cryptanalysis of the RSA Schemes with Short Secret Exponent from Asiacrypt'99. In: Okamoto, T. (ed.) ASIACRYPT 2000. LNCS, vol. 1976, pp. 14–29. Springer, Heidelberg (2000)
8. Howgrave-Graham, N.: Finding Small Roots of Univariate Modular Equations Revisited. In: IMA Int. Conf., pp. 131–142 (1997)
9. Howgrave-Graham, N.: Approximate Integer Common Divisors. In: Silverman, J.H. (ed.) CaLC 2001. LNCS, vol. 2146, pp. 51–66. Springer, Heidelberg (2001)
10. Itoh, K., Kunihiro, N., Kurosawa, K.: Small Secret Key Attack on a Variant of RSA (due to Takagi). In: Malkin, T.G. (ed.) CT-RSA 2008. LNCS, vol. 4964, pp. 387–406. Springer, Heidelberg (2008)
11. Jochemsz, E., May, A.: A Strategy for Finding Roots of Multivariate Polynomials with New Applications in Attacking RSA Variants. In: Lai, X., Chen, K. (eds.) ASIACRYPT 2006. LNCS, vol. 4284, pp. 267–282. Springer, Heidelberg (2006)
12. Kunihiro, N., Kurosawa, K.: Deterministic Polynomial Time Equivalence between Factoring and Key-Recovery Attack on Takagi's RSA. In: Okamoto, T., Wang, X. (eds.) PKC 2007. LNCS, vol. 4450, pp. 412–425. Springer, Heidelberg (2007)
13. Kunihiro, N.: Solving Generalized Small Inverse Problems. IACR eprint Archive: Report (2010)
14. Lenstra, A.K., Lenstra, H.W., Lovász, L.: Factoring polynomials with rational coefficients. Mathematische Annalen 261, 515–534 (1982)
15. May, A.: Cryptanalysis of Unbalanced RSA with Small CRT-Exponent. In: Yung, M. (ed.) CRYPTO 2002. LNCS, vol. 2442, pp. 242–256. Springer, Heidelberg (2002)
16. May, A.: Computing the RSA Secret Key Is Deterministic Polynomial Time Equivalent to Factoring. In: Franklin, M. (ed.) CRYPTO 2004. LNCS, vol. 3152, pp. 213–219. Springer, Heidelberg (2004)
17. May, A.: Chapter3.2, The univariate case, in New RSA Vulnerabilities Using Lattice Reduction Methods. Ph.D thesis, University of Paderborn (2003)
18. Rivest, R., Shamir, A., Adleman, L.: A Method for Obtaining Digital Signatures and Public-Key Cryptosystems. Communications of the ACM 21(2), 120–126 (1978)
19. Takagi, T.: Fast RSA-Type Cryptosystem Modulo $p^k q$. In: Krawczyk, H. (ed.) CRYPTO 1998. LNCS, vol. 1462, pp. 318–326. Springer, Heidelberg (1998)
20. Wiener, M.: Cryptanalysis of Short RSA Secret Exponents. IEEE Transactions on Information Theory 36, 553–558 (1990)

A   Proofs

A.1   Proof of Lemma 4

The determinant of the submatrix B_u is given by

$$\det B_u = M^{(m-u)w} X_0^{uw} \det B_h^{(u)}(M) = M^{mw} \left(\frac{X_0}{M}\right)^{uw} \det B_h^{(u)}(M).$$

Since the determinant $\det B$ for f is given by $\det B = \prod_{u=0}^{m} \det B_u$, we have the lemma.   □

A.2   Proof of Lemma 5

For Lemma 1, if the norm of $\boldsymbol{b}_{n+1}$ is less than $M^m/\sqrt{w}$, we can reduce the modular equations into integer equations. Combining Proposition 1, this condition can be transformed into

$$\det B < M^{mW}/\gamma, \tag{9}$$

where γ is a constant. Since this term is negligible compared to M^{mW}, we can ignore this term. By substituting Eq. (4) into Eq. (9), we have

$$(X_0/M)^{\sum uw(u)} < \prod_{u=0}^{m} (\det B_h^{(u)}(M))^{-1}. \tag{10}$$

It is important that M^{mW} in both hand sides are canceled. By transforming this inequality, we have the above condition.   □

A.3   Proof of Theorem 2

The function $l(m;t)$ is given by

$$l(m;t) = \frac{a_0 m t^2 + a_1 m^2 t/2 + a_2 m^3/3}{b_1 m^2 t/2 + b_2 m^3/3}. \tag{11}$$

By replacing $x = t/m$, we have $l(x) \equiv l(m;mx) = \frac{6a_0 x^2 + 3a_1 x + 2a_2}{3b_1 x + 2b_2}$. The value x minimizing $l(x)$ satisfies $3a_0 b_1 x^2 + 4a_0 b_2 x + (a_1 b_2 - a_2 b_1) = 0$. If $a_1 b_2 - a_2 b_1 < 0$, this equation has a positive solution. By solving the above equation, we have $x = \frac{\sqrt{4a_0^2 b_2^2 - 3a_0 a_1 b_1 b_2 + 3a_0 a_2 b_1^2} - 2a_0 b_2}{3a_0 b_1}$. Letting this value c' and plugging c' into Eq. (7), we have the following condition for X_0:

$$\log_M X_0 < 1 - \frac{4a_0}{b_1} c' - \frac{a_1}{b_1}.$$

In particular, if $b_1 = b_2$, we have simply $c' = \frac{\sqrt{4a_0^2 - 3a_0 a_1 + 3a_0 a_2}}{3a_0} - \frac{2}{3}$.

One-Time-Password-Authenticated Key Exchange

Kenneth G. Paterson[1] and Douglas Stebila[2]

[1] Information Security Group, Royal Holloway,
University of London, Egham, Surrey, UK
kenny.paterson@rhul.ac.uk
[2] Information Security Institute, Queensland University of Technology,
Brisbane, Australia
douglas@stebila.ca

Abstract. To reduce the damage of phishing and spyware attacks, banks, governments, and other security-sensitive industries are deploying *one-time password* systems, where users have many passwords and use each password only once. If a single password is compromised, it can be only be used to impersonate the user once, limiting the damage caused. However, existing practical approaches to one-time passwords have been susceptible to sophisticated phishing attacks.

We give a formal security treatment of this important practical problem. We consider the use of one-time passwords in the context of password-authenticated key exchange (PAKE), which allows for mutual authentication, session key agreement, and resistance to phishing attacks. We describe a security model for the use of one-time passwords, explicitly considering the compromise of past (and future) one-time passwords, and show a general technique for building a secure one-time-PAKE protocol from any secure PAKE protocol. Our techniques also allow for the secure use of pseudorandomly generated and time-dependent passwords.

Keywords: one-time passwords, key exchange, protocols, cryptography.

1   Introduction

Many security attacks on the Internet today, such as phishing and spyware, aim to compromise a user's password. As a result, some businesses and government agencies are deploying *one-time password* systems. In these systems, users carry a sheet of paper listing passwords or an electronic device that generates passwords, and use a different password each time they log in. Ideally, without obtaining this physical list of passwords (or the device generating them), an attacker should be unable to impersonate the user.

It is unfortunately too easy these days for passwords to be compromised. For example, users at an Internet café cannot trust that the café operator has not installed a key logger, yet they may still have an urgent need to login to a particular website. Many home users unknowingly have malware installed on

R. Steinfeld and P. Hawkes (Eds.): ACISP 2010, LNCS 6168, pp. 264–281, 2010.

their computer. One-time password systems can help reduce the damage from such compromises: although we cannot prevent the password from being stolen, it can only be used once, and reveals no information about future passwords. As a result, one-time password systems are being deployed by banks, governments, and corporate virtual private networks (VPNs).

However, most deployments of one-time passwords have not used them in the strongest way possible. In a typical usage, Alice visits a bank's website in her browser, views a challenge on the website indicating which one-time password to use, and enters that one-time password into her browser, which transmits the one-time password to the website. This type of usage remains susceptible to the same phishing attacks that threaten regular passwords today: if Alice did not really have an encrypted link with her actual bank, then an attacker may be able to learn the one-time password and impersonate Alice. Unfortunately, average users are not very good at telling if a SSL/TLS connection is really encrypted and authenticated.

More advanced cryptographic protocols, such as *password-authenticated key exchange* (PAKE), can allow us to use passwords in a secure way that reveals no useful information about the password to a phishing or man-in-the-middle attacker. These protocols can provide strong mutual authentication as well: not only does the bank learn whether Alice knows her password, but Alice learns whether the bank knows her password.

To date, one-time password schemes have not been formally studied using techniques from provable security. One existing work [1] presents a PAKE protocol that uses pseudorandom passwords, but does not consider how the security properties of one-time passwords or pseudorandom passwords differ from normal long-term passwords. The goal of this work is to describe and formalize security properties for one-time password systems, especially in the context of authenticated key exchange protocols.

We emphasize that one-time password schemes are practical, as numerous deployments [25,20,21,10,7] have shown. Businesses that have already deployed one-time passwords in the form of token cards or sheets of paper could benefit from the greater security offered by our techniques by upgrading their back end systems without needing to deploy new password data to users; however, clients would need to upgrade their browsers or VPN clients to support these new protocols.

Contributions. In this work, we aim to answer three questions on the security of one-time password schemes:

1. How should we model the security of one-time password schemes?
2. How should we build secure one-time password schemes?
3. Are existing one-time password schemes secure?

To answer the first question, we describe in Sect. 2 an extension to the Bellare-Pointcheval-Rogaway [3] PAKE security model that adds one-time passwords and handles the compromise of other past or future one-time passwords.

For the second question, we give a general construction in Sect. 3 for building a one-time-PAKE protocol from any PAKE protocol and show that this transformation preserves security. The transformation itself is straightforward and efficient, and allows for extensions to the basic functionality of one-time passwords: the secure use of pseudorandomly generated passwords (Sect. 4), time-dependent passwords (Sect. 5), and verifier-based one-time passwords, in which the server stores a one-way transformation of the passwords, not the passwords themselves (Sect. 2.2).

Existing uses of one-time passwords over TLS connections can be troublesome as they require a public key infrastructure and users often have difficulty validating public keys. To our knowledge, the only existing consideration of one-time passwords in PAKE is the OPKeyX protocol [1], which requires the one-time passwords be of a particular form (namely, a hash chain), and that future passwords not be revealed. We discuss the security of OPKeyX in Sect. 6, noting that our model is stronger and allows for arbitrary passwords to be revealed.

Outline. The rest of this paper is organized as follows. In Sect. 1.1, we describe related work. Section 2 deals with the security of one-time password protocols: it introduces the general properties we seek, and then presents a security model encompassing those properties. In Sect. 3, we give our central theoretical result that secure one-time-password-authenticated key exchange protocols can be built out of secure password-authenticated key exchange protocols. We then discuss the use of pseudorandom (Sect. 4) and time-dependent (Sect. 5) passwords. We conclude with a brief discussion of how this work relates to the existing OPKeyX protocol in Sect. 6 and some general conclusions in Sect. 7. Security proofs, details of verifier-based security definitions, and an example construction appear in the full version [22].

1.1   Related Work

Many businesses, especially banks, have adopted one-time passwords in their authentication procedures. One-time passwords can be efficiently deployed using electronic tokens [25], using a chip-and-pin card in combination with a reader device as some British banks are doing [20], or on sheets of paper as some European banks do [21]; interestingly, there have subsequently been phishing attacks specifically targeting these sheets of one-time passwords [10]. One-time passwords are also being used for stronger authentication in virtual economies such as World of Warcraft [7]. The Internet Engineering Task Force (IETF) has standardized various mechanisms for deriving [14,15] and using [16,23,17] one-time passwords. While all of these systems may generate and deploy one-time passwords securely, none of them proceed to use one-time passwords in cryptographically secure way.

Password-authenticated key exchange was first introduced by Bellovin and Merritt in 1992 [5] as a protocol in which the client and server share a plaintext password and exchange encrypted information to allow them to derive a shared session key. A later variant [6], often called *verifier-based*, removed the requirement that the server have the plaintext password, instead having a one-way transformation of the password.

The most extensively used model for the security of PAKE protocols is the Bellare-Pointcheval-Rogaway (BPR) model [3] and its extension [12] for verifier-based protocols. This model is the starting point of our model for the security of one-time-PAKE protocols. One particular such protocol is the PAK protocol [8,18], which is the basis of our construction in the full version of this paper.

Various authors have noted the value of using one-time passwords in authenticated key exchange protocols [1,11,27]. Abdalla *et al.* [1] (see also [9]) describe the OPKeyX protocol, a verifier-based one-time-PAKE protocol. It uses a hash chain to derive subsequent one-time passwords from a seed such that the server can verify but not compute the next password. We will discuss OPKeyX in greater detail in Sect. 6.

2   Security of One-Time-Password Protocols

The main security property that protocols employing one-time passwords should achieve is: strong mutual authentication based on knowledge of one-time passwords. Our work will address one-time passwords in the context of PAKE protocols, which provide an additional property: secure key exchange.

The motivation for using one-time passwords is that the compromise of one password should not affect the security of sessions involving another password. The one-time password serves to mutually authenticate the client and the server; there are no other long-term values like public keys or certificates. Authentication is based on knowledge of the shared password. Informally, a protocol will provide secure mutual authentication if no honest party $\hat{A}$ accepts a session as being with party $\hat{B}$ unless $\hat{B}$ participated in the protocol, and vice versa. We want a one-time-password protocol to give secure mutual authentication for the current session even if other one-time passwords have been revealed. Such passwords could be revealed accidentally by the user or obtained by an adversary who has installed malware on the user's computer, for example.

In addition to mutually authenticating two parties to each other, we want a protocol that will also output a *session key* that can be used to encrypt and protect the integrity of future communications between those two parties. This is a common feature required of many secure communication protocols.

The traditional use of one-time passwords – sending the password over a TLS connection – is not compatible with our approach. Using TLS to establish an authentic channel requires that the user can obtain and properly use an authentic public key for the server. In other words, it requires a public key infrastructure, whereas one-time-PAKE only needs shared passwords. We need not remove the TLS infrastructure, however: one-time password-authenticated key exchange could be provided as a new TLS cipher suite.

2.1   Security Model

In the most widely adopted security model for PAKE, that of Bellare, Pointcheval, and Rogaway [3], when the adversary corrupts a party it learns all of the party's authentication secrets at once. In the one-time password setting, we want to model the situation where users have multiple passwords and the attacker can learn the passwords one by one. This more closely models the functionality, design goals, and capabilities of the adversary in many one-time password scenarios.

Participants. An instance of the protocol takes place between two interacting parties, each of which is a member of the set Parties; each party is identified by a unique fixed length string. Each pair of distinct parties $\{\hat{A}, \hat{B}\}$ shares a set of one-time passwords $\{\mathsf{pw}_{\hat{A},\hat{B},\mathsf{ch}}\}$ indexed by $\mathsf{ch} \in$ Indices, the set Indices being publicly known (we use the notation ch to suggest that the one-time password may be selected in response to a challenge, although the model does not assume that need be the case). We note that $\mathsf{pw}_{\hat{A},\hat{B},\mathsf{ch}} = \mathsf{pw}_{\hat{B},\hat{A},\mathsf{ch}}$ (this is the symmetric setting; in Sect. 2.2, we discuss how to model verifier-based one-time passwords). The size of the set Indices determines the maximum number of passwords shared between each pair of parties. Each one-time password is chosen uniformly at random from the set Passwords.[1]

Protocol execution. The protocol is a message-driven protocol. During execution, a party $\hat{U}$ may have multiple instances of the protocol running; each instance is modelled as an oracle and is denoted by $\Pi^{\hat{U}}_{(\hat{U}',\mathsf{ch})}$: it is indexed by the values $(\hat{U}', \mathsf{ch}) \in$ Parties $\times$ Indices, where $\hat{U}'$ is its purported partner and ch is the one-time password index for that instance. A party $\hat{U}$ must be *activated* to act as an initiator or a responder with $\hat{U}'$ for a particular instance by having oracle $\Pi^{\hat{U}}_{(\hat{U}',\mathsf{ch})}$ be sent a message of the form "initiator" or "responder", respectively. An instance for a particular partner-index pair can only be activated once. This restriction can be achieved by having each party maintain a record of used one-time passwords. In practice, this is easy to achieve: for example, a user could cross out a one-time password on a piece of paper once it has been used, or increment a counter if pseudorandomly generated passwords are used.

There are distinguished instances $\Pi^{\hat{U}}_{(\hat{U}',\perp)}$ which can be sent messages of the form "initiator" or "responder"; $\hat{U}$ then picks an unused one-time password index ch and activates the corresponding instance $\Pi^{\hat{U}}_{(\hat{U}',\mathsf{ch})}$ with the given role.

There is a sequence of messages, or *flows*, specified by the protocol, starting with a flow from the initiator to the responder, then from the responder to the initiator, and so on. After some number of flows, an instance may *accept*, at which point it holds a *session key* sk, *partner id* pid, and *session id* sid, and, possibly after some additional flows, *terminate*. Alternatively, at any point in time, an instance may *reject* (note that instances that reject have not terminated; accepting is a precondition for terminating). Two instances $\Pi^{\hat{A}}_{(\mathsf{pid},\mathsf{ch})}$ and $\Pi^{\hat{B}}_{(\mathsf{pid}',\mathsf{ch}')}$ are said to be *partnered* if they both accept, hold $(\mathsf{pid}, \mathsf{sid}, \mathsf{sk})$ and $(\mathsf{pid}', \mathsf{sid}', \mathsf{sk}')$,

---

[1] One common complaint about models for PAKE protocols is the typical assumption that passwords are uniformly distributed. In practice, human-selected passwords are rarely uniformly distributed. By contrast, one-time passwords are more likely in practice to be uniformly distributed since they are often generated by a computer.

respectively, with $\mathsf{pid} = \hat{B}$, $\mathsf{pid}' = \hat{A}$, $\mathsf{sid} = \mathsf{sid}'$, $\mathsf{sk} = \mathsf{sk}'$, and $\mathsf{ch} = \mathsf{ch}'$, and no other instance accepts with session id equal to sid. It is likely that the session identifier will include the one-time password index ch.

Definition 1 (Correctness). *A protocol is said to be* correct *if, for all distinct* $\hat{A}, \hat{B} \in \mathsf{Parties}$ *and all* $\mathsf{ch} \in \mathsf{Indices}$, *whenever messages are faithfully relayed between* $\Pi^{\hat{A}}_{\hat{B},\mathsf{ch}}$ *and* $\Pi^{\hat{B}}_{\hat{A},\mathsf{ch}}$, *both instances are partnered and terminate with probability 1.*

Queries allowed. The protocol is determined by how participants respond to inputs from the environment, and the environment is considered to be controlled by the adversary, which is a probabilistic algorithm that issues queries to parties' oracle instances and receives responses. For a protocol P, the queries that the adversary can issue are as follows (where clear by the setting, we may omit the subscript P). The first two queries model normal operation of the protocol:

- $\mathsf{Execute}_P(\hat{A}, \hat{B}, \mathsf{ch})$: This query activates initiator instance $\Pi^{\hat{A}}_{(\hat{B},\mathsf{ch})}$ and responder instance $\Pi^{\hat{B}}_{(\hat{A},\mathsf{ch})}$ with one-time password indexed by ch, causes them to faithfully execute protocol P, and returns the resulting transcript.
- $\mathsf{Send}_P(\hat{U}, (\hat{U}', \mathsf{ch}), M)$: Send message M to user instance $\Pi^{\hat{U}}_{(\hat{U}',\mathsf{ch})}$, which performs the appropriate portion of protocol P based on its current state and the message M, updates its state, and returns any resulting messages.

The next two queries model the compromise of information by the adversary:

- $\mathsf{RevealSessionKey}_P(\hat{U}, \hat{U}', \mathsf{ch})$: If instance $\Pi^{\hat{U}}_{(\hat{U}',\mathsf{ch})}$ has accepted, then it returns the session key sk held by $\Pi^{\hat{U}}_{(\hat{U}',\mathsf{ch})}$.
- $\mathsf{RevealPW}_P(\hat{U}, \hat{U}', \mathsf{ch})$: Returns the one-time password $\mathsf{pw}_{\hat{U},\hat{U}',\mathsf{ch}}$.

The $\mathsf{RevealPW}$ query models the adversary learning the authentication secrets, which corresponds to weak corruption in the Bellare-Pointcheval-Rogaway model. The adversary cannot modify stored authentication secrets (also called strong corruption). We note that the $\mathsf{RevealPW}(\hat{U}, \hat{U}', \mathsf{ch})$ query allows the adversary to reveal any password, regardless of whether it has been used in a session.

The final query is used to define the task that the adversary has to achieve in order for the session key security of the protocol to be considered broken. To define security, the adversary will interact with a *challenger* who, simulating the parties, answers all the queries above, as well as this one:

- $\mathsf{Test}_P(\hat{U}, \hat{U}', \mathsf{ch})$: If instance $\Pi^{\hat{U}}_{(\hat{U}',\mathsf{ch})}$ has accepted, then the following happens: the challenger chooses $b \in_R \{0, 1\}$; if $b = 1$, then it returns the session key held by $\Pi^{U}_{(\hat{U}',\mathsf{ch})}$, otherwise it returns a random string of the same length as the session key. This query may only be asked once.

Freshness. We adapt the notion of freshness in the Bellare-Pointcheval-Rogaway model to allow the adversary to compromise one-time passwords from any session except the target session.

Definition 2 (Freshness). *In a one-time-PAKE protocol, an instance $\Pi^{\hat{U}}_{(\hat{U}',ch)}$ is fresh (with forward-secrecy) if and only if none of the following events occur:*

1. *a* RevealSessionKey$(\hat{U}, \hat{U}', ch)$ *query occurs;*
2. *a* RevealSessionKey$(\hat{U}', \hat{U}, ch)$ *query occurs;*
3. *either of the following queries occur before the* Test *query:*
   (a) RevealPW$(\hat{U}, \hat{U}', ch)$ *or (b)* RevealPW$(\hat{U}', \hat{U}, ch)$*;*
   and Send$(\hat{U}, (\hat{U}', ch), M)$ *occurs for some string M.*

We note that this definition of freshness allows the adversary considerable power in terms of revealed passwords. In particular, the adversary could reveal every one-time password – past and future – except the single password for the target session.

Adversary's goals. The adversary's goals are to break either the confidentiality of the session key or the security of the mutual authentication.

For confidentiality, the goal of an adversary is to guess the bit b used in the Test query of a fresh session: this corresponds to the ability of an adversary to distinguish the session key from a random string of the same length. Let $\mathsf{Succ}_P^{1\times\mathsf{ake}}(\mathcal{A})$ be the event that the adversary $\mathcal{A}$ makes a single Test query to some fresh instance $\Pi^{\hat{U}}_{(\hat{U}',ch)}$ that has accepted and $\mathcal{A}$ eventually outputs a bit b', where $b' = b$ and b is the randomly selected bit in the Test query. The $1\times$ake-*advantage* of $\mathcal{A}$ attacking P is defined to be

$$\mathsf{Adv}_P^{1\times\mathsf{ake}}(\mathcal{A}) = \left| 2\Pr\left(\mathsf{Succ}_P^{1\times\mathsf{ake}}(\mathcal{A})\right) - 1 \right| . \qquad (1)$$

We can define a similar notion for *mutual authentication*. Let $\mathsf{Succ}_P^{1\times\mathsf{ma}}(\mathcal{A})$ be the event that the adversary $\mathcal{A}$ causes a participant instance $\Pi^{\hat{U}}_{(\hat{U}',ch)}$ with partner id $\hat{U}'$ and one-time password index ch to terminate without a partnered instance, before either of the RevealPW queries in part 3 of Definition 2. The $1 \times$ ma-*advantage* of $\mathcal{A}$ attacking P is defined to be $\mathsf{Adv}_P^{1\times\mathsf{ma}}(\mathcal{A}) = \Pr\left(\mathsf{Succ}_P^{1\times\mathsf{ma}}(\mathcal{A})\right)$.

Definition 3 (Security). *Let λ be a security parameter. A protocol P is a secure one-time-password-authenticated key agreement protocol if, for all adversaries $\mathcal{A}$ running in time polynomial in λ and making at most q_{se} Send$_P$ queries, there exists a constant δ and a negligible $\epsilon(\lambda)$ such that*

$$\mathsf{Adv}_P^{1\times\mathsf{ake}}(\mathcal{A}) \leq \frac{\delta q_{\mathsf{se}}}{|\mathsf{Passwords}|} + \epsilon(\lambda) , \qquad (2)$$

and a similar bound applies for $\mathsf{Adv}_P^{1\times\mathsf{ma}}(\mathcal{A})$.

This notion of security says that no polynomially bounded adversary can do negligibly better than randomly guessing an unknown password in each online attempt and can gain no advantage by doing an offline dictionary attack.

This bound is of the same form as bounds for the security of PAKE. One might expect that we could do better in the one-time password setting, since passwords are not reused. However, the adversary *always* has a password guessing strategy each time it participates in the protocol, leading to the $q_{se}/|\mathsf{Passwords}|$ factor. Hence, this bound is effectively the best possible, up to making δ or $\epsilon(\lambda)$ smaller. The advantage of one-time password systems comes from their robustness in the face of richer models of compromise.

Remark. This security definition protects users from authentication and confidentiality failures. It offers no protection against denial of service attacks, especially attacks in which an attacker aims to exhaust a user's supply of one-time passwords. An adversary could keep a client and server "out of sync" on which password to use, preventing a connection from being established. Unless there is some additional form of server-to-client authentication – for example, the challenge being signed by a server certificate, which is outside the scope of this work since it would require a public key infrastructure – this appears to be unavoidable.

2.2   Verifier-Based One-Time Passwords

In the verifier-based model, the server stores a *verifier*, which is a one-way transformation of the client's password that cannot be used to impersonate the user. This offers increased security against server database compromise. The security of verifier-based PAKE protocols is defined by the extension of the BPR model given by Gentry *et al.* [12]. The main difference is that an instance can remain fresh even if either the password or the verifier (but not both) is compromised. This necessitates the introduction of a new query for revealing the verifier. Additionally, it allows for the separate definitions of client-to-server and server-to-client authentication.

The model we described in Sect. 2.1 can be extended in the natural way to use verifier-based one-time passwords by introducing a RevealV query to reveal one-time verifiers and adjusting the freshness definition appropriately; the details appear in the full version [22].

3   A Generic Construction for One-Time Password Protocols

We now describe a technique for building a one-time-PAKE protocol, $1(P)$, out of any PAKE protocol P, and then show that the one-time-password protocol is at least as secure as the password protocol out of which it is built. The basic idea is that a PAKE protocol in which passwords are used only once is also a good one-time-PAKE protocol.

3.1   Construction of $1(P)$ from P

The construction proceeds as follows. For each client-server-index combination $(\hat{C}, \hat{S}, \mathsf{ch})$ in the one-time-password protocol, we will construct a new pair of users with compound names $(\hat{C}, \hat{S}, \mathsf{ch})$ and $(\hat{S}, \hat{C}, \mathsf{ch})$ in the password protocol, and pass the queries against the session in the one-time-password protocol down to the new pair of users in the underlying password protocol. Since every PAKE protocol should be secure even if each pair of users is used only once, this constructed one-time-PAKE protocol should also be secure.

We now specify in detail the technique to construct a one-time-PAKE protocol $1(P)$ from a PAKE protocol P. There are two phases: the *registration phase*, in which pairs of clients and servers establish passwords, and the *login phase*, in which pairs of clients and servers attempt to establish a secure session.

Registration phase. The registration phase of the $1(P)$ protocol is specified in Fig. 1 below. For every client-server pair $(\hat{C}, \hat{S}) \in \mathsf{Parties} \times \mathsf{Parties}$, and for each one-time-password index $\mathsf{ch} \in \mathsf{Indices}$, initiate the registration phase of P with the users $(\hat{C}, \hat{S}, \mathsf{ch})$ and $(\hat{S}, \hat{C}, \mathsf{ch})$, and set $\mathsf{pw}_{\hat{C}, \hat{S}, \mathsf{ch}}$ in $1(P)$ equal to the corresponding password in P.

Although one might be concerned about the time that it takes to complete the registration phase if $\mathsf{Indices}$ is large, the registration phase of any one-time password protocol can not, in general, be completed in less time asymptotically if truly one-time passwords are used. In other words, this is effectively the same complexity as password establishment in currently deployed one-time password schemes, and hence is quite practical. Moreover, the registration for each challenge can be run in parallel to reduce the number of communication rounds.

Protocol $1(P)$ – Registration Phase	
Client $\hat{C}$	Server $\hat{S}$
for each $\mathsf{ch} \in \mathsf{Indices}$:	
1. run registration phase of protocol P with users $(\hat{C}, \hat{S}, \mathsf{ch})$ and $(\hat{S}, \hat{C}, \mathsf{ch})$	
2. $\mathsf{pw}_{\hat{C}, \hat{S}, \mathsf{ch}}$ in $1(P) \leftarrow \mathsf{pw}_{(\hat{C}, \hat{S}, \mathsf{ch}), (\hat{S}, \hat{C}, \mathsf{ch})}$ in P	
3. $\mathsf{used}_{\hat{C}}(\hat{S}, \mathsf{ch}) \leftarrow \mathsf{false}$ $\mathsf{used}_{\hat{S}}(\hat{C}, \mathsf{ch}) \leftarrow \mathsf{false}$	
end for each	

Fig. 1. Protocol $1(P)$ – Registration Phase; must use a private, authenticated channel

Login phase. The login phase of the $1(P)$ protocol is specified in Figure 2 below. Each party $\hat{U}$ maintains a set of tables $\mathsf{used}_{\hat{U}}(\hat{U}', \mathsf{ch})$, where each entry in the table is either true or false and indicates whether the one-time-password indexed by ch has been used by $\hat{U}$ with $\hat{U}'$.

To initiate the protocol, instance $\varPi_{(\hat{S}, \perp)}^{\hat{C}}$ of user $\hat{C}$ sends a message ("hello", $\hat{C}$) to instance $\varPi_{(\hat{C}, \perp)}^{\hat{S}}$ of party $\hat{S}$. When a party $\hat{S}$ receives a message ("hello", $\hat{C}$), it picks a one-time-password index ch from $\mathsf{Indices}$ such that $\mathsf{used}_{\hat{S}}(\hat{C}, \mathsf{ch}) = \mathsf{false}$.

Then it sets $\mathsf{used}_{\hat{S}}(\hat{C}, \mathsf{ch}) \leftarrow \mathsf{true}$ and activates $\Pi^{\hat{S}}_{(\hat{C},\mathsf{ch})}$. Finally, it sends "hello" to instance $\Pi^{\hat{C}}_{(\hat{S},\mathsf{ch})}$ of party $\hat{C}$. It then waits to engage in a single instance of protocol P acting as user $(\hat{S}, \hat{C}, \mathsf{ch})$ interacting with party $(\hat{C}, \hat{S}, \mathsf{ch})$. When the corresponding instantiation of protocol P accepts, the instance in $1(P)$ sets its session key to the session key in P and then accepts. When it rejects in P, it rejects in $1(P)$; when it terminates in P, it terminates in $1(P)$ as well.

When instance $\Pi^{\hat{C}}_{(\hat{S},\mathsf{ch})}$ of party $\hat{C}$ receives a message ("hello"), it checks to see if $\mathsf{used}_{\hat{C}}(\hat{S}, \mathsf{ch}) = \mathsf{true}$; if so, then it rejects; if not, then it sets $\mathsf{used}_{\hat{C}}(\hat{S}, \mathsf{ch}) \leftarrow \mathsf{true}$. It then initiates the login phase of protocol P acting as user $(\hat{C}, \hat{S}, \mathsf{ch})$ interacting with party $(\hat{S}, \hat{C}, \mathsf{ch})$. It follows protocol P until it accepts or rejects. When the corresponding instantiation of protocol P accepts, the instance in $1(P)$ sets its session key to the session key in P and then accepts. When it rejects in P, it rejects in $1(P)$; when it terminates in P, it terminates in $1(P)$ as well.

It follows easily from inspection that, if P is correct, $1(P)$ is also correct.

Protocol $1(P)$ – Login Phase	
Client $\hat{C}$	Server $\hat{S}$
1. $\xrightarrow{\text{"hello"}, \hat{C}}$	
2.	pick $\mathsf{ch} \in \mathsf{Indices}$ s.t.
	$\mathsf{used}_{\hat{S}}(\hat{C}, \mathsf{ch}) = \mathsf{false}$
3.	$\mathsf{used}_{\hat{S}}(\hat{C}, \mathsf{ch}) \leftarrow \mathsf{true}$
4. $\Pi^{\hat{C}}_{(\hat{S},\mathsf{ch})} \xleftarrow{\text{"hello"}}$	
5. if $(\mathsf{used}_{\hat{C}}(\hat{S}, \mathsf{ch}) = \mathsf{true})$ then reject	
6. $\mathsf{used}_{\hat{C}}(\hat{S}, \mathsf{ch}) \leftarrow \mathsf{true}$	
7. run protocol P with users $(\hat{C}, \hat{S}, \mathsf{ch})$ and $(\hat{S}, \hat{C}, \mathsf{ch})$ and password $\mathsf{pw}_{(\hat{C},\hat{S},\mathsf{ch}),(\hat{S},\hat{C},\mathsf{ch})}$	
8. if P accepts then	if P accepts then
8.a) $\mathsf{sid}_{1(P)} \leftarrow \mathsf{sid}_P$; $\mathsf{pid} \leftarrow \hat{S}$	$\mathsf{sid}_{1(P)} \leftarrow \mathsf{sid}_P$; $\mathsf{pid} \leftarrow \hat{C}$
8.b) $\mathsf{sk}_{1(P)} \leftarrow \mathsf{sk}_P$	$\mathsf{sk}_{1(P)} \leftarrow \mathsf{sk}_P$
8.c) accept in $1(P)$	accept in $1(P)$
9. if P terminates then terminate	if P termiantes then terminate
10. if P rejects then reject	if P rejects then reject

Fig. 2. Protocol $1(P)$ – Login Phase; can use a public, unauthenticated channel

3.2   Security of $1(P)$

Theorem 1. *Let P be a secure password-authenticated key exchange protocol. Then $1(P)$ is a secure one-time-password-authenticated key exchange protocol.*

Due to length restrictions, the security argument appears in the full version [22]. The basic idea of the argument is as follows. We will show that attacks against $1(P)$ correspond to attacks against P. We construct a $1(P)$ simulator in which the adversary's queries to $1(P)$ are translated into queries on a P challenger as follows:

- $\mathsf{Execute}_{1(P)}(\hat{A}, \hat{B}, \mathsf{ch})$:   Return   the   result   of $\mathsf{Execute}_P((\hat{A}, \hat{B}, \mathsf{ch}), 1, (\hat{B}, \hat{A}, \mathsf{ch}), 1)$.
- $\mathsf{Send}_{1(P)}(\hat{U}, (\hat{U}', \mathsf{ch}), M)$: If message M is for one of the two flows added by the $1(P)$ construction, then respond as indicated in Figure 2. If message M is for one of the flows from P, then return the result of $\mathsf{Send}_P((\hat{U}, \hat{U}', \mathsf{ch}), 1, M)$.
- $\mathsf{RevealSessionKey}_{1(P)}(\hat{U}, \hat{U}', \mathsf{ch})$: Return $\mathsf{RevealSessionKey}_P((\hat{U}, \hat{U}', \mathsf{ch}), 1)$.
- $\mathsf{RevealPW}_{1(P)}(\hat{A}, \hat{B}, \mathsf{ch})$: Return $\mathsf{RevealPW}_P((\hat{A}, \hat{B}, \mathsf{ch}), (\hat{B}, \hat{A}, \mathsf{ch}))$.
- $\mathsf{Test}_{1(P)}(\hat{U}, \hat{U}', \mathsf{ch})$: Return $\mathsf{Test}_P((\hat{U}, \hat{U}', \mathsf{ch}), 1)$.

Using this simulation, if an adversary could break $1(P)$, it could break P just as efficiently. But since P is a secure PAKE protocol, no adversary should be able to attack P, and hence no adversary should be able to attack $1(P)$. The argument is a straightforward simulation involving creating separate user instances in P for each instance of $1(P)$ by constructing users in P with identities that are the concatenation of the user name and one-time password index from $1(P)$, and assuring that fresh instances in $1(P)$ correspond to fresh instances in P. We note that the security reduction is tight.

Example instantiation. Suppose we were to construct a one-time-password-authenticated key exchange protocol using the $1(P)$ construction where the underlying password-authenticated key exchange protocol is the (symmetric, non-verifier-based) protocol PAK [8]. The 1(PAK) protocol is particularly interesting because, with an appropriate reordering of messages, it can be made to fit inside the message flow of the TLS handshake protocol. This makes it suitable for use as a new cipher suite in TLS. A full presentation of the 1(PAK) protocol is given in the full version [22].

In our example, we wish for an adversary to be able to break the one-time-password protocol with probability at most 2^{-20}, where the adversary runs in time at most 2^{60}, and can only make a limited number (2^{10}) of Send queries. Assuming the hardness of solving the elliptic curve computational Diffie-Hellman problem (using estimates in [2]), we can achieve this security level using 10-digit numerical passwords ($\mathsf{Passwords} = \{0, \ldots, 9\}^{10}$) and a 348-bit elliptic curve group. (See the full version [22] for the full analysis.)

3.3   Efficiency and Practicality of $1(P)$

Login phase and computational efficiency. During the login phase, the $1(P)$ construction provides no loss of efficiency in terms of the number of expensive operations (such as group exponentiations) or security level of P, since the reduction is tight. $1(P)$ does add two additional message flows to the length of the protocol, but depending on the message flow of protocol P it may be possible to combine some flows without affecting security.

One might think that designing a one-time-password protocol from scratch may lead to greater efficiency, since some of the effort in designing PAKE protocols goes to preventing the transcript of one session leaking information about the password and in a one-time-password protocol we may not have to worry as much about leaking information about passwords during a protocol run. However, many PAKE protocols are already highly efficient in terms of number of operations. For example, the Diffie-Hellman-based PAK [18] protocol can be run with just 2 group exponentiations on each side (plus a group inversion on the client side, which is inexpensive in many groups like elliptic curve groups), which is very close to the operation count of the basic, unauthenticated Diffie-Hellman protocol (2 group exponentiations for both parties). The main efficiency to be gained, then, would be in improving the tightness in the security reduction to the underlying Diffie-Hellman problem so as to allow smaller group sizes.

Registration phase. The registration phase of $1(P)$ obviously requires establishing many more passwords than a single instance of P, but *any* one-time password scheme requires establishing many more passwords than a long-term password scheme. The $1(P)$ registration phase calls the registration phase of P many times. Depending on the PAKE protocol P, the registration phase can be quite efficient: for example, in the PAK protocol [18], the registration phase can be optimized to consist of just one hash function evaluation.

Password storage. In practice it is important to consider how clients will store a list of one-time passwords, especially if they wish to log in to a site while away from their normal computer. One method is to provide a piece of paper with a list of one-time passwords; for example, the Swedish bank Nordea provides its customers with a "scratch sheet" of 120 one-time passwords [21]. Alternatively, one-time passwords could be delivered through an out-of-band channel such as an SMS message to the user's mobile phone (for example, [19]). Passwords can also be stored on or generated by an electronic token device, for example the RSA SecurID [25], or even in a smart card built into credit cards [24].

We can further reduce the complexity of the registration phase and password storage by using pseudorandom or time-based one-time passwords, which we describe in the following sections.

4   Using Pseudorandom Passwords

To improve the efficiency of password registration and storage, it may be desirable to pseudorandomly generate passwords instead of truly random ones. For example, users may be given a hardware token [25,7] with a preprogrammed private seed which iteratively generates one-time passwords, or the device may accept a challenge as an input and then output a response from a pseudorandom function based on the seed and that challenge. We show that pseudorandomly generated passwords can be safely used in one-time-PAKE protocols.

Suppose P is secure one-time-PAKE protocol. We construct a new protocol $\tilde{P}$ based on P that uses pseudorandomly generated passwords as follows.

We modify the registration phase of $\tilde{P}$ as follows. For each (unordered) pair of users $\{\hat{A}, \hat{B}\} \in$ Parties $\times$ Parties, choose a random seed $\mathsf{seed}_{\hat{A},\hat{B}} \in_R \{0,1\}^\lambda$, where λ is a security parameter. Let $\mathcal{F} = \{F_k\}$ be a family of pseudorandom functions [13]. For each one-time-password index $\mathsf{ch} \in$ Indices, set $\mathsf{pw}_{\hat{A},\hat{B},\mathsf{ch}} = F_{\mathsf{seed}_{\hat{A},\hat{B}}}(\mathsf{ch})$.

The login phase of $\tilde{P}$ is exactly as in P, except that the passwords chosen in the modified registration phase above are used. For the purposes of the security model in Sect. 2.1, the RevealPW queries work exactly as before and only reveal an individual password $\mathsf{pw}_{\hat{A},\hat{B},\mathsf{ch}}$. No query reveals $\mathsf{seed}_{\hat{A},\hat{B}}$.[2]

The only difference between $\tilde{P}$ and P is that pseudorandom passwords are being used instead of random passwords. It is then easy to see that any efficient adversary $\mathcal{A}$ that can defeat session key security or mutual authentication in $\tilde{P}$ can be used to build either an adversary $\mathcal{A}_1$ that breaks session key security or mutual authentication in P, or an algorithm $\mathcal{A}_2$ that acts as a polynomial-time distinguisher for the pseudorandom function family $\mathcal{F}$. Thus $\tilde{P}$ is secure if P is, and we see that pseudorandomly generated one-time passwords can be safely used in any secure one-time-PAKE protocol.

5   Using Time-Dependent Pseudorandom Passwords

A further refinement to the use of pseudorandomly generated passwords is to use passwords that also depend on the current time. This allows the client and server to agree upon a challenge – the current time – without any communication, while easily enforcing the one-time use of passwords.

For example, consider a hardware token for party $\hat{A}$ interacting with party $\hat{B}$ which has a pseudorandom function $F_{\mathsf{seed}_{\hat{A},\hat{B}}} \in \mathcal{F}$ and an onboard clock. It generates one-time passwords as follows. Let t be the hardware token's current time. Treat t as the one-time password index ch, and then compute

$$\mathsf{pw}_{\hat{A},\hat{B},t} = F_{\mathsf{seed}_{\hat{A},\hat{B}}}(t) \ . \tag{3}$$

User $\hat{A}$ then participates in the one-time-PAKE protocol using $\mathsf{pw}_{\hat{A},\hat{B},t}$.

Whenever clocks are used by two parties, one must consider the issue of *clock skew*, in which the two clocks may not be perfectly synchronized. For example, ordinary quartz clocks drift at a rate of approximately 10^{-6} seconds per second, or about 1 second every 12 days.

One solution is to have a common network time server that both parties use for synchronization. This is problematic for two reasons: (1) the network time server must be trusted (or at least dealt with in the security model); (2) all of

---

[2] We could add a further query, say RevealPWSeed($\hat{A}, \hat{B}$), that does reveal the value $\mathsf{seed}_{\hat{A},\hat{B}}$ and then add an additional constraint to the definitions of freshness and authentication so that an instance is not considered fresh if the relevant RevealPWSeed query is called before the Test query. This enhanced model would make it clear that corruption of one pair of users' pseudorandom seed should not affect the security of another pair of users. It is not hard to see that this construction would satisfy this enhanced security model, assuming independent random seeds.

the parties participating must have a way of synchronizing with the clock server; an inexpensive, credit-card-sized hardware token may not be connected to the network, making synchronization difficult or impossible.

Another method for dealing with clock skew is to have the server accept multiple passwords from a small window around the server's current time (say, plus or minus 60 seconds). However, this is a problem for PAKE protocols, as the server never receives the client's password directly. Rather, each party uses what it believes to be the password in the protocol, and at the end the two parties know that the same password was used if and only if they arrive at the same session key. This prevents the server from accepting multiple passwords as valid. (Traditional one-time password systems often avoid this problem by having the client send the password itself to the server over an existing encrypted but not mutually authenticated channel.)

A simpler alternative mechanism for dealing with clock skew is for one party (the initiator) to just tell the other party (the responder) what time t it used in the protocol. If the time used by the initiator is acceptable to the responder (say within plus or minus 60 seconds of the responder's clock) then the responder continues the protocol using the specified time. This provides a simple mechanism for ensuring both sides use the same time-dependent password while accommodating clock skew.

Adjusting the model. In order to accommodate this alternate mechanism in the security model described in Sect. 2.1, the definition of freshness would need to be adapted (in part 3.(a) and 3.(b) of Definition 2) so that a responder instance $\Pi^{\hat{U}}_{(\hat{U}',t)}$ is fresh provided that no RevealPW$(\hat{U}, \hat{U}', t)$ or RevealPW$(\hat{U}', \hat{U}, t)$ query was issued. This captures the notion that the authentication should be secure as long as the currently valid password has not been revealed.

With this modified security definition, and assuming $\mathcal{F}$ is a secure family of pseudorandom functions, one-time time-based passwords generated in equation (3) can be safely used in a secure one-time-PAKE protocol as a result of the discussion about pseudorandom passwords in Sect. 4.

6 Analysis of the **OPKeyX** Protocol

The OPKeyX protocol [1] is a PAKE protocol that uses a sequence of passwords derived via a hash chain from a single seed. The protocol is a verifier-based protocol, meaning that the compromise of the value stored on the server should not allow someone to impersonate the client. We note that [1] omits a complete analysis of OPKeyX: it gives a proof for a non-verifier-based PAKE protocol in the BPR model but no proof that OPKeyX, a verifier-based hash-chain variant of the protocol, is also secure.

The sequence of passwords in OPKeyX is as follows. Each client $\hat{C}$ picks, for each server $\hat{S}$, a seed password pw. Let $N_{\max}$ be the maximum number of login sessions for the seed pw. During the registration phase, the client gives the server its verifier $V_{N_{\max}} \leftarrow f^{N_{\max}+1}(\text{pw})$, where f is a random oracle [4] and f^i denotes the i-fold application of f. The parties each maintain internal counters n of the current login phase, starting from $n = N_{\max}$ and decreasing to 1. During login phase with internal counter equal to n, the client and server do an encrypted key exchange where the Diffie-Hellman ephemeral public keys are encrypted using a value derived from the verifier $V_n = f^{n+1}(\mathsf{pw})$. Then, the client encrypts $f^n(\mathsf{pw})$ under a value s derived from the shared Diffie-Hellman key (but distinct from the session key sk) and sends it to the client. The server decrypts to obtain V', verifies $V_n = f(V')$, and sets $V_{n-1} \leftarrow V'$ and $n \leftarrow n - 1$.

OPKeyX relies on the correct sequence of passwords being used. In the security model for verifier-based one-time-PAKE in Section 2.2, we allow the adversary to reveal one-time passwords in any order. As a result, OPKeyX cannot be a secure verifier-based one-time-PAKE protocol in that sense. For example, an adversary could reveal the password for session with counter i, which is $f^{i+1}(\mathsf{pw})$, and then be able to derive the password for the earlier session with counter $i+1$ (recalling that counters decrease as time passes), which is $f^{i+2}(\mathsf{pw}) = f(f^{i+1}(\mathsf{pw}))$. To describe the security of OPKeyX, we would need to further restrict our model so that a session is not fresh if the password or verifier of a subsequent session has been revealed which, although weaker from a theoretical perspective, still models a plausible practical scenario. The situation is even more complicated if RevealSessionKey is deemed to reveal the value s (which encrypted the next verifier V' and is in some sense a "session key") in addition to sk, in which case no earlier s value for the target users can have been revealed before the Test query.

7   Conclusions

One-time password systems are already being widely deployed by banks, governments, and corporate virtual private networks (VPNs) to reduce the effects of password compromise. Bank customers today are using sheets of paper with lists of one-time passwords. Online shoppers and gamers today are using hardware one-time password generators. The money being spent on deploying one-time passwords is wasted if these passwords are not being used safely and securely.

By using one-time passwords in one-time-PAKE protocols, as we have proposed in this paper, we can be assured that one-time passwords are being used in a more secure way. We have presented a model for the secure use of one-time passwords in PAKE protocols, taking into account the idea that such protocols should be secure even if previous or future one-time passwords have been compromised. We have given a generic technique for constructing secure one-time password protocols. Our construction can be used with pseudorandomly generated one-time passwords or time-based one-time passwords, providing greater efficiency in one-time password distribution.

An important open problem based on this work is the task of determining whether it is possible to construct one-time password protocols that are more efficient than regular password protocols, as discussed in Sect. 3.3.

As with all cryptographic protocols, an essential precondition to security is *getting users to use the protocol*. If an adversary can trick a user into entering their password in a non-secure manner so that the secure protocol is never used – a so-called *chosen protocol attack* – then the cryptographic countermeasures are bypassed. For any PAKE protocol to succeed, user training and user interface design will be very important.

Spyware remains a significant threat to password security. In the face of passive spyware, such as a keystroke logger which collects information and occasionally relays it back to the attacker, both traditional one-time password schemes and one-time-PAKE are useful since used one-time passwords are useless to an attacker. If the spyware is active – it captures a one-time password, terminates the user's connection, and immediately sends the password to the attacker – the captured password may still be useful to an attacker, and it seems that neither traditional one-time password schemes nor one-time-PAKE can do much unless time-dependent passwords are used with careful expiration procedures.

An additional challenge is widespread deployment of such secure protocols. Passwords, as they are used in HTTP and TLS on the Internet today, remain susceptible to phishing attacks. The huge installed base of web browsers and web servers has significantly slowed efforts to deploy PAKE. Our techniques, like PAKE, require some changes to TLS implementations. It may be possible to implement a large portion of a new security protocol as a browser add-on (like a Firefox extension), making deployment easier.

Our approach may see more immediate application in corporate virtual private network (VPN) software. Many corporate VPNs use one-time passwords now, albeit in a less secure way than we have proposed. Moreover, both endpoints – the user's computer and the VPN server – are often under control of the same organization and using software from the same vendor, making it easier to deploy enhancements. An interesting avenue of future research is the integration of secure PAKE and one-time-PAKE protocols into IPsec for use in corporate VPNs. Indeed, IKEv2 (one of the key exchange protocols for IPsec) notes the need for password authentication: after showing how to derive a shared key for authenticated Diffie-Hellman key exchange in IKEv2, the RFC goes on to say:

> "... deriving the shared secret from a password is not secure. This construction is used because it is anticipated that people will do it anyway" [16, p. 30].

One-time-password-authenticated key exchange is one way in which one-time passwords can be used more securely.

References

1. Abdalla, M., Chevassut, O., Pointcheval, D.: One-time verifier-based encrypted key exchange. In: Vaudenay, S. (ed.) PKC 2005. LNCS, vol. 3386, pp. 47–64. Springer, Heidelberg (2005)
2. Babbage, S., Catalano, D., Cid, C., Dunkelman, O., Gehrmann, C., Granboulan, L., Lange, T., Lenstra, A., Nguyen, P.Q., Paar, C., Pelzl, J., Pornin, T., Preneel, B., Rechberger, C., Rijmen, V., Robshaw, M., Rupp, A., Smart, N., Ward, M.: ECRYPT yearly report on algorithms and keysizes (2007–2008) (July 2008), http://www.ecrypt.eu.org/documents/D.SPA.28-1.1.pdf

3. Bellare, M., Pointcheval, D., Rogaway, P.: Authenticated key exchange secure against dictionary attacks. In: Preneel, B. (ed.) EUROCRYPT 2000. LNCS, vol. 1807, pp. 139–155. Springer, Heidelberg (2000)

4. Bellare, M., Rogaway, P.: Random oracles are practical: a paradigm for designing efficient protocols. In: Proc. 1st ACM Conference on Computer and Communications Security (CCS), pp. 62–73. ACM, New York (1993)

5. Bellovin, S.M., Merritt, M.: Encrypted key exchange: Password-based protocols secure against dictionary attacks. In: Proceedings of the 1992 IEEE Computer Society Conference on Research in Security and Privacy. IEEE, Los Alamitos (May 1992)

6. Bellovin, S.M., Merritt, M.: Augmented encrypted key exchange: a password-based protocol secure against dictionary attacks and password file compromise. In: Proc. 1st ACM Conference on Computer and Communications Security (CCS), pp. 244–250. ACM, New York (1993)

7. Blizzard Entertainment: Blizzard authenticator (2009), http://eu.blizzard.com/support/article.xml?locale=en_GB&articleId=28152

8. Boyko, V., MacKenzie, P., Patel, S.: Provably secure Password-Authenticated Key exchange using Diffie-Hellman. In: Preneel, B. (ed.) EUROCRYPT 2000. LNCS, vol. 1807, pp. 156–171. Springer, Heidelberg (2000)

9. Chevassut, O., Siebenlist, F., Helm, M.: Secure (one-time-) password authentication for the Globus toolkit. In: GlobusWorld Conference (February 2005), http://acs.lbl.gov/Projects/OPKeyX/Talks/GlobusWorld05/GlobusWorld05.html

10. F-Secure: Weblog: More on international phishing (October 2005), http://www.f-secure.com/weblog/archives/00000689.html

11. Fang, L., Meder, S., Chevassut, O., Siebenlist, F.: Secure password-based authenticated key exchange for web services. In: Proc. 2004 Workshop on Secure Web Service (SWS), pp. 9–15. ACM, New York (2004)

12. Gentry, C., MacKenzie, P., Ramzan, Z.: PAK-Z+ contribution to the IEEE P1363-2000 study group for Future PKC Standards (August 2005), http://grouper.ieee.org/groups/1363/WorkingGroup/presentations/pakzplusv2.pdf;

13. Goldreich, O., Goldwasser, S., Micali, S.: How to construct random functions. Journal of the ACM 33(4), 792–807 (1986)

14. Haller, N.: The S/KEY one-time password system. RFC 1760 (February 1995), http://www.ietf.org/rfc/rfc1760.txt

15. Haller, N., Metz, C., Nesser II, P.J., Straw, M.: A one-time password system. RFC 2289 (February 1998), http://www.ietf.org/rfc/rfc2289.txt

16. Kaufman, C.: Internet Key Exchange (IKEv2) protocol. RFC 4306 (2005), http://www.ietf.org/rfc/rfc4306.txt

17. Kumar, S., Sing, A.: One time password in IKE version 2, non-EAP based (November 2008) (internet-Draft), http://tools.ietf.org/id/draft-sunabhi-otp-ikev2-03.txt

18. MacKenzie, P.: The PAK suite: Protocols for password-authenticated key exchange. Tech. Rep. 2002-46, DIMACS Center, Rutgers University (2002), http://dimacs.rutgers.edu/TechnicalReports/abstracts/2002/2002-46.html

19. Mobile-OTP Project: Mobile one time passwords, http://motp.sourceforge.net/

20. Nationwide Building Society: Card reader security (May 2009), http://www.nationwide.co.uk/rca/

21. Nordea Bank: Netbank security (2009), http://www.nordea.ee/Private+customers/E-channels++Netbank/Netbank/Netbank+Security/936612.html

22. Paterson, K.G., Stebila, D.: One-time-password-authenticated key exchange (full version), http://eprint.iacr.org/2009/430

23. Nystroem, M.: The EAP protected one-time password protocol (EAP-POTP). RFC 4793 (February 2007), `http://www.ietf.org/rfc/rfc4793.txt`
24. Prigg, M.: The new credit card with keypad that promises to fight online fraud (November 2008) (The Daily Mail Online), `http://www.dailymail.co.uk/sciencetech/article-1085642/The-new-credit-card-keypad-promises-fight-online-fraud.html?ITO=1490`
25. RSA Security Inc.: RSA SecurID (2009), `http://www.rsa.com/node.aspx?id=1156`
26. Shoup, V.: On formal models for secure key exchange (version 4) (November 1999), `http://shoup.net/papers/skey.pdf`, earlier version appeared as Report RZ 3120, IBM Research (April 1999), `http://www.zurich.ibm.com/security/publications/1999/Shoup99.ps.gz`
27. Stebila, D.: Classical Authenticated Key Exchange and Quantum Cryptography. Ph.D. thesis, University of Waterloo (2009), `http://www.douglas.stebila.ca/research/papers/ste09/`

Predicate-Based Key Exchange

James Birkett and Douglas Stebila

Information Security Institute, Queensland University of Technology,
Brisbane, Australia
james.birkett@qut.edu.au, douglas@stebila.ca

Abstract. We provide the first description of and security model for
authenticated key exchange protocols with predicate-based authentication. In addition to the standard goal of session key security, our security
model also provides for credential privacy: a participating party learns
nothing more about the other party's credentials than whether they satisfy the given predicate. Our model also encompasses attribute-based key
exchange since it is a special case of predicate-based key exchange.

We demonstrate how to realize a secure predicate-based key exchange
protocol by combining any secure predicate-based signature scheme with
the basic Diffie-Hellman key exchange protocol, providing an efficient and
simple solution.

Keywords: predicate-based, attribute-based, key exchange, protocols,
security models, cryptography.

1 Introduction

Two of the fundamental goals of key exchange are authentication and confidentiality. Entity authentication inherently depends on some pre-established piece of
trusted information; the most common examples include a shared key, a shared
password, or a certified public key. Recently, cryptographers have developed ways
of providing more fine-grained access control in cryptographic operations.

Identity-based encryption allows a sender to encrypt a message for a recipient
based solely on the recipient's identity (and public parameters for the system); in
other words, without requiring a recipient-dependent public key. The identities
used in identity-based cryptography may be simple usernames, but they could
contain more structured information as well, for example by appending an expiry
date or security level. The utility of this idea is limited by the fact that identities
must be encoded as strings, and a trusted key generation centre must generate
decryption keys for each resulting string.

In attribute-based encryption, a message can be encrypted so that it can
only be decrypted by keys whose attributes satisfy a certain policy. Attributes
are boolean variables, such as "student=false", "CS_department=false", and
"Math_department=true", and policies are boolean functions. Decryption keys
are constructed based on the user's attributes, and decryption only succeeds if

R. Steinfeld and P. Hawkes (Eds.): ACISP 2010, LNCS 6168, pp. 282–299, 2010.

the user's attributes satisfy the policy encoded in the ciphertext.[1] Research in attribute-based cryptography has focused on encryption and signatures.

The subject of this paper, *predicate-based cryptography*, is a generalization of identity- and attribute-based cryptography. Like attribute-based cryptography, it allows for fine-grained access control based on whether the given credentials satisfy a certain policy. However, credentials and access policies can be more general than in the attribute-based case. Credentials can consist of name-value pairs, where the values can be from arbitrary sets, not just boolean values. Access policies are expressed as predicates over the set of credentials, and can for example involve equality, comparison, subset, AND, and OR gates. Existing work in predicate-based cryptography has focused on encryption, particularly on expanding the expressiveness of predicates.

Our goal in this work is to consider the use of predicate-based cryptography in a multi-user interactive network setting, specifically examining the cryptographic task of *predicate-based authenticated key exchange*.

1.1 Contributions

Predicate-based key exchange security model. We give the first security model for authenticated key exchange using predicate-based authentication. Our security model has two security experiments:

1. *Session key security*: The session key should be indistinguishable to an adversary. Unlike attribute-based encryption, attribute-based group key exchange, and predicate-based encryption, the session key should be secret even from other parties satisfying the same predicates as either of the two original parties in the key exchange.
2. *Credential privacy*: In a key exchange, it should not be possible for anyone – including the legitimate peer – to learn anything more about a user's credentials other than whether they satisfy the chosen predicate. We argue that this is an essential property for predicate-based key exchange: without it, we might as well return to identity- or public-key-based key exchange with certified lists of credentials.

When restricted to the special case of attribute-based credentials, our security model for predicate-based key exchange also serves as the first full security model for attribute-based key exchange.

A generic predicate-based key exchange protocol. We present a protocol for predicate-based key exchange that satisfies the two security properties above, session key security and attribute privacy. The protocol is a signed-Diffie-Hellman construction that can be used with any secure predicate-based signature scheme. Although our definition of predicate-based signature scheme is new, attribute-based signature schemes are a special case of predicate-based signatures, so attribute-based signatures can be employed in our protocol construction.

---

[1] We have described *ciphertext-policy* attribute-based encryption, in which keys have attributes and ciphertexts have policies. These can be switched to obtain *key-policy* attribute-based encryption.

Outline. The remainder of this paper is organized as follows. We begin in Sect. 2 with a motivating example. We review existing work in Sect. 3 and introduce notation in Sect. 4. In Sect. 5, we present our security model for predicate-based key exchange protocols – including session key security and attribute privacy – and comment on implementation issues. We define predicate-based signature schemes in Sect. 6, and show in Sect. 7 how to build a secure predicate-based key exchange protocol using predicate-based signatures and a Diffie-Hellman construction. We conclude in Sect. 8.

2   Motivation

When one party wishes to establish a shared secret key with another party, it may not be as simple as Alice saying that she wants to talk to Bob. Alice may in fact wish to talk a customer service supervisor in the international trading group of the Bank of Bob. In other words, Alice has an *policy* against which she checks the *credentials* of the other party. Predicate-based cryptography allows parties to specify fine-grained access control policies and has been used in the context of encryption. It is natural to consider the problem in the context of key exchange, which allow two parties to authentically establish a secure channel.

We begin with a motivating example, drawn from the health care industry. Imagine a patient who wishes to communicate with a psychologist about a mental illness issue. What are some security goals for each party? The goals of the patient are to ensure that she is communicating with a qualified registered psychologist, to use a confidential channel so that no one can eavesdrop, and to maintain her anonymity so her disclosures about her mental illness cannot be used prejudicially against her in another context. The goals of the psychologist are to verify that the patient has valid insurance coverage from an insurer and to ensure that no one else can eavesdrop on the conversation so as to maintain patient-doctor confidentiality.

There are four types of security goals seen in the example above. The first goal is *policy-based authentication*, where one party can be confident the other party's credentials satisfy some security policy, and moreover that multiple parties cannot *collude* to combine their credentials to satisfy a policy that none of them individually satisfies. The second goal is *confidentiality*, where the parties are ensured that no one except the other authenticated party is able to read their communications; this means only the party with whom we started communicating, not just any partner who satisfies the authentication policy, for we do not want all patients to be able to read messages sent to one patient. The third and fourth goals are interrelated: we seek *anonymity*, so an adversary cannot distinguish between two parties who have credentials satisfying the same policy, and *credential privacy*, meaning that no information is leaked about which precise combination of credentials were used to satisfy the policy.

We aim to achieve these security goals using predicate-based key exchange. The credentials held by a party can be expressed using name-value pairs assigned by one or more credential authorities. For example, a patient with medical insurance may have a private key with the credentials "Employer = Acme Widgets", "Coverage = Gold", "Expires = 2011/06/30", and "Insurer = Red Cross".

The policy used by party to evaluate credentials will be expressed as a *predicate* over credentials; the predicate may be composed of a variety of operations, such as equality and subset tests, AND, OR, and threshold gates, and comparisons. A natural example of a predicate is a *threshold access tree*. Leaves of a threshold access tree consist of boolean-valued functions such as equality tests and comparisons. Interior nodes of a threshold access tree indicate how many of the children nodes must be satisfied; for example, a node with threshold 1 having 4 children corresponds to an OR gate, while a node with threshold n having n children corresponds to an AND gate. An example threshold access tree for the case of a psychologist checking medical insurance is given in Fig. 1.

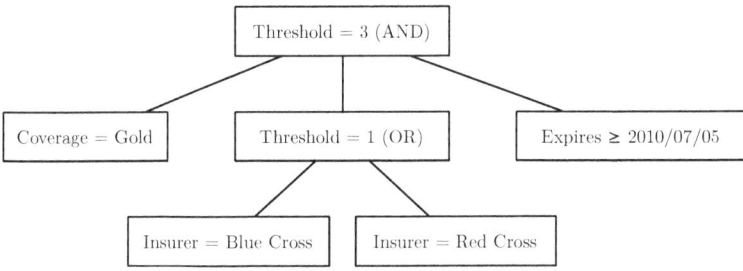

Fig. 1. A threshold access tree for checking medical insurance coverage

3   Related Work

Identity-, attribute-, and predicate-based encryption. Identity-based encryption, in which individual parties need not have public keys but only identity strings, was first proposed by Shamir [25] and has recently been the subject of much research. It was extended by Sahai and Waters [23] to *fuzzy identity-based encryption* in which parties must match at least a certain number – a threshold – of attributes. An *attribute*, usually labeled by a string, is a boolean variable: it is either present or absent. Goyal et al. [11] extended fuzzy identity-based encryption to *attribute-based encryption* supporting *boolean threshold access tree* predicates, which consist of boolean combinations of attributes using AND, OR, and threshold gates.

Boneh and Waters [5] extended credentials from boolean variable attributes to arbitrary values and supported encryption using predicates consisting of equality conjunctions, comparison conjunctions, and subset conjunctions; the support of arbitrary, not just boolean, values is what distinguishes *predicate-based* cryptography from attribute-based cryptography. Katz et al. [13] developed a technique for disjunctive predicates and inner products and Shen et al. [26] introduced the notion of predicate privacy for symmetric encryption. The improvement of predicate expressivity continues to be an active area of research.

Key exchange. The first protocol for identity-based key exchange was presented by Günther in 1989 [12] but it was not until 2003 that the first formal security model for identity-based key exchange protocols was proposed by Chen and Kudla [8]; their model was an extension of the public key authenticated key exchange security model of Blake-Wilson et al. [3] (itself based on the Bellare-Rogaway model [2]). Kudla and Paterson [18] subsequently created a generic key exchange security model to allow for modular security proofs which is also suitable for identity-based key exchange. A more refined security model for identity-based key exchange was proposed by Chen, Cheng, and Smart [7]. A common approach to designing secure key exchange protocols is using a signed-Diffie-Hellman construction (for example, [6]).

Wang, Xu, and Ban [27] and Wang, Xu, and Fu [28,29] have protocols for what they call attribute-based key agreement protocols (in the random oracle and standard models, respectively). The security proofs treat attributes as identification strings and then revert to the security model of Chen et al. [7] for identity-based authenticated key exchange. These two papers provide no mechanism for evaluating policy predicates and do not consider attribute privacy at all. As such, we consider these schemes to be merely identity-based. Ateniese et al. [1] provide a protocol for secret handshakes – key exchange where participating parties do not learn either the credentials or the predicate of the other party unless the protocol succeeds – using fuzzy attribute matching. Their protocol is secure in the fuzzy selective ID model for encryption [23].

Gorantla et al. [10] present a protocol for attribute-based group key exchange, which differs from our work in that all members of the group satisfying the predicate can compute the session key. In contrast, we allow each user to specify a predicate which the peer must satisfy, and these predicates need not be the same; moreover, in our approach the session key can only be computed by the two participants in the key-exchange protocol, not all parties that satisfy the predicate; this is related to the notion of forward-security.

Signature schemes. Attribute-based signatures were first introduced by Maji et al. [22], who provided a scheme that supported predicates containing threshold access trees, with a proof in the generic group model. Additional schemes supporting single threshold gates, in either the standard or random oracle models, have been proposed by Shahandashti and Safavi-Naini [24] and Li et al. [20], and a scheme with threshold access trees was given by Khader [16]. These schemes all achieve the goal of *attribute privacy*, in which the attributes used to satisfy a predicate are unknown the verifier. An attribute-based authentication scheme was proposed by Khader et al. [17] with some additional properties beyond signature schemes such as traceability by an authorized entity.

There are also a number of attribute-based group or ring signature schemes that provide lesser privacy guarantees, namely that the signer is anonymous among all signers possessing the same attributes [15,14,21].

4   Notation

We will use different typefaces to refer to *variables*, algorithms and oracles, and **constants**. The notation $a \leftarrow \mathsf{B}(c)$ indicates that algorithm B is run on input c and the output is assigned to a, and $a \xleftarrow{\text{R}} X$ denotes a value x being chosen uniformly at random from the set X. We use the notation $\mathsf{B}(c) \rightarrow a$ and $\mathsf{B}(c) \xrightarrow{\text{R}} a$ when defining deterministic and probabilistic algorithms, respectively, with input c and output a. We let $\lambda \in \mathbb{Z}_+$ denote a security parameter. We typically use $\mathcal{A}$ to denote the adversary; $\mathcal{A}^{\mathsf{B}(\cdot)}$ denotes $\mathcal{A}$ run with oracle access to B. Suppose $\mathbb{A}$ is a finite set of size n and $A \in \mathbb{A}$; I_A denotes the binary indicator vector of length n for the set A (assuming a canonical ordering). $\perp$ denotes a null value. We use $\mathbb{G}$ to denote a finite cyclic group, typically of order q and generated by g. A function f is *negligible* if, for sufficiently large x, $|f(x)|$ is smaller than the inverse of any polynomial in x.

Credentials and predicates. Let $\mathbb{C}$ be a finite set; we will call $\mathbb{C}$ the set of *credentials*. A *predicate* is a function $\varPhi : \mathbb{C} \rightarrow \{\mathbf{true}, \mathbf{false}\}$. We say that a credential $C \in \mathbb{C}$ *satisfies* a predicate $\varPhi$ if $\varPhi(C) = \mathbf{true}$. Let $\mathbb{P} \subseteq \{\mathbf{true}, \mathbf{false}\}^{\mathbb{C}}$ denote a set of predicates.

5   Predicate-Based Key Exchange

In this section, we define the functionality and security of a predicate-based key exchange protocol.

Definition 1 (Predicate-based key exchange protocol). *Let λ be a security parameter. A predicate-based key exchange protocol $\varPi$ consists of the following algorithms:*

- Setup$(1^\lambda) \xrightarrow{\text{R}} (MPK, MSK)$: *Returns public parameters MPK and a master secret MSK. The public parameters must uniquely define the key space $\mathbb{K}$, the set $\mathbb{C}$ of credentials used in the system and a set $\mathbb{P}$ of predicates over $\mathbb{C}$; we implicitly assume MPK is an input to all subsequent algorithms.*
- KeyGen$(MSK, C \in \mathbb{C}) \xrightarrow{\text{R}} sk$: *The credential issuing authority generates a secret key sk corresponding to the credentials $C \in \mathbb{C}$*
- Initiate$(sk, role \in \{\mathbf{init}, \mathbf{resp}\}, \varPhi \in \mathbb{P}) \xrightarrow{\text{R}} state$: *The user initiates a new session with the given role and predicate $\varPhi$.*
- Action$(sk, m, state) \xrightarrow{\text{R}} (m', state, status, k)$: *This is the core of the protocol: it takes a secret key, an incoming message (or the empty string if no messages have yet been exchanged) and the corresponding session state as input and returns the next message in the the protocol, an updated session state, the status of the session (either $\mathbf{Incomplete}$, $\mathbf{Established}$, or $\mathbf{Failed}$), and a session key $k \in \mathbb{K}$, which should be set to $\perp$ until the session is $\mathbf{Established}$.*

We have defined predicate-based key exchange in terms of non-interactive algorithms so that it is independent of any networking layer for message delivery. In particular, we deliberately do not specify how the user determines what predicates to use or to which session an incoming message belongs. For example, when using TCP over the Internet, messages may be directed to an IP address (specifying the user) and a port number (specifying the session), but a key-exchange protocol should be substrate-neutral: whether messages are delivered by carrier pigeon or pneumatic tube, the protocol actions are the same. In the case of predicate-based key exchange, these implementation issues have important implications for the security properties we desire, and any application making use of predicate-based key exchange *must* take them into consideration. We will discuss problems that arise from these networking details further in Sect. 5.3.

5.1 Correctness

A predicate-based key exchange is correct if, whenever two users who each satisfy their peer's predicate run the protocol over a benign network which faithfully delivers their messages unaltered, both parties complete the session in state **Established** and they agree on a key.

Let $role(j) = R$ if j is even and $role(j) = I$ if j is odd. Let $\mathsf{Correct}(MSK, C_I, C_R, \Phi_I, \Phi_R)$ be as follows: Set $sk_I \leftarrow \mathsf{KeyGen}(MSK, C_I)$ and $sk_R \leftarrow \mathsf{KeyGen}(MSK, C_R)$. Let $state_I \leftarrow \mathsf{Initiate}(sk_I, \mathbf{init}, \Phi_I)$, and $state_R \leftarrow \mathsf{Initiate}(sk_R, \mathbf{resp}, \Phi_R)$. Set $(m_1, state_I, status_I, k) \leftarrow \mathsf{Action}(sk_I, \perp, state_I)$. For $j = 1, \ldots, r-1$, set $(m_{j+1}, state_{role(j+1)}, status_{role(j+1)}, k_{role(j+1)}) \leftarrow \mathsf{Action}(sk_{role(j)}, m_j, state_{role(j)})$. If $status_I = \mathbf{Established} = status_R$ and $k_I = k_R$, then return **true**, otherwise return **false**.

Definition 2 (Correctness). *A predicate-based key-exchange protocol is said to be* correct *if, for* $(MPK, MSK) \leftarrow \mathsf{Setup}(1^k)$, *for all* $\Phi_I, \Phi_R \in \mathbb{P}$ *and for all* $C_I, C_R \in \mathbb{C}$ *such that* $\Phi_R(C_I) = \mathbf{true} = \Phi_I(C_R)$,

$$\Pr(\mathsf{Correct}(MSK, C_I, C_R, \Phi_I, \Phi_R) = \mathbf{true}) = 1 \ .$$

5.2 Security Model

We require a predicate-based key exchange protocol to satisfy two security properties: session-key security and credential privacy. Our security model combines aspects of the Bellare-Rogaway [2] model for key exchange, the Maji et al. model for attribute-based signature schemes [22], and aspects of predicate-based encryption from Boneh and Waters [5]. We define these properties using two security experiments, each played by an adversary against a challenger.

In both security experiments, the challenger maintains a list of users $U_1, \ldots, U_N$, which is not fixed, but is under the control of the adversary. Each user U_u has credentials C_u and a secret key sk_u, and the challenger maintains a numbered list of sessions, $s_{u,1}, \ldots, s_{u,n_u}$, with the following associated variables:

- $m_{u,\ell,1}, \ldots, m_{u,\ell,i}$: The protocol messages exchanged in session $s_{u,\ell}$.
- $state_{u,\ell}$: The private session state information.

- $status_{u,\ell} \in \{\mathbf{Established}, \mathbf{Incomplete}, \mathbf{Failed}\}$: The status of the session.
- $k_{u,\ell} \in \mathbb{K}$: The session key.
- $\Phi_{u,\ell} \in \mathbb{P}$: The predicate which the peer of the session must satisfy.
- $\Phi'_{u,\ell} \in \mathbb{P}$: The predicate which the owner of the session must satisfy; in our example construction, this value is sent to the peer as part of the first protocol message, but it could in principle be specified by some other means.
- $role_{u,\ell} \in \{\mathbf{init}, \mathbf{resp}\}$: The role (initiator or responder) played by the user U_u in session ℓ.

We now present the queries available to the adversary in both games:

- Create($C \in \mathbb{C}$): The challenger increments N, the number of users, sets $C_N \leftarrow C$, computes $sk_N \leftarrow \mathsf{KeyGen}(MSK, C_N)$ and returns N.
- Activate($u, role, \Phi \in \mathbb{P}$): The challenger increments n_u, sets $state_{u,n_u} \leftarrow \mathsf{Initiate}(sk_u, role, \Phi)$, and returns n_u.
- Send($u, \ell, m_{u,\ell,i}$): The challenger sets $(m_{u,\ell,i+1}, state_{u,\ell}, status_{u,\ell}, k_{u,\ell}) \leftarrow \mathsf{Action}(sk_u, m_{u,\ell}, state_{u,\ell})$ and returns $(m_{u,\ell,i+1}, status_{u,\ell})$. If $role_{u,\ell} = \mathbf{init}$ and $i = 0$, then $m_{u,\ell,i}$ must be $\bot$.
- SKReveal(u, l): Returns $k_{u,\ell}$.
- Corrupt(u): Returns sk_u.

Session Key Security. The definition of session key security is based on the idea that an adversary should not be able to distinguish the session key of a sufficiently uncompromised session from a random string, except with negligible probability. First, we adapt the Bellare-Rogaway definition of a matching conversation [2] to our setting as follows.

Definition 3 (Matching session). *A session $s_{u',\ell'}$ is a matching session of a session $s_{u,\ell}$ if $\Phi_{u,\ell} = \Phi'_{u',\ell'}$, $\Phi'_{u,\ell} = \Phi_{u',\ell'}$, and any of the following rules hold.*

- *For protocols where r, the number of rounds, is odd:*
  - *$role_{u,\ell} = \mathbf{init}$, $role_{u',\ell'} = \mathbf{resp}$, and $(m_{u,\ell,1}, \ldots, m_{u,\ell,r-1}) = (m_{u',\ell',1}, \ldots, m_{u',\ell',r-1})$;*
  - *$role_{u,\ell} = \mathbf{resp}$, $role_{u',\ell'} = \mathbf{init}$, and $(m_{u,\ell,1}, \ldots, m_{u,\ell,r}) = (m_{u',\ell',1}, \ldots, m_{u',\ell',r})$.*
- *For protocols where r is even:*
  - *$role_{u,\ell} = \mathbf{init}$, $role_{u',\ell'} = \mathbf{resp}$, and $(m_{u,\ell,1}, \ldots, m_{u,\ell,r}) = (m_{u',\ell',1}, \ldots, m_{u',\ell',r})$;*
  - *$role_{u,\ell} = \mathbf{resp}$, $role_{u',\ell'} = \mathbf{init}$, and $(m_{u,\ell,1}, \ldots, m_{u,\ell,r-1}) = (m_{u',\ell',1}, \ldots, m_{u',\ell',r-1})$.*

This captures the idea that the owner and the peer in the matching session must satisfy each other's predicates and agree on all of the messages exchanged, except perhaps if the owner of the session $s_{u,\ell}$ sent the final message. In this case the owner of the session completes the protocol without knowing if the final message was delivered, or if a different message was delivered instead, so we do not require that the final messages are equal in this case. Note that the relation "is a matching session of" is *not* symmetric!

Definition 4 (Session key security). *Let λ be a security parameter and let $\mathcal{A}$ be a polynomial-time (in λ) probabilistic algorithm. A predicate-based key exchange protocol Π is* session-key-secure *if*

$$\mathsf{Adv}_{\Pi,\mathcal{A}}^{\mathrm{PB\text{-}SK}}(\lambda) := \left| \Pr\left(\mathsf{Expt}_{\Pi,\mathcal{A}}^{\mathrm{PB\text{-}SK}}(\lambda) = \mathbf{true} \right) - \frac{1}{2} \right|$$

is negligible, where $\mathsf{Expt}_{\Pi,\mathcal{A}}^{\mathrm{PB\text{-}SK}}(\lambda)$ is the following algorithm:

1. *Set $(MPK, MSK) \leftarrow \mathsf{Setup}(1^\lambda)$.*
2. *Let $\mathsf{Test}(u, \ell)$ be the following algorithm. Choose a bit $b \overset{\mathrm{R}}{\leftarrow} \{0,1\}$ at random. If $b = 0$, then return $k_{u,\ell}$, otherwise return $k \overset{\mathrm{R}}{\leftarrow} \mathbb{K}$.*
3. *Set $b' \leftarrow \mathcal{A}(MPK)$, where $\mathcal{A}$ has oracle access to Create, $\mathsf{Activate}$, Send, $\mathsf{SKReveal}$, $\mathsf{Corrupt}$, and Test. $\mathcal{A}$ is restricted as follows:*
   - *$\mathcal{A}$ may make a single query to the Test oracle; let u, ℓ be the arguments to that query.*
   - *$\mathcal{A}$ must not have made any query of the form $\mathsf{Corrupt}(u')$ for any u' such that $\Phi_{u,\ell}(C_{u'}) = \mathbf{true}$ prior to the Test query.*
   - *When the Test query is made, it must be that $status_{u,\ell} = \mathbf{Established}$.*
   - *$\mathcal{A}$ may not query $\mathsf{SKReveal}(u, \ell)$ or $\mathsf{SKReveal}(u', \ell')$ for any (u', ℓ') such that $s_{u',\ell'}$ is a matching session of $s_{u,\ell}$, even after the Test query is made.*
4. *If $b' = b$, then return $\mathbf{true}$, otherwise return $\mathbf{false}$.*

Collusion resistance. This definition of session key security also implies collusion resistance, since the adversary may perform $\mathsf{Corrupt}$ queries for multiple users with credentials that collectively, but not individually, satisfy the predicate.

Credential Privacy. For the credential privacy experiment, the adversary should not be able to distinguish between two users whose credentials satisfy the same predicate, even if they have different credentials.

Definition 5 (Credential privacy). *Let λ be a security parameter and let $\mathcal{A}$ be a polynomial-time (in λ) probabilistic algorithm. A predicate-based key exchange protocol Π is* credential-private *if*

$$\mathsf{Adv}_{\Pi,\mathcal{A}}^{\mathrm{PB\text{-}Priv}}(\lambda) := \left| \Pr\left(\mathsf{Expt}_{\Pi,\mathcal{A}}^{\mathrm{PB\text{-}Priv}}(\lambda) = \mathbf{true} \right) - \frac{1}{2} \right|$$

is negligible, where $\mathsf{Expt}_{\Pi,\mathcal{A}}^{\mathrm{PB\text{-}Priv}}(\lambda)$ is the following algorithm:

1. *Set $(MPK, MSK) \leftarrow \mathsf{Setup}(1^\lambda)$.*
2. *Let $\mathsf{TestActivate}(u_0, u_1, role, \Phi \in \mathbb{P})$ be the following algorithm. Choose a bit $b \overset{\mathrm{R}}{\leftarrow} \{0,1\}$ at random. Set $state^* \leftarrow \mathsf{Initiate}(sk_{u_b}, role, \Phi)$ and return $\bot$.*
3. *Let $\mathsf{Send}^*(m_i^*)$ be the following algorithm. Set $(m_{i+1}^*, state^*, status^*, k^*) \leftarrow \mathsf{Action}(sk_{u_b}, m_i^*, state^*)$ and return m_{i+1}^*.*
4. *Set $b' \leftarrow \mathcal{A}(MPK)$, where $\mathcal{A}$ has oracle access to Create, $\mathsf{Activate}$, Send, Send^*, $\mathsf{SKReveal}$, $\mathsf{Corrupt}$, and $\mathsf{TestActivate}$. $\mathcal{A}$ is restricted as follows:*

– $\mathcal{A}$ *may make a single query to the* TestActivate *oracle.*
– *The predicate Φ'^* which C_{u_b} has to satisfy (which is determined by the* Send$^*(\cdot)$ *queries made by the adversary) must be chosen so that $\Phi'^*(C_{u_0})$* $= \Phi'^*(C_{u_1})$. *(If this were not the case then the adversary could trivially distinguish U_{u_0} from U_{u_1}.)*
5. *If $b' = b$, then return* **true**, *otherwise return* **false**.

Credential privacy captures the notion of anonymity: the adversary cannot distinguish between two users satisfying the same predicate. It also ensures that the adversary cannot tell whether two sessions with the same predicate are owned by the same user; we call this property *unlinkability*. To see why this holds, suppose that an adversary executes a session with U_{u_0}, and the test session with U_{u_b} using the same predicate. If the adversary *could* tell whether those two sessions are owned by the same user, then it can discover the identity of U_{u_b} and win the credential privacy experiment.

5.3   Implementation Issues

Credential privacy is an essential feature of any predicate-based key exchange protocol. If an application does not need credential privacy, then standard public key or identity-based systems may be used in combination with a credential-issuing authority which simply issues a certificate on the users public key declaring that they hold a given credential. This shows that there is simply no need for predicate-based key exchange unless credential-privacy is desired.

Our definition of credential privacy ensures that the contents of the protocol messages exchanged reveal no information about either party's credentials, except whether they satisfy their peer's chosen predicate. Unlike predicate-based encryption or signatures, predicate-based key exchange faces an additional challenge: users need to be identified by some means in order to deliver messages. It seems unavoidable that this should leak some information about a user's credentials, but we will discuss some approaches that may be fruitful.

Suppose that a predicate-based key exchange protocol is used on an IP network, with each user having a fixed IP address. An adversary may initiate multiple sessions with the same user using different predicates to exhaustively search the credential space. A user initiating a session may mitigate this problem if she is able to obtain a new IP address for each session, for example by using tunnelling, or an anonymising service such as Tor [9]. Unfortunately, a user acting as a responder cannot use this solution, since the initiator must know an address to initiate a session. Depending on the application, it may be that only the initiator needs credential privacy. In the example from Sect. 2, the patient desires to remain anonymous when discussing their mental-health problems, but it seems unlikely that the psychologist has the same requirement. However, a society of secretive psychologists acting together could preserve some degree of anonymity by operating a trusted proxy which knows their individual credentials, and could choose a psychologist who satisfies a given predicate at random from among the society.

6   Predicate-Based Signature Schemes

Our definition of predicate-based signature schemes is a natural extension from the definition of attribute-based signature schemes [22].

Definition 6 (Predicate-based signature scheme). *Let λ be a security parameter. A predicate-based signature scheme $\mathcal{S}$ is a tuple consisting of the following polynomial-time (in λ) algorithms:*

- Setup$(1^\lambda) \xrightarrow{\text{R}} (mpk, msk)$: *The credential authority obtains a master private key msk and public parameters mpk. The public parameters must uniquely define the set $\mathbb{C}$ of credentials and a set $\mathbb{P}$ of predicates over $\mathbb{C}$; we assume mpk is an implicit input to all subsequent algorithms.*
- KeyGen$(msk, C \in \mathbb{C}) \xrightarrow{\text{R}} sk$: *The authority generates a signing key sk for credentials C.*
- Sign$(sk, m, \Phi \in \mathbb{P}) \xrightarrow{\text{R}} \sigma$: *The signer generates a signature σ for a message m and predicate Φ, provided sk was generated with C such that $\Phi(C) = \textbf{true}$.*
- Verify$(m, \Phi \in \mathbb{P}, \sigma) \rightarrow \{\textbf{true}, \textbf{false}\}$: *The verifier checks if σ is a valid signature on m for predicate Φ.*

Definition 7 (Correctness). *A predicate-based signature scheme $\mathcal{S}$ is correct if, for $(mpk, msk) \leftarrow$ Setup(1^λ), all messages m, all credentials $C \in \mathbb{C}$, all signing keys $sk \leftarrow$ KeyGen(msk, C), and all predicates $\Phi \in \mathbb{P}$ such that $\Phi(C) = \textbf{true}$, we have $\Pr\left(\text{Verify}\left(m, \Phi, \text{Sign}\left(sk, m, \Phi\right)\right) = \textbf{true}\right) = 1$.*

Definition 8 (Perfect privacy). *A predicate-based signature scheme $\mathcal{S}$ is perfectly private if, for $(mpk, msk) \leftarrow$ Setup(1^λ), all messages m, all credentials $C_1, C_2 \in \mathbb{C}$, all signing keys $sk_1 \leftarrow$ KeyGen(msk, C_1), $sk_2 \leftarrow$ KeyGen(msk, C_2), and all predicates $\Phi \in \mathbb{P}$ such that $\Phi(C_1) = \Phi(C_2) = \textbf{true}$, the distributions Sign$(sk_1, m, \Phi)$ and Sign(sk_2, m, Φ) are equal.*

A perfectly private predicate-based signature scheme does not leak any information about which credentials or secret keys were used in signing.

Definition 9 (Unforgeability). *Let λ be a security parameter and let $\mathcal{A}$ be a polynomial-time (in λ) probabilistic algorithm. A perfectly private predicate-based signature scheme $\mathcal{S}$ is unforgeable if*

$$\text{Adv}_{\mathcal{S},\mathcal{A}}^{\text{PB-Forge}}(\lambda) := \Pr\left(\text{Expt}_{\mathcal{S},\mathcal{A}}^{\text{PB-Forge}}(\lambda) = \textbf{true}\right)$$

is negligible, where $\text{Expt}_{\mathcal{S},\mathcal{A}}^{\text{PB-Forge}}(\lambda)$ is the following algorithm:

1. *Set $(mpk, msk) \leftarrow$ Setup(1^λ).*
2. *Let AltSign$(msk, m, C \in \mathbb{C}, \Phi \in \mathbb{P})$ be an algorithm that, provided $\Phi(C) = \textbf{true}$, sets $sk \leftarrow$ KeyGen(msk, C), and returns Sign(sk, m, Φ).*
3. *Set $(m, \Phi, \sigma) \leftarrow \mathcal{A}^{\text{KeyGen}(msk,\cdot),\text{AltSign}(msk,\cdot,\cdot)}(mpk)$.*
4. *If Verify$(m, \Phi, \sigma) = \textbf{true}$, $\mathcal{B}$ never queried AltSign$(m, \cdot, \Phi)$, and $\mathcal{B}$ never queried KeyGen(C) for any $C \in \mathbb{C}$ such that $\Phi(C) = \textbf{true}$, then return $\textbf{true}$, otherwise return $\textbf{false}$.*

The security experiment for unforgeability is slightly different than is typical for signature schemes, because the signing oracle generates a new key for each signature rather than using an existing key. However, for a predicate-based signature scheme with perfect privacy, the signature depends on the predicate used, but not the specific credentials (or secret key), so the definition is appropriate.

An example instantiation. Attribute-based signature schemes are a special case of predicate-based signature schemes. We can rewrite the notation of attribute-based signature schemes in terms of the more expressive notation of predicate-based schemes, as indicated in Fig. 2. Thus, all attribute-based schemes are predicate-based schemes, but in general predicate-based schemes are more expressive than attribute-based schemes. It follows that existing secure attribute-based schemes [22,24,17] are also secure predicate-based signature schemes.

	Attribute-based [22]	Predicate-based (Sect. 4)
Credential universe	$\mathbb{A}$, $|\mathbb{A}| = n$	$\mathbb{C} = \{0,1\}^{|\mathbb{A}|}$
Credentials	$A \subseteq \mathbb{A}$	$C \in \mathbb{C}$, $C = I_A$
Predicate	$\Upsilon : \{0,1\}^n \to \{\mathbf{true}, \mathbf{false}\}$	$\Phi : \mathbb{C} \to \{\mathbf{true}, \mathbf{false}\}$
	A satisfies Υ iff $\Upsilon(I_A) = \mathbf{true}$	C satisfies Φ iff $\Phi(C) = \mathbf{true}$

Fig. 2. Representation of attribute-based notation in predicate-based notation

7 A Signed Diffie-Hellman Construction

We present a simple signed-Diffie-Hellman protocol using a secure predicate-based signature scheme and a group in which the Decisional Diffie-Hellman (DDH) problem is hard.

Definition 10 (Decisional Diffie-Hellman problem [4]). *Let $(\mathbb{G}_\lambda)_{\lambda \in \mathbb{N}}$ be a family of multiplicatively written cyclic groups of prime order q_λ, indexed by a security parameter λ. Fix a security parameter λ; let g be a generator of $\mathbb{G}_\lambda$ and let $x, y, z \xleftarrow{\mathrm{R}} \mathbb{Z}_{q_\lambda}$. For any probabilistic polynomial-time algorithm $\mathcal{A}$, we define*

$$\mathsf{Adv}_{\mathbb{G}_\lambda, \mathcal{A}}^{\mathrm{DDH}}(\lambda) = |\Pr\left(\mathcal{A}(g, g^x, g^y, g^z) = 1\right) - \Pr\left(\mathcal{A}(g, g^x, g^y, g^{xy}) = 1\right)| \ .$$

The DDH problem is hard if, for any probabilistic polynomial-time algorithm $\mathcal{A}$, $\mathsf{Adv}_{\mathbb{G}_\lambda, \mathcal{A}}^{\mathrm{DDH}}(\lambda)$ is negligible.

7.1 Protocol Definition

Let $\mathcal{S} = (\mathsf{Setup}_\mathcal{S}, \mathsf{KeyGen}_\mathcal{S}, \mathsf{Sign}, \mathsf{Verify})$ be a predicate-based signature scheme. We define the protocol $\Pi_{\mathcal{S}, \mathbb{G}}$ as the following tuple of algorithms:

- $\mathsf{Setup}(1^\lambda)$: Set $(mpk, msk) \leftarrow \mathsf{Setup}_\mathcal{S}(1^\lambda)$; recall that mpk defines a set $\mathbb{C}$ of credentials and a set $\mathbb{P}$ of predicates over $\mathbb{C}$. Let $\mathbb{G} = \mathbb{G}_\lambda$ be a finite cyclic group of order $q = q_\lambda$ generated by g. Set $MPK \leftarrow (mpk, \mathbb{G}, g, q)$ and $MSK \leftarrow msk$. Return (MPK, MSK).

- KeyGen($MSK, C \in \mathbb{C}$): Return KeyGen$_\mathcal{S}(msk, C)$.
- Initiate($sk, \mathbf{init}, \Phi_I$): Return $state \leftarrow \Phi_I$.
- Initiate($sk, \mathbf{resp}, \Phi_R$): Return $state \leftarrow \Phi_R$.
- Action($sk, m, state$): For clarity, we write the protocol action as four separate algorithms which may be combined in the natural way. We also present the protocol diagrammatically in Fig. 3.
  - InitiatorAction1($sk, \perp, \Phi_I$): Set $x \xleftarrow{\text{R}} \mathbb{Z}_q$ and $X \leftarrow g^x$. Set $m' \leftarrow (X, \Phi_I)$ and $state' \leftarrow (\Phi_I, x)$. Return $(m', state', \mathbf{Incomplete}, \perp)$.
  - ResponderAction1($sk, (X, \Phi_I), \Phi_R$): If $\Phi_I(C_R) = \mathbf{false}$, then return $(\perp, \perp, \mathbf{Failed}, \perp)$. Otherwise, set $y \xleftarrow{\text{R}} \mathbb{Z}_q$ and $Y \leftarrow g^y$. Set $\sigma_R \leftarrow$ Sign($sk, (\mathbf{resp}, X, \Phi_I, Y, \Phi_R), \Phi_I$). Set $m' \leftarrow (Y, \Phi_R, \sigma_R)$ and $state' \leftarrow (X, \Phi_I, Y, y, \Phi_R, \sigma_R)$. Return $(m', state', \mathbf{Incomplete}, \perp)$.
  - InitiatorAction2($sk, (Y, \Phi_R, \sigma_R), (\Phi_I, x)$): If Verify($(\mathbf{resp}, X, \Phi_I, Y, \Phi_R)$, Φ_I, σ_R) $= \mathbf{false}$ or $\Phi_R(C_I) = \mathbf{false}$, then return $(\perp, \perp, \mathbf{Failed}, \perp)$. Set $\sigma_I \leftarrow$ Sign($sk, (\mathbf{init}, X, \Phi_I, Y, \Phi_R, \sigma_R), \Phi_R$). Set $k \leftarrow Y^x$. Return $(\sigma_I, \perp, \mathbf{Established}, k)$.
  - ResponderAction2($sk, \sigma_I, (X, \Phi_I, Y, y, \Phi_R, \sigma_R)$): If Verify($(\mathbf{init}, X, \Phi_I, Y, \Phi_R, \sigma_R), \Phi_R, \sigma_I$) $\neq \mathbf{true}$, then return $(\perp, \perp, \mathbf{Failed}, \perp)$. Set $k \leftarrow X^y$. Return $(\perp, \perp, \mathbf{Established}, k)$.

It is easy to see that the $\Pi_{\mathcal{S},\mathbb{G}}$ is correct when the signature scheme is correct.

7.2   Credential Privacy

Theorem 1. *If $\mathcal{S}$ is a perfectly-credential-private signature scheme, then $\Pi_{\mathcal{S},\mathbb{G}}$ is credential-private.*

Proof (sketch). Consider the test session in the credential privacy experiment for the predicate-based key exchange protocol. If u_b does not satisfy the chosen predicate Φ'^*, specified by the adversary – that is, if $\Phi'^*(C_{u_b}) = \mathbf{false}$ – then the session terminates with status **Failed**, by definition of the protocol. However, the choice of Φ'^* is restricted so that $\Phi'^*(C_{u_0}) = \Phi'^*(C_{u_1})$, so in this case the responses of the challenger are independent of the bit b. Similarly, if $\Phi'^*(C_{u_b}) = \mathbf{true}$, the distribution of the signature returned to the adversary does not depend on the bit b by the perfect privacy of $\mathcal{S}$. Since the bit b is not used in answering any other queries, we now see that the responses to the adversary's queries are all independent of b, so $\Pr(b' = b) = \frac{1}{2}$ and $\mathsf{Adv}^{\text{PB-Priv}}_{\Pi_{\mathcal{S},\mathbb{G}},\mathcal{A}}(\lambda) = 0$. $\square$

7.3   Session Key Security

Theorem 2. *If $\mathcal{S}$ is an unforgeable signature scheme and the DDH problem is hard in $\mathbb{G}$, then $\Pi_{\mathcal{S},\mathbb{G}}$ is session-key secure.*

Proof. Let $\mathcal{A}$ be an adversary against the session key security of $\Pi_{\mathcal{S},\mathbb{G}}$ and consider the experiment $\mathsf{Expt}^{\text{PB-SK}}_{\Pi_{\mathcal{S},\mathbb{G}}}(\lambda)$. Let u^*, ℓ^* be the test session. Define M to be the event that a matching session $s_{u',\ell'}$ of s_{u^*,ℓ^*} exists.

$\Pi_{\mathcal{S},\mathbb{G}}$ – Protocol flow	
Initiator	**Responder**
secret key sk_I	secret key sk_R
responder predicate Φ_I	initiator predicate Φ_R

InitiatorAction1		
$x \xleftarrow{\text{R}} \mathbb{Z}_q, X \leftarrow g^x$	$\xrightarrow{\quad X, \Phi_I \quad}$	ResponderAction1
		$y \xleftarrow{\text{R}} \mathbb{Z}_q, Y \leftarrow g^y$
		$\sigma_R \leftarrow \mathsf{Sign}(sk_R, (\mathbf{resp}, X,$
InitiatorAction2	$\xleftarrow{\quad Y, \Phi_R, \sigma_R \quad}$	$\Phi_I, Y, \Phi_R), \Phi_I)$
If $\neg\mathsf{Verify}((\mathbf{resp}, X, \Phi_I, Y,$		
$\Phi_R), \Phi_I, \sigma_R)$ then		
$\qquad status \leftarrow \mathbf{Failed}$		
$\qquad$ Abort		
$\sigma_I \leftarrow \mathsf{Sign}(sk_I, (\mathbf{init}, X, \Phi_I, Y,$		
$\Phi_R, \sigma_R), \Phi_R)$		
$k \leftarrow Y^x$		
$status \leftarrow \mathbf{Established}$	$\xrightarrow{\quad \sigma_I \quad}$	ResponderAction2
		If $\neg\mathsf{Verify}((\mathbf{init}, X, \Phi_I, Y,$
		$\Phi_R, \sigma_R), \Phi_R, \sigma_I)$ then
		$\qquad status \leftarrow \mathbf{Failed}$
		$\qquad$ Abort
		$k \leftarrow X^y$
		$status \leftarrow \mathbf{Established}$

Fig. 3. Protocol flow of $\Pi_{\mathcal{S},\mathbb{G}}$

Case 1: No session matching s_{u^*,ℓ^*} exists (event $\neg M$). We construct an adversary $\mathcal{B}$ against the unforgability of $\mathcal{S}$ as follows. $\mathcal{B}$ runs $\mathcal{A}(mpk)$ and simulates the challenger's responses according to the definition of the $\mathsf{Expt}^{\text{PB-SK}}_{\Pi_{\mathcal{S},\mathbb{G}},\lambda}$, with the following modifications: whenever the challenger would compute $\mathsf{Sign}(sk_u, m, \Phi)$ (while responding to a Send query), $\mathcal{B}$ instead queries the $\mathsf{AltSign}$ oracle on input (msk, m, C_u, Φ). Whenever $\mathcal{A}$ makes a $\mathsf{Corrupt}(u)$ query, $\mathcal{B}$ responds by querying $\mathsf{KeyGen}_{\mathcal{S}}(C_u)$ and returning the result.

Now consider the test session s_{u^*,ℓ^*}. By the definition of $\Pi_{\mathcal{S},\mathbb{G}}$, $m_{u^*,\ell^*,1} = (X, \Phi_{u^*,\ell^*})$ for some $X \in \mathbb{G}$, $m_{u^*,\ell^*,2} = (Y, \Phi'_{u^*,\ell^*}, \sigma_R)$ for some $Y \in \mathbb{G}$, and $m_{u^*,\ell^*,3} = \sigma_I$. When $\mathcal{A}$ terminates, if $role_{u^*,\ell^*} = \mathbf{init}$, then $\mathcal{B}$ chooses $m^* \leftarrow (\mathbf{resp}, X, \Phi_{u^*,\ell^*}, Y, \Phi'_{u^*,\ell^*})$ as the message to forge a signature on and returns $(m^*, \Phi_{u^*,\ell^*}, \sigma_R)$ as the forgery. If $role_{u^*,\ell^*} = \mathbf{resp}$, $\mathcal{B}$ chooses $m^* \leftarrow (\mathbf{init}, X, \Phi'_{u^*,\ell^*}, Y, \Phi_{u^*,\ell^*}, \sigma_R)$ and returns $(m^*, \Phi_{u^*,\ell^*}, \sigma_I)$ as the forgery.

We must now show that if the test session has no matching session, then $\mathcal{B}$ satisfies the requirements of Definition 9, namely that $\mathsf{Verify}(m, \Phi, \sigma) = \mathbf{true}$, $\mathcal{B}$ never queried $\mathsf{AltSign}(msk, m, \cdot, \Phi)$ and $\mathcal{B}$ never queried $\mathsf{KeyGen}_{\mathcal{S}}(C)$ for any credential C such that $\Phi(C) = \mathbf{true}$.

Since the test session must be an **Established** session, it follows that $\mathsf{Verify}(m, \Phi_{u^*,\ell^*}, \sigma_R) = \mathbf{true}$. Because of the constraints on $\mathcal{A}$ concerning the test session, it follows that $\mathcal{A}$ never queried $\mathsf{Corrupt}(u)$ for any u satisfying $\Phi_{u^*,\ell^*}(C_u) =$

true, which implies that $\mathcal{B}$ never queried $\mathsf{KeyGen}_{\mathcal{S}}(C)$ for any credential C such that $\Phi_{u^*,\ell^*}(C) = \textbf{true}$.

Finally, suppose $\mathcal{A}$ made a query of the form $\mathsf{Send}(u', \ell', m_{u',\ell',i})$ which caused $\mathcal{B}$ to query $\mathsf{AltSign}(m^*, C, \Phi_{u^*,\ell^*})$, where m^* is the forged message defined above. If $role_{u^*,\ell^*} = \textbf{init}$, then $m^* = (\textbf{resp}, X, \Phi_{u^*,\ell^*}, Y, \Phi'_{u^*,\ell^*})$, and the only circumstances where $\mathcal{B}$ could query $\mathsf{AltSign}(msk, m^*, C, \Phi_{u^*,\ell^*})$ are if $\Phi_{u',\ell'} = \Phi'_{u^*,\ell^*}$, $\Phi'_{u',\ell'} = \Phi_{u^*,\ell^*}$, $m_{u',\ell',1} = (X, \Phi'_{u',\ell'})$, and $m_{u',\ell',2} = (Y, \Phi_{u',\ell'}, \sigma_R)$: in other words, when $s_{u',\ell'}$ is a matching session of s_{u^*,ℓ^*}, contradicting our original assumption. Conversely, if $role_{u^*,\ell^*} = \textbf{resp}$, then $m^* = (\textbf{init}, X, \Phi'_{u^*,\ell^*}, Y, \Phi_{u^*,\ell^*}, \sigma_R)$, and if $\mathcal{B}$ queried $\mathsf{AltSign}(m^*, C, \Phi_{u^*,\ell^*})$ then $\Phi_{u',\ell'} = \Phi'_{u^*,\ell^*}$, $\Phi'_{u',\ell'} = \Phi_{u^*,\ell^*}$, $m_{u',\ell',1} = (X, \Phi_{u',\ell'})$, $m_{u',\ell',2} = (Y, \Phi'_{u',\ell'}, \sigma_R)$ and $m_{u',\ell',3} = \sigma_I$. Once again this implies that $s_{u',\ell'}$ is a matching session of s_{u^*,ℓ^*} contradicting our original assumption.

Therefore $\mathcal{B}$ wins the forgery game whenever $\mathcal{A}$ selects a test session with no matching session, so $\Pr(\neg M) = \mathsf{Adv}_{\mathcal{S},\mathcal{B}}^{\text{PB-Forge}}(\lambda)$, which is negligible.

Case 2: There is a session $s_{u',\ell'}$ which matches s_{u^*,ℓ^*} (event M). Since s_{u^*,ℓ^*} is required to be **Established**, and $s_{u',\ell'}$ matches s_{u^*,ℓ^*} by assumption, we see that $m_{u^*,\ell^*,1} = (X, \Phi_{u^*,\ell^*}) = (X, \Phi'_{u',\ell'}) = m_{u',\ell',1}$, $m_{u^*,\ell^*,2} = (Y, \Phi_{u',\ell'}, \sigma_R) = (Y, \Phi'_{u^*,\ell^*}, \sigma_R) = m_{u',\ell',2}$.

In particular, this shows that both X and Y were chosen by the challenger in response to the corresponding Send queries. This allows us to construct a DDH adversary $\mathcal{C}$ as follows. Let $q_{\mathsf{Activate}}(\lambda)$ be an upper bound on the number of $\mathsf{Activate}$ queries that an adversary in the PB-SK experiment makes. The adversary $\mathcal{C}$ takes a DDH tuple (g, X^*, Y^*, Z^*) as input and chooses $i, j \xleftarrow{\text{R}} \{1, \ldots, q_{\mathsf{Activate}}(\lambda)\}$. It then generates a key pair $(mpk, msk) \leftarrow \mathsf{KeyGen}_{\mathcal{S}}(1^\lambda)$ and runs $\mathcal{A}(msk)$. $\mathcal{C}$ responds to all of $\mathcal{A}$'s queries according to the rules of $\mathsf{Expt}_{\Pi_{\mathcal{S},\mathbb{G}},\mathcal{A}}^{\text{PB-SK}}(\lambda)$, except that it inserts the Diffie-Hellman values X^* and Y^* into the ith and j^{th} sessions instead of generating a random group element. We refer to these session as s_i and s_j. If $\mathcal{A}$ queries $\mathsf{SKReveal}(s_i)$ or $\mathsf{SKReveal}(s_j)$, $\mathcal{C}$ aborts. When $\mathcal{A}$ queries $\mathsf{Test}(s_{u^*,\ell^*})$, $\mathcal{C}$ aborts *unless* $s_{u^*,\ell^*} = s_i$ and $s_{u',\ell'} = s_j$. Assuming it does not abort, $\mathcal{C}$ sets $k \leftarrow Z^*$. When $\mathcal{A}$ terminates and returns a guess b', $\mathcal{C}$ returns b' as its guess for the DDH problem.

Since the test session s_{u^*,ℓ^*} and its matching session $s_{u',\ell'}$ are chosen by the adversary $\mathcal{A}$ independently of the choice of i and j, $\Pr(\mathcal{C}$ does not abort$) \geq \frac{1}{q_{\mathsf{Activate}}^2}$. Whenever it does not abort, $\mathcal{C}$ wins the DDH game if and only if $\mathcal{A}$ wins the PB-SK experiment.

Combining results from Case 1 and Case 2, we see that

$$\mathsf{Adv}_{\Pi_{\mathcal{S},\mathbb{G}},\mathcal{A}}^{\text{PB-SK}}(\lambda) = \Pr(b' = b) = \Pr(b' = b|M)\Pr(M) + \Pr(b' = b|\neg M)\Pr(\neg M)$$

$$\leq \frac{1}{q_{\mathsf{Activate}}^2(\lambda)}\mathsf{Adv}_{\mathbb{G},\mathcal{C}}^{\text{DDH}}(\lambda)\Pr(M) + \Pr(b' = b|\neg M)\mathsf{Adv}_{\mathcal{S},\mathcal{B}}^{\text{PB-Forge}}(\lambda)$$

$$\leq \frac{1}{q_{\mathsf{Activate}}^2(\lambda)}\mathsf{Adv}_{\mathbb{G},\mathcal{C}}^{\text{DDH}}(\lambda) + \mathsf{Adv}_{\mathcal{S},\mathcal{B}}^{\text{PB-Forge}}(\lambda)$$

which is negligible as required.   □

8   Conclusions

We have introduced the notion of predicate-based key exchange, given a security model, and presented a secure protocol satisfying the security definitions. Our security model for predicate-based key exchange can also be specialized to attribute-based key exchange, a cryptographic task for which there was previously no rigourous security definition.

Our security model incorporates two notions of security: session key security and credential privacy. We have argued that credential privacy is an essential property of predicate-based key exchange; without it, we might as well use certificates to link public keys and a list of credentials. However, achieving credential privacy requires careful consideration of the networking layer over which the protocol runs, as the addressing information of messages – the packet headers – may leak information. In practice, then, a secure deployment of predicate-based key exchange may rely on an anonymising network such as Tor.

The protocol we have presented is a generic protocol that combines any secure predicate-based signature scheme with a Diffie-Hellman construction, providing efficiency and simplicity.

Future work. The major security models for public-key-based authenticated key exchange have an additional query to allow revealing some of the session variables: either a SessionStateReveal query [6], which reveals the session state variables stored during the protocol, or an EphemeralKeyReveal query [19] which reveals all randomness used during the run of a protocol. Adding either of these queries to our security model would be a natural way to improve its security guarantees. Our generic protocol construction may still be secure with a SessionStateReveal query, but cannot be secure with an EphemeralKeyReveal query unless the underlying signature scheme is secure against revealing the randomness used in signing. No existing schemes have been shown to have this property, at least in the case of attribute-based or predicate-based signatures.

Our definition of credential privacy for predicate-based key exchange is computational in nature, but our proof for the generic construction relies on the perfect privacy of the underlying signature scheme, as defined by Maji et al. [22]. However, it seems plausible that a suitably defined computational notion of credential privacy would suffice. It may also be possible to give alternative constructions based on ciphertext-policy predicate-based encryption schemes, though as yet only ciphertext-policy attribute-based encryption schemes exist.

Finally, predicate-based key exchange could be extended to support multiple, independent, mutually distrusting, potentially corrupt, credential authorities, as in multiple attribute authorities for attribute-based signature schemes [22, §4].

Acknowledgements

The authors are grateful for discussions with Juanma González Nieto. J.B. was supported by Australian Research Council (ARC) Discovery Project DP0666065.

References

1. Ateniese, G., Kirsch, J., Blanton, M.: Secret handshakes with dynamic and fuzzy matching. In: Proc. Internet Society Network and Distributed System Security Symposium (NDSS 2007). Internet Society (2007)
2. Bellare, M., Rogaway, P.: Entity authentication and key distribution. In: Stinson, D.R. (ed.) CRYPTO 1993. LNCS, vol. 773, pp. 232–249. Springer, Heidelberg (1994)
3. Blake-Wilson, S., Johnson, D., Menezes, A.: Key agreement protocols and their security analysis. In: Darnell, M.J. (ed.) Cryptography and Coding 1997. LNCS, vol. 1355. Springer, Heidelberg (1997)
4. Boneh, D.: The decision Diffie-Hellman problem. In: Buhler, J.P. (ed.) ANTS 1998. LNCS, vol. 1423, pp. 48–63. Springer, Heidelberg (1998)
5. Boneh, D., Waters, B.: Conjunctive, subset, and range queries on encrypted data. In: Vadhan, S.P. (ed.) TCC 2007. LNCS, vol. 4392, pp. 535–554. Springer, Heidelberg (2007)
6. Canetti, R., Krawczyk, H.: Analysis of key-exchange protocols and their use for building secure channels. In: Pfitzmann, B. (ed.) EUROCRYPT 2001. LNCS, vol. 2045, pp. 453–474. Springer, Heidelberg (2001)
7. Chen, L., Cheng, Z., Smart, N.P.: Identity-based key agreement protocols from pairings. International Journal of Information Security 6(4), 213–241 (2007)
8. Chen, L., Kudla, C.: Identity based authenticated key agreement protocols from pairings. In: Proceedings 16th IEEE Computer Security Foundations Workshop (CSWF-16), pp. 219–233. IEEE, Los Alamitos (2003)
9. Dingledine, R., Mathewson, N., Syverson, P.: Tor: The second-generation onion router. In: Proc. 13th USENIX Security Symposium. The USENIX Association (2004)
10. Gorantla, M.C., Boyd, C., González Nieto, J.: Attribute-based authenticated key exchange (2010) (unpublished manuscript)
11. Goyal, V., Pandey, O., Sahai, A., Waters, B.: Attribute-based encryption for fine-grained access control of encrypted data. In: Wright, R., De Capitani de Vimercati, S., Shmatikov, V. (eds.) Proc. 13th ACM Conference on Computer and Communications Security (CCS), pp. 89–98. ACM, New York (2006)
12. Günther, C.G.: An identity-based key-exchange protocol. In: Quisquater, J.-J., Vandewalle, J. (eds.) EUROCRYPT 1989. LNCS, vol. 434, pp. 29–37. Springer, Heidelberg (1990)
13. Katz, J., Sahai, A., Waters, B.: Predicate encryption supporting disjunctions, polynomial equations, and inner products. In: Smart, N.P. (ed.) EUROCRYPT 2008. LNCS, vol. 4965, pp. 146–162. Springer, Heidelberg (2008)
14. Khader, D.: Attribute based group signature with revocation, Cryptology ePrint Archive, Report 2007/241 (2007)
15. Khader, D.: Attribute based group signatures, Cryptology ePrint Archive, Report 2007/159 (2007)
16. Khader, D.: Authenticating with attributes, Cryptology ePrint Archive, Report 2008/031 (2008)
17. Khader, D., Chen, L., Davenport, J.H.: Certificate-free attribute authentication. In: Parker, M.G. (ed.) Cryptography and Coding 2009. LNCS, vol. 5921, pp. 301–325. Springer, Heidelberg (2009)
18. Kudla, C., Paterson, K.G.: Modular security proofs for key agreement protocols. In: Roy, B. (ed.) ASIACRYPT 2005. LNCS, vol. 3788, pp. 549–565. Springer, Heidelberg (2005)

19. LaMacchia, B., Lauter, K., Mityagin, A.: Stronger security of authenticated key exchange. In: Susilo, W., Liu, J.K., Mu, Y. (eds.) ProvSec 2007. LNCS, vol. 4784, pp. 1–16. Springer, Heidelberg (2007)
20. Li, J., Au, M.H., Susilo, W., Xie, D., Ren, K.: Attribute-based signature and its applications. In: Proc. 2010 ACM Symposium on Information, Computer and Communications Security (ASIACCS 2010). ACM Press, New York (2010)
21. Li, J., Kim, K.: Attribute-based ring signatures, Cryptology ePrint Archive, Report 2008/394 (2008)
22. Maji: H., Prabhakaran, M., Rosulek, M.: Attribute-based signatures: Achieving attribute-privacy and collusion-resistance, Cryptology ePrint Archive, Report 2008/328 (2008)
23. Sahai, A., Waters, B.: Fuzzy identity-based encryption. In: Cramer, R. (ed.) EUROCRYPT 2005. LNCS, vol. 3494, pp. 457–473. Springer, Heidelberg (2005)
24. Shahandashti, S.F., Safavi-Naini, R.: Threshold attribute-based signatures and their application to anonymous credential systems. In: Preneel, B. (ed.) AFRICACRYPT 2009. LNCS, vol. 5580, pp. 198–216. Springer, Heidelberg (2009)
25. Shamir, A.: Identity-based cryptosystems and signature schemes. In: Blakely, G.R., Chaum, D. (eds.) CRYPTO 1984. LNCS, vol. 196, pp. 47–53. Springer, Heidelberg (1985)
26. Shen, E., Shi, E., Waters, B.: Predicate privacy in encryption systems. In: Reingold, O. (ed.) TCC 2009. LNCS, vol. 5444, pp. 457–473. Springer, Heidelberg (2009)
27. Wang, H., Xu, Q., Ban, T.: A provably secure two-party attribute-based key agreement protocol. In: Proceedings of the 2009 Fifth International Conference on Intelligent Information Hiding and Multimedia Signal Processing, pp. 1042–1045 (2009)
28. Wang, H., Xu, Q., Fu, X.: Revocable attribute-based key agreement protocol without random oracles. Journal of Networks 4(8), 787–794 (2009)
29. Wang, H., Xu, Q., Fu, X.: Two-party attribute-based key agreement protocol in the standard model. In: Yu, F., Shu, J., Yue, G. (eds.) Proceedings of the 2009 International Symposium on Information Processing (ISIP 2009), pp. 325–328. Academy Publisher (2009)

Attribute-Based Authenticated Key Exchange[*]

M. Choudary Gorantla, Colin Boyd, and Juan Manuel González Nieto

Information Security Institute, Faculty of IT, Queensland University of Technology
GPO Box 2434, Brisbane, QLD 4001, Australia
mc.gorantla@gmail.com, {c.boyd,j.gonzaleznieto}@qut.edu.au

Abstract. We introduce the concept of attribute-based authenticated key exchange (AB-AKE) within the framework of ciphertext-policy attribute-based systems. A notion of AKE-security for AB-AKE is presented based on the security models for group key exchange protocols and also taking into account the security requirements generally considered in the ciphertext-policy attribute-based setting. We also introduce a new primitive called encapsulation policy attribute-based key encapsulation mechanism (EP-AB-KEM) and then define a notion of chosen ciphertext security for EP-AB-KEMs. A generic one-round AB-AKE protocol that satisfies our AKE-security notion is then presented. The protocol is generically constructed from any EP-AB-KEM that achieves chosen ciphertext security. Finally, we propose an EP-AB-KEM from an existing attribute-based encryption scheme and show that it achieves chosen ciphertext security in the generic group and random oracle models. Instantiating our AB-AKE protocol with this EP-AB-KEM will result in a concrete one-round AB-AKE protocol also secure in the generic group and random oracle models.

Keywords: Attribute-based Key Exchange, Attribute-based KEM, Group Key Exchange.

1 Introduction

In a distributed collaborative system, it is often convenient for the members to communicate with the others in the system using attributes that describe their roles or responsibilities. These attributes are highly desirable if the members join/leave the system dynamically. Consider an Internet forum where the members are organized into user groups based on the members' skills or privileges. It is a natural requirement that the members of a user group should be able to establish secure communication with the other members belonging to particular user groups. The communication in these forums is generally carried out through initiating a thread or by posting messages within an existing thread. To enable authentic and confidential communication, the forum administrator may specify an access policy with the user groups being attributes. Obviously, only the

---

[*] This work has been supported in part by the Australian Research Council through Discovery Project DP0666065.

R. Steinfeld and P. Hawkes (Eds.): ACISP 2010, LNCS 6168, pp. 300–317, 2010.

members of the forum whose attributes (e.g. membership to user groups) satisfy the policy should be able to have read and/or write access to the thread.

In the above scenario, the members do not necessarily have to know the identity of the other members with whom they want to communicate. In fact, the administrator may be requested not to disclose the identity of a member to the others for privacy reasons. Any member whose attributes satisfy the policy specified by the administrator should be able to participate in the communication. Note that the communication can naturally be among a group of more than two members, since the defined policy may be satisfied by attributes of more than two members. Hence, an authenticated group key exchange protocol that facilitates attributes usage can be employed in this setting. We call such a protocol, an attribute-based authenticated key exchange (AB-AKE) protocol. Once a session key among the willing participants has been established via the key exchange protocol, it can be used for establishing secure communication among the participants.

We can further envisage applications for AB-AKE in interactive chat rooms and also in organizations with strict hierarchy like the military. In interactive chat rooms, each room may be associated with a policy defined with a set of interests being the attributes. Any member whose interests satisfy the policy of a chat room can have read and/or write access to it. Similarly, a policy over ranks (e.g., Sergeant, Lieutenant, Major, Colonel etc.) as attributes can be specified for the units in the military by another unit at a higher level in the hierarchy. All the units whose attributes satisfy the policy can establish secure communication among themselves through an AB-AKE protocol.

ATTRIBUTE-BASED ENCRYPTION. Sahai and Waters [26] introduced the concept of attribute based encryption (ABE) as an extension to ID-based encryption [6], in which a set of descriptive attributes is regarded as an identity. Goyal et al. [19] further extended ABE and introduced two variants: key policy attribute based encryption (KP-ABE) and ciphertext policy attribute based encryption (CP-ABE). In a KP-ABE system, the private key of a party is associated with an access policy defined over a set of attributes while the ciphertext is associated with a set of attributes. A ciphertext can be decrypted by a party if the attributes associated with the ciphertext satisfy the policy associated the user's private key. A CP-ABE system can be seen as a complementary form to KP-ABE system, wherein the private key is associated with a set of attributes, while a policy defined over a set of attributes is attached to the ciphertext. A ciphertext can be decrypted by a party if the attributes associated with its private key satisfy the ciphertext's policy.

1.1   Contributions

In this paper, we introduce the concept of AB-AKE. We assume that each member willing to participate in an AB-AKE protocol is issued a private key for a set of attributes that he/she possesses. Our modelling of AB-AKE follows the framework of CP-ABE in that the attributes are associated with the private keys. We assume that the members are given an access policy which their attributes have to satisfy for them to participate in the protocol. Alternatively, a common policy may be negotiated by the group members themselves. The protocol takes the access policy as input and computes messages for the other parties. Similar to CP-ABE systems, we may assume that the policy is attached to the protocol messages in an AB-AKE protocol, although this assumption is not necessary since each member knows the policy at the outset of the protocol. A member whose attributes satisfy the given policy can compute the session key from the incoming messages and (if exists) its own contribution.

While a complementary flavour of AB-AKE can be conceptualized based on KP-ABE systems, we do not explore this direction in this work. For the type of applications that we have discussed earlier, AB-AKE protocols based on CP-ABE systems suit well. AB-AKE can be seen as an extension of group key exchange (GKE) [9,23,22] with the additional expressiveness provided by the ciphertext-policy attribute-based systems. We define a notion of authenticated key exchange security (AKE-security) for AB-AKE by adapting a corresponding notion for GKE to the attribute-based setting. The property of collusion resistance considered by attribute-based systems [19,3,28] is naturally embedded into our AKE-security notion.

We then propose a generic one-round AB-AKE protocol that satisfies our AKE-security notion. The protocol is based on a type of attribute-based key encapsulation mechanism (KEM) that we call *encapsulation-policy attribute based KEM* (EP-AB-KEM). In an EP-AB-KEM, the attributes are associated to the private key of a party and access policy is attached to the encapsulation. We define a notion of chosen ciphertext security for EP-AB-KEM based on a corresponding notion considered for CP-ABE schemes.

Our AB-AKE protocol is generic in the sense that it can be instantiated using any EP-AB-KEM that satisfies chosen ciphertext security. We propose a chosen-ciphertext secure EP-AB-KEM based on the CP-ABE scheme of Bethencourt et al. [3] and using the generic technique of Boneh et al. [7]. While we apply the technique of Boneh et al. to the chosen plaintext secure EP-AB-KEM implicit in Bethencourt et al.'s scheme, we also make some non-trivial changes to adapt it to the attribute-based setting. The proposed EP-AB-KEM is then proven secure in the generic group and random oracle models. Incidentally, we are the first to model and construct EP-AB-KEMs, which are of independent interest.

An AB-AKE protocol satisfying our AKE-security provides implicit authentication that is similar to the corresponding notion considered for normal key exchange protocols. Particularly, our AKE-security notion ensures each protocol participant that no other party apart from parties who satisfy the given policy can possibly learn the value of the session key. Note that an EP-AB-KEM cannot achieve this property since it does not provide any sender authentication. Consequently, the receivers in EP-AB-KEM whose attributes satisfy the policy have no way of knowing whether the sender actually satisfies the same policy or not. For example, if we use an EP-AB-KEM in a user group, any one can post a message that is encrypted with the symmetric of the EP-AB-KEM. Alternatively, if the message is encrypted with a session key derived from an AB-AKE protocol the readers will get the assurance that only someone with valid attribute set has posted the message.

Our generic construction of AB-AKE can be seen as an extension of the protocols of Boyd et al. [8] and Gorantla et al. [17] to the attribute-based setting. One disadvantage of our protocol is that it cannot provide forward secrecy. However, for some of the applications that we have discussed forward secrecy may not be necessary. For example, in an Internet forum the administrator may like to moderate the content posted in the user groups or in the military a unit at a higher rank would like to monitor the communication among the units at the same or a lower rank. In such scenarios, an AB-AKE protocol without forward secrecy will be useful since any party with the right attribute set will be able to recover the session key and consequently the messages encrypted with it. Nevertheless, forward secrecy is generally a highly desirable property for key exchange protocols. Hence, we also sketch constructions of AB-AKE protocols that can achieve forward secrecy.

1.2 Related Work

The concept of fuzzy secret handshake proposed by Ateniese et al. [1] seems closely related to our modelling of AB-AKE. However, there are a few important differences: In AB-AKE, we allow policies specified by the members to be very expressive consisting of several threshold gates, while fuzzy secret handshake only considers a single threshold gate. In a (fuzzy) secret handshake protocol, if a member do not satisfy the attributes specified by another member, the attributes of none of the members can be learned by the other member. On the other hand, in an AB-AKE protocol, if a member does not satisfy the policy specified by the other members, the members do not know anything about the attributes of the other members except what can be inferred by the policies attached to the protocol messages. Although both the properties look similar, we emphasize that an AB-AKE protocol would not hide the affiliation of the members even if the protocol was not successful [20]. Note that the property of "affiliation hiding" is the main requirement for (fuzzy) secret handshakes. Finally, the fuzzy secret handshake protocol of Ateniese et al. considers only two party setting, while our protocol naturally operates in a group setting.

In independent work, Steinwandt and Corona [27] proposed a two-round attribute-based group key exchange protocol that achieves forward secrecy. Their protocol uses the GKE protocol of Bohli et al. [5] as the base protocol and replaces the public key signature scheme in Bohli et al. with an attribute-based signcryption scheme to authenticate the protocol messages. Recently, Birkett and Stebila [4] introduced the concept of predicate-based key exchange which encompasses key policy attribute-based key exchange. However, their security model considers key exchange between only two parties.

1.3 Organization

Section 2 presents a security model for EP-AB-KEM and also proposes a chosen ciphertext secure EP-AB-KEM. We define a security model for AB-AKE in

Section 3 and present a generic one-round AB-AKE protocol in Section 4. In Section 5, we outline how to construct AB-AKE protocols with forward secrecy.

2 Encapsulation Policy Attribute-Based KEM

We first give a formal definition of security for EP-AB-KEM. As in the earlier attribute-based systems [19,3], we review the definition of an access structure and use it in the security model. Later, we present a concrete EP-AB-KEM based on the CP-ABE scheme of Bethencourt et al. [3].

Definition 1 (Access Structure [2]). Let $\{U_1, \cdots, U_n\}$ be a set of parties. A collection $\mathbb{A} \subseteq 2^{\{U_1, \cdots, U_n\}}$ is monotone if $\forall B, C$: if $B \in \mathbb{A}$ and $B \subseteq C$ then $C \in \mathbb{A}$. An access structure (respectively, monotone access structure) is a collection (respectively, monotone collection) $\mathbb{A}$ of non-empty subsets of $\{U_1, \cdots, U_n\}$, i.e., $\mathbb{A} \subseteq 2^{\{U_1, \cdots, U_n\}} \setminus \{\phi\}$. The sets in $\mathbb{A}$ are called authorized sets, and the sets not in $\mathbb{A}$ are called the unauthorized sets.

In our EP-AB-KEM and later in the protocol, each party is assumed to possess a set of attributes. A policy over a set of attributes is specified through an access structure $\mathbb{A}$. Hence, $\mathbb{A}$ contains the authorized sets of attributes i.e., $\mathbb{A} \subseteq 2^{\{S_1, \cdots, S_n\}} \setminus \{\phi\}$ for a given set of attributes $\{S_1, \cdots, S_n\}$. As in the CP-ABE of Bethencourt et al., we consider only monotonic access structures. In the rest of the paper, by an access structure we mean a monotonic one.

A EP-AB-KEM consists of five polynomial-time algorithms:

Setup: takes the security parameter k and the attribute universe description $\mathbb{U}$ as inputs. The public parameters PK and the master key MK are the outputs.

Encapsulation: takes as input the public parameters PK and an access structure $\mathbb{A}$ over the attribute universe $\mathbb{U}$. It outputs an encapsulation C and a symmetric key K such that only a user with attributes satisfying $\mathbb{A}$ can recover K from C. Similar to the CP-ABE schemes, we assume that the encapsulation implicitly contains $\mathbb{A}$.

KeyGen: takes as input the master key MK, the public parameters PK and a set of attributes S of a user that gives a description of the user's private key. The output is the user's private key SK.

Decapsulation: takes as input the public parameters PK, an encapsulation C and a private key SK corresponding to a set of attributes S. The algorithm outputs either a symmetric key K or $\bot$.

We also define an optional delegation algorithm, which allows a user with attribute sets S and a corresponding secret key SK to derive a secret key for another set of attributes $\tilde{S}$ such that $\tilde{S} \subseteq S$.

Delegate: takes as input the public parameters PK, a secret key SK corresponding to a set of attributes S and a set $\tilde{S} \subseteq S$. It outputs a secret key $\tilde{SK}$ for the attribute set $\tilde{S}$.

For an EP-AB-KEM to be considered valid, it is required that for any key SK corresponding to an attribute set S, if S satisfies $\mathbb{A}$ and if $(K, C) \leftarrow$ Encapsulation$(PK, \mathbb{A})$, then Decapsulation$(PK, C, SK) = K$.

2.1 Security Model

Bethencourt et al. [3] defined the notion of indistinguishability under chosen plaintext attack (IND-CPA) for CP-ABE schemes. In this section, we adapt their notion and extend it to define a notion of indistinguishability under chosen ciphertext attacks (IND-CCA) for EP-AB-KEM. The security notion is formally defined as follows.

Definition 2. An EP-AB-KEM is IND-CCA secure if the advantage of any probabilistic polynomial time adversary $\mathcal{A}^{\text{cca}}$ in the following game is negligible in the security parameter k.

Setup: The challenger runs the Setup algorithm and returns PK to $\mathcal{A}^{\text{cca}}$.

Phase 1: $\mathcal{A}^{\text{cca}}$ issues Extract and Decap queries as follows:

  Extract: This query can be issued multiple times with sets of attributes $S_1, \cdots, S_{q_1}$ as input. The challenger returns a private key corresponding to each input attribute set. We do not require the input attribute sets to be distinct.

  Decap: This query is issued with an encapsulation C and an attribute set S as inputs. Note that C implicitly contains an access structure $\mathbb{A}$ defined over the attribute universe $\mathbb{U}$. The challenger executes the Decapsulation algorithm on C using a private key corresponding to S and returns the output of Decapsulation to $\mathcal{A}^{\text{cca}}$.

Challenge: At the end of **Phase 1**, $\mathcal{A}^{\text{cca}}$ gives an access structure $\mathbb{A}^*$ defined over $\mathbb{U}$ to the challenger. The challenger first chooses a bit b. It then runs the Encapsulation algorithm with $\mathbb{A}^*$ as input and generates a symmetric key–encapsulation pair (K_1, C^*). It then sets K_0 to be a random key drawn from the probability distribution of the symmetric key. The tuple (K_b, C^*) is returned to $\mathcal{A}^{\text{cca}}$ as the challenge. A trivial restriction on the adversary's choice of $\mathbb{A}^*$ is that none of the attributes sets $S_1, \cdots, S_{q_1}$ passed as input to Extract queries in **Phase 1** should satisfy $\mathbb{A}^*$.

Phase 2: $\mathcal{A}^{\text{cca}}$ is allowed to execute in the same way as in **Phase 1** with the following restrictions: (1) none of the attribute sets $S_{q_1+1}, \cdots, S_q$ passed as input to Extract queries in **Phase 2** satisfy $\mathbb{A}^*$ and (2) a Decap query with C^* as input in combination with an attribute set S^* that satisfies $\mathbb{A}^*$ is not allowed.

Guess: The goal of $\mathcal{A}^{\text{cca}}$ is to guess whether the key K_b is encapsulated within C^* or not. $\mathcal{A}^{\text{cca}}$ finally outputs a guess bit b'. It wins the game if $b' = b$. The advantage of $\mathcal{A}^{\text{cca}}$ is given as $Adv_{\mathcal{A}^{\text{cca}}} = |2 \cdot \Pr[b' = b] - 1|$.

Existing security notions for CP-ABE schemes also consider the weaker *selective model* where $\mathcal{A}^{\text{cca}}$ declares the challenge access structure $\mathbb{A}^*$ before the **Setup** phase. Similarly, a corresponding model for EP-AB-KEMs can be defined.

Similar to earlier CP-ABE schemes [3,11,28], we have not explicitly modelled the delegation mechanism in the security model for EP-AB-KEMs. However, we require that for a given set of attributes, a secret key output by the Delegate algorithm will have identical distribution to the one output by the KeyGen algorithm. In particular, the Decapsulation algorithm using a private key SK should work in the same way irrespective of SK being an output of KeyGen or Delegate. Our security model for EP-AB-KEMs suffices in the presence of an adversary who may obtain delegated private keys since such queries can be simulated using Extract queries.

Remark 1. In Definition 2, $\mathcal{A}^{cca}$ is allowed to issue multiple Extract queries with attribute sets as input such that none of the individual sets S_i satisfy the challenge access structure $\mathbb{A}^*$. Hence, similar to earlier definitions of attribute-based encryption schemes, our definition also takes care of collusion resistance. An EP-AB-KEM satisfying the above definition ensures that from the private keys of S_i's, $\mathcal{A}^{cca}$ cannot construct a private key corresponding to another attribute set S^* such that S^* satisfies $\mathbb{A}^*$.

HYBRID CP-ABE. An EP-AB-KEM satisfying the above IND-CCA security notion can be combined with any IND-CCA secure data encapsulation mechanism to construct an IND-CCA secure CP-ABE scheme [12,13]. We describe the hybrid construction and prove its security in the full version of this paper [16].

2.2   A Chosen Ciphertext Secure EP-AB-KEM

Bethencourt et al. [3] first proposed a construction of a CP-ABE scheme. Their scheme was shown IND-CPA secure assuming generic group and random oracle models. Later, many CP-ABE schemes [18,11,28] have been proposed and shown IND-CPA secure without assuming generic group or random oracle models, but analyzed only in the selective model of security. Recently, Lewko et al. [24] proposed a fully secure CP-ABE scheme in the standard model using composite order bilinear groups.

We now construct an IND-CCA secure EP-AB-KEM based on the CP-ABE scheme of Bethencourt et al. The idea is to enhance the security of the IND-CPA secure EP-AB-KEM that is implicit in Bethencourt et al.'s CP-ABE scheme. For this purpose, the techniques of Fujisaki and Okamoto [15,14] and Canetti et al. (CHK) [10] can be applied in the random oracle and standard models respectively. As remarked by Bethencourt et al., IND-CCA security for CP-ABE (and correspondingly for EP-AB-KEM) schemes can be achieved by a straightforward application of the Fujisaki-Okamoto technique.

Bethencourt et al. also suggested that the delegation mechanism of their CP-ABE scheme can be leveraged to achieve IND-CCA security using the CHK transform. However, we observe that applying the CHK transform to CP-ABE schemes (similarly to EP-AB-KEMs) is slightly more involved. Specifically, contrary to the approach followed by KP-ABE schemes, IND-CCA security for CP-ABE schemes cannot be achieved by directly leveraging the delegation mechanism. We later discuss why this is so and then present an IND-CCA secure

EP-AB-KEM by making a few changes to the Setup and Encapsulation algorithms derived from Bethencourt et al.'s CP-ABE scheme. Although the CHK technique can be used to achieve IND-CCA security in the standard model, our EP-AB-KEM will only be secure assuming generic groups and random oracles since the base CP-ABE scheme also assumes the same. Finally, we choose the scheme of Bethencourt et al. because it is secure in the fully adaptive model (i.e., non-selective model). Alternatively, one could derive an EP-AB-KEM secure in the fully adaptive model from the CP-ABE scheme of Lewko et al. [24]. In the full version [16], we discuss the necessity of an EP-AB-KEM to be secure in the adaptive model for constructing AB-AKE protocols.

The IND-CCA secure scheme first generates a one-time key pair (sk, vk) for a signature scheme with the condition that the verification key is of the same length as the length of an attribute in the attribute universe $\mathbb{U}$. Let $\mathbb{A}$ be the access structure given as input to the EP-AB-KEM. We now construct a more restrictive access structure $\mathbb{A}' = \mathbb{A}$ AND vk and execute the CPA-secure EP-AB-KEM under $\mathbb{A}'$. The resulting encapsulation is then signed using the one-time signing key sk. The encapsulation of the CCA-secure EP-AB-KEM contains the encapsulation generated by the underlying CPA-secure EP-AB-KEM, the signature generated on it and the verification key vk. The recipient first checks the signature using vk and then executes the CPA-secure KEM's decapsulation algorithm under $\mathbb{A}'$ to extract the symmetric key.

While the above informal description of our construction directly follows the CHK technique, the tricky part in the context of EP-AB-KEM (or CP-ABE) is to empower the recipient with a private key corresponding to the attributes that satisfy the modified access structure $\mathbb{A}'$. The recipient may already possess attributes that satisfy $\mathbb{A}$. However, since the verification key vk is one-time and chosen randomly for each execution of EP-AB-KEM, the recipient cannot be issued with a private key that can decrypt messages encrypted under $\mathbb{A}' = \mathbb{A}$ AND vk. This problem cannot be addressed by the delegation mechanism in an EP-AB-KEM (or CP-ABE) scheme since it can be used to derive private key corresponding to an attribute set S' from the one corresponding to S only if $S' \subseteq S$. But, we have an additional attribute in the form of vk. Note that this is not a problem in the KP-ABE system since it naturally allows a party with a private key corresponding to an access structure $\mathbb{A}$ to derive private keys corresponding to access structures that are more restrictive than $\mathbb{A}$.

To address the above problem, we make modifications to the Setup and Encapsulation algorithms derived from the CP-ABE scheme of Bethencourt et al. [3]. Our EP-AB-KEM now enables a recipient with private key for attributes that satisfy $\mathbb{A}$ to decapsulate an encapsulation created under $\mathbb{A}'$, irrespective of the choice of vk by the sender. As in the CP-ABE scheme of Bethencourt et al., an access structure $\mathbb{A}$ is represented in the form of an access tree $\mathcal{T}$.

Access Tree. Let $\mathcal{T}$ be a tree representing an access structure. Each interior node of $\mathcal{T}$ represents a threshold gate, while each leaf node is described by an attribute. Let num_x be the number of children of a node x and let k_x be its threshold value. We have $0 \leq k_x \leq num_x$. A threshold gate associated to an internal node with threshold value k_x outputs true if at least k_x of its children output true. If the threshold gate represented by an interior node is an AND gate then $k_x = num_x$ and if the gate is OR, $k_x = 1$. The threshold value for each leaf node x is defined to be $k_x = 1$. The parent of a node x in the tree $\mathcal{T}$ is denoted by the function $\mathsf{parent}(x)$, while the attribute of a leaf node x is denoted by $\mathsf{att}(x)$. The children of each interior node are numbered from 1 to num_x. The function $\mathsf{index}(x)$ returns such a number associated with a node x. We assume that the index values are uniquely assigned in an arbitrary manner for a given access structure.

Satisfying an access tree. Let r be the root of an access tree $\mathcal{T}$. The subtree of $\mathcal{T}$ rooted at a node x is denoted by $\mathcal{T}_x$. If a set of attributes γ satisfy the access tree $\mathcal{T}_x$, it is denoted as $\mathcal{T}_x(\gamma) = 1$. The function $\mathcal{T}_x(\gamma)$ is computed recursively as follows: If x is an interior node, for each children x' of x, $\mathcal{T}_{x'}(\gamma)$ is evaluated. $\mathcal{T}_x(\gamma)$ returns 1 if and only if at least k_x children of x return 1. If x is a leaf node, $\mathcal{T}_x(\gamma)$ returns 1 if and only if $\mathsf{att}(x) \in \gamma$.

Let $\mathbb{G}_0$ and $\mathbb{G}_1$ be two multiplicative groups of prime order p and g be an arbitrary generator of $\mathbb{G}_0$. Let $e : \mathbb{G}_0 \times \mathbb{G}_0 \to \mathbb{G}_1$ be an admissible bilinear map. The Lagrange's coefficient $\Delta_{i,S}$ for $i \in \mathbb{Z}_p$ and a set S of elements in $\mathbb{Z}_p$ is defined as: $\Delta_{i,S} = \Pi_{j \in S, j \neq i} \frac{x-j}{i-j}$.

$\mathsf{Setup}(k)$. It chooses the groups $\mathbb{G}_0$, $\mathbb{G}_1$ and defines a bilinear map $e : \mathbb{G}_0 \times \mathbb{G}_0 \to \mathbb{G}_1$. It also selects $\alpha, \beta_1, \beta_2 \in \mathbb{Z}_p$ such that $\beta_1 \neq \beta_2$, $\beta_1 \neq 0$ and $\beta_2 \neq 0$. The public key is

$$PK = \left(\mathbb{G}_0, \mathbb{G}_1, e, g, h_1 = g^{\beta_1}, f_1 = g^{1/\beta_1}, h_2 = g^{\beta_2}, f_2 = g^{1/\beta_2}, e(g, g)^\alpha \right).$$

The master key MK is $(\beta_1, \beta_2, g^\alpha)$.

$\mathsf{Encapsulation}(PK, \mathcal{T})$. This algorithm generates an encapsulation and a symmetric key under the access tree $\mathcal{T}$ using the public key PK. It first executes the KeyGen algorithm of the signature scheme and obtains a one-time key pair (sk, vk). Let $\mathbb{A}$ be the access structure represented by $\mathcal{T}$. The algorithm now constructs a new access tree $\mathcal{T}'$ for the access structure $(\mathbb{A} \text{ AND } vk)$ as follows: Let R be the root node of $\mathcal{T}$. The root node R' of the new tree $\mathcal{T}'$ is set as the AND gate with $\mathcal{T}$ as its subtree and the verification key vk as a leaf node attached to R'.

The algorithm now generates a polynomial q_x for each node x in the tree $\mathcal{T}'$ in a top-down approach as follows: Starting from the root node R', for each node x in the tree set the degree d_x of the polynomial associated with x to be $k_x - 1$ i.e., the degree of the polynomial is one less than the threshold value associated with the node x. The algorithm starts from the root node and first chooses a random $s \in \mathbb{Z}_p$. Then it chooses $d_{R'}$ other points randomly to define the polynomial $q(R')$. For any node x other than the root, it sets $q_x(0) = q_{\mathsf{parent}(x)}(\mathsf{index}(x))$ and chooses d_x other points randomly to define the polynomial $q(x)$.

Let Y be the set of leaf nodes in the subtree $\mathcal{T}$ rooted at R. The only other leaf node in the tree $\mathcal{T}'$ is the one that describes the verification key vk. The algorithm proceeds as follows:

1. $K = e(g, g)^{\alpha s}$.
2. $C_1 = h_1^s$.
3. $\forall y \in Y : \; C_y = g^{q_y(0)}, C_y' = H(\mathsf{att}(y))^{q_y(0)}$.
4. $C_{vk} = h_2^{q_{vk}(0)}, \; C_{vk}' = H(vk)^{q_{vk}(0)}$.
5. Let $\mathcal{C} = (\mathcal{T}', C_1, C_y, C_y', C_{vk}, C_{vk}'), \; \forall y \in Y$. Compute a signature $\sigma = \mathsf{Sig}_{sk}(\mathcal{C})$.

The final encapsulation $C = (\mathcal{C}, vk, \sigma)$.

KeyGen(MK,PK,S). The key generation algorithm takes as input the master key MK and a set of attributes S and outputs a private key corresponding to S. It chooses $r, r_{vk} \in \mathbb{Z}_p$ and $r_j \in \mathbb{Z}_p$ for each $j \in S$. The private key is computed as:

$$SK = (D = g^{(\alpha+r)/\beta_1}, \; E = g^{r/\beta_2}, \; \forall j \in S : \; D_j = g^r \cdot H(j)^{r_j}, \; D_j' = g^{r_j}).$$

Delegate($SK, PK, \tilde{S}$). It takes as input a secret key SK corresponding to a set of attributes S and another set $\tilde{S} \subseteq S$. The key SK is of the form $SK = (D, E, \forall j \in S : D_j, D_j')$. The algorithm chooses $\tilde{r}$ and $\tilde{r}_k \forall k \in \tilde{S}$. The new key for $\tilde{S}$ is generated as:

$$\tilde{SK} = (\tilde{D} = Df_1^{\tilde{r}}, \; \tilde{E} = Ef_2^{\tilde{r}}, \; \forall k \in \tilde{S} : \; \tilde{D}_k = D_k g^{\tilde{r}} H(k)^{\tilde{r}_k}, \; \tilde{D}_k' = D_k' g^{\tilde{r}_k}).$$

Decapsulation(SK, PK, C). Upon receiving an encapsulation C, the decryptor first parses the access tree $\mathcal{T}'$. It then extracts the subtree $\mathcal{T}$ rooted at R from $\mathcal{T}'$. Note that this can be easily done since the node that describes the verification key as an attribute can be identified with the help of the verification key vk sent in the encapsulation. The algorithm first verifies the signature σ on C using the verification key vk. If the verification succeeds, it proceeds as follows:

$$F_{vk} = \frac{e(C_{vk}, H(vk) \cdot g^{r/\beta_2})}{e(C_{vk}', h_2)} = \frac{e(C_{vk}, g^{r/\beta_2}) \cdot e(C_{vk}, H(vk))}{e(C_{vk}', h_2)} \tag{1}$$

$$= \frac{e(h_2^{q_{vk}(0)}, g^{r/\beta_2}) \cdot e(h_2^{q_{vk}(0)}, H(vk))}{e(H(vk)^{q_{vk}(0)}, h_2)}$$

$$= e(g^{\beta_2 \cdot q_{vk}(0)}, g^{r/\beta_2}) = e(g, g)^{r q_{vk}(0)}.$$

A recursive algorithm DecryptNode($\mathcal{C}, SK, x$) that takes as input $\mathcal{C}$, a private key SK associated with a set of attributes S and a node x from the subtree $\mathcal{T}$ is then executed as below:

If x is a leaf node, then let $i = \mathsf{att}(x)$. If $i \notin S$, then DecryptNode($\mathcal{C}, SK, x$) = $\perp$. Otherwise it is defined as follows:

$$\mathsf{DecryptNode}(\mathcal{C}, SK, x) = \frac{e(D_i, C_x)}{e(D_i', C_x')} = \frac{e(g^r \cdot H(i)^{r_i}, g^{q_x(0)})}{e(g^{r_i}, H(i)^{q_x(0)})} = e(g,g)^{r q_x(0)}.$$

If x is an interior node then $\mathsf{DecryptNode}(\mathcal{C}, SK, x)$ proceeds as follows: For all nodes z that are children of x, the algorithm $\mathsf{DecryptNode}(\mathcal{C}, sk, z)$ is called. The output is stored as F_z. Let S_x be an arbitrary k_x-sized set of child nodes z such that $F_z \neq \perp$. If no such set exists, the function returns $\perp$. Otherwise, the decapsulation algorithm proceeds as follows:

$$
\begin{aligned}
F_x &= \prod_{z \in S_x} F_z^{\Delta_{i, S_x'}(0)}, \text{ where } i = \mathsf{index}(z), S_x' = \{\mathsf{index}(z) : z \in S_x\} \\
&= \prod_{z \in S_x} (e(g,g)^{r \cdot q_z(0)})^{\Delta_{i, S_x'}(0)} \\
&= \prod_{z \in S_x} (e(g,g)^{r \cdot q_{\mathsf{parent}(z)}(\mathsf{index}(z))})^{\Delta_{i, S_x'}(0)} \\
&= \prod_{z \in S_x} (e(g,g)^{r \cdot q_x(i) \cdot \Delta_{i, S_x'}(0)}) \\
&= (e(g,g)^{r \cdot q_x(0)}).
\end{aligned}
$$

Finally, the decapsulation algorithm calls the $\mathsf{DecryptNode}$ algorithm on the node R, which is the root of the subtree $\mathcal{T}$. If $\mathcal{T}$ is satisfied by the attribute set S, then we have $F_R = \mathsf{DecryptNode}(\mathcal{C}, SK, R) = e(g,g)^{r \cdot q_R(0)}$. We now compute $F_{R'}$ from F_{vk} and F_R using polynomial interpolation as follows:

$$
\begin{aligned}
F_{R'} &= \prod_{x \in \{R, vk\}} F_x^{\Delta_{\mathsf{index}(x), \{R, vk\}}} \\
&= e(g,g)^{r \cdot q_{R'}(0)} \\
&= e(g,g)^{rs}.
\end{aligned}
$$

Let $A = e(g,g)^{rs}$. The symmetric key is recovered as

$$\frac{e(C_1, D)}{A} = \frac{e(h_1^s, g^{(\alpha+r)\beta_1})}{e(g,g)^{rs}} = \frac{e(g,g)^{s(\alpha+r)}}{e(g,g)^{rs}} = e(g,g)^{\alpha s} = K. \qquad (2)$$

Note that in Equation 1, we implicitly verify that the one-time verification key has not been replaced. If vk was replaced the symmetric key computed in Equation 2 would be $\perp$. Alternatively, the verification check can be done explicitly at the cost of an additional pairing operation. In the full version [16], we show that the proposed EP-AB-KEM is IND-CCA secure in the generic group and random oracle models.

3   Attribute-Based Authenticated Key Exchange

An AB-AKE protocol consists of three polynomial-time algorithms: Setup, Key-Gen and KeyExchange. The Setup and KeyGen algorithms are identical to those defined for EP-AB-KEM in Section 2. Each party in the AB-AKE protocol executes the KeyExchange algorithm which initially takes as input PK, an access structure $\mathbb{A}$ and a private key for a set of attributes S. If S satisfies $\mathbb{A}$, KeyExchange proceeds as per specification and may generate outgoing messages and also accept incoming messages from other parties as inputs. The output of KeyExchange is either a session key κ or $\perp$.

Communication Model. Let $\mathcal{U} = \{U_1, \cdots, U_n\}$ be a set of n users. The protocol may be executed among any subset $\tilde{\mathcal{U}} \subseteq \mathcal{U}$ of size $\tilde{n} \geq 2$. We assume that each user has a set of descriptive attributes. Let SK_i be the private key corresponding to an attribute set S_i of user U_i. We assume that an access structure $\mathbb{A}$ is given as input to all the users. Note that this $\mathbb{A}$ may be specified by a higher level protocol. Alternatively, the users can run an interactive protocol to negotiate a common access structure $\mathbb{A}$. We also assume that all the users execute the protocol honestly. If a user U_i wants to establish a session key with respect to an access structure $\mathbb{A}$, it first checks whether its attribute set S_i satisfies $\mathbb{A}$ or not i.e., checks if $S_i \in \mathbb{A}$. U_i proceeds with the protocol execution only if S_i satisfies $\mathbb{A}$. Thus, any user U_j with attribute set S_j that satisfies $\mathbb{A}$ is a potential participant in the key exchange protocol. The set of parties whose individual attributes satisfy $\mathbb{A}$ can compute a common session key.

An AB-AKE protocol π executed among $\tilde{n} \leq n$ users is modelled as a collection of $\tilde{n}$ programs running at the $\tilde{n}$ parties. Each instance of π within a party is defined as a session and each party may have multiple such sessions running concurrently. Let π_i^j be the j-th run of the protocol π at party $U_i \in \tilde{\mathcal{U}}$. Each protocol instance at a party is identified by a unique session ID. We assume that the session ID is derived during the run of the protocol. The session ID of an instance π_i^j is denoted by sid_i^j. An instance π_i^j enters an *accepted* state when it computes a session key sk_i^j. Note that an instance may terminate without ever entering into an accepted state. The information of whether an instance has terminated with acceptance or without acceptance is assumed to be public.

Note that there may be more than one party whose attributes satisfy $\mathbb{A}$, hence we consider group setting for AB-AKE. We define partnership in AB-AKE protocol as follows: A set of $\tilde{n}$ instances at $\tilde{n}$ different parties $\tilde{\mathcal{U}} \subseteq \mathcal{U}$ are called partners if

1. they all have the same session ID; **and**
2. the attributes of each $U_i \in \tilde{\mathcal{U}}$ satisfy $\mathbb{A}$.

An AB-AKE protocol is called *correct* if the instances at the parties in $\tilde{\mathcal{U}}$ are partnered and output identical session keys in the presence of a passive adversary.

Adversarial Model. The communication network is assumed to be fully controlled by the adversary, which schedules and mediates the sessions among all the parties. The adversary is allowed to insert, delete or modify the protocol messages. We also assume that it is the adversary that may select the protocol participants from the set $\mathcal{U}$. While the adversary may not know the attribute set that a user possesses, it can initiate an instance of the AB-AKE protocol with an access structure of its choice. In addition to controlling the message transmission, the adversary is allowed to ask the following queries.

- Send(π_i^j,m) sends a message m to the instance π_i^j. If the message is $\mathbb{A}$, the instance π_i^j is initiated with the access structure $\mathbb{A}$. Otherwise, the message is processed as per the protocol specification. The response of π_i^j to any Send query is returned to the adversary.
- RevealKey(π_i^j) If π_i^j has accepted, the adversary is given the session key sk_i^j established at π_i^j.
- Corrupt(S_i) This query returns a private key SK_i corresponding to the attribute set S_i.
- Test(π_i^j) A random bit b is secretly chosen. If $b = 1$, the adversary is given sk_i^j established at π_i^j. Otherwise, a random value chosen from the session key probability distribution is given. Note that a Test query is allowed only on an accepted instance.

Definition 3 (Freshness). Let $\mathbb{A}$ be the access structure for an instance π_i^j. π_i^j is called fresh if the following conditions hold: (1) the instance π_i^j or any of its partners has not been asked a RevealKey query **and** (2) there has not been a Corrupt query on an input S_i such that S_i satisfies $\mathbb{A}$.

Definition 4 (AKE-security). An adversary $\mathcal{A}_{\mathsf{ake}}$ against the AKE-security notion is allowed to make Send, RevealKey and Corrupt queries in Stage 1. $\mathcal{A}_{\mathsf{ake}}$ makes a Test query to an instance π_i^j at the end of Stage 1 and is given a challenge key K_b as described above. It can continue asking queries in Stage 2. Finally, $\mathcal{A}_{\mathsf{ake}}$ outputs a bit b' and wins the AKE-security game if (1) $b' = b$ **and** (2) the Test instance π_i^j remains fresh till the end of $\mathcal{A}_{\mathsf{ake}}$'s execution. Let $\mathsf{Succ}_{\mathcal{A}_{\mathsf{ake}}}$ be the event that $\mathcal{A}_{\mathsf{ake}}$ wins the AKE-security game. The advantage of $\mathcal{A}_{\mathsf{ake}}$ in winning this game is $\mathsf{Adv}_{\mathcal{A}_{\mathsf{ake}}} = |2 \cdot \Pr[\mathsf{Succ}_{\mathcal{A}_{\mathsf{ake}}}] - 1|$. A protocol is called AKE-secure if $\mathsf{Adv}_{\mathcal{A}_{\mathsf{ake}}}$ is negligible in the security parameter k for any polynomial time $\mathcal{A}_{\mathsf{ake}}$.

Remark 2. By allowing the adversary to reveal the private keys corresponding to attribute sets which individually do not satisfy the given access structure $\mathbb{A}^*$ in the test session, our definition naturally considers collusion resistance. In other words, any number of parties whose individual attribute sets do not satisfy $\mathbb{A}^*$ may collude among themselves and try to violate the AKE-security of the protocol. An AB-AKE protocol satisfying our AKE-security notion will still remain secure against such collusion attacks.

4   A Generic One-Round AB-AKE Protocol

We now present a simple generic AB-AKE protocol based on IND-CCA secure EP-AB-KEM. Informally, each party executes an EP-AB-KEM in parallel and combines the symmetric key it has generated with the symmetric keys extracted

---

Computation

      Each U_i executes an EP-AB-KEM on the input $(PK, \mathcal{T})$ where PK is the public parameters and $\mathcal{T}$ is the access tree that represents an access structure $\mathbb{A}$. As a result, a symmetric key and encapsulation pair (K_i, C_i) is obtained.

$$(K_i, C_i) \leftarrow \mathsf{Encapsulation}(PK, \mathcal{T}).$$

Broadcast

      Each U_i broadcasts the generated encapsulation C_i.

$$U_i \rightarrow * : \quad C_i.$$

Key Computation

  1. Each U_i executes the decapsulation algorithm using its private key SK_i on each of the incoming encapsulations C_j and obtains the symmetric keys K_j, for $j \neq i$.

$$K_j \leftarrow \mathsf{Decapsulation}(sk_i, C_j) \text{ for each } j \neq i.$$

  2. Each U_i then computes the session ID as the concatenation of all the outgoing and incoming messages exchanged i.e. $\mathsf{sid} = (C_1 \| \cdots \| C_{\tilde{n}})$, where $\tilde{n}$ is the number of protocol participants.

  3. The session key κ is then computed as

$$\kappa = f_{K_1}(\mathsf{sid}) \oplus f_{K_2}(\mathsf{sid}) \oplus \cdots \oplus f_{K_{\tilde{n}}}(\mathsf{sid})$$

      where f is a pseudorandom function.

Fig. 1. A Generic One-round AB-AKE Protocol

from the incoming messages to establish a common session key. Our construction is an extension of the one-round protocols of Boyd et al. [8] and Gorantla et al. [17] to the attribute-based setting. Figure 1 presents our generic one-round AB-AKE protocol.

At the beginning of the protocol each party is given an access structure $\mathbb{A}$ represented via an access tree $\mathcal{T}$. The protocol uses an EP-AB-KEM scheme (Setup, Encapsulation, KeyGen, Decapsulation). Each U_i is issued a private key SK_i corresponding to the attributes set S_i that it possesses. Each party U_i who has attribute set S_i satisfying the access structure $\mathbb{A}$ runs the Encapsulation algorithm and obtains a symmetric key-encapsulation pair (K_i, C_i). The parties broadcast the encapsulations to the other parties. Upon receiving the encapsulations, each party runs the Decapsulation algorithm using its private key on each of the incoming encapsulations and extracts the symmetric keys. The number of protocol participants $\tilde{n}$ can be derived based on the number of input messages received within a prescribed time period. The session key is finally computed by each party from the symmetric key that it has generated and all the symmetric keys decapsulated from the incoming encapsulations.

A pseudorandom function f is applied to derive the session key. We assume that the output of the Decapsulation algorithm can be directly used as a seed for f. Otherwise, we will have to extract and then expand the randomness from the output of the Decapsulation algorithm as done by Boyd et al. [8].

Theorem 1. *The AB-AKE protocol in Fig. 1 is AKE-secure as per Definition 4 assuming that the underlying EP-AB-KEM is IND-CCA secure. The advantage of $\mathcal{A}_{\mathsf{ake}}$ is*

$$Adv_{\mathcal{A}_{\mathsf{ake}}} \leq \tilde{n} \cdot \frac{q_s^2}{|C|} + q_s \cdot (\tilde{n} \cdot Adv_{\mathcal{A}^{\mathsf{prf}}} + Adv_{\mathcal{A}^{\mathsf{cca}}})$$

where $\tilde{n}$ is the number of parties in the protocol, q_s is the number of sessions $\mathcal{A}_{\mathsf{ake}}$ is allowed to activate, $|C|$ is the size of the ciphertext space, $Adv_{\mathcal{A}^{\mathsf{cca}}}$ is the advantage of a polynomial adversary $\mathcal{A}^{\mathsf{cca}}$ against the IND-CCA security of the underlying EP-AB-KEM and $Adv_{\mathcal{A}^{\mathsf{prf}}}$ is the advantage of a polynomial adversary $\mathcal{A}^{\mathsf{prf}}$ against the pseudorandomness of the pseudorandom function f.

The proof of the above theorem is given in the full version [16].

Concrete Instantiation. From the EP-AB-KEM proposed in Section 2.2, a concrete AB-AKE protocol can be directly realized. It follows from the security of the EP-AB-KEM and the generic AB-AKE protocol that the instantiated protocol is AKE-secure in the generic group and the random oracle models.

5   Extensions

The security model in Section 3 is concerned only about the basic notion of AKE-security without forward secrecy. Forward secrecy is one of the most important security attributes for key exchange protocols since it limits the damage of long-term key exposure. A key exchange protocol with forward secrecy ensures that even if the long-term key of a party is exposed, all the past session keys established using that long-term key will remain uncompromised.

Forward secrecy seems to be more important for AB-AKE protocols than in the case of normal key exchange protocols. To see why, let us assume that the adversary obtains the private key of a user U_i who possesses a set of attributes S_i. If an AB-AKE protocol does not achieve forward secrecy, then the adversary can compromise all the protocol sessions which have been established with access structures that can be satisfied by S_i. Note that the party U_i does not even have to participate in any of these sessions. We now define a notion of freshness that takes forward secrecy into account.

5.1   AKE-Security with Forward Secrecy

Definition 5 (FS-Freshness). Let $\mathbb{A}$ be the access structure for an instance π_i^j. π_i^j is called fs-fresh if the following the conditions hold: (1) the instance π_i^j or any its partners has not been asked a RevealKey query **and** (2) there has not been a Corrupt query on an input S_i before π_i^j or its partner instances have terminated, such that S_i satisfies $\mathbb{A}$.

Definition 5 can be coupled with the AKE-security notion in Definition 4 to arrive at AKE-security notion with forward secrecy for AB-AKE protocols.

5.2 Constructing AB-AKE Protocols with Forward Secrecy

Our one-round AB-AKE protocol can be modified to achieve AKE-security with forward secrecy for two-party and three-party settings using known techniques. For a two-party AB-AKE protocol with forward secrecy, one can use the technique of Boyd et al. [8] where ephemeral Diffie-Hellman public keys are appended with the encapsulations. Similarly, for a three-party AB-AKE protocol with forward secrecy, the protocol of Joux [21] can be executed in the same round with our EP-AB-KEM based protocol. The session keys in both the protocols will include ephemeral Diffie-Hellman key components, which ensure forward secrecy. However, the protocols will achieve *weak forward secrecy*, wherein the adversary has to remain passive during protocol execution. The security of the resulting two-party and three-party AB-AKE protocols will depend on the hardness of the computational Diffie-Hellman and bilinear Diffie-Hellman problems respectively along with the security of the underlying AB-AKE protocol (the security of the latter has been proven already).

Constructing AB-AKE protocols in the more general group setting needs more than one round. The compiler of Katz and Yung (KY) [23] turns an unauthenticated group key exchange protocol into an authenticated one. The compiler uses a public key based signature as an "authenticator" for this purpose. One may adapt the KY compiler to the attribute-based setting by replacing the normal public key based signature with an attribute-based signature [25]. The resulting compiler can then be applied to the two-round unauthenticated Burmester and Desmedt (BD) protocol [9] to achieve a three-round AB-AKE protocol with forward secrecy. Since the session key established by the BD protocol is ephemeral it achieves forward secrecy, where as the attribute-based KY compiler provides authentication. Although the attribute-based version of the KY compiler can be constructed with necessary changes to the KY compiler, it may not be straightforward. We leave this construction for future work.

6 Conclusion

We have initiated the concept of AB-AKE in the ciphertext-policy attribute-based system. Our modelling of AB-AKE assumes that each party has a set of attributes and a corresponding private key. A policy is defined (or negotiated) for each execution of the protocol and the parties satisfying the policy can establish a common shared key by executing the protocol. In the security model for AB-AKE, we have considered only outsider adversaries. Our security model can be extended by considering insider attackers who try to impersonate other protocol participants [22].

We have also introduced the concept of EP-AB-KEM. We then proposed a one-round generic AB-AKE protocol based on IND-CCA secure EP-AB-KEMs. For concrete instantiation of this protocol, we have presented an EP-AB-KEM and shown it secure under the IND-CCA notion in the generic group and random oracle models. As a consequence, a concrete AB-AKE protocol based on this EP-AB-KEM would also be secure in the generic group and random oracle models.

References

1. Ateniese, G., Kirsch, J., Blanton, M.: Secret Handshakes with Dynamic and Fuzzy Matching. In: Proceedings of the Network and Distributed System Security Symposium–NDSS 2007. The Internet Society (2007)
2. Beimel, A.: Secure Schemes for Secret Sharing and Key Distribution. PhD thesis, Israel Institute of Technology, Technion, Haifa, Israel (1996)
3. Bethencourt, J., Sahai, A., Waters, B.: Ciphertext-Policy Attribute-Based Encryption. In: IEEE Symposium on Security and Privacy, pp. 321–334. IEEE Computer Society, Los Alamitos (2007)
4. Birkett, J., Stebila, D.: Predicate-Based Key Exchange. Cryptology ePrint Archive, Report 2010/082 (2010); To appear at ACISP 2010,
http://eprint.iacr.org/2010/082
5. Bohli, J.M., Gonzalez Vasco, M.I., Steinwandt, R.: Secure group key establishment revisited. Int. J. Inf. Sec. 6(4), 243–254 (2007)
6. Boneh, D., Franklin, M.: Identity-Based Encryption from the Weil Pairing. In: Kilian, J. (ed.) CRYPTO 2001. LNCS, vol. 2139, pp. 213–229. Springer, Heidelberg (2001)
7. Boneh, D., Canetti, R., Halevi, S., Katz, J.: Chosen-Ciphertext Security from Identity-Based Encryption. SIAM J. Comput. 36(5), 1301–1328 (2007)
8. Boyd, C., Cliff, Y., González Nieto, J.M., Paterson, K.G.: One-Round Key Exchange in the Standard Model. International Journal of Applied Cryptography 1(3), 181–199 (2009)
9. Burmester, M., Desmedt, Y.: A Secure and Efficient Conference Key Distribution System (Extended Abstract). In: De Santis, A. (ed.) EUROCRYPT 1994. LNCS, vol. 950, pp. 275–286. Springer, Heidelberg (1995)
10. Canetti, R., Halevi, S., Katz, J.: Chosen-Ciphertext Security from Identity-Based Encryption. In: Cachin, C., Camenisch, J.L. (eds.) EUROCRYPT 2004. LNCS, vol. 3027, pp. 207–222. Springer, Heidelberg (2004)
11. Cheung, L., Newport, C.: Provably secure ciphertext policy ABE. In: CCS 2007: Proceedings of the 14th ACM conference on Computer and communications security, pp. 456–465. ACM, New York (2007)
12. Cramer, R., Shoup, V.: Design and Analysis of Practical Public-Key Encryption Schemes Secure against Adaptive Chosen Ciphertext Attack. SIAM J. Comput. 33(1), 167–226 (2004)
13. Dent, A.W.: A Designer's Guide to KEMs. In: Paterson, K.G. (ed.) Cryptography and Coding 2003. LNCS, vol. 2898, pp. 133–151. Springer, Heidelberg (2003)
14. Fujisaki, E., Okamoto, T.: How to Enhance the Security of Public-Key Encryption at Minimum Cost. In: Imai, H., Zheng, Y. (eds.) PKC 1999. LNCS, vol. 1560, pp. 53–68. Springer, Heidelberg (1999)
15. Fujisaki, E., Okamoto, T.: Secure Integration of Asymmetric and Symmetric Encryption Schemes. In: Wiener, M.J. (ed.) CRYPTO 1999. LNCS, vol. 1666, pp. 537–554. Springer, Heidelberg (1999)
16. Gorantla, M.C., Boyd, C., González Nieto, J.M.: Attribute-based Authenticated Key Exchange. Cryptology ePrint Archive, Report 2010/084 (2010),
http://eprint.iacr.org/2010/084

17. Gorantla, M.C., Boyd, C., González Nieto, J.M., Manulis, M.: Generic One Round Group Key Exchange in the Standard Model. In: 12th International Conference on Information Security and Cryptology–ICISC 2009. Springer, Heidelberg (2009)
18. Goyal, V., Jain, A., Pandey, O., Sahai, A.: Bounded Ciphertext Policy Attribute Based Encryption. In: Aceto, L., Damgård, I., Goldberg, L.A., Halldórsson, M.M., Ingólfsdóttir, A., Walukiewicz, I. (eds.) ICALP 2008, Part II. LNCS, vol. 5126, pp. 579–591. Springer, Heidelberg (2008)
19. Goyal, V., Pandey, O., Sahai, A., Waters, B.: Attribute-based encryption for fine-grained access control of encrypted data. In: Proceedings of the 13th ACM Conference on Computer and Communications Security–CCS 2006, pp. 89–98. ACM, New York (2006)
20. Jarecki, S., Liu, X.: Private Mutual Authentication and Conditional Oblivious Transfer. In: Halevi, S. (ed.) CRYPTO 2009. LNCS, vol. 5677, pp. 90–107. Springer, Heidelberg (2009)
21. Joux, A.: A One Round Protocol for Tripartite Diffie-Hellman. In: Bosma, W. (ed.) ANTS 2000. LNCS, vol. 1838, pp. 385–394. Springer, Heidelberg (2000)
22. Katz, J., Shin, J.S.: Modeling insider attacks on group key-exchange protocols. In: Proceedings of the 12th ACM Conference on Computer and Communications Security–CCS 2005, pp. 180–189. ACM, New York (2005)
23. Katz, J., Yung, M.: Scalable Protocols for Authenticated Group Key Exchange. In: Boneh, D. (ed.) CRYPTO 2003. LNCS, vol. 2729, pp. 110–125. Springer, Heidelberg (2003)
24. Lewko, A., Okamoto, T., Sahai, A., Takashima, K., Waters, B.: Fully Secure Functional Encryption: Attribute-Based Encryption and (Hierarchical) Inner Product Encryption. Cryptology ePrint Archive, Report 2010/100 (2010); To appear at EUROCRYPT 2010, http://eprint.iacr.org/2010/110
25. Maji, H., Prabhakaran, M., Rosulek, M.: Attribute-based signatures: Achieving attribute-privacy and collusion-resistance. Cryptology ePrint Archive, Report 2008/328 (2008), http://eprint.iacr.org/2008/328
26. Sahai, A., Waters, B.: Fuzzy Identity-Based Encryption. In: Cramer, R. (ed.) EUROCRYPT 2005. LNCS, vol. 3494, pp. 457–473. Springer, Heidelberg (2005)
27. Steinwandt, R., Corona, A.S.: Attribute-based group key establishment (unpublished manuscript)
28. Waters, B.: Ciphertext-Policy Attribute-Based Encryption: An Expressive, Efficient, and Provably Secure Realization. Cryptology ePrint Archive, Report 2008/290 (2008), http://eprint.iacr.org/

Optimally Tight Security Proofs for Hash-Then-Publish Time-Stamping

Ahto Buldas[1,2,3,*] and Margus Niitsoo[1,3,**]

[1] Cybernetica AS, Akadeemia tee 21, 12618 Tallinn, Estonia
[2] Tallinn University of Technology, Raja 15, 12618 Tallinn, Estonia
[3] University of Tartu, Liivi 2, 50409 Tartu, Estonia

Abstract. We study the security of hash-then-publish time-stamping schemes and concentrate on the tightness of security reductions from the collision-resistance of the underlying hash functions. While the previous security reductions create a quadratic loss in the security in terms of time-success ratio of the adversary being protected against, this paper achieves a notably smaller loss of power 1.5. This is significant for two reasons. Firstly, the reduction is asymptotically optimally tight, as the lower bound of 1.5 on the power was proven recently by the authors in ACISP 2009 and this is the first application for which optimality in this sense can be demonstrated. Secondly, the new reduction is the first one efficient enough to allow meaningful security guarantees to be given for a global-scale time-stamping service based on 256 bit hash functions, which considerably increases the efficiency of possible practical solutions.

1 Introduction

Time stamps are proofs that electronic data was created at certain time. Time stamps support rights protection as well as extending the lifetime of public key digital signatures considering the possible revocation of public-key certificates.

Before 1990, it was believed that the only possible way to achieve secure time-stamping is to use a trusted third party who adds time-readings to electronic data and then signs the data by using a public-key digital signature scheme. Although this scheme has been in use, it does have drawbacks. The assumption of a trusted third party is rather strong and often not feasible in the global corporate scale as nearly everyone has their own interests. Even when such a trusted party could be found, it is generally impossible to guarantee absolute security of the private signature keys. It would therefore be desirable to use time-stamping schemes that are free of secret keys and do not assume ultimate trustworthiness of third parties.

The so-called hash-then-publish time-stamping schemes were first introduced in 1990 by Haber and Stornetta [6] in connection with attempts to eliminate secret-based cryptography and trusted third parties from time-stamping schemes. In such a scheme,

---

* Supported by the European Regional Development Fund through the Estonian Center of Excellence in Comp. Sci., by Estonian SF grant no. 6944, and by EU FP6-15964: "AEOLUS".
** Supported by Estonian SF grant no. 6944 and the Tiger University Program of the Estonian Information Technology Foundation.

R. Steinfeld and P. Hawkes (Eds.): ACISP 2010, LNCS 6168, pp. 318–335, 2010.

a collection of N documents is hashed down to a single digest of few dozen bytes that is then published in a widely available medium such as a newspaper. Using Merkle hash trees [9] as a hashing scheme provides a possibility of creating compact certificates (of size $O(\log N)$) for each one of the N documents. To create such a certificate, it is sufficient to store all sibling hash values in the corresponding path in the hash tree from a document to the root of the tree. The sibling hash values are sufficient to re-compute the root hash value from the document and as such they can be used as a *proof of membership*. Based on this idea, Haber and Stornetta then drafted a large-scale time-stamping scheme [1] where a giant Merkle tree is created co-operatively by numerous servers all over the world and the root value is published in newspapers as the hash value of this particular unit of time. In such schemes N is potentially very large.

It might seem obvious that the security of hash-then-publish time-stamping schemes can be reduced to the collision-resistance of the hash function. However, the first correct security proof of such a scheme was published as late as 2004 [5]. It then became evident that the number N of time-stamps explicitly affects the efficiency (security guarantee) of the security proof. In the very first security proof [5] it was shown that if there is an adversary with running time t that is able to backdate a document with probability δ, then there is also a collision-finding adversary that works in time $t' \approx 2t$ and succeeds with probability $\delta' \approx \frac{\delta^2}{N}$. When measuring security in terms of time-success ratio introduced by Luby [8] we have to use $2N \cdot \frac{t}{\delta^2}$-collision resistant hash functions to have a $\frac{t}{\delta}$-secure time-stamping scheme. This means that the hash function must be roughly $\frac{2N}{\delta}$ times more secure against collisions than the time-stamping system constructed from it is against backdating. As N could be very large, the security requirements for the hash function may grow unreasonably large. Indeed, it is mentioned in [5] that such a security proof is practical only for hash functions with 400 or more output bits.

In [4], a more efficient security proof was given, where $\frac{t'}{\delta'} \approx 48\sqrt{N} \cdot \frac{t}{\delta^2}$. This was a considerable improvement because it allowed for much larger values of N. In this paper, we propose a new security reduction, where $\frac{t'}{\delta'} \approx 14\sqrt{N} \cdot \frac{t}{\delta^{1.5}}$, i.e. we get a power 1.5 reduction instead a quadratic one in terms of time-success ratio. This allows us to use shorter hash functions in practical applications while still maintaining good security guarantees. Based on a recently proved separation result [2] we also argue why the exponent 1.5 is the least achievable.

2   Notation

By $x \leftarrow \mathcal{D}$ we mean that x is chosen randomly according to a distribution $\mathcal{D}$. By $\mathbf{E}[X]$ we mean the average of a random variable X. If A is a probabilistic function or a Turing machine, then $x \leftarrow \mathsf{A}(y)$ means that x is chosen according to the output distribution of A on an input y. If $\mathcal{D}_1, \ldots, \mathcal{D}_m$ are distributions and $F(x_1, \ldots, x_m)$ is a predicate, then $\Pr[x_1 \leftarrow \mathcal{D}_1, \ldots, x_m \leftarrow \mathcal{D}_m : F(x_1, \ldots, x_m)]$ is the probability that $F(x_1, \ldots, x_m)$ is true after the ordered assignment of $x_1, \ldots, x_m$. For functions $f, g \colon \mathbb{N} \to \mathbb{R}$, we write $f(k) = O(g(k))$ if there are $c, k_0 \in \mathbb{R}$, so that $f(k) \le cg(k)$ $(\forall k > k_0)$. We write $f(k) = \omega(g(k))$ if $\lim\limits_{k \to \infty} \frac{g(k)}{f(k)} = 0$. If $f(k) = k^{-\omega(1)}$, then f is *negligible*. For every two functions $f(k)$ and $g(k)$, we will write $f \gtrsim g$ iff $f(k) \ge g(k) - k^{-\omega(1)}$. A Turing machine M is *poly-time* if it runs in time $k^{O(1)}$, where k is the input size.

Let $\mathcal{F} = \{\mathcal{F}_k\}_{k \in \mathbb{N}}$ be a function family such that every $h \leftarrow \mathcal{F}_k$ is a function $h\colon \{0,1\}^{\ell(k)} \to \{0,1\}^k$, where $\ell(k) = k^{O(1)}$ and $\ell(k) > k$ for every $k \geq 0$. We say that $\mathcal{F}$ is *collision-free* if for every poly-time (non-uniform) Turing machine A:

$$\Pr\left[h \leftarrow \mathcal{F}_k, (x, x') \leftarrow \mathsf{A}(1^k, h)\colon x \neq x', h(x) = h(x')\right] = k^{-\omega(1)} .$$

3   Hash-then-Publish Time-Stamping

A time-stamping procedure consists of the following two general steps:

1. Client sends a request $x \in \{0,1\}^k$ to Server.
2. Server binds x with a time value t and sends Client a time-certificate c.

Time-stamping protocols process requests in batches $\mathcal{X}_1, \mathcal{X}_2, \mathcal{X}_3 \ldots$ that we call *rounds*. The rounds correspond to time periods of fixed duration (one hour, one day, etc.) After the t-th period, a short commitment $r_t = \mathsf{Com}(\mathcal{X}_t)$ of $\mathcal{X}_t$ is published. A request $x \in \mathcal{X}_t$ precedes another request $x' \in \mathcal{X}_{t'}$ if $t < t'$. The requests of the same batch are considered simultaneous. For this scheme to be efficient there must be an efficient way to prove inclusions $x \in \mathcal{X}_t$, i.e. there is a verification algorithm Ver that on input a request x, a certificate c and a commitment r_t returns true if $x \in \mathcal{X}_t$. On the one hand, it should be easy to create certificates for the members $x \in \mathcal{X}_t$, i.e. there has to be an efficient certificate generation algorithm Cert that outputs a certificate $c = \mathsf{Cert}(x, \mathcal{X}_t)$. On the other hand, for security, it must be infeasible to create such proofs for non-members $y \notin \mathcal{X}_t$, i.e. it is hard to find a certificate c' so that $\mathsf{Ver}(y, c', r_t) = \mathsf{true}$.

Definition 1. *A time-stamping scheme is a triple* $\mathsf{T} = (\mathsf{Com}, \mathsf{Cert}, \mathsf{Ver})$ *of efficient algorithms, where:*

- Com *is a commitment algorithm which, on input a set* $\mathcal{X}$ *of requests, outputs a commitment* $r = \mathsf{Com}(\mathcal{X})$.
- Cert *is a certificate generation algorithm which, on input a set* $\mathcal{X}$ *and an element* $x \in \mathcal{X}$, *generates a certificate* $c = \mathsf{Cert}(x, \mathcal{X})$.
- Ver *is a verification algorithm which, on input a request* x, *a certificate* c *and a commitment* r, *outputs* yes *or* no, *depending on whether* x *is a member of* $\mathcal{X}$ *(the set that corresponds to the commitment* r). *It is assumed that for every set* $\mathcal{X}$ *of requests and every member-request* $x \in \mathcal{X}$ *the following correctness condition holds:*

$$\mathsf{Ver}(x, \mathsf{Cert}(x, \mathcal{X}), \mathsf{Com}(\mathcal{X})) = \mathsf{yes} . \tag{1}$$

3.1   Security Condition for Time-Stamping Schemes

It was shown in [5] that giving a consistent security definition for *hash-then-publish* time-stamping schemes is not an easy task. Intuitively, a time-stamping adversary *back-dates a document that never existed before*, but the "existence" itself is not that easy to capture in formal definitions. In this paper, we use the so-called *entropy-based security* condition [3] that models the "fresh" documents by using high-entropy distributions.

Such approach has been the most common in this line of research. This security condition is inspired by the following attack-scenario with a malicious Server:

1. Server computes a commitment r and publishes it. Server is potentially malicious, so there are no guarantees that r is created by applying Com to a set X of requests.
2. Alice creates an invention $\mathcal{D}_A \in \{0, 1\}^*$ and protects it by obtaining a time stamp.
3. Some time later, $\mathcal{D}_A$ is disclosed to the public and Server tries to steal it by showing that the invention was known to Server long before Alice time-stamped it. He creates a slightly modified version $\mathcal{D}'_A$ of A, i.e. changes the invertor's name, modifies the creation time, and possibly rewords the document in a suitable way.
4. Finally, Server back-dates a hash value x of $\mathcal{D}'_A$, by finding a certificate c, so that $\mathsf{Ver}(x, c, r) = \mathsf{yes}$. It is shown in [3] that the hash function that computes x from $\mathcal{D}'_A$ must convert poly-sampleable high entropy input distributions to high entropy output distributions, and this is in fact also a sufficient condition.

Security definitions for time-stamping are usually based on this scenario. However, to our knowledge, there have been no academic discussions whether such a scenario is sufficient for the security level we really expect. One major assumption that has been made here is that before creating and publishing the commitment r, Server has no information about the invention $\mathcal{D}_A$. For example, if Alice obtains a time stamp for $\mathcal{D}_A$ from malicious Server *before* r is published (i.e. steps 1 and 2 are exchanged) then during the computation of r, Server knows the time stamp request x which is partial information about $\mathcal{D}_A$. So, one may imagine that Server tries to extract useful information from x about $\mathcal{D}_A$, create a request x' for a similar document $\mathcal{D}'_A$ that describes the same invention, and then refuse to issue a time stamp for Alice. If such an attack succeeds, Server has the earliest time stamp for Alice's invention. But there are many practical objections against such an attack:

– Time stamp requests only contain a relatively short hash value of $\mathcal{D}_A$ which (in practice) can hardly contain any useful information about the invention.
– It is improbable that all time-stamping servers could be simultaneously corrupted and Alice is usually free to commit to several of them at the same time. This means that malicious servers who try to delay the publishing of r in order to have more time for creating x' based on x will "lose the race" against honest servers who create their time stamps earlier.

So, the assumption that server has no information about $\mathcal{D}_A$ when publishing r is heuristic but still justified in practice and hence it is reasonable to study the security of time-stamping schemes under such assumption.

To formalize such an attack, a two-staged adversary $A = (A_1, A_2)$ is used. The first stage A_1 computes and outputs r and an advice string a, which contains useful information for the second stage A_2. Note that a may contain all random coins of A_1, which makes all useful information that A_1 gathered available to A_2. After that, the second stage A_2 finds a new x (which is assumed to be a random variable with a sufficient amount of entropy) and a certificate c such that $\mathsf{Ver}(x, c, r) = \mathsf{yes}$. Note that x must be unpredictable because otherwise x could have been pre-computed by A_1 and there would be nothing wrong in proving that x existed before r was computed and published.

Hence, for defining the security of time-stamping schemes, the class of possible adversaries is restricted. Only adversaries that produce unpredictable x are considered [3]. A poly-time adversary (A_1, A_2) is *unpredictable* if for every poly-time predictor Π:

$$\Pr\left[(r, a) \leftarrow A_1(1^k), x' \leftarrow \Pi(r, a), (x, c) \leftarrow A_2(r, a)\colon x' = x\right] = k^{-\omega(1)} . \quad (2)$$

It is reasonable to assume that a contains all internal random coins of A_1 (see [3] for more details). An equivalent definition for the unpredictability of A is that the probability $\Pr[\mathsf{Equ}]$ that $A_2(r, a)$ outputs the same x twice is negligible. We can also say that x should have large (super-logarithmic in k) conditional min entropy $H_\infty(x \mid A_1(1^k))$.

Definition 2. *A time-stamping scheme is secure if for every unpredictable* (A_1, A_2):

$$\Pr\left[(r, a) \leftarrow A_1(1^k), (x, c) \leftarrow A_2(r, a)\colon \mathsf{Ver}(x, c, r) = \mathsf{yes}\right] = k^{-\omega(1)} . \quad (3)$$

3.2   Hash Tree Time-Stamping Schemes

The commitments r_t are computed as the root hash values of Merkle hash trees [9]. To make the paper more self-contained, we outline the basic facts about hash-chains and how they are used in time-stamping. We use the notation and definitions introduced in [3]. By () we mean an empty list.

Definition 3 (Hash-Chain). *Let* $h\colon \{0,1\}^{2k} \to \{0,1\}^k$ *be a twice-compressing hash function and* $x, y \in \{0,1\}^k$. *By an* h-link *from* x *to* y *we mean a pair* (s, b), *where* $s \in \{0,1\}^k$ *and* $b \in \{0,1\}$, *such that either* $b = 0$ *and* $y = h(x\|s)$, *or* $b = 1$ *and* $y = h(s\|x)$. *By an* h-chain *from* x *to* y *we mean a (possibly empty) list* $c = ((s_1, b_1), \ldots, (s_\ell, b_\ell))$ *of* h-links, *such that either* $c = ()$ *and* $x = y$; *or* (2) *there is a sequence* $y_0, y_1, \ldots, y_\ell$ *of* k-bit strings, *such that* $x = y_0$, $y = y_\ell$, *and* (s_i, b_i) *is an* h-link *from* y_{i-1} *to* y_i *for every* $i \in \{1, \ldots, \ell\}$. *We denote by* $\mathsf{Chain}_h(x, c) = y$ *the proposition that* c *is an* h-chain *from* x *to* y. *Note that* $\mathsf{Chain}_h(x, ()) = x$ *for every* $x \in \{0,1\}^k$. *By the* shape $\rho(c)$ *of* c *we mean the* ℓ-bit string $b_1 b_2 \ldots b_\ell$.

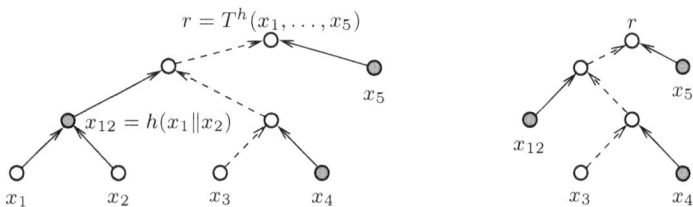

Fig. 1. A hash tree for $\mathcal{X} = \{x_1, \ldots, x_5\}$ and the hash chain $c = ((x_4, 0), (x_{12}, 1), (x_5, 0))$ with shape $\rho(c) = 010$ for x_3

Hash-tree time-stamping schemes use Merkle trees to compute the commitments r_t for batches $\mathcal{X}_t$. The commitment $\mathsf{Com}(\mathcal{X}_t)$ of a batch $\mathcal{X}_t = \{x_1, \ldots, x_N\}$ is $r_t = T^h(x_1, \ldots, x_N) \in \{0,1\}^k$, where T^h is a tree-shaped hashing scheme. A certificate for $x \in \mathcal{X}_t$ is a hash chain c such that $\mathsf{Chain}_h(x, c) = r_t$. The verification procedure $\mathsf{Ver}(x, c, r_t)$ returns yes if $\mathsf{Chain}_h(x, c) = r_t$. In this work, we denote the hash-tree time-stamping scheme by T^h. An example of a hash-tree scheme is depicted in Fig. 1.

Hash-forest time-stamping schemes are obvious generalizations of hash tree schemes. Input for these schemes is a sequence of batches $\mathcal{X}_1, \mathcal{X}_2, \ldots, \mathcal{X}_m$ and the commitments are sequences $r = (r_1, r_2, \ldots, r_m)$ of hash values, where every $r_i = \mathsf{Com}(\mathcal{X}_i)$ is computed by using a hash-tree scheme. A certificate for $x \in \mathcal{X}_t$ is a pair $c = (c', t)$ where c' is a hash chain such that $\mathsf{Chain}_h(x, c') = r_t$. The verification procedure $\mathsf{Ver}(x, c, r)$, having as input a request x, a certificate $c = (c', t)$, and a commitment $r = (r_1, r_2, \ldots, r_m)$ returns yes whenever there is $t \in \{1, \ldots, m\}$ and $\mathsf{Chain}_h(x, c) = r_t$. By the *shape* $\rho(c)$ of $c = (c', t)$ we mean the pair $(\rho(c'), t)$.

4   Existing Security Proofs

It was shown in [5] that this scheme cannot be proved secure in a traditional black-box way by assuming only the one-wayness and collision-resistance of h. In [5] they also define a restricted scheme, with a modified verification procedure that uses a set N of *allowed shapes* with size $|N| = N$ and the verification procedure Ver was completed with an additional check for $\rho(c) \in N$. Note that N can be considered as the total capacity of the time-stamping system, i.e. the total number of time-stamps that can be securely issued in the system. All the known security proofs for hash-tree or hash-forest schemes use the following general collision-extraction property:

Definition 4 (Collision-Extraction Property). *If* $\mathsf{Ver}^h(x_1, c_1, r) = \mathsf{Ver}^h(x_2, c_2, r) =$ *yes,* $\rho(c_1) = \rho(c_2)$, *and* $(x_1, c_1) \neq (x_2, c_2)$, *then the h-calls of* $\mathsf{Ver}^h(x_i, c_i, r)$ *$(i = 1, 2)$ comprise an h-collision.*

Essentially, this means that given two certificates of the same shape, we can always find a collision. For hash trees or hash forests it is rather easy to see: if two different hash chains c and c' have the same shape and the same root value, there must be an index l such that $c_l \neq c'_l$ but $h(c_l) = h(c'_l)$ which gives the collision that we need. Note that this property also implies that the number of different time-stamp requests per round is limited to N, for otherwise we would have a collision to the hash function we use.

We now proceed to describe the reduction itself. However, in order to give a better intuition to the results we use an iterative process of proving increasingly more precise bounds. All security reductions we illustrate use the following general schema. Having an adversary $\mathsf{A} = (\mathsf{A}_1, \mathsf{A}_2)$ for a time-stamping scheme T^h with success

$$\delta(k) = \mathsf{Pr}\left[h \leftarrow \mathcal{F}_k, (r, a) \leftarrow \mathsf{A}_1(1^k, h), (x, c) \leftarrow \mathsf{A}_2(r, a) \colon \mathsf{Ver}^h(x, c, r) = \mathsf{yes}\right] . \quad (4)$$

and running time $t = t(k)$, we construct a collision finder $\mathsf{CF}_k^{h, \mathsf{A}, \mathsf{T}}(m)$ (Fig. 2) with approximate running time $t' \approx m \cdot t$, where m is a reduction-specific parameter and then analyze the success δ' of the collision finder. Although the running time t and the success δ of A depend on the security parameter k, we will use the shorthand notations t and δ instead of $t(k)$ and $\delta(k)$. Let Equ denote the event that $x_i = x_j$ for some

1. Compute $(r, a) \leftarrow A_1(1^k, h)$.
2. Generate m independent samples: $(x_1, c_1) \leftarrow A_2(r, a), \ldots, (x_m, c_m) \leftarrow A_2(r, a)$.
3. Find $x_i \neq x_j$ such that $\mathsf{Ver}^h(x_i, c_i, r) = \mathsf{Ver}^h(x_j, c_j, r) = $ yes and $\rho(c_i) = \rho(c_j)$.
4. If such a pair was found, use it to extract a collision and output it. Otherwise, output $\perp$.

Fig. 2. Generic collision finder $\mathsf{CF}_k^{h, A, T}(m)$

$i \neq j$ and $\overline{\mathsf{Equ}}$ denote the opposite event, i.e. that all the x_i-s are different. Considering the collision-extraction property, it would be good if all the successfully back-dated bit-strings were different because then it would be sufficient to find two back-dating certificates of the same shape. Let Coll denote the event that $\mathsf{CF}_k^{h, A, T}(m)$ finds a collision for h. A general estimate for the success of the collision finder $\mathsf{CF}_k^{h, A, T}(m)$ is:

$$\Pr\left[\mathsf{Coll}\right] \geq \Pr\left[\mathsf{Coll} \cap \overline{\mathsf{Equ}}\right] = \Pr\left[\mathsf{Coll} \mid \overline{\mathsf{Equ}}\right] \cdot (1 - \Pr\left[\mathsf{Equ}\right]) \gtrsim \Pr\left[\mathsf{Coll} \mid \overline{\mathsf{Equ}}\right] ,$$

because $\Pr\left[\mathsf{Equ}\right] = k^{-\omega(1)}$ due to the unpredictability of (A_1, A_2). We can therefore neglect the event Equ in the analysis on the security reductions, i.e. we can just assume that all $x_1, \ldots, x_m$ are different, and use the fact that the success probability of the collision-finder is $\delta' \gtrsim \Pr\left[\mathsf{Coll} \mid \overline{\mathsf{Equ}}\right]$. Let

$$\Pr[h, r, a] = \Pr\left[H \leftarrow \mathcal{F}_k, (R, A) \leftarrow A_1(1^k): H = h, R = r, A = a\right] ,$$

$$\delta_{h,r,a}^{(n)} = \Pr\left[(x, c) \leftarrow A_2(r, a): \mathsf{Ver}^h(x, c, r) = \text{yes}, \rho(c) = n\right] ,$$

$$\delta^{(n)} = \mathop{\mathbf{E}}_{h,r,a}\left[\delta_{h,r,a}^{(n)}\right], \text{ and} \tag{5}$$

$$\delta_{h,r,a} = \delta_{h,r,a}^{(1)} + \ldots + \delta_{h,r,a}^{(N)} .$$

We have $\delta = \sum_{h,r,a} \Pr[h, r, a] \cdot \delta_{h,r,a} = \mathop{\mathbf{E}}_{h,r,a}[\delta_{h,r,a}]$ and $\delta = \delta^{(1)} + \ldots + \delta^{(N)}$. The success probability of the collision finder is:

$$\delta' \gtrsim \sum_{h,r,a} \Pr[h, r, a] \cdot f(m; \delta_{h,r,a}^{(1)}, \ldots, \delta_{h,r,a}^{(N)}) = \mathop{\mathbf{E}}_{h,r,a}\left[f(m; \delta_{h,r,a}^{(1)}, \ldots, \delta_{h,r,a}^{(N)})\right] ,$$

where $f(m; \delta_1, \ldots, \delta_N)$ is a function that computes the probability that $\mathsf{CF}_k^{h, A, T}(m)$ made at least two successive A_2-calls (among the total m) with the same certificate shape (Tab. 1). For example, if $N = 1$ (we have only one shape) and $m = 2$ then $f(m, \delta) = \delta^2$ and by the Jensen inequality $\delta' \gtrsim \mathop{\mathbf{E}}_{h,r,a}\left[\delta_{h,r,a}^2\right] \geq \left(\mathop{\mathbf{E}}_{h,r,a}[\delta_{h,r,a}]\right)^2 = \delta^2$.

4.1   Tightness Measure for Security Reductions

In order to compare the efficiency of adversaries with different running time and success probability, Luby [8] introduced *time-success ratio* $\frac{t}{\delta}$, where t is the running time and δ

Table 1. The success function $f(m; \delta_1, \ldots, \delta_N)$ and its special cases

	$N = 1$	Arbitrary N
$m = 2$	$f(2; \delta) = \delta^2$	$f(2; \delta_1, \ldots, \delta_N) = \delta_1^2 + \ldots + \delta_N^2$
Arbitrary m	$f(m; \delta) =$ $1 - m\delta(1-\delta)^{m-1} - (1-\delta)^m$	$f(m; \delta_1, \ldots, \delta_N) =$ $1 - \sum_{j=0}^{m} \binom{m}{j} j! \sigma_j(\delta_1, \ldots, \delta_N)(1-\delta)^{m-j}$

is the success of the adversary. A cryptographic primitive is said to be S-secure if every adversary has time-success ratio $\frac{t}{\delta} \geq S$. In terms of exact security, this means that the primitive is (t, δ)-secure for every t and δ with $\frac{t}{\delta} \geq S$. Time-success ratio provides a general measure for the tightness of cryptographic reductions. If the time-success ratio $\frac{t'}{\delta'}$ of the constructed adversary (i.e. $\mathsf{CF}_k^{h,\mathsf{A},\mathsf{T}}(m)$) is represented as a function $\frac{t'}{\delta'} = F(t, \frac{1}{\delta})$, where t and δ are the running time and the success of the assumed adversary (i.e. $(\mathsf{A}_1, \mathsf{A}_2)$), then the reduction is *tight* if F grows slowly and *loose* if the growth is faster. The reduction is said to be *linear* if $F(a, b) = O(a) \cdot O(b)$, *quadratic* if $F(a, b) = O(a^2) \cdot O(b^2)$, and *polynomial* if $F(a, b) = a^{O(1)} \cdot b^{O(1)}$. The equation $\frac{t'}{\delta'} = F(t, \frac{1}{\delta})$ is also called as the *security loss (formula)* of the reduction.

4.2   Reduction with Quadratic Security Loss

To get a security reduction with quadratic security loss, we take $m = 2$ and use the estimate [1] $f(2; \delta_1, \ldots, \delta_N) \geq N \cdot f(2; \frac{\delta_1 + \ldots + \delta_N}{N})$, and hence by using Jensen inequality

$$\delta' \gtrsim \mathop{\mathbf{E}}_{h,r,a}\left[f(2; \delta_{h,r,a}^{(1)}, \ldots, \delta_{h,r,a}^{(N)})\right] \geq \mathop{\mathbf{E}}_{h,r,a}\left[N \cdot f\left(2; \frac{\delta_{h,r,a}}{N}\right)\right] = N \cdot \mathop{\mathbf{E}}_{h,r,a}\left[\left(\frac{\delta_{h,r,a}}{N}\right)^2\right] \geq \frac{\delta^2}{N} .$$

Such a reduction has the security loss formula $\frac{t'}{\delta'} \approx 2N \cdot \frac{t}{\delta^2}$ and was given in [5].

4.3   Reducing the Power of N

Buldas and Laur [4] used the birthday bound to improve the efficiency of the reduction. Their main idea was to use the collision-finder $\mathsf{CF}_k^{h,\mathsf{A},\mathsf{T}}(m)$ with $m = \frac{\sqrt{N}}{\delta}$ instead of $\mathsf{CF}_k^{h,\mathsf{A},\mathsf{T}}(2)$. After generating the samples $(x_1, c_1), \ldots, (x_m, c_m)$ and verifying them with Ver_k, the collision finder $\mathsf{CF}_k^{h,\mathsf{A},\mathsf{T}}(m)$ has on average $\delta m = \sqrt{N}$ successfully back-dated bit-strings on average. The birthday bound implies that with a probability of roughly $\frac{1}{2}$ we then have two successfully back-dated bit strings with the same shape n.

---

[1] This holds because $\frac{\delta_1^2 + \ldots + \delta_N^2}{N} \geq \left(\frac{\delta_1 + \ldots + \delta_N}{N}\right)^2$ due to the convexity of the square function.

These can then be used to extract a collision. This idea was made precise in [4] and resulted in a reduction with security loss [2] $\frac{t'}{\delta'} \approx 48\sqrt{N} \cdot \frac{t}{\delta^2}$. Their reduction was the best known for this problem so far.

5 New Reduction

We now establish a power 1.5 reduction by first showing an inefficient reduction and then using combinatorial counting arguments to make it considerably more efficient. Finally, we obtain a reduction with security loss $\frac{t'}{\delta'} = 14\sqrt{N} \cdot \frac{t}{\delta^{1.5}}$. For this, we use $\mathsf{CF}_k^{h,\mathsf{A},\mathsf{T}}(m)$ with $m = \Theta\left(\sqrt{\frac{N}{\delta}}\right)$. We start from the case $N = 1$ when all certificates have the same shape and we only need two successful A_2 calls to get a collision. If the success of A_2 is δ, the success of $\mathsf{CF}_k^{h,\mathsf{A},\mathsf{T}}(m)$ is:

$$f(m, \delta) = 1 - m\delta(1 - \delta)^{m-1} - (1 - \delta)^m , \tag{6}$$

where the first negative term is the probability that only one call is successful and the second negative term is the probability that no call was successful. To explain the theoretical obstacles we will meet when going from power 2.0 to power 1.5 reductions, we first show why it is not trivial to construct a linear reduction even for the case $N = 1$.

5.1 Problems with Establishing a Linear Reduction

It might seem that when $N = 1$, it is nearly trivial to construct a linear reduction with security loss $\frac{t'}{\delta'} = c \cdot \frac{t}{\delta}$. One could just take $m = \max\left\{2, \lceil\frac{1}{\delta}\rceil\right\}$, where δ is the success of the back-dating adversary $(\mathsf{A}_1, \mathsf{A}_2)$, and the success δ' of C will be:

$$\delta' \approx f\left(\frac{1}{\delta}, \delta\right) = \begin{cases} 1 - (1 - \delta)^{\frac{1-\delta}{\delta}} - (1 - \delta)^{\frac{1}{\delta}} & \text{if } \delta < \frac{1}{2} \\ 1 - 2\delta(1 - \delta) - (1 - \delta)^2 = \delta^2 & \text{if } \delta \geq \frac{1}{2} \end{cases} .$$

It is easy to see that $\lim_{\delta\to 0} f\left(\frac{1}{\delta}, \delta\right) = 1 - 2e^{-1} \approx 0.26424 \geq \frac{1}{4}$ and if the running time of A_2 is t, we seemingly have that the time success ratio of C is $\frac{t'}{\delta'} \approx 4 \cdot \frac{t}{\delta}$.

However, this approach overlooks the fact that h is randomly chosen and therefore the probability δ in (6) depends on particular choices of h and also on the output (r, a) of A_1. This means that the success of C is the mathematical expectation $\underset{h,r,a}{\mathbf{E}}\left[f(m, \delta_{h,r,a})\right]$. As f turns out not to be convex, Jensen's inequality cannot be used and the averaging becomes a nontrivial task in which the power of δ necessarily has to increase.

5.2 Tightness Bounds for Security Reductions

It is easy to see that any hash function used in hash-then-publish time-stamping schemes[3] must be *division-resistant* [2], i.e. any poly-time adversary $\mathsf{A} = (\mathsf{A}_1, \mathsf{A}_2)$ has success:

$$\Pr\left[h \leftarrow \mathcal{F}_k, (y,a) \leftarrow \mathsf{A}_1(h), x_1 \leftarrow \{0,1\}^k, x_2 \leftarrow \mathsf{A}_2(y,a,x_1): h(x_1\|x_2) = y\right] = k^{-\omega(1)} .$$

---

[2] The larger constant is due to technical reasons and could probably be reduced somewhat.

[3] More precisely, in schemes where the set $\mathcal{N}$ of allowed shapes contains at least one shape that begins with a 0-bit. In all schemes that are used in practice, this is indeed the case. If for some reasons, all allowed shapes begin with a 1-bit, then we can show in a similar way that h must satisfy a dual condition with success predicate $h(x_2\|x_1) = y$ instead of $h(x_1\|x_2) = y$.

Indeed, if there is $A = (A_1, A_2)$, such that $\text{ADV}_k(A) = \delta$, then we construct (A'_1, A'_2) so that A'_1 first calls $(r, a) \leftarrow A_1$, creates an h-chain $c' = ((s_1, b_1), \ldots, (s_\ell, b_\ell))$ such that $0b_1 \ldots b_\ell \in \mathcal{N}$, and with output $\text{Chain}_r(c', =)r'$, and outputs (r', a') where $a' = (a, r, c')$. The second stage $A_2(r', a')$ first generates a random $x \leftarrow \{0,1\}^k$, then executes $x_2 \leftarrow A_2(r, a, x)$ and outputs (x, c), where $c = ((x_2, 0), (s_1, b_1), \ldots (s_\ell, b_\ell))$. It is easy to see that the modified adversary is unpredictable (because x_1 is chosen independent of y and uniformly at random) and breaks the h-based time-stamping scheme in terms of (3) with success δ.

By using oracle separation techniques it has been proved [2] that every black-box security reduction that derives division-resistance from the collision-resistance of the same function is at least a power-1.5 reduction. Hence, power-1.5 black-box reductions are also the best we can get when proving entropy-based security of a hash-then-publish time-stamping scheme from the collision-resistance of the underlying hash function.

5.3 New Reduction: The Case $N = 1$

If $m = \max\{\frac{1}{\sqrt{\delta}}, 2\}$, the success of the generic collision-finder $\text{CF}_k^{h,A,T}(m)$ is:

$$\delta' \gtrsim \mathop{\mathbf{E}}_{h,r,a} [f(m, \delta_{h,r,a})] = \sum_{h,r,a} \Pr[h, r, a] \cdot f(m, \delta_{h,r,a}) \ . \tag{7}$$

Note that, in general, $\delta' \not\gtrsim f(m, \mathop{\mathbf{E}}_{h,r,a}[\delta_{h,r,a}]) = f(m, \delta)$ because f is not convex and we cannot apply the Jensen inequality directly. However, $f(m, \delta)$ is convex in the interval $\left[0 \ldots \frac{1}{m-1}\right]$ (Lemma 2 in Appendix A) and lower bounded by the identity function in the interval $\left[\frac{1}{m-1} \ldots 1\right]$ (Lemma 4 in Appendix A). Defining

$$p = \sum_{\substack{h,r,a \\ \delta_{h,r,a} \geq \frac{1}{m-1}}} \Pr[h, r, a] \cdot \delta_{h,r,a} \ ,$$

we estimate the success δ' of the collision-finder as follows:

$$\delta' \gtrsim \sum_{\substack{h,r,a \\ \delta_{h,r,a} < \frac{1}{m-1}}} \Pr[h, r, a] \cdot f(m, \delta_{h,r,a}) + \sum_{\substack{h,r,a \\ \delta_{h,r,a} \geq \frac{1}{m-1}}} \Pr[h, r, a] \cdot f(m, \delta_{h,r,a}) \geq f(m, \delta - p) + p \ ,$$

where the first sum is lower-bounded by using Lemma 3 of Appendix A. From the observation that $p \geq \frac{\delta}{6}$ or $\delta - p \geq \frac{5\delta}{6}$, and that $f\left(m, \frac{5\delta}{6}\right) \geq \frac{\delta}{6}$ (Appendix B), it follows that $\delta' \gtrsim \frac{\delta}{6}$. The security loss of the reduction is $\frac{t'}{\delta'} \approx 6 \cdot \frac{t}{\delta^{1.5}}$.

5.4 New Reduction: General Case

We simply use the fact that from $\delta = \delta^{(1)} + \ldots + \delta^{(N)}$ it follows that there is $n \in \mathcal{N}$ such that $\delta^{(n)} \geq \frac{\delta}{N}$. We now take $m = \max\{\frac{1}{\sqrt{\delta^{(n)}}}, 2\}$ and modify the adversary A_2 so that it only outputs (x, c) if $\rho(c) = n$. The success of A is $\delta^{(n)}$ by the defining equation (5). Hence, we reduced the general case to the case $N = 1$ and the success of the collision finder $\mathsf{CF}_k^{h,\mathsf{A},\mathsf{T}}(m)$ is $\delta' \gtrsim \frac{\delta^{(n)}}{6} \geq \frac{\delta}{6N}$ and the security loss of the reduction is $\frac{t'}{\delta'} \approx 6 \cdot N^{1.5} \cdot \frac{t}{\delta^{1.5}}$. In the next section, we show that $N^{1.5}$ can actually be reduced to $\sqrt{N}$ which makes our reduction strictly better than the one given in [4].

5.5 New Reduction: Reducing the Power of N

The adversary previously considered only used collisions for the most probable certificate shape. We can get significantly better bounds if we try to take advantage of all possible collisions. We again use $\mathsf{CF}_k^{h,\mathsf{A},\mathsf{T}}(m)$ as our adversary construction. However, we try to bound the success probability δ' of the collision-finder tighter than before. It is clear that the adversary can fail to find a collision only when all the certificates returned by the time-stamping adversary are of different shapes or when two certificates coincide completely. The readers who are not interested in mathematical details of the proof may skip this subsection.

We analyze what happens if the advice a, the hash function h, and the root hash value r for A_1 have been fixed already. Then the probability of all the successfully back-dated certificates having different shapes after m tries is

$$\sum_{k=0}^{m}\binom{m}{k}k!\sigma_k(\delta^{(1)}\ldots\delta^{(N)})(1-\delta)^{m-k} = \sum_{k=0}^{m}\binom{m}{k}\binom{N}{k}k!S_k(\delta^{(1)}\ldots\delta^{(N)})(1-\delta)^{m-k}$$

$$\leq \sum_{k=0}^{m}\binom{m}{k}\binom{N}{k}k!S_1(\delta^{(1)}\ldots\delta^{(N)})^k(1-\delta)^{m-k} = \sum_{k=0}^{m}\binom{m}{k}\frac{N^{\underline{k}}}{N^k}\delta^k(1-\delta)^{m-k} \quad , \quad (8)$$

where σ_k is the k-th *elementary symmetric polynomial*, $S_k = \sigma_k/\binom{N}{k}$ and $N^{\underline{k}} = N\cdot(N-1)\cdot\ldots\cdot(N-k+1)$ is the *falling factorial power*. The *MacLaurin's inequality* says that $\sqrt[k]{S_k} \leq \sqrt[l]{S_l}$ whenever $k \geq l$ and $\delta_i \geq 0$. Now note that

$$\delta^k(1-\delta)^{m-k} = \sum_{i=0}^{m-k}(-1)^i\binom{m-k}{i}\delta^{i+k} = \sum_{j=k}^{m}(-1)^{j-k}\binom{m-k}{j-k}\delta^j \quad .$$

We plug this into (8), change the order of summation and use $\binom{m}{k}\binom{m-k}{j-k} = \binom{m}{j}\binom{j}{k}$ to get

$$S = \sum_{k=0}^{m}\binom{m}{k}\frac{N^{\underline{k}}}{N^k}\left(\sum_{j=k}^{m}(-1)^{j-k}\binom{m-k}{j-k}\delta^j\right) = \sum_{k=0}^{m}\sum_{j=k}^{m}(-1)^{j+k}\frac{N^{\underline{k}}}{N^k}\binom{m}{k}\binom{m-k}{j-k}\delta^j =$$

$$= \sum_{j=0}^{m}\sum_{k=j}^{m}(-1)^{j+k}\frac{N^{\underline{k}}}{N^k}\binom{m}{j}\binom{j}{k}\delta^j = \sum_{j=0}^{m}(-1)^j\binom{m}{j}\left(\sum_{k=0}^{j}(-1)^k\frac{N^{\underline{k}}}{N^k}\binom{j}{k}\right)\delta^j \quad .$$

Computing the first few terms we get $1 - \frac{1}{N}\binom{m}{2}\delta^2 + \frac{2}{N^2}\binom{m}{3}\delta^3 + \frac{3N-6}{N^3}\binom{m}{4}\delta^4 + \ldots$. Denote $\phi_n = \sum_{k=0}^{n}(-1)^k\frac{N^{\underline{k}}}{N^k}\binom{n}{k}$ and $\psi_n = \binom{m}{n}|\phi_n|\delta^n$. It turns out that ϕ_n satisfy the recurrence[4] $\phi_{k+1} = \frac{k}{N}(\phi_k - \phi_{k-1})$. Assuming $c_1\sqrt{\frac{N}{\delta}} + 1 \leq m \leq c_2\sqrt{\frac{N}{\delta}}$, we get

---

[4] This recurrence was found using Zeilberger's algorithm [10]. See Appendix C for a proof.

$$\psi_{k+1} = \binom{m}{k+1}\frac{k}{N}|\phi_{k-1}-\phi_k|\,\delta^{k+1} \le \binom{m}{k+1}\frac{2k}{N}\max(|\phi_{k-1}|,|\phi_k|)\delta^{k+1}$$

$$= \max\left(\frac{2(m-k)(m-k-1)}{(k+1)N}\binom{m}{k-1}|\phi_{k-1}|,\;\frac{2(m-k)k}{(k+1)N}\binom{m}{k}|\phi_k|\right)\delta^{k+1}$$

$$\le \max\left(\frac{c_2^2 N}{N\delta}\psi_{k-1}\delta^2,\;\frac{2c_2\sqrt{N}}{N\sqrt{\delta}}\psi_k\delta\right) = c_2\sqrt{\delta}\max\left(\frac{2}{\sqrt{N}}\psi_k, c_2\sqrt{\delta}\psi_{k-1}\right).$$

To simplify further analysis we assume that $N \ge 4$. By noting that $\psi_1 = 0$, we get that $\psi_3 \le c_2\sqrt{\delta}\psi_2$, $\psi_4 \le c_2^2\delta\psi_2$ and in general, $\psi_k \le (c_2\sqrt{\delta})^{k-2}\psi_2$ for all $k \ge 2$ which can be easily verified by induction. Using this, we get a simple bound on the sum of the remaining elements if we assume $c_2\sqrt{\delta} < 1$:

$$\left|\sum_{k=3}^{m}(-1)^k\binom{m}{k}\phi_k\delta^k\right| \le \sum_{k=3}^{m}\psi_k \le \sum_{k=1}^{m-2}(c_2\sqrt{\delta})^k\psi_2 \le \frac{c_2\sqrt{\delta}\psi_2}{1-c_2\sqrt{\delta}}.$$

We thus know that the success of the adversary for fixed h, r and a is at least

$$f(N,\delta) \ge \left(1-\frac{c_2\sqrt{\delta}}{1-c_2\sqrt{\delta}}\right)\frac{1}{N}\binom{m}{2}\delta^2 \ge \frac{1-2c_2\sqrt{\delta}}{N(1-c_2\sqrt{\delta})}\frac{c_1^2 N}{2\delta}\delta^2 = \frac{c_1^2(1-2c_2\sqrt{\delta})}{2(1-c_2\sqrt{\delta})}\delta.$$

We analyze the lower bound described for convexity. Assuming $\frac{c_1}{c_2} = const.$ we can substitute $c_2\sqrt{\delta} = x$ and disregard a constant multiplier to get $x^2\frac{1-2x}{1-x}$ which is easily seen to be convex whenever $x < 1 - \frac{1}{\sqrt[3]{2}} \approx 0.2$. In order to guarantee the convexity of the approximation for f we need to have $c_2\sqrt{\delta} \le 1 - \frac{1}{\sqrt[3]{2}}$ for all possible δ. As $\delta \le 1$, this can easily be achieved by taking $c_2 \le 1 - \frac{1}{\sqrt[3]{2}}$.

Let $\delta_{h,r,a}$ denote the success when h, r and a are fixed and let $\delta = \mathop{\mathbb{E}}\limits_{h,r,a}[\delta_{h,r,a}]$ be the average success. Since f is convex for δ when we fix c_2 as described, we can use the Jensen inequality to get $\bar{f}(N,\delta) = \mathop{\mathbb{E}}\limits_{h,r,a}[f(N,\delta_{h,r,a})] \ge f\left(N, \mathop{\mathbb{E}}\limits_{h,r,a}[\delta_{h,r,a}]\right)$. Thus,

$$\frac{t'}{\delta'} \approx \frac{mt}{\bar{f}(N,\delta)} \le \frac{c_1\sqrt{\frac{N}{\delta}}t}{\frac{c_1^2(1-2c_2\sqrt{\delta})}{2(1-c_2\sqrt{\delta})}\delta} = \frac{2(1-c_2\sqrt{\delta})\sqrt{N}t}{c_1(1-2c_2\sqrt{\delta})\delta^{1.5}}.$$

We want to make the bound. Again, assuming $\frac{c_1}{c_2} = const.$ and also that $\delta = N = const.$ we can see that the problem we are facing is equivalent to maximizing $\frac{1-x\sqrt{\delta}}{x(1-2x\sqrt{\delta})}$. The derivative of that function is positive whenever $(\sqrt{\delta}x)^2 - 2\sqrt{\delta}x + 0.5 > 0$. Since $\sqrt{\delta} \le 1$ and $x = c_2 \le 1 - \frac{1}{\sqrt[3]{2}}$ and both are also greater than 0, the derivative is always positive and as such the maximum is achieved when we take $c_2 = 1 - \frac{1}{\sqrt[3]{2}}$.

We now upper bound $\frac{1-c_2\sqrt{\delta}}{(1-2c_2\sqrt{\delta})}$. As the function is strictly increasing for c_2 fixed to $1 - \frac{1}{\sqrt[3]{2}}$ and $\sqrt{\delta} \leq 1$, the upper bound is achieved when $\delta = 1$ when the result is $\frac{1}{2-\sqrt[3]{2}} < 1.4$. Taking $c_1 = 0.2$ then gives $\frac{t'}{\delta'} \approx 14\frac{\sqrt{N}t}{\delta^{1.5}}$.

6 Practical Implications

In order to show the practical consequences of the new reduction we will compare three reductions: the reduction given by Buldas and Saarepera in Asiacrypt 2004 [5], the reduction by Buldas and Laur in PKC 2007 [4], and the new reduction given in this article. We study a hypothetic global scale time-stamping system capable of issuing 67 million (about 2^{26}) time stamps per second and with lifetime at least 34 years (about 2^{30} seconds), i.e. we need to take $N = 2^{56}$. Systems of that scale are indeed in practical use. Our security proof is the first practical statement about the security of such systems if a 256 bit hash function (such as SHA2-256) is used. This is because we want the system to be secure against back-dating adversaries with time-success ratio $t/\delta = 2^{64}$. We study adversaries with three different time-success profiles: $(t, \delta) \in \{(1, 2^{-64}), (2^{32}, 2^{-32}), (2^{64}, 1)\}$. For each profile and reduction we compute the necessary output length of the hash function that is used in the time-stamping system considering that the hash function's security is near the birthday barrier, i.e. hash functions of output size k are $2^{k/2}$-secure. The results are presented in Table 2.

Table 2. Efficiency of reductions. The numbers denote hash function's output size in bits.

Reduction	Formula	$t = 1, \delta = 2^{-64}$	$t = 2^{32}, \delta = 2^{-32}$	$t = 2^{64}, \delta = 1$
Asiacrypt 2004	$\frac{t'}{\delta'} \approx 2N\frac{t}{\delta^2}$	370	306	242
PKC 2007	$\frac{t'}{\delta'} \approx 48\sqrt{N}\frac{t}{\delta^2}$	324	260	196
This paper	$\frac{t'}{\delta'} \approx 14\sqrt{N}\frac{t}{\delta^{1.5}}$	256	224	190

We see that a 256-bit hash function is indeed sufficient for such a time-stamping scheme though the previously proposed reductions were incapable of showing this.

It is also interesting to analyze how the hash-function output size k depends on the capacity N and the required security of the time-stamping system against back-dating. We study two levels of security: against 2^{64}-adversaries and against 2^{80}-adversaries. The results are summarized in Table 3. For example, in order to construct a 2^{64}-secure

Table 3. Efficiency of reductions. How hash function output size k depends on the capacity N.

Reduction	Formula	2^{64}-security	2^{80}-security
Asiacrypt 2004	$\frac{t'}{\delta'} \approx 2N\frac{t}{\delta^2}$	$k = 2\log_2 N + 258$	$k = 2\log_2 N + 322$
PKC 2007	$\frac{t'}{\delta'} \approx 48\sqrt{N}\frac{t}{\delta^2}$	$k = \log_2 N + 268$	$k = \log_2 N + 332$
This paper	$\frac{t'}{\delta'} \approx 14\sqrt{N}\frac{t}{\delta^{1.5}}$	$k = \log_2 N + 200$	$k = \log_2 N + 248$

time-stamping system with total capacity $N = 2^{56}$, we need a 256-bit hash function. Unfortunately, for achieving 2^{80}-security with a 256-bit hash function the capacity should be $N \leq 2^8 = 256$, which is clearly insufficient for a global scale time-stamping system. As the reduction we have is asymptotically tight, we have almost no hope of improving the efficiency of the reduction. Hence, in order to draw practical security conclusions about the large time-stamping systems that use a 256-bit hash function, we are forced to use security assumptions stronger than collision-freeness, even if the function is assumed to be collision-free to the birthday barrier.

References

1. Bayer, D., Haber, S., Stornetta, W.-S.: Improving the efficiency and reliability of digital timestamping. In: Sequences II: Methods in Communication, Security, and Computer Science, pp. 329–334. Springer, Heidelberg (1993)
2. Buldas, A., Jürgenson, A., Niitsoo, M.: Efficiency bounds for adversary constructions in black-box reductions. In: Boyd, C., González Nieto, J. (eds.) ACISP 2009. LNCS, vol. 5594, pp. 264–275. Springer, Heidelberg (2009)
3. Buldas, A., Laur, S.: Do broken hash functions affect the security of time-stamping schemes? In: Zhou, J., Yung, M., Bao, F. (eds.) ACNS 2006. LNCS, vol. 3989, pp. 50–65. Springer, Heidelberg (2006)
4. Buldas, A., Laur, S.: Knowledge-binding commitments with applications in time-stamping. In: Okamoto, T., Wang, X. (eds.) PKC 2007. LNCS, vol. 4450, pp. 150–165. Springer, Heidelberg (2007)
5. Buldas, A., Saarepera, M.: On provably secure time-stamping schemes. In: Lee, P.J. (ed.) ASIACRYPT 2004. LNCS, vol. 3329, pp. 500–514. Springer, Heidelberg (2004)
6. Haber, S., Stornetta, W.-S.: How to time-stamp a digital document. Journal of Cryptology 3(2), 99–111 (1991)
7. Haber, S., Stornetta, W.-S.: Secure names for bit-strings. In: ACM Conference on Computer and Communications Security, pp. 28–35 (1997)
8. Luby, M.: Pseudorandomness and Cryptographic Applications. Princeton University Press, Princeton (1996)
9. Merkle, R.C.: Protocols for public-key cryptosystems. In: Proceedings of the 1980 IEEE Symposium on Security and Privacy, pp. 122–134 (1980)
10. Petkovšek, M., Wilf, H.S., Zeilberger, D.: A=B. A.K. Peters, Ltd, Wellesley (1996)
11. Simon, D.: Finding Collisions on a One-Way Street: Can secure hash functions be based on general assumptions? In: Nyberg, K. (ed.) EUROCRYPT 1998. LNCS, vol. 1403, pp. 334–345. Springer, Heidelberg (1998)

A   Properties of the Success Function f

We prove some useful properties of $f(m, x) = 1 - mx(1 - x)^{m-1} - (1 - x)^m$.

Lemma 1. *If $m \geq 2$, then the function $f(m, x)$ is increasing in $[0 \ldots 1]$.*

Proof. This follows from the observation that $\frac{d}{dx} f(m, x) = m(m - 1)x(1 - x)^{m-2}$ is always positive in $x \in [0 \ldots 1]$.   □

Lemma 2. *If $m \geq 2$, then the function $f(m, x)$ is convex in $\left[0 \ldots \frac{1}{m-1}\right]$ and concave in $\left[\frac{1}{m-1} \ldots 1\right]$.*

Proof. We use zeroes of the second derivative of $f(m, x)$. The equation

$$\frac{d^2}{dx^2} f(m, x) = -m(m-1)(1-x)^{m-3}[(m-1)x - 1] = 0$$

implies that $x \in \left\{\frac{1}{m-1}, 1\right\}$. It is easy to see by using direct computations that the second derivative is positive if $0 \leq x \leq \frac{1}{m-1}$ and negative if $\frac{1}{m-1} \leq x \leq 1$. □

Lemma 3. *For every $m \geq 2$, for every collection of points $x_1, \ldots, x_n \in \left[0 \ldots \frac{1}{m-1}\right]$ and coefficients $p_1, \ldots, p_n$ so that $\sum_i p_i \leq 1$ we have*

$$\sum_{i=1}^{n} p_i \cdot f(m, x_i) \geq f\left(m, \sum_{i=1}^{n} p_i \cdot x_i\right).$$

Proof. We use the fact that $f(m, 0) = 0$, add an artificial term to the sum, and use the convexity of $f(m, x)$ and apply the Jensen's inequality. Let $p_0 = 1 - \sum_i p_i$ and $x_0 = 0$. Then we have:

$$\sum_{i=1}^{n} p_i \cdot f(m, x_i) = p_0 \cdot f(m, x_0) + \sum_{i=1}^{n} p_i \cdot f(m, x_i) \geq f\left(m, \ p_0 \cdot x_0 + \sum_{i=1}^{n} p_i \cdot x_i\right)$$

$$= f\left(m, \sum_{i=1}^{n} p_i \cdot x_i\right),$$

which proves the claim. □

Lemma 4. *For every $m \geq 2$ and $x \geq \frac{1}{m-1}$ we have $f(m, x) \geq x$.*

Proof. It is sufficient to prove that $f\left(m, \frac{1}{m-1}\right) \geq \frac{1}{m-1}$ for every $m \geq 2$ and then use the fact that $f(m, x)$ is concave in $\left[\frac{1}{m-1} \ldots 1\right]$. Indeed, $f\left(2, \frac{1}{1}\right) = \frac{1}{1}$, $f\left(3, \frac{1}{2}\right) = \frac{1}{2}$, and $f\left(4, \frac{1}{3}\right) = \frac{11}{27} \geq \frac{1}{3}$. If $m \geq 5$ then

$$f\left(m, \frac{1}{m-1}\right) = 1 - \frac{m}{m-1} \cdot \left(1 - \frac{1}{m-1}\right)^{m-1} - \left(1 - \frac{1}{m-1}\right)^{m}$$

$$= 1 - \left(\frac{m+1}{m-1} - 1\right) \cdot \left(1 - \frac{1}{m-1}\right)^{m-1} \geq 1 - 2 \cdot \left(1 - \frac{1}{m-1}\right)^{m-1}$$

$$= 1 - 2e^{-1} \geq \frac{1}{4} \geq \frac{1}{m-1}.$$

As $f(m, 1) = 1$ and f is concave in $\left[\frac{1}{m-1} \ldots 1\right]$, we have $f(m, x) \geq x$, $\forall x \in \left[\frac{1}{m-1} \ldots 1\right]$. □

B Lower Bound for $f\left(\max\left\{2,\frac{1}{\sqrt{\delta}}\right\},\frac{5\delta}{6}\right)$

Theorem 1. *For every* $0 \le \delta \le \frac{1}{4}$ *we have* $f(\max\left\{2,\frac{1}{\sqrt{\delta}}\right\},\frac{5\delta}{6}) \ge \frac{\delta}{6}$.

Lemma 5. *If* $0 \le x \le \frac{1}{m-1}$, *then* $f(m,x) \ge \frac{m(m-1)}{2}x^2 - \frac{m(m-1)(m-2)}{3}x^3$.

Proof. First, we expand $f(m,x)$ as follows:

$$f(m,x) = 1 - mx(1-x)^{m-1} - (1-x)^m = 1 - mx\sum_{i=0}^{m-1}(-1)^i\binom{m-1}{i}x^i - \sum_{i=0}^{m}(-1)^i\binom{m}{i}x^i$$

$$= 1 + \sum_{i=0}^{m-1}(-1)^{i+1}m\binom{m-1}{i}x^{i+1} - \sum_{i=0}^{m}(-1)^i\binom{m}{i}x^i$$

$$= 1 + \sum_{i=1}^{m}(-1)^i m\binom{m-1}{i-1}x^i - \sum_{i=0}^{m}(-1)^i\binom{m}{i}x^i = \sum_{i=1}^{m}(-1)^i\left[m\binom{m-1}{i-1}-\binom{m}{i}\right]x^i$$

$$= \sum_{i=1}^{m}(-1)^i m\left(1-\frac{1}{i}\right)\binom{m-1}{i-1}x^i = \sum_{i=2}^{m}(-1)^i m\left(1-\frac{1}{i}\right)\binom{m-1}{i-1}x^i$$

Obviously, $a_i = m\left(1-\frac{1}{i}\right)\binom{m-1}{i-1} > 0$ and if $x < \frac{1}{m-1}$ and $2 \le i < m$ then

$$\frac{a_i x^i}{a_{i+1}x^{i+1}} = \frac{1}{x}\frac{\left(1-\frac{1}{i}\right)\binom{m-1}{i-1}}{\left(1-\frac{1}{i+1}\right)\binom{m-1}{i}} = \frac{1}{x}\frac{\frac{i-1}{i}\binom{m-1}{i-1}}{\frac{i}{i+1}\binom{m-1}{i}} = \frac{1}{x}\frac{i^2-1}{i^2}\frac{\binom{m-1}{i-1}}{\binom{m-1}{i}}$$

$$= \frac{1}{x}\frac{i^2-1}{i^2}\frac{\frac{(m-1)!}{(i-1)!(m-i)!}}{\frac{(m-1)!}{i!(m-i-1)!}} = \frac{1}{x}\frac{i^2-1}{i^2}\frac{i!(m-i-1)!}{(i-1)!(m-i)!} = \frac{1}{x}\frac{i^2-1}{i^2}\frac{i}{m-i}$$

$$= \frac{1}{x}\frac{i^2-1}{i(m-i)} = \frac{1}{x}\cdot\left(1+\frac{1}{i}\right)\frac{i-1}{m-i} \ge \frac{m-1}{m-i}\cdot(i-1) > 1 \;.$$

Therefore, the expansion of $f(m,x)$ when $x \le \frac{1}{m-1}$ is an alternating sum of decreasing terms. This means that

$$f(m,x) \ge p(m,x) = m\left(1-\frac{1}{2}\right)\binom{m-1}{1}x^2 - m\left(1-\frac{1}{3}\right)\binom{m-1}{2}x^3$$

$$= \frac{m(m-1)}{2}x^2 - \frac{m(m-1)(m-2)}{3}x^3 \;.$$

□

Lemma 6. *If* $m = \frac{1}{\sqrt{\delta}}$ *and* $0 < \delta < 1$ *then* $\frac{5\delta}{6} \le \frac{1}{m-1}$.

Proof. $\frac{1}{m-1} = \frac{1}{\frac{1}{\sqrt{\delta}}-1} = \frac{\sqrt{\delta}}{1-\sqrt{\delta}} = \frac{\delta}{\sqrt{\delta}-\delta} \ge \frac{\delta}{1} > \frac{5\delta}{6}$.   □

Lemma 7. *The polynomial* $h(\delta) = \frac{1}{\delta}\cdot p(\frac{1}{\sqrt{\delta}},\frac{5\delta}{6})$ *is decreasing in* $[0\ldots 1]$.

Proof. As $h(\delta) = \frac{25}{72} - \frac{175}{324}\sqrt{\delta} + \frac{125}{216}\delta - \frac{125}{324}\delta^{3/2}$ and the equation $\frac{d}{dx}h(x) = 0$ has no real solutions and $\lim_{x\to\infty} h(x) = -\infty$ we conclude that $h(x)$ is decreasing in $[0\ldots\infty)$ and $h(\delta)$ is decreasing in $[0\ldots1]$. □

Therefore, the global minimum of $h(\delta)$ in $[0\ldots1/4]$ is $h(1/4) = \frac{25}{144} > \frac{1}{6}$. The function $\frac{1}{\delta}\cdot f(\max\left\{2, \frac{1}{\sqrt{\delta}}\right\}, \frac{5\delta}{6})$ is increasing in $[1/4\ldots1]$ because then $\max\left\{2, \frac{1}{\sqrt{\delta}}\right\} = 2$ and $\frac{1}{\delta}\cdot f(2, \delta) = \delta$ is increasing. Hence, $f(\max\left\{2, \frac{1}{\sqrt{\delta}}\right\}, \frac{5\delta}{6})$ is lower-bounded by $\frac{\delta}{6}$.

C   Proof of the Recurrence Relation

Lemma 8. *Stirling numbers of first kind $s(n, m)$ satisfy the following identity ($\forall m, n$):*

$$\sum_{k=0}^{n+1}(-1)^k s(k, m + k - n - 1)\binom{n+1}{k} = \sum_{k=0}^{n}(-1)^k k \cdot s(k, m + k - n)\binom{n}{k} . \quad (9)$$

Proof. We use the recurrence relation $s(a, b-1) - s(a+1, b) = a \cdot s(a, b)$ and transform the left hand side sum ℓ of (9) as follows:

$$\ell = \sum_{k=0}^{n+1}(-1)^k s(k, m + k - n - 1)\binom{n+1}{k}$$

$$= s(0, m - n - 1)\binom{n+1}{0} + \sum_{k=1}^{n}(-1)^k s(k, m + k - n - 1)\binom{n+1}{k} +$$

$$+ (-1)^{n+1}s(n + 1, m)\binom{n+1}{n+1}$$

$$= s(0, m - n - 1)\binom{n}{0} + \sum_{k=1}^{n}(-1)^k s(k, m + k - n - 1)\left[\binom{n}{k} + \binom{n}{k-1}\right] +$$

$$+ (-1)^{n+1}s(n + 1, m)\binom{n}{n}$$

$$= s(0, m - n - 1)\binom{n}{0} + \sum_{k=1}^{n}(-1)^k s(k, m + k - n - 1)\binom{n}{k} +$$

$$+ \sum_{k=1}^{n}(-1)^k s(k, m + k - 1 - n)\binom{n}{k-1} + (-1)^{n+1}s(n + 1, m)\binom{n}{n}$$

$$= \sum_{k=0}^{n}(-1)^k s(k, m + k - n - 1)\binom{n}{k} + \sum_{k=0}^{n}(-1)^{k+1}s(k + 1, m + k - n)\binom{n}{k}$$

$$= \sum_{k=0}^{n}(-1)^k [s(k, m + k - n - 1) - s(k + 1, m + k - n)]\binom{n}{k}$$

$$= \sum_{k=0}^{n}(-1)^k k \cdot s(k, m + k - n)\binom{n}{k} ,$$

which is equal to the right hand side of (9). □

Theorem 2. *The sequence* $\phi_n = \sum_{k=0}^{n}(-1)^k \frac{N^{\underline{k}}}{N^k}\binom{n}{k}$ *satisfies the recurrence relation:*

$$\phi_{n+1} = \frac{n}{N}(\phi_n - \phi_{n-1}) \ .$$

Proof. It is sufficient to show that $A(N) = N^{n+1}\phi_{n+1}$ and $B(N) = nN^n(\phi_n - \phi_{n-1})$ are identical as polynomials with variable N, i.e. all their coefficients coincide. We use the formula $N^{\underline{m}} = \sum_{j=0}^{m} s(m,j) \cdot N^j$, where $s(m,j)$ are Stirling numbers of the first kind. So, we have:

$$A(N) = \sum_{k=0}^{n+1}(-1)^k N^{\underline{k}}N^{n+1-k}\binom{n+1}{k} = \sum_{k=0}^{n+1}\sum_{j=0}^{k}(-1)^k s(k,j)\binom{n+1}{k}N^{n+1+j-k} \ ,$$

from which it follows that the coefficient $\mathsf{coef}_m(A)$ of N^m is:

$$\mathsf{coef}_m(A) = \sum_{k=0}^{n+1}(-1)^k s(k, m + k - n - 1)\binom{n+1}{k} \ ,$$

which is equal to the left hand side of identity (9). Similarly, for $B(N)$ we obtain:

$$
\begin{aligned}
B(N) &= nN^n(\phi_n - \phi_{n-1})\\
&= \sum_{k=0}^{n}(-1)^k N^{\underline{k}}N^{n-k}n\binom{n}{k} - \sum_{k=0}^{n-1}(-1)^k N^{\underline{k}}N^{n-k}n\binom{n-1}{k}\\
&= \sum_{k=0}^{n-1}(-1)^k N^{\underline{k}}N^{n-k}n\left[\binom{n}{k} - \binom{n-1}{k}\right] + (-1)^n N^{\underline{n}}N^0 n\binom{n}{n}\\
&= \sum_{k=0}^{n-1}(-1)^k N^{\underline{k}}N^{n-k}n\binom{n-1}{k-1} + (-1)^n N^{\underline{n}}N^0 n\binom{n}{n}\\
&= \sum_{k=0}^{n-1}(-1)^k N^{\underline{k}}N^{n-k}k\binom{n}{k} + (-1)^n N^{\underline{n}}N^0 n\binom{n}{n}\\
&= \sum_{k=0}^{n}(-1)^k N^{\underline{k}}N^{n-k}k\binom{n}{k} = \sum_{k=0}^{n}\sum_{j=0}^{k}(-1)^k k \cdot s(k,j)\binom{n}{k} \cdot N^{n-k+j}
\end{aligned}
$$

and

$$\mathsf{coef}_m(B) = \sum_{k=0}^{n}(-1)^k k \cdot s(k, m + k - n)\binom{n}{k} \ ,$$

which coincides with the right hand side of (9). Hence, $\mathsf{coef}_m(A) = \mathsf{coef}_m(B)$ for every $m > 0$, and by Lemma 8 the statement follows. $\qquad\square$

Additive Combinatorics and Discrete Logarithm Based Range Protocols

Rafik Chaabouni[1], Helger Lipmaa[2,3], and Abhi Shelat[4]

[1] EPFL, Switzerland
[2] Cybernetica AS, Estonia
[3] Tallinn University, Estonia
[4] University of Virginia, USA

Abstract. We show how to express an arbitrary integer interval $\mathcal{I} = [0, H]$ as a sumset $\mathcal{I} = \sum_{i=1}^{\ell} G_i * [0, u-1] + [0, H']$ of smaller integer intervals for some small values ℓ, u, and $H' < u - 1$, where $b * A = \{ba : a \in A\}$ and $A + B = \{a + b : a \in A \wedge b \in B\}$. We show how to derive such expression of $\mathcal{I}$ as a sumset for any value of $1 < u < H$, and in particular, how the coefficients G_i can be found by using a nontrivial but efficient algorithm. This result may be interesting by itself in the context of additive combinatorics. Given the sumset-representation of $\mathcal{I}$, we show how to decrease both the communication complexity and the computational complexity of the recent pairing-based range proof of Camenisch, Chaabouni and shelat from ASIACRYPT 2008 by a factor of 2. Our results are important in applications like e-voting where a voting server has to verify thousands of proofs of e-vote correctness per hour. Therefore, our new result in additive combinatorics has direct relevance in practice.

Keywords: Additive combinatorics, cryptographic range proof, sumset, zero knowledge.

1 Introduction

In a cryptographic range proof, the prover proves in zero knowledge that for given C and H, C is a commitment of some element $\sigma \in [0, H]$. (Modifying it to general ranges $[L, H]$ is trivial when one uses a homomorphic commitment scheme.) Range proofs are needed in various applications like e-voting [10,11] (where usually $H + 1$ is the number of candidates, that is, relatively small — though in the case of certain elections, there may be thousands of candidates), e-auctions [16] (where $H + 1$ is the number of number of possible bids, that is, relatively large), e-cash, etc. Range proofs with communication complexity $O(1)$ were introduced in [4,15].

However, such proofs work under very specific security assumptions, and thus there is still interest in protocols that are based on the discrete logarithm scenario. There exists a well-known folklore cryptographic range proof, see for example [11], in the special case when $H = u^{\ell} - 1$ for some integers $u, \ell > 0$. In this

R. Steinfeld and P. Hawkes (Eds.): ACISP 2010, LNCS 6168, pp. 336–351, 2010.

protocol, the prover writes σ as $\sigma = \sum \sigma_j u^j$, commits—by using a homomorphic commitment scheme—to all values σ_j, and then proves in zero-knowledge (using say a Σ-protocol) that $\sigma_j \in [0, u-1]$ for all j. The asymptotic communication complexity of this folklore range proof is $\Theta(\log H)$ times the complexity of the range proof of smaller interval $[0, u-1]$.

Recently, Camenisch, Chaabouni and shelat [5] presented a new range proof that works in the non-binary case. Assuming again $H = u^\ell - 1$, the verifier in their range proof first publishes signatures on all integers in $[0, u-1]$. The prover gives a proof of knowledge on signatures of ℓ committed elements σ_j. Analogously to the folklore protocol, this shows that the prover knows elements $\sigma_j \in [0, u-1]$ such that $\sigma = \sum \sigma_j u^j$ (the latter part is trivial with a public homomorphic commitment to σ).

However, if $H \neq u^\ell - 1$ then both the folklore protocol and the protocol of [5] get more complicated, and require up to 2 times more communication. In a nutshell, this is because they show that $\sigma \in [0, H]$ by using an AND composition of two range proofs, $\sigma' \in [0, u^\ell - 1]$ and $\sigma' \in [H - (u^\ell - 1), H]$. While such an AND composition is standard [9], it requires roughly two times more resources than the non-composed protocol for the case $H = u^\ell - 1$.

In the special case $u = 2$, an efficient modification of the folklore protocol for general ranges was proposed (though its correctness was not proven) in [16]. There it was noted that for any $H \geq 1$, $\sigma \in [0, H]$ if and only if $\sigma = \sum_{j=0}^{\lfloor \log_2 H \rfloor} G_j \sigma_j$, where $\sigma_j \in \{0, 1\}$ and $G_j := \lfloor (H + 2^j)/2^{j+1} \rfloor$. For example, $\sigma \in [0, 11]$ iff $\sigma = 6\sigma_0 + 3\sigma_1 + \sigma_2 + \sigma_3$ for $\sigma_j \in \{0, 1\}$. Thus the folklore protocol can be extended to arbitrary values of H with virtually no efficiency loss. In particular, there is no need for the AND composition. No improvement upon the folklore protocol in the general case $u > 2$ and $H \neq u^\ell - 1$ is known.

New Result in Additive Combinatorics. The principal contribution of this paper is to show that for *any* integer interval $\mathcal{I} = [0, H]$ and for *any* $1 < u < H$, there is a sumset-representation

$$\mathcal{I} = \sum_{j=0}^{\ell-1} G_j * [0, u-1] + [0, H'] \tag{1}$$

for some $\ell \leq \lceil \log_u(H+1) \rceil$ and $H' \in [0, u-2]$, where $b * A = \{ba : a \in A\}$ and $A + B = \{a + b : a \in A \wedge b \in B\}$. We first derive a recursive formula for computing G_j for any $u > 1$. As an interesting technical contribution, we then show a semi-closed form for G_j, that is, we show how to compute G_j given only H, j and u. This algorithm is efficient and only requires simple arithmetic. More precisely, we show that G_j is equal to the sum of $\lfloor H/u^{j+1} \rfloor$ and a simple (but nontrivial) function of the $j + 1$ lowest u-ary digits of H. We think that the presented algorithm may be interesting by itself say in the general context of additive combinatorics [17]: decompositions of sets as sumsets are common in additive combinatorics, but our concrete result differs significantly from existing results in that field.[1]

Note that in the language of additive combinatorics, the result of [16] says that

$$[0, H] = \sum_{j=0}^{\lfloor \log_2 H \rfloor} \lfloor (H + 2^j)/2^{j+1} \rfloor * [0, 1] . \tag{2}$$

Eq. (2) does not straightforwardly generalize to the case where we are interested in a larger range $[0, u - 1]$. In fact, [16] did not even present a proof that Eq. (2) holds. As a straightforward corollary of our sumset-representation of $[0, H]$, we obtain a proof that the presentation of Eq. (2) is correct.

Application of the Sumset-Representation in Range Proofs. We show how to use the sumset-representation Eq. (1) to modify the pairing-based range proof of [5] so that it will become at least 50% more communication-efficient in practice (and so that it is *always* more efficient than the folklore protocol). For this we use a simple corollary of our general sumset-representation that $[0, H] = \sum_{j=0}^{\ell-1} G_j * [0, u - 1]$ whenever $(u - 1) \mid H$. Moreover, if we set $u = O(\log H / \log \log H)$, then the total communication of the range proof is $\Theta(\log H / \log \log H)$. We also point out some mistakes in [5], namely, that the so called OR composition proposed there does not work in most of the cases, and thus their protocols are somewhat less efficient than claimed. In addition, the new protocol is also about 2 times more computation-efficient than the protocol from [5]. A factor of 2 times improvement in communication and computation is extremely relevant to practical applications like e-voting where a voting server may have to verify thousands of proofs of e-vote correctness per hour.[2] Moreover, the used sumset-representation is optimal, so the achieved speedup is optimal for this kind of range proofs. (Note that in applications like e-voting, one requires non-interactive zero-knowledge proofs. The latter can be efficiently constructed from Σ-protocols using the well-known Fiat-Shamir heuristic [13].)

Finally, we hope there will be more applications of the new sumset-representation in cryptography.

2   Preliminaries

We summarize and copy some of the notation and definitions from [5] for consistency and to make it easier for the reader to follow.

---

[1] Recall that typical questions of additive combinatorics are of type how large or small can sumsets of type $A \pm A$ be, and how is the cardinality of this set related to the cardinalities of A_i. Note that our question can be reworded as follows: we are asking for the maximal cardinality of $\mathcal{I} = \sum G_j * [0, u - 1] + [0, H']$ for fixed u and H' but variable G_j.

[2] Such e-voting servers are currently running at least in Estonia, http://www.vvk.ee/index.php?id=11178, and will hopefully be more widespread in the near future. For example, in the last e-voting in Estonia, 4 500 votes were cast during the peak hour.

Notation. PPT means probabilistic polynomial-time. k is the security parameter. In all protocols, prover and verifier send elements from $\mathbb{G}_1$, $\mathbb{G}_T$ and $\mathbb{Z}_p$. We denote the length of representation (which may differ from the logarithm of the cardinality of the groups) of such elements by $\mathsf{rlen}(\mathbb{G}_1)$, $\mathsf{rlen}(\mathbb{G}_T)$ and $\mathsf{rlen}(\mathbb{Z}_p)$ respectively.

Additive Combinatorics. For any two integers $L \leq H$, let

$$[L, H] := \{x \in \mathbb{Z} : L \leq x \leq H\} \ .$$

We use the usual "set-theoretic" arithmetic notation. For example, if A and B are sets then $A + B = \{a + b : a \in A \wedge b \in B\}$. Moreover, for an integer b and $A \subset \mathbb{Z}$, $b * A = \{ba : a \in A\}$, this is also called the b-dilate of A [17].

Commitment Schemes. A (string) commitment scheme is a triple of algorithms $C = (\mathsf{Gen}, \mathsf{Com}, \mathsf{Open})$ representing the generation, the commit and the open algorithm. The Gen algorithm generates parameters p for a scheme. The Com algorithm runs on input (p, m, r) where m is a string, and r is a random tape, and produces a pair of values (c, o) representing respectively the committed string and an opening string. The Open algorithm runs on input (c, m, o) and outputs 0 or 1. The scheme should have a "hiding" property and a "binding" property which informally require it to be difficult (or impossible) for the adversary to determine the message m from c or to open the value of a commitment c to two different messages m_1, m_2.

Zero-Knowledge Proofs and Σ-Protocols. We use definitions from [1,8]. A pair of interacting algorithms (P, V) is a proof of knowledge (PK) for a relation $R = \{(\alpha, \beta)\} \subseteq \{0,1\}^* \times \{0,1\}^*$ with knowledge error $\kappa \in [0, 1]$ if (1) for all $(\alpha, \beta) \in R$, $\mathsf{V}(\alpha)$ accepts a conversation with $\mathsf{P}(\beta)$ with probability 1; and (2) there exists an expected polynomial-time algorithm E, called the *knowledge extractor*, such that if a cheating prover P^* has probability ϵ of convincing V to accept α, then E, when given rewindable black-box access to P^*, outputs a witness β for α with probability $\epsilon - \kappa$.

A proof system (P, V) is *computational honest-verifier zero-knowledge* if there exists a PPT algorithm Sim, called the *simulator*, such that for any $(\alpha, \beta) \in R$, the outputs of $V(\alpha)$ after interacting with $\mathsf{P}(\beta)$ and that of $\mathsf{Sim}(\alpha)$ are computationally indistinguishable. When we will talk about *honest-verifier zero-knowledge* we will assume the computational case.

Note that standard techniques can be used to transform an honest-verifier zero-knowledge proof system into a general zero-knowledge one [8]. This is especially true of special Σ-protocols that will be presented later in the paper. Thus, for the remainder of the paper, our proofs will be honest-verifier zero-knowledge. (This also allows us to make more accurate comparisons with the other proof techniques since they are usually also presented as honest-verifier protocols.)

A Σ-protocol for language $\mathcal{L}$ is a proof system (P, V) where the conversation is of the form (a, c, z), where a and z are computed by P, and c is a challenge randomly chosen by V. The verifier accepts if $\phi(\alpha, a, c, z) = 1$ for some efficiently computable predicate ϕ. A Σ-protocol must satisfy three security requirements: correctness, special soundness and special honest-verifier zero knowledge (SHVZK). A Σ-protocol is correct when a honest prover convinces honest verifier with probability $1 - k^{-\omega(1)}$. A Σ-protocol has the special soundness property when from two accepting views (a, c, z) and (a, c', z'), where $c \neq c'$, one can efficiently recover a witness w such that $w \Rightarrow x \in \mathcal{L}$. A Σ-protocol has the SHVZK property if there exists a PPT simulator Sim that can first randomly pick c^*, z^* (from some fixed sets) and then compute an a^* such that the view (a^*, c^*, z^*) is accepting and the distribution (a^*, c^*, z^*) is computationally indistinguishable from the distribution of accepting views between honest prover and honest verifier.

We use the notation introduced by Camenisch and Stadler [6] for various zero-knowledge proofs of knowledge of discrete logarithms and proofs of the validity of statements about discrete logarithms. For instance, $PK\{(\alpha, \beta, \gamma) : y = g^\alpha h^\beta \wedge \mathfrak{y} = \mathfrak{g}^\alpha \mathfrak{h}^\gamma \wedge (u \leq \alpha \leq v)\}$ denotes a *"zero-knowledge Proof of Knowledge of integers α, β, and γ such that $y = g^\alpha h^\beta$ and $\mathfrak{y} = \mathfrak{g}^\alpha \mathfrak{h}^\gamma$ holds, where $u \leq \alpha \leq v$,"* where $y, g, h, \mathfrak{y}, \mathfrak{g}$, and $\mathfrak{h}$ are elements of some groups $G = \langle g \rangle = \langle h \rangle$ and $\mathfrak{G} = \langle \mathfrak{g} \rangle = \langle \mathfrak{h} \rangle$. The convention is that Greek letters denote quantities the knowledge of which is being proved, while all other parameters are known to the verifier. Using this notation, a proof-protocol can be described by just pointing out its aim while hiding all details. We note that all of the protocols we present in this notation can be easily instantiated as Σ-protocols.

Definition 1 (Proof of Set Membership [5]). *Let $C = (\mathrm{Gen}, \mathrm{Com}, \mathrm{Open})$ be the generation, the commit and the open algorithm of a string commitment scheme. For an instance c, a proof of set membership with respect to commitment scheme C and set Φ is a proof of knowledge for the following statement: $PK\{(\sigma, \rho) : c \leftarrow \mathrm{Com}(\sigma; \rho) \wedge \sigma \in \Phi\}$.*

Definition 2 (Range Proof [5]). *A range proof with respect to a commitment scheme C is a special case of a proof of set membership in which the set Φ is a continuous sequence of integers $\Phi = [a, b]$ for $a, b \in \mathbb{N}$.*

As discussed in the introduction, some efficient range proofs were proposed in [4,16,15,5]. We will give a precise description of the proof from [5] in Sect. 4.

Any secure Σ-protocol can be efficiently transferred into a non-interactive zero-knowledge proof (in the random oracle model) by using the Fiat-Shamir heuristic [13]. In many applications, the Σ-protocol needs to have nontransferability properties. In all such cases (like e-voting), one uses the corresponding non-interactive zero-knowledge proof. Since the Fiat-Shamir heuristic is well-known and its use is standard in say e-voting literature [10,11], we will omit any explicit mention of it in what follows.

3   Sumset-Representation of Integer Intervals

The goal of this section is to derive a sumset-representation $[0, H] = \sum_{i=0}^{\ell-1} G_i * [0, u - 1] + [0, H']$, where $1 < H' < u \ll H$, of an arbitrary integral interval

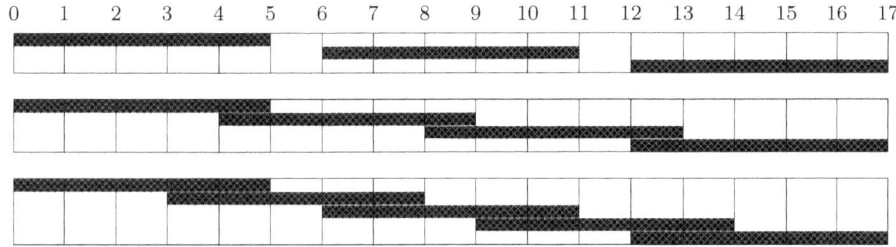

Fig. 1. Illustration of the first recursive step of Thm. 1. Here $H = H_0 = 17$, and $u \in \{3, 4, 5\}$. For example, in the top graph, $u = 3$, $G_0 = \lfloor (17 + 1)/3 \rfloor = 6$, and $H_1 = 17 - 2 \cdot 6 = 5$.

$[0, H]$. (Integral means that all involved parameters H, H', u and G_i are positive integers.) Moreover, we aim to find minimal ℓ for any fixed value of u.

First we give an intuitive derivation of our result. (See also Fig. 1.) Fix H and u. Let $H_0 = H$. Then clearly $[0, H_0] = G_0 * [0, u - 1] + [0, H_1]$, where $G_0 := \lfloor (H_0 + 1)/u \rfloor$ and $H_1 = H_0 - (u - 1) \cdot G_0$. This can be derived as follows: we want to divide $[0, H_0]$ into u smaller (possibly overlapping) intervals of equal size H_1 such that H_1 is minimal. The sub-intervals should start at periodic positions jG_0, for some G_0. Because all elements from $[0, H_0]$ must belong to at least one of those subareas, it must be the case that $H_1 \geq G_0 - 1$ and $(u - 1)G_0 + H_1 \geq H_0$. Thus, in the optimal case when $H_1 = G_0 - 1$, we get $uG_0 - 1 = H_0$ or $G_0 = (H_0 + 1)/u$. Since G_0 has to be an integer, we set $G_0 = \lfloor (H_0 + 1)/u \rfloor$. Finally, $H_1 = H_0 - (u - 1)G_0$ as stated.

These formulas reduce the case $[0, H_0]$ to a smaller case $[0, H_1]$ that can be solved similarly. Recursively, $[0, H] = [0, u - 1] \cdot \sum_j G_j + [0, H']$, where

$$G_j := \left\lfloor \frac{H_j + 1}{u} \right\rfloor . \tag{3}$$

and $H_{j+1} := H_j - (u - 1) \cdot G_j = H_j - (u - 1) \cdot \left\lfloor \frac{H_j + 1}{u} \right\rfloor$.

This process stops when the interval $[0, H_{j+1}]$ is small enough so that it cannot be covered by u different non-empty intervals, that is, if $H_{j+1} \leq u - 1$. Then we define $\ell(u, H) := j + 1$ to be the number of steps in this recursive process. Clearly, after we are done with the recursive process,

$$H' := H_{j+1} = H - \left\lfloor \frac{H}{u - 1} \right\rfloor \cdot (u - 1) .$$

This means in particular that if $(u - 1) \mid H$ then $H' = 0$.

Example 1. For example, with $H = 57$ and $u = 4$, one can verify that $[0, 57] = 14 * [0, 3] + 4 * [0, 3] + [0, 3]$. As another example, $[0, 160] = 40 * [0, 3] + [0, 40] = 40 * [0, 3] + 10 * [0, 3] + [0, 10] = 40 * [0, 3] + 10 * [0, 3] + 2 * [0, 3] + [0, 4] = 40 * [0, 3] + 10 * [0, 3] + 2 * [0, 3] + 1 * [0, 3] + [0, 1]$. Now we are done since $1 < u - 1 = 3$.

(Another, though non-recursive, example with $H = 17$ was already depicted by Fig. 1.)

Finally, the sequence $(\ldots, G_j, \ldots)$ clearly decreases the slowest when for all j, $u \mid (H_{j+1} + 1)$, since then the floor operation is not applied. But $u \mid (H_{j+1} + 1)$ iff $u \mid (H_j - (u-1)G_j + 1)$ iff $u \mid (H_j + G_j + 1)$ iff (because also $u \mid (H_j + 1)$) $u \mid G_j$. Thus, the sequence is slowest to decrease if $H + 1 = u^\ell$ for some ℓ. This means, that the process is guaranteed to stop in $\ell(u, H) \leq \log_u(H+1)$ steps.

This leads us to the following theorem.

Theorem 1. *Let $u \geq 2$, $H \geq u$. Let G_j, H_j and H' be defined as before. Denote $\ell = \ell(u, H) \leq \lceil \log_u(H+1) \rceil$ as above. Then $[0, H] = \sum_{j=0}^{\ell-1} G_j * [0, u-1] + [0, H']$. If $(u-1) \mid H$ then $H' = 0$.*

Proof. Clear from above.   □

Semi-Closed Form for G_j. While the presented recursive formulas for G_j and H_{j+1} are efficient, it is desirable to have a closed form for G_j. In the following we construct a semi-closed form, that is, a formula for G_j that only depends on u, j and H.

Assume that $H = \sum h_j u^j$ with $h_j \in \{0, \ldots, u-1\}$. For any j, write $h_{j+} := \lfloor H/u^j \rfloor$, that is, $H = u^j h_{j+} + \sum_{i=0}^{j-1} u^i h_i$. In particular, $h_{j+} = u h_{(j+1)+} + h_j$. Define $[\![x]\!] := x \pmod{u-1}$. Our proof is built up on the initial observation that:
$G_0 = h_{1+} + \lfloor \frac{h_0+1}{u} \rfloor$, and $H_1 = h_{1+} + h_0 - (u-1)\lfloor \frac{1+h_0}{u} \rfloor = h_{1+} + [\![h_0]\!]$.
The latter equation is obvious: if $h_0 < u-1$ then $h_0 - (u-1)\lfloor \frac{1+h_0}{u} \rfloor = h_0 = [\![h_0]\!]$ and if $h_0 = u - 1$ then $h_0 - (u-1)\lfloor \frac{1+h_0}{u} \rfloor = u-1 - (u-1) = 0 = [\![h_0]\!]$. We can now prove that

Theorem 2. $G_j = h_{(j+1)+} + \left\lfloor \dfrac{h_j + [\![\sum_{i=0}^{j-1} h_i]\!] + 1}{u} \right\rfloor.$

Proof. By induction. We prove that $H_j = h_{j+} + [\![\sum_{i=0}^{j-1} h_i]\!]$, from this the claim for G_j is obvious. Induction basis $(j = 0)$ is obvious since $H_0 = h_{0+}$.

Induction step $(j > 0)$. Assume that $H_j = h_{j+} + [\![\sum_{i=0}^{j-1} h_i]\!] = u h_{(j+1)+} + h_j + [\![\sum_{i=0}^{j-1} h_i]\!]$ and $G_j = h_{(j+1)+} + \left\lfloor \frac{h_j + [\![\sum_{i=0}^{j-1} h_i]\!] + 1}{u} \right\rfloor$. Then

$$H_{j+1} = H_j - (u-1)G_j = h_{(j+1)+} + h_j + [\![\sum_{i=0}^{j-1} h_i]\!] - (u-1) \cdot \left\lfloor \frac{h_j + [\![\sum_{i=0}^{j-1} h_i]\!] + 1}{u} \right\rfloor.$$

Thus to finish the proof we only have to show that

$$h_j + [\![\sum_{i=0}^{j-1} h_i]\!] - (u-1) \cdot \left\lfloor \frac{h_j + [\![\sum_{i=0}^{j-1} h_i]\!] + 1}{u} \right\rfloor = [\![\sum_{i=0}^{j} h_i]\!] \qquad (4)$$

for any $h_i \in \{0, \ldots, u-1\}$. We consider the next cases.

Case 1, $\left[\!\left[\sum_{i=0}^{j-1} h_i\right]\!\right] = 0$. Then the left hand side of Eq. (4) is $h_j - (u-1) \cdot \lfloor(1 + h_j)/u\rfloor = [\![h_j]\!]$ and the right hand side is equal to the same value.

Case 2, $\left[\!\left[\sum_{i=0}^{j-1} h_i\right]\!\right] \neq 0$ and $h_j + \left[\!\left[\sum_{i=0}^{j-1} h_i\right]\!\right] + 1 < u$. Then the left hand side of Eq. (4) is $h_j + \left[\!\left[\sum_{i=0}^{j-1} h_i\right]\!\right]$ and the right hand side is $\left[\!\left[\sum_{i=0}^{j-1} h_i + h_j\right]\!\right] = \left[\!\left[\sum_{i=0}^{j-1} h_i\right]\!\right] + h_j$.

Case 3, $\left[\!\left[\sum_{i=0}^{j-1} h_i\right]\!\right] \neq 0$ and $h_j + \left[\!\left[\sum_{i=0}^{j-1} h_i\right]\!\right] + 1 \geq u$. Then the left hand side of Eq. (4) is $h_j + \left[\!\left[\sum_{i=0}^{j-1} h_i\right]\!\right] - (u-1)$ and the right hand side is $\left[\!\left[\sum_{i=0}^{j-1} h_i + h_j\right]\!\right] = \left[\!\left[\sum_{i=0}^{j-1} h_i\right]\!\right] + h_j - (u-1)$.   □

In the binary case $u = 2$, a formula like this was already given in [16]. However, while [16] stated the closed form, they did not prove it. Fortunately, their formula follows straightforwardly from the general result.

Corollary 1 (Binary case, [16]). *If $u = 2$ then $G_j = h_{(j+1)+} + \left\lfloor \frac{h_j+1}{u} \right\rfloor = \left\lfloor \frac{H+2^j}{2^{j+1}} \right\rfloor$.*

Proof. Straightforward corollary.

4   Preliminaries: CCS Range Proof

Computational Assumptions. The following protocols require bilinear groups and associated hardness assumptions. These assumptions are summarized from [5].

Let PG be a pairing group generator that on input 1^k outputs descriptions of multiplicative groups $\mathbb{G}_1$ and $\mathbb{G}_T$ of prime order p where $|p| = k$. Let $\mathbb{G}_1^* = \mathbb{G}_1 \setminus \{1\}$ and let $g \in \mathbb{G}_1^*$. The generated groups are such that there exists an admissible bilinear map $e : \mathbb{G}_1 \times \mathbb{G}_1 \to \mathbb{G}_T$, meaning that (1) for all $a, b \in \mathbb{Z}_p$ it holds that $e(g^a, g^b) = e(g, g)^{ab}$; (2) $e(g, g) \neq 1$; and (3) the bilinear map is efficiently computable.

Definition 3 (Strong Diffie-Hellman Assumption [3]). *We say that the q-SDH assumption associated to a pairing generator PG holds if for all PPT adversaries A, the probability that $A(g, g^x, \ldots, g^{x^q})$ where $(\mathbb{G}_1, \mathbb{G}_T) \leftarrow \mathsf{PG}(1^k)$, $g \leftarrow \mathbb{G}_1^*$ and $x \leftarrow \mathbb{Z}_p$, outputs a pair $(c, g^{1/(x+c)})$ where $c \in \mathbb{Z}_p$ is negligible in k.*

As noted by [5], Cheon's [7] attack against this type of assumption is not relevant if $q \leq 50$ as it is in this protocol.

Boneh-Boyen Signatures. Our scheme relies on the elegant Boneh-Boyen short signature scheme [3] which we briefly summarize. The signer's secret key is $x \leftarrow \mathbb{Z}_p$, the corresponding public key is $y = g^x$. The signature on a message m is $\sigma \leftarrow g^{1/(x+m)}$; verification is done by checking that $e(\sigma, y \cdot g^m) = e(g, g)$. This scheme is similar to the Dodis and Yampolskiy verifiable random function [12].

Security under ℓ-weak chosen-message attack is defined through the following game. The adversary begins by outputting ℓ messages $m_1, \ldots, m_\ell$. The challenger generates a fresh key pair and gives the public key to the adversary, together with signatures $\sigma_1, \ldots, \sigma_\ell$ on $m_1, \ldots, m_\ell$. The adversary wins if it succeeds in outputting a valid signature σ on a message $m \notin \{m_1, \ldots, m_\ell\}$. The scheme is said to be unforgeable under an ℓ-weak chosen-message attack if no PPT adversary A has non-negligible probability of winning this game. Our scheme relies on the following property of the Boneh-Boyen short signature [3] which we paraphrase below:

Lemma 1 ([3]). *Suppose the q-Strong Diffie Hellman assumption holds in $(\mathbb{G}_1, \mathbb{G}_T)$. Then the basic Boneh-Boyen signature scheme is secure against an existential forgery under a q-weak chosen message attack.*

The Camenisch-Chaabouni-shelat range proof in the case when $H = u^\ell - 1$ is depicted by Protocol 1. In particular, $e : \mathbb{G}_1 \times \mathbb{G}_1 \to \mathbb{G}_T$ is an admissible bilinear map for some multiplicative groups $\mathbb{G}_1, \mathbb{G}_T$, and g is a generator of $\mathbb{G}_1$ with $h \in \langle g \rangle$.

Communication of CCS Range Proof for "Nice" H. The CCS range proof for nice H requires the prover to compute 3ℓ exponentiations and 2ℓ pairings (in [5], this was summed up as 5ℓ exponentiations). It requires non-interactive (static) communication of

$$\mathsf{NICom}_{\mathsf{ccs}}(u, \ell) := (1 + u) \cdot \mathsf{rlen}(\mathbb{G}_1) \text{ bits}$$

(signatures and public keys that can be shared between different protocol runs), and interactive communication (which is unique for every protocol run) of

$$\mathsf{ICom}_{\mathsf{ccs}}(u, \ell) := (1 + \ell) \cdot \mathsf{rlen}(\mathbb{G}_1) + \ell \cdot \mathsf{rlen}(\mathbb{G}_T) + (2 + 2\ell) \cdot \mathsf{rlen}(\mathbb{Z}_p) \text{ bits}.$$

Communication of CCS for Arbitrary Range $[L, H]$. As noted in [5], to prove that $\sigma \in [L, H]$ for arbitrary L and H, one can use an AND composition. More precisely, suppose that $u^{\ell-1} < H < u^\ell$. Then to show that $\sigma \in [L, H]$, it suffices to show that $\sigma \in [L, L + u^\ell)$ and $\sigma \in [H - u^\ell, H)$. Equivalently, one has to show that $\sigma - L \in [0, u^\ell)$ and $\sigma - H + u^\ell \in [0, u^\ell)$.

For this, one uses the standard AND composition of Protocol 1 with itself. Recall that an AND composition of two Σ-protocols A_1 and A_2 is a Σ protocol where the first message is a composition of the first messages of A_1 and A_2, the second message is a single challenge c, and the third message is a composition of the third messages of A_1 and A_2 that correspond to the first messages and the single challenge c. Moreover, static information (the public key y and all signatures) and also the values V_j are only sent once. Thus, in the AND composition of the CCS protocol, there are two versions of a_j, D, $z_j^{(\sigma)}$, $z_j^{(v)}$ and $z^{(m)}$, which makes the (static) communication of the AND composition of Protocol 1 with

---

Assume $\sigma = \sum_{j=0}^{\lfloor \log_u(H+1) \rfloor} \sigma_j u^j$.

Common input: g, h, u, ℓ, and a commitment C.

Prover's input: σ, r such that $C = g^\sigma h^r$ and $\sigma \in [0, H]$.

1. **The verifier does:** generate a random $x \leftarrow \mathbb{Z}_p$, and set $y \leftarrow g^x$. For $i \in [0, u-1]$, set $A_i \leftarrow g^{1/(x+i)} \in \mathbb{G}_1$. She sends $(y, A_0, \dots, A_{u-1})$ to the prover.

2. **The prover does:** For all $j \in [0, \ell-1]$, generate random $v_j \leftarrow \mathbb{Z}_p$, set $V_j \leftarrow A_{\sigma_j}^{v_j} \in \mathbb{G}_1$. He sends $(V_0, \dots, V_{\ell-1})$ to the verifier.

3. **The prover** uses the following Σ-protocol to prove to the verifier that $C = h^r \cdot g^{\sum \sigma_j u^j}$, and $V_j = g^{v_j/(x+\sigma_j)}$ for all j:

   (a) **The prover** picks $s_j, t_j, m_j \leftarrow \mathbb{Z}_p$ for $j \in [0, \ell-1]$. He sets $a_j \leftarrow e(V_j, g)^{-s_j} e(g, g)^{t_j} \in \mathbb{G}_T$, for $j \in [0, \ell-1]$, and $D \leftarrow g^{\sum_j u^j s_j} \cdot h^{\sum_j m_j} \in \mathbb{G}_1$. He sends $(a_0, \dots, a_{\ell-1}, D)$ to the verifier.

   (b) **The verifier** sends a random challenge $c \leftarrow \mathbb{Z}_p$ to the prover.

   (c) **The prover** sets $z_j^{(\sigma)} \leftarrow s_j - \sigma_j c \mod p$, $z_j^{(v)} \leftarrow t_j - v_j c \mod p$, for $j \in [0, \ell-1]$. He sets $z^{(m)} \leftarrow m - rc \mod p$, where $m = \sum_{j=0}^{\ell-1} m_j$. He sends $(z_0^{(\sigma)}, \dots, z_{\ell-1}^{(\sigma)}, z_0^{(v)}, \dots, z_{\ell-1}^{(v)}, z^{(m)})$ to the verifier.

   (d) **The verifier** checks that $D = C^c h^{z^{(m)}} g^{\sum_j u^j \cdot z_j^{(\sigma)}}$ and $a_j = e(V_j, y)^c \cdot e(V_j, g)^{-z_j^{(\sigma)}} \cdot e(g, g)^{z_j^{(v)}}$ for every $j \in [0, \ell-1]$.

Protocol 1. The CCS cryptographic range proof for range $[0, u^\ell - 1]$

itself equal to $\mathsf{NICom}_{\mathsf{ccsand}}(u, \ell) = \mathsf{NICom}_{\mathsf{ccs}}(u, \ell) = (1 + u) \cdot \mathsf{rlen}(\mathbb{G}_1)$, and the dynamic communication is equal to

$$\mathsf{ICom}_{\mathsf{ccsand}}(u, \ell)$$
$$= \mathsf{ICom}_{\mathsf{ccs}}(u, \ell) + \ell \cdot \mathsf{rlen}(\mathbb{G}_T) + \mathsf{rlen}(\mathbb{G}_1) + (2\ell + 1) \cdot \mathsf{rlen}(\mathbb{Z}_p)$$
$$= (\ell + 2) \cdot \mathsf{rlen}(\mathbb{G}_1) + 2\ell \cdot \mathsf{rlen}(\mathbb{G}_T) + (4\ell + 3) \cdot \mathsf{rlen}(\mathbb{Z}_p) \ .$$

Remark on OR Composition. [5] also considered the OR composition. The communication of an OR composition is twice the communication of the single protocol, but with $\ell - 1$ instead of ℓ, and thus the OR composition has the potential to be more efficient than the AND composition. In our case, for the OR composition to work, we have to assume that u is such that $u^{\ell-1} < H \leq 2 \cdot u^{\ell-1}$. In this case, $\sigma \in [0, H]$ iff $\sigma \in [0, u^{\ell-1}]$ or $\sigma \in [H - u^{\ell-1}, H]$. This differs slightly from the misstated requirement of [5], where it was said that one just needs that $u^{\ell-1} < H$. In particular, this means that the OR composition does not work for values, considered in Sect. 4.4 of [5], and thus the communication-efficiency of their range proof is (in most of the cases) slightly worse than claimed in [5].

Communication Analysis. Let us assume that $\mathsf{rlen}(\mathbb{G}_T) \approx 12 \cdot \mathsf{rlen}(\mathbb{G}_1) \approx 12 \cdot \mathsf{rlen}(\mathbb{Z}_p)$ [14]. Following [5] and plugging in parameters in terms of the $\mathsf{rlen}(\mathbb{G}_1)$, the communication can then be minimized by solving the following system min $(6 + u + 29\ell)$ s.t. $u^\ell \geq H$. Setting $u = \frac{\log H}{\log \log H}$ then we get a total asymptotic communication complexity of

$$Com(u, \ell) = O\left(\frac{\log H}{\log \log H - \log \log \log H}\right)$$

which is asymptotically smaller than $O(\log H)$. For concrete parameters, we substitute the constraint that $u^\ell \approx H$ into the equation $u + \ell$ above, set the derivative with respect to u to 0 and attempt to solve the equation:

$$1 - \frac{29 \log H}{u \log^2 u} = 0$$

which simplifies to

$$u \log^2 u = 29 \log H. \tag{5}$$

This equation cannot be solved analytically. However, given H, we can use numerical methods to find a good u as described in [2].

5   Modified Range Proof: New

The idea of the next proof follows from Thm. 1. We can assume that $u > 1$. Clearly, $\sigma \in [0, H]$ iff $(u - 1)\sigma \in [0, (u - 1)H]$ iff, because of Thm. 1,

$$(u - 1)\sigma = \sum_{j=0}^{\ell(u,(u-1)H)-1} \sigma_j G_j$$

for some $\sigma_j \in [0, u - 1]$, and G_j are defined as in Thm. 1 with $H_0 = (u - 1)H$.

Thus, we can propose a new range proof where we prove that C^{u-1} commits a value in $(0, (u - 1)H]$ by using the CCS protocol for "nice" H, see Protocol 2. Note that changing 0 to any meaningful L, $1 \leq L < (u - 1)H$, is trivial. In the description of the protocol, see Protocol 2, new parts (compared to the CCS protocol) have been **bolded** for easy parsing.

Rationale for multiplying by $u - 1$. If $(u - 1)$ divides H, then it is not necessary to multiply the commitment by $(u - 1)$. Recall that if $u - 1$ does not divide H, then $H' < u$ (the leftover value) defines some small range $[0, H']$. In this case, one could (instead of multiplying by $u - 1$) add an extra step to the range proof that shows that some new committed element belongs to the range $[0, H']$. Doing this would require an extra $H' + 1$ elements from $\mathsf{rlen}(\mathbb{G}_1)$, one extra element from $\mathsf{rlen}(\mathbb{G}_T)$, and one extra element from $\mathsf{rlen}(\mathbb{Z}_p)$ to be transmitted. Thus, it will always be more expensive to add this extra step. Another idea might be to use a simple OR-proof to handle the last $[0, H']$ elements. This would require extra communication of $H' \cdot \mathsf{rlen}(\mathbb{G}_1) + (1 + H') \cdot \mathsf{rlen}(\mathbb{Z}_p)$ bits. Since one element of $\mathsf{rlen}(\mathbb{G}_T)$ is roughly 12 times larger than the size of one element from either $\mathbb{G}_1$ or $\mathbb{Z}_p$, this approach is favorable when $H' < 6$.

Assume $(\mathbf{u}-1)\cdot\sigma = \sum_{j=0}^{\ell-1}\sigma_j\cdot\mathbf{G_j}$ for $\ell = \ell(\mathbf{u},(\mathbf{u}-1)H) \leq \lceil\log_u((\mathbf{u}-1)\cdot H+1)\rceil$.
Common input: g, h, u, ℓ, and a commitment C.
Prover's input: σ, r such that $C = g^\sigma h^r$ and $\sigma \in [0, H]$.

1. **The verifier does:** she generates a random $x \leftarrow \mathbb{Z}_p$, and sets $y \leftarrow g^x$. For $i \in [0, u-1]$, she sets $A_i \leftarrow g^{1/(x+i)} \in \mathbb{G}_1$. She sends $(y, A_0, \ldots, A_{u-1})$ to the prover.
2. **The prover does:** For all $j \in [0, \ell-1]$, generate random $v_j \leftarrow \mathbb{Z}_p$, set $V_j \leftarrow A_{\sigma_j}^{v_j} \in \mathbb{G}_1$. He sends $(V_1, \ldots, V_{\ell-1})$ to the verifier.
3. **The prover** uses the following Σ-protocol to prove to the verifier that $C^{\mathbf{u}-1} = h^{(\mathbf{u}-1)\cdot r}\cdot g^{\sum\sigma_j\mathbf{G_j}}$, and $V_j = g^{v_j/(x+\sigma_j)}$ for all $j \in [0, \ell-1]$:
   (a) **The prover** picks $s_j, t_j, m_j \leftarrow \mathbb{Z}_p$ for $j \in [0, \ell-1]$. He sets $a_j \leftarrow e(V_j, g)^{-s_j}e(g,g)^{t_j} \in \mathbb{G}_T$, for $j \in [0, \ell-1]$, and $D \leftarrow g^{\sum_j s_j\cdot\mathbf{G_j}}\cdot h^{(\mathbf{u}-1)\cdot\sum_j m_j} \in \mathbb{G}_1$. He sends $(a_0, \ldots, a_{\ell-1}, D)$ to the verifier.
   (b) **The verifier** sends a random challenge $c \leftarrow \mathbb{Z}_p$ to the prover.
   (c) **The prover** sets $z_j^{(\sigma)} \leftarrow s_j - \sigma_j c$, $z_j^{(v)} \leftarrow t_j - v_j c$ for $j \in [0, \ell-1]$ and $z^{(m)} \leftarrow m - rc$ for $m = \sum_j m_j$. He sends $(z_0^{(\sigma)}, \ldots, z_{\ell-1}^{(\sigma)}, z_0^{(v)}, z_{\ell-1}^{(v)}, z^{(m)})$ to the verifier.
   (d) **The verifier** checks that $D = C^{c(\mathbf{u}-1)}\cdot h^{(\mathbf{u}-1)\cdot z^{(m)}}\cdot g^{\sum_j z_j^{(\sigma)}\cdot\mathbf{G_j}}$ and $a_j = e(V_j, y)^c\cdot e(V_j, g)^{-z_j^{(\sigma)}}\cdot e(g,g)^{z_j^{(v)}}$ for every $j \in [0, \ell-1]$.

Protocol 2. New, generalization of CCS protocol for arbitrary range $[0, H]$

Theorem 3. *Assuming the q-SDH assumption, Protocol 2 is correct and has the property of special soundness and SHVZK.*

Proof (Sketch.). The proof is a straightforward extension of the security proof from [5]. □

Concrete Efficiency of New. Clearly, both the static and dynamic communication of New is related to communication of the CCS protocol in the following simple way:

$$\mathsf{NICom}_{\mathsf{New}}(u, \ell) := \mathsf{NICom}_{\mathsf{ccs}}(u, \ell(u, (u-1)(H+1))) \text{ and}$$
$$\mathsf{ICom}_{\mathsf{New}}(u, \ell) := \mathsf{ICom}_{\mathsf{ccs}}(u, \ell(u, (u-1)(H+1))).$$

This is easily seen to be a factor of 2 more efficient than having to use two proofs to handle an arbitrary range H.

Efficiency of New. Asymptotically, the *total* communication of $\mathsf{NICom}_{\mathsf{New}} + \mathsf{ICom}_{\mathsf{New}}$ remains the same:

$$u + \ell(u, (u-1)(H+1)) \leq u + \log_u((u-1)(H+1))$$
$$= u + \log_u(u-1) + \log_u(H+1)$$
$$\leq u + 1 + \log_u(H) + \frac{1}{H}$$

As before, this value is (approximately) minimized when we set $u \leftarrow \frac{\log_2 H}{\log_2 \log_2 H}$. Concretely, there is a factor of two difference. The communication can be minimized by solving

$$\min(4 + u + 15\ell) \quad \text{such that} \quad \ell > \log_u((u-1)(H+1))$$

As mentioned before, in some cases when $u - 1$ already divides H, it is not necessary to multiply by $u - 1$; even when $(u - 1)$ does not evenly divide H, a standard OR-proof can sometimes be used to handle H'. We take this fact in account when computing the protocol's efficiency for a given range below. In the graph below, we show how the complexity of our new protocol compares with that of [5] for ranges $[0, H]$ where H varies from 1000 to $2 \cdot 10^8$.

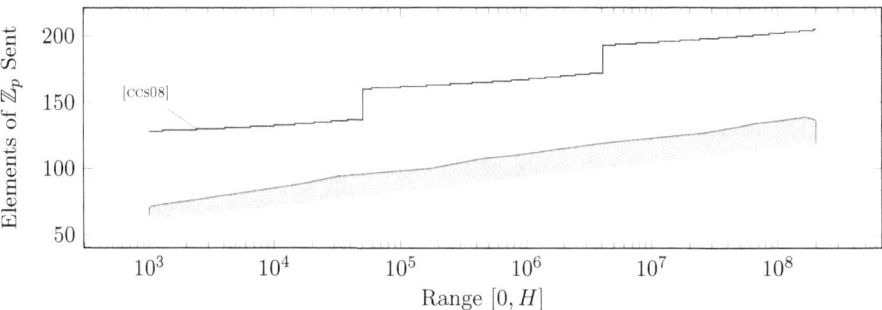

Fig. 2. Relative Efficiency of the New Protocol vs. [5]. The number of group elements are computed under the assumption that $\mathsf{rlen}(\mathbb{G}_T) \approx 12 \cdot \mathsf{rlen}(\mathbb{G}_1) \approx 12 \cdot \mathsf{rlen}(\mathbb{Z}_p)$. The complexity of our new protocol depends more sensitively on the exact value of H; therefore the shaded area represents the convex hull of the values for our new protocol. The vertical gaps in the curve for [5] are a result of the ratio 12 used above.

5.1 Comparison of Case Analysis

As a second way to compare the new protocol with protocol from [5] and other previous work, we use the same numbers as in Sect. 4 of [5]. In particular we assume that the size of $\mathbb{G}_1$ is 256 bits, the size of $\mathbb{G}_T$ is 3072 bits and the size of $\mathbb{Z}_p$ is upper-bounded by 256 bits. We also use the range $\mathcal{R} = [L, H) = [347184000, 599644800)$ as in [5]. Also, clearly, the new protocol (as in the CCS protocol) for $\mathcal{R}$ is exactly as efficient as protocol for range $[0, H']$, where $H' = H - L - 1$. That is, $H' = 252460799$.

The values of $\mathsf{NICom}_{\mathsf{ccsand}}$, $\mathsf{ICom}_{\mathsf{ccsand}}$, $\mathsf{NICom}_{\mathsf{New}}$ and $\mathsf{ICom}_{\mathsf{New}}$ for a few different choices of u and ℓ are shown in the following two tables. Note that the optimal choice of u depends on how many times the range proof is going to be reused: the larger is the number w of reuses, the larger should be u, and for w reuses, one should choose a value of u for which $\mathsf{NICom}_{\mathsf{New}}(u, \ell) + (w-1)\mathsf{ICom}_{\mathsf{New}}(u, \ell)$ is minimal.

The values of $\mathsf{NICom_{ccsand}}$ and $\mathsf{ICom_{ccsand}}$ for some chosen values of u, ℓ are given below. (Here we only use the AND composition. As mentioned above, the OR composition is sometimes more efficient but only under certain restrictions.) The numbers in Tbl. 1 show that the CCS protocol is less efficient than claimed but still more efficient than the previous range proofs.

Table 1. Communication of the CCS protocol with some chosen values of u (and implicitly chosen optimal ℓ)

u	ℓ	$\mathsf{NICom_{ccsand}}$	$\mathsf{ICom_{ccsand}}$	Comments
48	5	12 544	38 400	Minimal $\mathsf{NICom_{ccsand}} + \mathsf{ICom_{ccsand}} * [1,2]$
57	5	14 848	38 400	Same parameters as in [5]
633	3	162 304	23 552	Minimal $\mathsf{NICom_{ccsand}} + 10000 \cdot \mathsf{ICom_{ccsand}}$, $\mathsf{ICom_{ccsand}}$

Communication of New for some concrete choices of u and ℓ is given in Tbl. 2. Recall that we need to show that $(u-1)(\sigma+1) - 1 \in [0, (u-1)(H'+1)] = [0, 252460800 \cdot (u-1)]$. We have calculated ℓ according to the point where the recursions of Thm. 1 end, and we note that sometimes its value differs from the predicted value $\lfloor \log_u((u-1)(H+1)) \rfloor$.

Table 2. Communication of New with some chosen values of u (and implicitly chosen optimal ℓ)

u	ℓ	$\mathsf{NICom_{New}}(u, \ell)$	$\mathsf{ICom_{New}}(u, \ell)$	Comments
25	6	6 656	27 648	Minimal $\mathsf{NICom_{New}} + \mathsf{ICom_{New}}$
48	5	12 544	23 808	Minimal $\mathsf{NICom_{New}} + \mathsf{ICom_{New}}$
57	5	14 848	23 808	Same parameters as in [5]
632	4	162 048	16 128	Minimal $\mathsf{NICom_{New}} + 10000 \cdot \mathsf{ICom_{New}}$, $\mathsf{ICom_{New}}$

6   Conclusions

We showed that for any H and $1 < u < H$, the interval $[0, H]$ is equal to a sum $\sum G_i * [0, u-1] + [0, H']$, where $0 \le H' < H$, and both u and ℓ are "small" in terms of H. We gave efficient (closed form) algorithms for computing the coefficients G_i. This result may be interesting by itself in the context of additive combinatorics.

We then used this decomposition to show how to derive efficient range proofs for arbitrary intervals $[0, H]$. Compared to the previous work [5], we thus avoided the use of AND composition of Σ-protocols. In addition, (1) we showed also that an earlier result from [16] (that only considered the case $u = 2$) is correct, though it was left unproven in [16], and (2) we pointed out that the range proof from [5] is (in most of the cases) less efficient than claimed there. In addition, the new protocol is also about 2 times more computation-efficient than the protocol from [5]. While 2 times is not much, it is important in practical applications like e-voting where a voting server may have to verify thousands of proofs of e-vote correctness per hour.

Finally, we hope that our techiques can be extended to construct other efficient cryptographic protocols that use results from additive combinatorics.

Acknowledgments. The first author was partially supported by the Swiss National Science Foundation, 200021-124575, and by the European Commission through the ICT programme under contract ICT-2007-216676 ECRYPT II. The second author was supported by Estonian Science Foundation, grant #8058, and European Union through the European Regional Development Fund. The third author was partly supported by NSF CAREER Award CNS-0447808.

References

1. Bellare, M., Goldreich, O.: On Defining Proofs of Knowledge. In: Brickell, E.F. (ed.) CRYPTO 1992. LNCS, vol. 740, pp. 390–420. Springer, Heidelberg (1993)
2. Black, K.: Classroom Note: Putting Constraints in Optimization for First-Year Calculus Students. SIAM Rev. 39(2), 310–312 (1997)
3. Boneh, D., Boyen, X.: Short Signatures without Random Oracles. In: Cachin, C., Camenisch, J.L. (eds.) EUROCRYPT 2004. LNCS, vol. 3027, pp. 56–73. Springer, Heidelberg (2004)
4. Boudot, F.: Efficient Proofs That a Committed Number Lies in an Interval. In: Preneel, B. (ed.) EUROCRYPT 2000. LNCS, vol. 1807, pp. 431–444. Springer, Heidelberg (2000)
5. Camenisch, J., Chaabouni, R., Shelat, A.: Efficient Protocols for Set Membership and Range Proofs. In: Pieprzyk, J. (ed.) ASIACRYPT 2008. LNCS, vol. 5350, pp. 234–252. Springer, Heidelberg (2008)
6. Camenisch, J., Stadler, M.: Efficient Group Signature Schemes for Large Groups. In: Kaliski Jr., B.S. (ed.) CRYPTO 1997. LNCS, vol. 1294, pp. 410–424. Springer, Heidelberg (1997)
7. Cheon, J.H.: Security Analysis of the Strong Diffie-Hellman Problem. In: Vaudenay, S. (ed.) EUROCRYPT 2006. LNCS, vol. 4004, pp. 1–11. Springer, Heidelberg (2006)
8. Cramer, R., Damgård, I., MacKenzie, P.D.: Efficient Zero-Knowledge Proofs of Knowledge without Intractability Assumptions. In: Imai, H., Zheng, Y. (eds.) PKC 2000. LNCS, vol. 1751, pp. 354–373. Springer, Heidelberg (2000)
9. Cramer, R., Damgård, I., Schoenmakers, B.: Proofs of Partial Knowledge and Simplified Design of Witness Hiding Protocols. In: Desmedt, Y.G. (ed.) CRYPTO 1994. LNCS, vol. 839, pp. 174–187. Springer, Heidelberg (1994)
10. Cramer, R., Gennaro, R., Schoenmakers, B.: A Secure and Optimally Efficient Multi-Authority Election Scheme. In: Fumy, W. (ed.) EUROCRYPT 1997. LNCS, vol. 1233, pp. 103–118. Springer, Heidelberg (1997)
11. Damgård, I., Jurik, M.: A Generalisation, a Simplification and Some Applications of Paillier's Probabilistic Public-Key System. In: Kim, K.-c. (ed.) PKC 2001. LNCS, vol. 1992, pp. 119–136. Springer, Heidelberg (2001)
12. Dodis, Y., Yampolskiy, A.: A Verifiable Random Function with Short Proofs and Keys. In: Vaudenay, S. (ed.) PKC 2005. LNCS, vol. 3386, pp. 416–431. Springer, Heidelberg (2005)

13. Fiat, A., Shamir, A.: How to Prove Yourself: Practical Solutions to Identification and Signature Problems. In: Odlyzko, A.M. (ed.) CRYPTO 1986. LNCS, vol. 263, pp. 186–194. Springer, Heidelberg (1987)
14. Galbraith, S.D., Paterson, K.G., Smart, N.P.: Pairings for Cryptographers. Discrete Applied Mathematics 156(16), 3113–3121 (2008)
15. Lipmaa, H.: On Diophantine Complexity and Statistical Zero-Knowledge Arguments. In: Laih, C.-S. (ed.) ASIACRYPT 2003. LNCS, vol. 2894, pp. 398–415. Springer, Heidelberg (2003)
16. Lipmaa, H., Asokan, N., Niemi, V.: Secure Vickrey Auctions without Threshold Trust. In: Blaze, M. (ed.) FC 2002. LNCS, vol. 2357, pp. 87–101. Springer, Heidelberg (2003)
17. Tao, T., Vu, V.: Additive Combinatorics. Cambridge Studies in Advanced Mathematics. Cambridge University Press, Cambridge (2006)

Proof-of-Knowledge of Representation of Committed Value and Its Applications

Man Ho Au, Willy Susilo, and Yi Mu

Centre for Computer and Information Security Research
School of Computer Science and Software Engineering
University of Wollongong, Australia
{aau,wsusilo,ymu}@uow.edu.au

Abstract. We present a zero-knowledge argument system of representation of a committed value. Specifically, for commitments $C = \mathsf{Commit}_1(y)$, $D = \mathsf{Commit}_2(x)$, of value y and a tuple $x = (x_1, \ldots, x_L)$, respectively, our argument system allows one to demonstrate the knowledge of (x, y) such that x is a representation of y to bases $h_1, \ldots, h_L$. That is, $y = h_1^{x_1} \cdots h_L^{x_L}$. Our argument system is zero-knowledge and hence, it does not reveal anything such as x or y. We note that applications of our argument system are enormous. In particular, we show how round-optimal cryptography systems, where privacy is of a great concern, can be achieved. We select three interesting applications with the aim to demonstrate the significance our argument system. First, we present a concrete instantiation of two-move concurrently-secure blind signature without interactive assumptions. Second, we present the first compact e-cash with concurrently-secure withdrawal protocol. Finally, we construct two-move traceable signature with concurrently-secure join. On the side note, we present a framing attack against the original traceable signature scheme within the original model.

1 Introduction

The notion of zero-knowledge proof protocol was put forth by Goldwasser, Micali and Rackoff in [34]. In a zero-knowledge proof protocol, a prover convinces a verifier that a statement is true, while the verifier learns nothing except the validity of the assertion. A proof-of-knowledge [7] is a protocol such that the verifier is convinced that the prover knows a certain quantity w satisfying some kinds of relation R with respect to a commonly known string x. That is, the prover convinces the verifier that he knows some w such that $(w, x) \in R$. If it can be done in such a way that the verifier learns nothing besides the validity of the statement, this protocol is called a zero-knowledge proof-of-knowledge (ZKPoK) protocol. Various efficient ZKPoK protocols about knowledge of discrete logarithms and their relations have been proposed in the literature. For instance, knowledge of discrete logarithm [46], polynomial relations of discrete logarithms [15,27], inequality of discrete logarithms [18], range of discrete logarithms [13] and double discrete logarithm [19].

ZKPoK protocols have been used extensively as building blocks of many cryptosystems. In this paper, we present a ZKPoK protocol for the knowledge of representation

R. Steinfeld and P. Hawkes (Eds.): ACISP 2010, LNCS 6168, pp. 352–369, 2010.

of a committed value. We demonstrate that our protocol can be used to construct round-optimal cryptosystems, including blind signatures, traceable signatures and compact e-cash.

1.1 Related Work

ZKPoK of Double-Discrete Logarithm. Our protocol generalizes the ZKPoK protocol of double discrete logarithm ,introduced by Stadler [47], when it is used to construct a verifiable secret sharing scheme. Roughly speaking, a double discrete logarithm of an element y to base g and h is an element x such that $y = g^{h^x}$. Stadler introduces a ZKPoK protocol to demonstrate the knowledge of such x with respect to y. This protocol was employed in the construction of group signatures [19,2] and a divisible e-cash scheme [20]. Looking ahead, our zero-knowledge protocol further extends Stadler's protocol in which it allows the prover to demonstrate the knowledge of a set of values $(x_1, \ldots, x_L, r)$ such that $y = g^{h_1^{x_1} \cdots h_L^{x_L}} g_0^r$. We would like to stress that there is a *subtle difference* between Stadler's protocol and ours when $L = 1$. Specifically, with the introduction of the variable r, no information about x is leaked to the verifier. This turns out to be very useful when the prover wishes to demonstrate the same x, without being linked, to different verifiers.

Blind signatures. Introduced by Chaum [23], blind signature schemes allow a user to obtain interactively a signature on message m from a signer in such a way that the signer learns nothing about m (*blindness*) while at the same time, the user cannot output more signatures than the ones produced from the interaction with the signer (*unforgeability*). The formal definition of blind signatures was first proposed in [45], with the requirement that any user executing the protocol ℓ times with the signer cannot output $\ell + 1$ valid signatures on $\ell + 1$ distinct messages. One important feature of security offered by any blind signature construction is whether the execution of the signing protocol can be performed concurrently, that is, in an arbitrarily-interleaved manner. As pointed out in [31], a notable exception to the problems of constructing schemes secure against interleaving executions are those with an optimal two-move signing protocol, of which the problem of concurrency is solved immediately.

Table 1 summarizes existing schemes that are secure under concurrent execution. Note that [36], [31] and [35] provide generic construction only. [31] relies on generic NIZK while [35] utilizes ZAP. On the other hand, as pointed out in [35], [36] makes use of generic concurrently-secure 2-party computation and constructing such a protocol without random oracle or trusted setup is currently an open problem. Lindell's result [40] states that it is impossible to construct concurrently-secure blind signatures in the plain model if simulation-based definitions are used. Hazay *et al.* [35] overcome this limitation by employing a game-based definition. A construction achieving all properties is proposed in [32] recently.

Traceable Signatures. Introduced by Chaum and van Heyst [24], group signatures allow a group member to sign anonymously on behalf of the group. Whenever required, the identity of the signature's originator can be revealed only by the designated party. Traceable signatures, introduced in [37], are group signatures with added functionality

Table 1. Summary of Existing Blind Signatures Secure under Concurrent Signature Generation

Schemes	Round-Optimal?	W/o RO?	Non-Interactive Assumption?	Instantiation?
[35]	×	✓	✓	?
[36]	×	✓	✓	×
[31]	✓	✓	✓	?
[8]	✓	×	×	✓
[10]	✓	×	×	✓
[42]	×	✓	✓	✓
[32]	✓	✓	✓	✓
Our Scheme	✓	×	✓	✓

in which a designated party could output some tracing information on a certain user that allows the bearer to trace *all* signatures generated by that user. Subsequently, another traceable signature is propose in [25]. We discover a flaw in the security proof of [37] and are able to develop a concrete attack against their scheme under their model. Table 2 summarizes existing traceable signatures. Note that *none* of the existing schemes is secure when the join protocol is executed concurrently. In contrast, group signature scheme with concurrent join has been proposed in [39] and can also be constructed based on group encryption [22].

Table 2. Summary of Existing Traceable Signatures

Schemes	Round-Optimal?	W/o RO?	Support Concurrent-Join?	Secure?
[37]	×	×	×	×
[25]	×	×	×	✓
Our Scheme	✓	×	✓	✓

Compact E-Cash Invented by Chaum [23], electronic cash (E-Cash) is the digital counterpart of paper cash. In an e-cash scheme, a user withdraws an electronic coin from the bank and the user can spend it to any merchant, who will deposit the coin back to the bank. Compact e-cash, introduced in [16], aims at improving bandwidth efficiency. In compact e-cash, users can withdraw efficiently a wallet containing K coins. These coins, however, must be spent one by one. Other constructions of compact e-cash include [5,3,21]. Table 3 summarizes existing compact e-cash. Note that *none* of the existing schemes is secure when the withdrawal protocol is executed concurrently.

Table 3. Summary of Existing Compact E-Cash Systems

Schemes	Round-Optimal?	W/o RO?	Support Concurrent-Withdrawal?
[16]	×	×	×
[5]	×	×	×
[3]	×	×	×
[21]	×	×	×
Our Scheme	✓	×	✓

1.2   Overview of Our Approach

As discussed in [39], the most efficient and conceptually simple joining procedure for a group signature is for the user to choose a one way function f and compute $x = f(x')$ for some user secret x'. Next, the user sends x to the group manager (GM) and obtains a signature σ on x. A group signature from the user will then consist of a probabilistic encryption of x into ψ under the GM's public key, and a signature-of-knowledge of (1) the correctness of ψ as an encryption of some value x, (2) knowledge of x', a pre-image of x, and (3) knowledge of σ which is a valid signature on x. This approach is suggested by Camenisch and Stadler [19], and is given the name "single-message and signature-response paradigm" in [39]. Nonetheless, it turns out that a concrete instantiation of this approach is not as simple as it looks, since it is hard to choose a suitable signature scheme and function f so that efficient and secure proof is possible.

It turns out that our argument system together with the Boneh-Boyen signature [11] fits in perfectly with the above paradigm. In our construction, f is chosen to be a perfectly hiding malleable commitment scheme which allows the commitment of a block of values. This expands the flexibility of the paradigm and allows the construction of traceable signatures, compact e-cash as well as blind signature. Taking traceable signature as an example, a user first computes a commitment $f(x)$ of a secret value x. Due to the malleability of the commitment scheme, the group manager changes it to a commitment of a block of values $f(x, t)$ and issues a signature σ on this commitment. To generate a traceable signature, the user computes a probabilistic encryption of $f(x, t)$ into ψ^1, a random base $\tilde{g} = g^r$ and a tracing tag $T = \tilde{g}^t$. Next, the user generates a signature-of-knowledge of (1) the correctness of ψ, $\tilde{g}$ and T with respect to x and t, (2) knowledge of x, t, a pre-image of $f(x, t)$, and (3) knowledge of σ which is a valid signature on $f(x, t)$. To trace the user, the GM simply outputs t and everyone can test whether the tracing tag T and the random base $\tilde{g}$ associated with each group signature satisfies $T = \tilde{g}^t$.

1.3   Organization of the Paper

The rest of this paper is organized as follows. In Section 2, we review preliminaries that will be used throughout this paper. We then present our argument system, its security and efficiency analysis in Section 3. Then, we apply our argument system in constructing blind signatures, traceable signatures and compact e-cash. Those constructions are presented in Section 4, 5 and 6, respectively. Finally, we conclude the paper in Section 7.

2   Preliminaries

2.1   Notations

We employ the following notation throughout this paper. Let $\mathbb{G}_1$ be a cyclic group of prime order p. Let $\mathbb{G}_q \subset \mathbb{Z}_p^*$ be a cyclic group of prime order q. This can be generated

---

[1] In fact, this is for revealing signer's identity and encryption of either $f(x)$, x or σ also serves the purpose.

by setting p to be a prime of the form $p = \gamma q + 1$ for some integer γ and set $\mathbb{G}_q$ to be the group generated by an element of order q in $\mathbb{Z}_p^*$.

Let $g, g_0, g_1, g_2 \in_R \mathbb{G}_1$ be random elements of $\mathbb{G}_1$ and $h, h_0, h_1, \ldots, h_L \in_R \mathbb{G}_q$ be random elements of $\mathbb{G}_q$ (with the requirement that none of them being the identity element of their respective group). Since $\mathbb{G}_1$ and $\mathbb{G}_q$ are of prime order, those elements are generators of their respective groups.

We say that a function $\mathsf{negl}(\lambda)$ is a negligible function [6], if for all polynomials $f(\lambda)$, for all sufficiently large λ, $\mathsf{negl}(\lambda) < 1/f(\lambda)$.

2.2 Bilinear Map

A pairing is a bilinear mapping from a pair of group elements to a group element. Specifically, let $\mathbb{G}_T$ be cyclic group of prime order p. A function $\hat{e} : \mathbb{G}_1 \times \mathbb{G}_1 \to \mathbb{G}_T$ is said to be a pairing if it satisfies the following properties:

- (Bilinearity.) $\hat{e}(u^x, v^y) = \hat{e}(u, v)^{xy}$ for all $u, v \in \mathbb{G}_1$ and $x, y \in \mathbb{Z}_p$.
- (Non-Degeneracy.) $\hat{e}(g, g) \neq 1_{\mathbb{G}_T}$, where $1_{\mathbb{G}_T}$ is the identity element in $\mathbb{G}_T$.
- (Efficient Computability.) $\hat{e}(u, v)$ is efficiently computable for all $u, v \in \mathbb{G}_1$.
- (Unique Representation.) All elements in $\mathbb{G}_1$, $\mathbb{G}_T$ have unique binary representation.

Looking ahead, while we are assuming $\mathbb{G}_1$ is equipped with a bilinear map, it is not necessary for our zero-knowledge proof of knowledge of representation of committed value. Its presence is mainly for the many applications associated with our protocol.

2.3 Number-Theoretic Assumptions

We present below the number-theoretic problems related to the schemes presented in this paper. The respective assumptions state that no PPT algorithm has non-negligible advantage in security parameter in solving the corresponding problems. Let $\mathbb{G} = \langle g \rangle = \langle g_1 \rangle = \cdots = \langle g_k \rangle$ be a cyclic group.

- The *Discrete Logarithm Problem* (DLP) in $\mathbb{G}$ is to output x such that $Y = g^x$ on input $Y \in \mathbb{G}$.
- The *Representation Problem* (RP) [14] in $\mathbb{G}$ is to compute a k-tuple $(x_1, \ldots, x_k)$ such that $Y = g_1^{x_1} \cdots g_k^{x_k}$ on input Y. RP is as hard as DLP if the relative discrete logarithm of any of the g_i's are not known.
- The *Decisional Diffie-Hellman Problem* (DDHP) $\in \mathbb{G}$ is to decide if $z = xy$ on input a tuple (g^x, g^y, g^z).
- The *Decisional Linear Diffie-Hellman Problem* (DLDH problem) [12] in $\mathbb{G}$ is to decide if $z = x + y$ on input a tuple (g_1^x, g_2^y, g_3^z). The DLDH problem is strictly harder than the DDH problem.
- The *q-Strong Diffie-Hellman Problem* (q-SDH problem) [11] in $\mathbb{G}$ is to compute a pair (A, e) such that $A^{x+e} = g$ on input $(g^x, g^{x^2}, \ldots, g^{x^q})$.
- The *y-Decisional Diffie-Hellman Inversion Problem* (y-DDHI problem) [29,16] in $\mathbb{G}$ is to decide if $z = 1/x$ on input $(g^x, g^{x^2}, \ldots, g^{x^y}, g^z)$.

2.4   Cryptographic Tools

Commitment Schemes. A commitment scheme is a protocol between two parties, namely, committer Alice and receiver Bob. It consists of two stages: the *Commit* stage and the *Reveal* stage. In the *Commit* stage, Alice receives a value x as input, which is revealed to Bob at the *Reveal* stage. Informally speaking, a commitment scheme is secure if at the end of the *Commit* stage, Bob cannot learn anything about the committed value (a.k.a. hiding) while at the *Reveal* stage, Alice can only reveal one value, that is x (a.k.a. binding). Formally, we review the security notion from [33].

Definition 1. *A commitment scheme* (**Gen**, **Commit**)[2] *is secure if holding the following two properties:*

1. *(Perfect Hiding.) For all algorithm* $\mathcal{A}$ *(even computationally unbounded one), we require that*

$$Pr\left[\begin{array}{l} param \leftarrow \textbf{Gen}(1^\lambda); (x_0, x_1) \leftarrow \mathcal{A}(param); \\ b \in_R \{0,1\}; r \in_R \{0,1\}^\lambda; \\ C = \textbf{Commit}(param, x_b; r); b' \leftarrow \mathcal{A}(C); \end{array} : b' = b\right] \leq \frac{1}{2} + \textsf{negl}(\lambda).$$

2. *(Binding.) No PPT adversary* $\mathcal{A}$ *can open a commitment in two different ways. Specifically,*

$$Pr\left[\begin{array}{l} param \leftarrow \textbf{Gen}(1^\lambda); (x_0, x_1, r_0, r_1) \leftarrow \mathcal{A}(param) : \\ x_0 \neq x_1 \wedge \\ \textbf{Commit}(param, x_0; r_0) = \textbf{Commit}(param, x_1; r_1) \end{array}\right] = \textsf{negl}(\lambda).$$

In this paper, we restrict ourselves to a well-known non-interactive commitment scheme, the Pedersen Commitment [43], which is reviewed very briefly here. On input a value $x \in \mathbb{Z}_p$, the committer randomly chooses $r \in \mathbb{Z}_p$, computes and outputs commitment $C = g_0^x g^r$ as the commitment of value x. To reveal commitment C, the committer outputs (x, r). Everyone can test if $C = g_0^x g^r$. Sometimes (x, r) is referred to as an opening of the commitment C.

Recall that Pedersen Commitment is perfect hiding and computationally binding provided that the g_0 and g are randomly and independently generated and that relative discrete logarithm of g_0 to base g is unknown. One can easily extend the scheme to allow commitment of a block of values, say, $\boldsymbol{x} = (x_0, x_1, \ldots, x_k)$ by setting the commitment $C = g_0^{x_0} g_1^{x_1} \cdots g_k^{x_k} g^r$ with additional random generators $g_1, \ldots, g_k$ of $\mathbb{G}_1$.

Boneh-Boyen Short Signature. Boneh and Boyen introduced a short signature scheme in [11], which, is used extensively in the applications of our argument system. Hereafter, we shall refer to this scheme as BB-signature.

KeyGen. Let $\alpha, \beta \in_R \mathbb{Z}_p^*$ and $u = g^\alpha$ and $v = g^\beta$. The secret key sk is (α, β) while the public key pk is $(\hat{e}, \mathbb{G}_1, \mathbb{G}_T, p, g, u, v)$.

Sign. Given message $m \in \mathbb{Z}_p^*$, pick a random $e \in_R \mathbb{Z}_p$ and compute $A = g^{\frac{1}{\alpha+m+\beta e}}$. The term $\alpha + m + \beta e$ is computed modulo p. In case it is zero, choose another e. The signature σ on m is (A, e).

---

[2] With **Gen** being the parameter generation function.

Verify. Given a message m and signature $\sigma = (A, r)$, verify that

$$\hat{e}(A, ug^m v^e) = \hat{e}(g, g)$$

If the equality holds, output **valid**. Otherwise, output **invalid**.

Σ-Protocol. We restrict ourselves to a special class of ZKPoK protocol called Σ-protocol which is defined below. Informally speaking, Σ-protocols only guarantee zero-knowledgeness when the verifier is honest. We are interested in Σ-protocol since they can be transformed to 4-move perfect zero-knowledge ZKPoK protocol [26]. They can also be transformed to 3-move concurrent zero-knowledge protocol in the auxiliary string model using trapdoor commitment schemes [28].

Definition 2. *A Σ-protocol for a binary relation $\mathcal{R}$ is a 3-round ZKPoK protocol between two parties, namely, a prover $\mathcal{P}$ and a verifier $\mathcal{V}$. For every input $(w, x) \in \mathcal{R}$ to $\mathcal{P}$ and x to $\mathcal{V}$, the first round of the protocol consists of $\mathcal{P}$ sending a commitment t to $\mathcal{V}$. $\mathcal{V}$ then replies with a challenge c in the second round and $\mathcal{P}$ concludes by sending a response z in the last round. At the end of the protocol, $\mathcal{V}$ outputs accept or reject. We say a protocol transcript (t, c, z) is valid if the output of an honest verifier $\mathcal{V}$ is accept. A Σ-protocol has to satisfy the following two properties:*

- *(Special Soundness.) A cheating prover can at most answer one of the many possible challenges. Specifically, there exists an efficient algorithm KE, called knowledge extractor, that on input x, a pair of valid transcripts (t, c, z) and (t, c', z') with $c \neq c'$, outputs w such that $(w, x) \in \mathcal{R}$.*
- *(Special Honest-Verifier Zero-Knowledgeness(HVZK).) There exists an efficient algorithm KS, called zero-knowledge simulator, that on input x and a challenge c, outputs a pair (t, z) such that (t, c, z) is a valid transcript having the same distribution as a real protocol transcript resulted from the interaction between $\mathcal{P}$ with input $(w, x) \in \mathcal{R}$ and an honest $\mathcal{V}$.*

Signature of Knowledge. Any Σ-protocol can be turned into non-interactive form, called signature of knowledge [19], by setting the challenge to the hash value of the commitment together with the message to be signed [30]. Pointcheval and Stern [44] showed that any signature scheme obtained this way is secure in the random oracle model [9].

3   A Zero-Knowledge Proof-of-Knowledge Protocol for RCV

We present the main result of this paper, namely, a zero-knowledge proof-of-knowledge protocol of **R**epresentation of **C**ommitted **V**alue, RCV. Specifically, let $C = g_0^x g_1^r \in \mathbb{G}_1$ be a commitment of x with randomness r. Let $D = h_1^{m_1} \cdots h_L^{m_L} h^s \in \mathbb{G}_q$ be the commitment of x's representation (to bases $h_1, \ldots, h_L$, denoted as $\boldsymbol{m}$) with randomness $s \in_R \mathbb{Z}_q$. We construct a ZKPoK protocol of $(x, \boldsymbol{m})$, denoted as PK_{RCV}. Technically speaking, our protocol is an *argument* system rather than a *proof* system in the sense that soundness in our system only holds against a PPT cheating prover. This is sufficient for all our purposes when adversaries in the applications of our PK_{RCV} are modeled as PPT algorithms. PK_{RCV} for C, D can be abstracted as follows.

$$\mathsf{PK_{RCV}}\Big\{\,(x, r, s, m_1, \ldots, m_L):$$
$$C = g_0^x g^r \,\wedge\, D = \mathrm{h}_1^{m_1} \cdots \mathrm{h}_L^{m_L} \mathrm{h}^s \,\wedge\, x = \mathrm{h}_1^{m_1} \cdots \mathrm{h}_L^{m_L} \Big\}$$

The construction of $\mathsf{PK_{RCV}}$ consists of two parts. Note that while we describe them separately, they can be executed in parallel in its actual implementation.

3.1 The Actual Protocol

We construct a Σ-Protocol of $\mathsf{PK_{RCV}}$. Let λ_k be a security parameter. In practice, we suggest λ_k should be at least 80. The first part of $\mathsf{PK_{RCV}}$ is a zero-knowledge proof-of-knowledge of representation of an element, and we adapt the protocol from [41].

(Commitment.) The prover randomly generates $\rho_x, \rho_r \in_R \mathbb{Z}_p$, computes and sends $T = g_0^{\rho_x} g^{\rho_r}$ to the verifier.
(Challenge.) The verifier returns a random challenge $c \in_R \{0,1\}^{\lambda_k}$.
(Response.) The prover, treating c as an element in $\mathbb{Z}_p{}^3$, computes $z_x = \rho_x - cx \in \mathbb{Z}_p$, $z_r = \rho_r - cr \in \mathbb{Z}_p$ and returns (z_x, z_r) to the verifier.
(Verify.) Verifier accepts if and only if $T = C^c g_0^{z_x} g^{z_r}$.

The second part is more involved and can be thought of as the extension of the ZKPoK of double-discrete logarithm in combination with ZKPoK of equality of discrete logarithm.

(Commitment.) For $i = 1$ to λ_k, the prover randomly generates $\rho_{m_1,i}$, $\ldots$, $\rho_{m_L,i}$, $\rho_{s,i} \in_R \mathbb{Z}_q$ and $\rho_{r,i} \in_R \mathbb{Z}_p$. Then the prover computes $T_{1,i} = g_0^{\mathrm{h}_1^{\rho_{m_1,i}} \cdots \mathrm{h}_L^{\rho_{m_L,i}}} g^{\rho_{r,i}} \in \mathbb{G}_1$ and $T_{2,i} = \mathrm{h}_1^{\rho_{m_1,i}} \cdots \mathrm{h}_L^{\rho_{m_L,i}} h^{\rho_{s,i}} \in \mathbb{G}_q$. After that, the prover sends $(T_{1,i}, T_{2,i})_{i=1}^{\lambda_k}$ to the verifier.
(Challenge.) The verifier returns a random challenge $c \in_R \{0,1\}^{\lambda_k}$.
(Response.) Denote $c[i]$ as the i-th bit of c. That is, $c[i] \in \{0,1\}$. For $i = 1$ to λ_k, the prover computes $z_{m_1,i} = \rho_{m_1,i} - c[i]m_1 \in \mathbb{Z}_q, \ldots, z_{m_L,i} = \rho_{m_L,i} - c[i]m_L \in \mathbb{Z}_q$, $z_{s,i} = \rho_{s,i} - c[i]s \in \mathbb{Z}_q$ and $z_{r,i} = \rho_{r,i} - c[i]\mathrm{h}_1^{z_{m_1,i}} \cdots \mathrm{h}_L^{z_{m_L,i}} r \in \mathbb{Z}_p$. The prover sends $\big(z_{m_1,i}, \ldots, z_{m_L,i}, z_{s,i}, z_{r,i}\big)_{i=1}^{\lambda_k}$ to the verifier.
(Verify.) The verifier accepts if the following equations hold for $i = 1$ to λ_k.

$$T_{2,i} \overset{?}{=} \mathrm{D}^{c[i]} \mathrm{h}_1^{z_{m_1,i}} \cdots \mathrm{h}_L^{z_{m_L,i}} h^{z_{s,i}}$$
$$T_{1,i} \overset{?}{=} g_0^{\mathrm{h}_1^{z_{m_1,i}} \cdots \mathrm{h}_L^{z_{m_L,i}}} g^{z_{r,i}} \text{ if } c[i] = 0$$
$$T_{1,i} \overset{?}{=} C^{\mathrm{h}_1^{z_{m_1,i}} \cdots \mathrm{h}_L^{z_{m_L,i}}} g^{z_{r,i}} \text{ if } c[i] = 1$$

The two parts should be executed in parallel using the same challenge. Regarding the security of $\mathsf{PK_{RCV}}$, we have the following theorem whose proof can be found in the full version of the paper [4].

Theorem 1. $\mathsf{PK_{RCV}}$ *is a Σ-Protocol.*

---

[3] Consequently, the bit-length of p should be longer than λ_k.

3.2 Efficiency Analysis of PK$_{\mathsf{RCV}}$

Table 4 summarizes the time and space complexities of PK$_{\mathsf{RCV}}$. We breakdown the time complexity of the protocol into the number of multi-exponentiations (multi-EXPs)[4] in various groups. Note that with pre-processing, prover's online computation is minimal and does not involve any exponentiations. As for the bandwidth requirement, the non-interactive version is more space-efficient since the prover does not need to include the commitment using the technique of [1].

In practice, we can take $\lambda_k = 80$ and p (resp. q) to be a 1024-bit (resp. 160-bit) prime. Thus, $\mathbb{Z}_p$, $\mathbb{Z}_q$ and $\mathbb{G}_1$ will take 1024, 160 and roughly 1024 bit, respectively. The non-interactive form (of which our applications employ) takes up around $(12+1.5L)$kB. Looking ahead, L is 1, 3 and 3 in our construction of blind signature, traceable signature and compact e-cash, respectively. The most dominant operation in our applications is the Multi-EXPs in group $\mathbb{G}_1$ since we are using the elliptic curve group equipped with pairing. As a preliminary analysis, we find out that one multi-EXP in $\mathbb{G}_1$ takes about 25ms. The timing is obtained on a Dell GX620 with an Intel Pentium 4 3.0 GHz CPU and 2GB RAM running Windows XP Professional SP2 as the host. We used Sun xVM VirtualBox 2.0.0 to emulate a guest machine of 1GB RAM running Ubuntu 7.04. Our implementation is written in C and relies on the Pairing-Based Cryptography (PBC) library (version 0.4.18). $\mathbb{G}_1$ is taken to be an elliptic curve group equipped with type A1 pairing and the prime p is 1048 bits. In a nutshell, the verifier takes around 2 seconds in verifying the proof PK$_{\mathsf{RCV}}$.

Table 4. Time and Space Complexities of PK$_{\mathsf{RCV}}$

Time Complexities			
	Prover		Verifier
	w/o Preproc.	w/ Preproc.	
$\mathbb{G}_1$ multi-EXP	$\lambda_k + 1$	0	$\lambda_k + 1$
$\mathbb{G}_q$ multi-EXP	$\lambda_k(\lceil L/3 \rceil + 1) + 1$	0	$\lambda_k(\lceil L/3 \rceil + 2)$
Bandwidth Requirement			
	Interactive Form		Non-Interactive Form
$\mathbb{G}_1$	$2\lambda_k + 1$		0
$\mathbb{Z}_p$	$\lambda_k + 2$		$\lambda_k + 2$
$\mathbb{Z}_q$	$\lambda_k(L + 1)$		$\lambda_k(L + 1)$

4 Application to Round-Optimal Concurrently-Secure Blind Signature without Interactive Assumptions

4.1 Syntax

We review the definition of blind signature from Hazay *et al.* [35].

---

[4] A multi-EXP computes the product of exponentiations faster than performing the exponentiations separately. Normally, a multi-based exponentiation takes only 10% more time compared with a single-based exponentiation. We assume that one multi-EXP operation multiplies up to 3 exponentiations.

Definition 3. *A blind signature scheme is a tuple of PPT algorithms* **BGen**, **BVer** *and an interactive protocol* **BSign** *between a user and a signer such that:*

- **BGen**: *On input security parameter* 1^λ, *this algorithm outputs a key pair* (pk, sk).
- **BSign**: *Signer, with private input* sk *interacts with a user having input* pk *and a message* m *in the protocol. At the end of the execution, user obtains a signature* σ_m *on the message* m, *assuming neither party abort.*
- **BVer**: *On input* pk, m, σ_m, *outputs* **valid** *or* **invalid**.

As usual, correctness requires that for all (pk, sk) *output by* **BGen**(1^λ), *and for all* σ_m *which is the output of the user upon successful completion of the protocol run of* **BSign** *with appropriate inputs* $((pk, m)$ *and* sk *for user and signer respectively) to both parties,* **BVer** *with input* pk, m, σ_m *outputs* **valid**.

Definition 4. *Blind signature scheme* (**BGen**, **BSign**, **BVer**) *is unforgeable if the winning probability for any PPT adversary* $\mathcal{A}$ *in the following game is negligible:*

- **BGen** *outputs* (pk, sk) *and* pk *is given to* $\mathcal{A}$.
- $\mathcal{A}$ *interact concurrently with* ℓ *signer clones with input* sk *in* **BSign** *protocol.*
- $\mathcal{A}$ *outputs* $\ell + 1$ *signatures* σ_i *on* $\ell + 1$ *distinct messages* m_i.

$\mathcal{A}$ *wins the game if all* m_i *are distinct and* **BVer**$(pk, m_i, \sigma_i) = 1$ *for all* $i = 1$ *to* $\ell + 1$.

Definition 5. *Blind signature scheme* (**BGen**, **BSign**, **BVer**) *satisfies blindness if the advantage for any PPT adversary* $\mathcal{A}$ *in the following game is negligible:*

- $\mathcal{A}$ *outputs an arbitrary public key* pk *and two equal-length messages* m_0, m_1.
- *A random bit* $b \in_R \{0, 1\}$ *is chosen, and* $\mathcal{A}$ *interacts concurrently with two user clones, say* U_0 *and* U_1, *with input* (pk, m_b) *and* (pk, m_{1-b}) *respectively. Upon completion of both protocols, define* σ_0 *and* σ_1 *as follows:*
  - *If either of the* U_0 *or* U_1 *aborts, set* $(\sigma_0, \sigma_1) = (\bot, \bot)$.
  - *Otherwise, define* σ_i *be the output of* U_i *for* $i = 0$ *and* 1.
  (σ_0, σ_1) *are given to* $\mathcal{A}$.
- $\mathcal{A}$ *outputs a guess bit* $b' \in \{0, 1\}$.

$\mathcal{A}$ *wins the game if all* $b' = b$. *The advantage of* $\mathcal{A}$ *is defined as* $|Pr[b' = b] - 1/2|$.

4.2 Construction

BGen. Let $\alpha, \beta \in_R \mathbb{Z}_p$ and $u = g^\alpha$ and $v = g^\beta$. Let $H : \{0, 1\}^* \to \mathbb{Z}_q$ be a collision-resistant hash function. The signer's secret key sk is (α, β) while its public key pk is $(\mathbb{G}_1, \mathbb{G}_T, \hat{e}, \mathbb{G}_q, p, q, g, u, v, h, h_0, h_1, H)$.

BSign. On input message $m \in \mathbb{Z}_q$, the user computes $x = h_0^m h^s$ for some randomly generated $s \in_R \mathbb{Z}_q$. The user sends x to the signer. The signer selects $e \in_R \mathbb{Z}_p$ and computes $A = g^{\frac{1}{\alpha + x + \beta e}}$. The signer returns (A, e) to the user.

The user computes Π_m as an non-interactive zero-knowledge proof-of-knowledge of a BB signature (A, e) on a hidden value x, and that x is a commitment of m and output Π_m as the signature of m.

Specifically, denote $y = h^s$. The user computes $\mathfrak{A}_1 = Ag_2^{r_1}$, $\mathfrak{A}_2 = g_1^{r_1}g_2^{r_2}$, $\mathfrak{A}_3 = g_1^y g_2^{r_3}$ for some randomly generated $r_1, r_2, r_3 \in_R \mathbb{Z}_p$ and $A_4 = h_0^s h^t$ for some randomly generated $t \in_R \mathbb{Z}_q$. Parse $M = \mathfrak{A}_1 || \mathfrak{A}_2 || \mathfrak{A}_3 || A_4$. The user computes the following non-interactive zero-knowledge proof-of-knowledge Π_m comprising two parts, namely, SPK_1 and SPK_2. SPK_1 can be computed using standard techniques, while SPK_2 is computed using our newly constructed PK_{RCV}. Finally, parse Π_m as ($\mathfrak{A}_1$, $\mathfrak{A}_2$, $\mathfrak{A}_3$, A_4, SPK_1, SPK_2).

$$
\Pi_m : \begin{cases} SPK_1 \Big\{ (r_1, r_2, r_3, y, e, \beta_1, \beta_2, \beta_3, \beta_4) : \\ \qquad \mathfrak{A}_2 = g_1^{r_1}g_2^{r_2} \ \wedge \ 1 = \mathfrak{A}_2^{-e}g_1^{\beta_1}g_2^{\beta_2} \ \wedge \\ \qquad 1 = \mathfrak{A}_2^{-y}g_1^{\beta_3}g_2^{\beta_4} \ \wedge \ \mathfrak{A}_3 = g_1^y g_2^{r_3} \ \wedge \ \frac{\hat{e}(\mathfrak{A}_1, u)}{\hat{e}(g,g)} = \\ \qquad \hat{e}(g_2, u)^{r_1}\hat{e}(\mathfrak{A}_1, v)^{-e}\hat{e}(g_2, v)^{\beta_1}\hat{e}(\mathfrak{A}_1, h_0^{h_0^m})^{-y}\hat{e}(g_2, h_0^{h_0^m})^{\beta_3} \Big\}(M) \\ SPK_2 \Big\{ (r_3, y, s, t) : \mathfrak{A}_3 = g_1^y g_2^{r_3} \ \wedge A_4 = h_0^s h^t \ \wedge \ y = h_0^s \Big\}(M) \end{cases}
$$

BVer. On input message m and its signature Π_m, parse Π_m as ($\mathfrak{A}_1$, $\mathfrak{A}_2$, $\mathfrak{A}_3$, A_4, SPK_1, SPK_2) and verify that SPK_1 and SPK_2 are valid.

Regarding the security of our construction, we have the following theorems whose proofs can be found in the full version of the paper [4].

Theorem 2. *Our blind signature is unforgeable under the q-SDH assumption in $\mathbb{G}_1$ and DL assumption in $\mathbb{G}_q$ in the random oracle model.*

Theorem 3. *Our blind signature satisfies blindness unconditionally in the random oracle model.*

5   Application to Traceable Signatures with Concurrent Join

We describe the construction of our traceable signatures. Since traceable signatures are group signatures with added functionalities, it is easy to modify our scheme into a 'regular' group signature. An attack to the traceable signature due to [37] is given in Appendix A.

5.1   Syntax

We review briefly the definition of traceable signature from Choi *et al.* [25] which is an adaptation of the definition of traceable identification from Kiayias *et al.* [37]. Note that Traceable identifications can be turned into traceable signatures using the Fiat-Shamir Heuristics [30].

Definition 6. *A traceable signature scheme is a tuple of nine PPT algorithms / protocols (GGen, Join, GSign, GVer, Open, Trace, Claim, ClaimVer) between three entities, namely group manager (GM), users and tracing agents:*

- *GGen: On input security parameter 1^λ, this algorithm outputs a key pair (pk, sk) for the group manager.*

- *Join:* This is a protocol between a user and GM. Upon successful completion of the protocol, user U_i obtains a membership certificate **cert**$_i$. The GM stores the whole protocol transcript **Jtrans**$_i$.
- *GSign:* User U_i with membership certificate **cert**$_i$ signs a message m and produces a group signature σ_m.
- *GVer:* On input pk, m, σ_m, outputs **valid** or **invalid**.
- *Open:* On input m, σ_m, the group manager outputs the identity of the signer.
- *Reveal:* On input **Jtrans**$_i$, the group manager outputs tracing information **tr**$_i$, which is the tracing trapdoor that allows party to identity signatures generated by user U_i.
- *Trace:* On input a signature σ and a tracing information **tr**$_i$, output $0/1$ indicating the signature is generated by user U_i or not.
- *Claim:* On input a signature σ and a membership certificate **cert**$_i$, user U_i produces a proof τ to prove that he is the originator of the signature.
- *ClaimVer:* On input a signature σ, a proof τ, output $0/1$ indicating the signature is generated by claimer or not.

Security Requirements. We informally review the security notion of a traceable signature. Due to page limitation, please refer to [37,25] for formal definition. A traceable signature should be secure against three types of attack.

(Misidentification.) The adversary is allowed to observe the operation of the system while users are engaged with GM during the joining protocol. It is also allowed to obtain a signature from existing users on any messages of its choice. They are also allowed to introduce users into the system. The adversary's goal is to produce a valid signature on new message that is not open to users controlled by the adversary.

(Anonymity.) The adversary is allowed to observe the operation of the system while users are engaged with GM during the joining protocol. It is also allowed to obtain signature from existing users on any messages of its choice. They are also allowed to introduce users into the system. Finally, the adversary chooses a message and two target users he does not control, and then receives a signature of the message he returned from one of these two target users. The adversary's goal is to guess which of the two target users produced the signature.

(Framing.) The adversary plays the role of a malicious GM. It is considered successful with the following scenarios. Firstly, the adversary may construct a signature that opens to an honest user. Secondly, it may construct a signature, output some tracing information and that when traced, this maliciously-constructed signature will be traced to be from an honest user. Thirdly, it may claim a signature that was generated by an honest user as its own.

5.2   Construction

GGen. Let $\alpha, \beta \in_R \mathbb{Z}_p^*$ and $u = g^\alpha$ and $v = g^\beta$. $H : \{0,1\}^* \to \mathbb{Z}_q$ be a collision-resistant hash function. Further, randomly generate $\gamma_1, \gamma_2 \in_R \mathbb{Z}_p$, $w_3 \in_R \mathbb{G}_1$ and compute $w_1 = w_3^{\frac{1}{\gamma_1}}$ and $w_2 = w_3^{\frac{1}{\gamma_2}}$. GM's secret key sk is $(\alpha, \beta, \gamma_1, \gamma_2)$ while its public key pk is $(\hat{e}, \mathbb{G}_1, \mathbb{G}_T, \mathbb{G}_q, p, q, g, u, v, w_1, w_2, w_3, h, h_0, h_1, \ldots, h_4, H)$.

Join. A user U_i randomly selects $s, x \in_R \mathbb{Z}_q$ and sends $C' = h_0^s h_1^x \in \mathbb{G}_q$ to GM. GM computes $t = H(C') \in \mathbb{Z}_q$. It then computes $C = C' h_2^t$ and selects $e \in_R \mathbb{Z}_p$. Next, it computes $A = g^{\frac{1}{\alpha + C + \beta e}}$. The GM returns (A, e, t) to the user. User checks if $\hat{e}(A, uv^e g^{h_0^s h_1^x h_2^t}) = \hat{e}(g, g)$ and $t = H(C')$. He then stores (A, e, s, t, x) as his membership certificate cert_i. GM records t as the tracing information tr_i for this user. GM also stores the whole communication transcript.

GSign. Let the user membership certificate be (A, e, s, t, x). The user computes $S = h_3^k$, $U = h_3^{k'}$ for some randomly generated $k, k', k'' \in_R \mathbb{Z}_q$ and $T_1 = S^t$, $T_2 = S^{k''}$, $T_3 = h_0^s h_1^x T_1^{k''}$, $V = U^x$. Denote $y = h_0^s h_1^x h_2^t$. The user then randomly generates $r_1, r_2, r_3 \in_R \mathbb{Z}_p$, computes $\mathfrak{A}_1 = A w_3^{r_1 + r_2}$, $\mathfrak{A}_2 = w_1^{r_1}$, $\mathfrak{A}_3 = w_2^{r_2}$, $\mathfrak{A}_4 = g_1^y g_2^{r_3}$ and $A_5 = h^r h_0^s h_1^x h_2^t$ for some randomly generated $r \in_R \mathbb{Z}_q$. To generate a traceable signature for message m, parse $M = m||S||U||T_1||T_2||T_3||V||\mathfrak{A}_1||\mathfrak{A}_2||\mathfrak{A}_3||\mathfrak{A}_4||A_5$.

The user computes the following non-interactive zero-knowledge proof-of-knowledge Π_{grp} comprising two parts, namely, SPK_3 and SPK_4. SPK_3 can be computed using standard techniques, while SPK_4 is computed using $\mathrm{PK}_{\mathsf{RCV}}$. Finally, parse Π_{grp} as $(\mathfrak{A}_1, \mathfrak{A}_2, \mathfrak{A}_3, \mathfrak{A}_4, A_5, \mathrm{SPK}_3, \mathrm{SPK}_4)$ and the signature σ_m as $(\Pi_{\mathsf{grp}}, S, T_1, T_2, T_3, U, V)$.

$$\Pi_{\mathsf{grp}} : \begin{cases} \mathrm{SPK}_3\Big\{(r_1, r_2, r_3, \mathsf{y}, e, \beta_1, \beta_2, \beta_3, \beta_4, r, s, t, x, k, k', k'') : \\ \qquad \mathfrak{A}_2 = w_1^{r_1} \wedge 1 = \mathfrak{A}_2^{-e} w_1^{\beta_1} \wedge 1 = \mathfrak{A}_2^{-\mathsf{y}} w_1^{\beta_2} \wedge \\ \qquad \mathfrak{A}_3 = w_2^{r_2} \wedge 1 = \mathfrak{A}_3^{-e} w_1^{\beta_3} \wedge 1 = \mathfrak{A}_3^{-\mathsf{y}} w_1^{\beta_4} \wedge \\ \qquad \mathfrak{A}_4 = g_1^{\mathsf{y}} g_2^{r_3} \wedge A_5 = h^r h_0^s h_1^x h_2^t \wedge \\ \qquad S = h_3^k \wedge T_1 = S^t \wedge T_2 = S^{k''} \wedge T_3 = h_0^s h_1^x T_1^{k''} \wedge \\ \qquad U = h_3^{k'} \wedge V = U^x \wedge \frac{\hat{e}(\mathfrak{A}_1, u)}{\hat{e}(g, g)} = \\ \hat{e}(w_3, u)^{r_1 + r_2} \hat{e}(\mathfrak{A}_1, v)^{-e} \hat{e}(w_2, v)^{\beta_1 + \beta_3} \hat{e}(\mathfrak{A}_1, g)^{-\mathsf{y}} \hat{e}(w_3, g)^{\beta_2 + \beta_4} \Big\}(M) \\ \mathrm{SPK}_4\Big\{(r_3, \mathsf{y}, r, s, t, x) : \\ \qquad \mathfrak{A}_4 = g_1^{\mathsf{y}} g_2^{r_3} \wedge A_5 = h^r h_0^s h_1^x h_2^t \wedge \mathsf{y} = h_0^s h_1^x h_2^t \Big\}(M) \end{cases}$$

Basically, $\mathfrak{A}_1$, $\mathfrak{A}_2$ and $\mathfrak{A}_3$ is the linear encryption of A (part of the membership certificate), T_1, T_2, T_3 is the El-Gamal encryption of $h_0^s h_1^x$ (under the public key S^t), while the rest of the proof is to assure the verifier that the encryptions are properly done and that values U, V, S are correctly formed with respective to values s, t, x, r.

Open. On input a signature σ_m, GM computes $A := \frac{\mathfrak{A}_1}{\mathfrak{A}_2^{\gamma_1} \mathfrak{A}_3^{\gamma_2}}$. From A, GM looks up its list of join transcripts and identify the underlying user.

Reveal. To allow tracing of user U_i, the GM outputs tracing information tr_i.

Trace. Given a valid signature $\sigma_m = (\Pi_{\mathsf{grp}}, S, T_1, T_2, T_3, U, V)$ and tracing information tr_i, everyone can test if the signature is from user U_i by testing $T_1 \overset{?}{=} S^{\mathsf{tr}_i}$ and $\mathsf{tr}_i \overset{?}{=} H(\frac{T_3}{T_2^{\mathsf{tr}_i}})$.

Claim. On input a message $\sigma_m = (\Pi_{\mathsf{grp}}, S, T_1, T_2, T_3, U, V)$, the originator can produce an non-interactive proof τ as

$$\tau : \mathrm{SPK}_\tau\{(x) : V = U^x\}(\sigma_m)$$

ClaimVer. Given a signature σ_m and τ, everyone can verify τ.

Regarding the security of traceable signature, we have the following theorem whose proof can be found in the full version of the paper [4].

Theorem 4. *Our traceable signature is secure under the q-SDH assumption, the DLDH assumption in $\mathbb{G}_1$ and DL assumption in $\mathbb{G}_q$ in the random oracle model.*

6 Compact E-Cash with Concurrent Withdrawal

Our technique can also be applied to construct compact e-cash systems with concurrently-secure withdrawal protocol. Due to page limitation, only high-level description is given here. Its detail, together with definitions shall be found in the full version of the paper [4]. Roughly speaking, there are three entities, namely, the bank, users and merchants, in a compact e-cash system. To withdraw a wallet of K coins, user obtains a BB signature cert on commitment of values (s, t, x), in a similar manner as user obtains a membership certificate in our construction of traceable signatures. Note that the major difference being in this case, none of the values are known to the bank (with s being a random number jointly generated by the bank and user).

To spend a electronic coin to a merchant, user computes a serial number $S = \mathrm{h}_3^{\frac{1}{s+J+1}}$, a tracing tag $T = \mathrm{h}_0^s \mathrm{h}_1^t \mathrm{h}_2^x \mathrm{h}_3^{\frac{R}{t+J+1}}$, where J is the counter of the number of times the user has spent his wallet and R is a random challenge issued by the merchant. User sends the pair (S, T) to the merchant, along with a signature of knowledge $\Pi_\$$, stating that S and T are correctly formed. Specifically, the proof assures the merchant that (1)user is in possession of a valid BB signature from the bank on values (s, t, x); (2)counter $0 \leq J < K$; (3)S and T are correctly formed with respect to (s, t, x).

In the deposit protocol, merchant sends the coin $(\Pi_\$, S, T, R)$ to the bank. Since counter J runs from 0 to $K - 1$, user can at most spend his wallet for K times. If the user uses the counter for a second time, the serial number S of the double-spent coins will be the same and will thus be identified. Next, the bank can compute a value $C := (\frac{T^{R'}}{T'^R})^{1/(R'-R)}$, the commitment of (s, t, x) which allows the bank to identify the underlying double-spender.

7 Conclusion

We constructed a new zero-knowledge argument system and illustrated its significance with applications to blind signatures, traceable signatures and compact e-cash systems. We believe this system is useful in other cryptographic applications.

References

1. Ateniese, G., Camenisch, J., Joye, M., Tsudik, G.: A practical and provably secure coalition-resistant group signature scheme. In: Bellare, M. (ed.) CRYPTO 2000. LNCS, vol. 1880, pp. 255–270. Springer, Heidelberg (2000)
2. Ateniese, G., Song, D.X., Tsudik, G.: Quasi-efficient revocation in group signatures. In: Blaze, M. (ed.) FC 2002. LNCS, vol. 2357, pp. 183–197. Springer, Heidelberg (2003)

3. Au, M.H., Susilo, W., Mu, Y.: Practical compact e-cash. In: Pieprzyk, J., Ghodosi, H., Dawson, E. (eds.) ACISP 2007. LNCS, vol. 4586, pp. 431–445. Springer, Heidelberg (2007)
4. Au, M.H., Susilo, W., Yiu, S.-M.: Proof-of-knowledge of representation of committed value and its applications (2010), Full version available at,
http://uow.academia.edu/ManHoAu
5. Au, M.H., Wu, Q., Susilo, W., Mu, Y.: Compact e-cash from bounded accumulator. In: Abe, M. (ed.) CT-RSA 2007. LNCS, vol. 4377, pp. 178–195. Springer, Heidelberg (2006)
6. Bellare, M.: A note on negligible functions. J. Cryptology 15(4), 271–284 (2002)
7. Bellare, M., Goldreich, O.: On defining proofs of knowledge. In: Brickell, E.F. (ed.) CRYPTO 1992. LNCS, vol. 740, pp. 390–420. Springer, Heidelberg (1993)
8. Bellare, M., Namprempre, C., Pointcheval, D., Semanko, M.: The Power of RSA Inversion Oracles and the Security of Chaum's RSA-Based Blind Signature Scheme. In: Syverson, P.F. (ed.) FC 2001. LNCS, vol. 2339, pp. 309–328. Springer, Heidelberg (2002)
9. Bellare, M., Rogaway, P.: Random oracles are practical: A paradigm for designing efficient protocols. In: ACMCCS 1993, pp. 62–73 (1993)
10. Boldyreva, A.: Threshold signatures, multisignatures and blind signatures based on the gap-diffie-hellman-group signature scheme. In: Desmedt, Y.G. (ed.) PKC 2003. LNCS, vol. 2567, pp. 31–46. Springer, Heidelberg (2002)
11. Boneh, D., Boyen, X.: Short signatures without random oracles. In: Cachin, C., Camenisch, J.L. (eds.) EUROCRYPT 2004. LNCS, vol. 3027, pp. 56–73. Springer, Heidelberg (2004)
12. Boneh, D., Boyen, X., Shacham, H.: Short group signatures. In: Franklin, M. (ed.) CRYPTO 2004. LNCS, vol. 3152, pp. 41–55. Springer, Heidelberg (2004)
13. Boudot, F.: Efficient proofs that a committed number lies in an interval. In: Preneel, B. (ed.) EUROCRYPT 2000. LNCS, vol. 1807, pp. 431–444. Springer, Heidelberg (2000)
14. Brands, S.: An Efficient Off-Line Electronic Cash System Based on the Representation Problem. Technical report, CWI (1993)
15. Camenisch, J.: Group signature schemes and payment systems based on the discrete logarithm problem. PhD Thesis, ETH Zürich, 1998. Diss. ETH No. 12520. Hartung Gorre Verlag, Konstanz (1998)
16. Camenisch, J., Hohenberger, S., Lysyanskaya, A.: Compact e-cash. In: Cramer, R. (ed.) EUROCRYPT 2005. LNCS, vol. 3494, pp. 302–321. Springer, Heidelberg (2005)
17. Camenisch, J., Lysyanskaya, A.: A Signature Scheme with Efficient Protocols. In: Cimato, S., Galdi, C., Persiano, G. (eds.) SCN 2002. LNCS, vol. 2576, pp. 268–289. Springer, Heidelberg (2003)
18. Camenisch, J., Shoup, V.: Practical verifiable encryption and decryption of discrete logarithms. In: Boneh, D. (ed.) CRYPTO 2003. LNCS, vol. 2729, pp. 126–144. Springer, Heidelberg (2003)
19. Camenisch, J., Stadler, M.: Efficient group signature schemes for large groups (extended abstract). In: Kaliski Jr., B.S. (ed.) CRYPTO 1997. LNCS, vol. 1294, pp. 410–424. Springer, Heidelberg (1997)
20. Canard, S., Gouget, A.: Divisible e-cash systems can be truly anonymous. In: Naor, M. (ed.) EUROCRYPT 2007. LNCS, vol. 4515, pp. 482–497. Springer, Heidelberg (2007)
21. Canard, S., Gouget, A., Hufschmitt, E.: Handy Compact E-cash System. In: Proceedings of the 2nd Conference on Security in Network Architectures and Information Systems - SAR-SSI 2007 (2007)
22. Cathalo, J., Libert, B., Yung, M.: Group encryption: Non-interactive realization in the standard model. In: matsui, M. (ed.) ASIACRYPT 2009. LNCS, vol. 5912, pp. 179–196. Springer, Heidelberg (2009)
23. Chaum, D.: Blind Signatures for Untraceable Payments. In: Advances in Cryptology: Proceedings of CRYPTO 1982, pp. 199–203. Plenum, New York (1983)

24. Chaum, D., van Heyst, E.: Group signatures. In: Davies, D.W. (ed.) EUROCRYPT 1991. LNCS, vol. 547, pp. 257–265. Springer, Heidelberg (1991)

25. Choi, S.G., Park, K., Yung, M.: Short traceable signatures based on bilinear pairings. In: Yoshiura, H., Sakurai, K., Rannenberg, K., Murayama, Y., Kawamura, S.-i. (eds.) IWSEC 2006. LNCS, vol. 4266, pp. 88–103. Springer, Heidelberg (2006)

26. Cramer, R., Damgård, I., MacKenzie, P.D.: Efficient zero-knowledge proofs of knowledge without intractability assumptions. In: Imai, H., Zheng, Y. (eds.) PKC 2000. LNCS, vol. 1751, pp. 354–373. Springer, Heidelberg (2000)

27. Cramer, R., Damgård, I., Schoenmakers, B.: Proofs of partial knowledge and simplified design of witness hiding protocols. In: Desmedt, Y.G. (ed.) CRYPTO 1994. LNCS, vol. 839, pp. 174–187. Springer, Heidelberg (1994)

28. Damgård, I.: Efficient concurrent zero-knowledge in the auxiliary string model. In: Preneel, B. (ed.) EUROCRYPT 2000. LNCS, vol. 1807, pp. 418–430. Springer, Heidelberg (2000)

29. Dodis, Y., Yampolskiy, A.: A verifiable random function with short proofs and keys. In: Vaudenay, S. (ed.) PKC 2005. LNCS, vol. 3386, pp. 416–431. Springer, Heidelberg (2005)

30. Fiat, A., Shamir, A.: How to Prove Yourself: Practical Solutions to Identification and Signature Problems. In: Odlyzko, A.M. (ed.) CRYPTO 1986. LNCS, vol. 263, pp. 186–194. Springer, Heidelberg (1987)

31. Fischlin, M.: Round-optimal composable blind signatures in the common reference string model. In: Dwork, C. (ed.) CRYPTO 2006. LNCS, vol. 4117, pp. 60–77. Springer, Heidelberg (2006)

32. Fuchsbauer, G.: Automorphic signatures in bilinear groups and an application to round-optimal blind signatures. Cryptology ePrint Archive, Report 2009/320 (2009)

33. Goldwasser, S., Micali, S.: Probabilistic Encryption. J. Comput. Syst. Sci. 28(2), 270–299 (1984)

34. Goldwasser, S., Micali, S., Rackoff, C.: The Knowledge Complexity of Interactive Proof-Systems (Extended Abstract). In: STOC 1985, pp. 291–304 (1985)

35. Hazay, C., Katz, J., Koo, C.-Y., Lindell, Y.: Concurrently-secure blind signatures without random oracles or setup assumptions. In: Vadhan, S.P. (ed.) TCC 2007. LNCS, vol. 4392, pp. 323–341. Springer, Heidelberg (2007)

36. Juels, A., Luby, M., Ostrovsky, R.: Security of blind digital signatures (extended abstract). In: Kaliski Jr., B.S. (ed.) CRYPTO 1997. LNCS, vol. 1294, pp. 150–164. Springer, Heidelberg (1997)

37. Kiayias, A., Tsiounis, Y., Yung, M.: Traceable signatures. In: Cachin, C., Camenisch, J.L. (eds.) EUROCRYPT 2004. LNCS, vol. 3027, pp. 571–589. Springer, Heidelberg (2004)

38. Kiayias, A., Tsiounis, Y., Yung, M.: Traceable signatures. Cryptology ePrint Archive, Report 2004/007 (2004), http://eprint.iacr.org/

39. Kiayias, A., Yung, M.: Group signatures with efficient concurrent join. In: Cramer, R. (ed.) EUROCRYPT 2005. LNCS, vol. 3494, pp. 198–214. Springer, Heidelberg (2005)

40. Lindell, Y.: Bounded-concurrent secure two-party computation without setup assumptions. In: STOC 2003, pp. 683–692 (2003)

41. Okamoto, T.: Provably secure and practical identification schemes and corresponding signature schemes. In: Brickell, E.F. (ed.) CRYPTO 1992. LNCS, vol. 740, pp. 31–53. Springer, Heidelberg (1993)

42. Okamoto, T.: Efficient blind and partially blind signatures without random oracles. Cryptology ePrint Archive, Report 2006/102 (2006), http://eprint.iacr.org/

43. Pedersen, T.P.: Non-interactive and information-theoretic secure verifiable secret sharing. In: Feigenbaum, J. (ed.) CRYPTO 1991. LNCS, vol. 576, pp. 129–140. Springer, Heidelberg (1992)

44. Pointcheval, D., Stern, J.: Security proofs for signature schemes. In: Maurer, U.M. (ed.) EUROCRYPT 1996. LNCS, vol. 1070, pp. 387–398. Springer, Heidelberg (1996)

45. Pointcheval, D., Stern, J.: Security arguments for digital signatures and blind signatures. J. Cryptology 13(3), 361–396 (2000)
46. Schnorr, C.-P.: Efficient Signature Generation by Smart Cards. J. Cryptology 4(3), 161–174 (1991)
47. Stadler, M.: Cryptographic protocols for revocable privacy. PhD Thesis, ETH Zürich, 1996. Diss. ETH No. 11651 (1996)

A   A Framing Attack on KTY Traceable Signatures

In this section, we present a high level description of the traceable signatures from [37] (KTY) and a concrete attack within their security model.

Overview of the KTY Traceable Signature

GGen: The group manager chooses a signature scheme. The signature scheme in KTY is in fact a variant of the CL signature [17].

Join: User chooses a random number x' and obtains a CL signature (denoted as cert) from the GM on values x', x using the signature generation protocol of CL signature. In particular, x' is unknown to GM while x is known. The value x is stored as the tracing information tr of the user. User stores cert as his membership certificate.

GSign: To sign a message m, user with membership certificate cert on values x', x first computes:

    1. a tuple (T_1, T_2, T_3), which is the El-Gamal encryption of part of cert.
    2. a tuple (T_4, T_5) such that $T_5 = g^k$ and $T_4 = T_5^x$ for some random number k.
    3. a tuple (T_6, T_7) such that $T_7 = g^{k'}$ and $T_6 = T_7^{x'}$ for some random number k'.

The traceable signature is a signature of knowledge σ_m such that $(T_1, \ldots, T_7)$ are correctly formed.

GVer: The verifier simply verifies the signature-of-knowledge σ_m.

Open: On input m, σ_m, the group manager outputs the identity of the signer by decrypting T_1, T_2, T_3 and obtains cert of the user.

Reveal: On input Jtrans$_i$, the group manager outputs tracing information tr $= x$.

Trace: On input a signature σ_m and a tracing information tr, test whether $T_4 \stackrel{?}{=} T_5^x$.

Claim/ClaimVer: To claim a signature, the signer produces a non-interactive proof-of-knowledge of discrete logarithm of T_6 to base T_7 (which is x').

The Framing Attack. The framing attack is considered successful if the attacker can generate a signature that traces to an honest user. Specifically, the adversary is considered successful if it can output a signature σ_m^* such that Trace(Reveal(Jtrans$_i$), σ_m^*) = 1 and that user U_i is an honest user who has not generated σ_m^* himself. The attack is based on the fact that σ_m^* does not need to open to U_i, and the attacker knows the corresponding tracing information, that is, x, of an honest user. To frame an honest user, the adversary generates another membership certificate cert* on values x^*, x and uses it to produce a signature σ_m^*. Obviously, this signature will trace to the honest user.

    The attack is possible due to a flaw in the security proof [38] (full version of [37] , Section 9.3), in which it is stated that "Then if the adversary outputs an identification transcript that either opens to user j traces to the user j, it is clear that we can rewind the adversary and obtain a witness for that transcript that will reveal the logarithm of C base b, and thus solving the discrete-logarithm problem." The argument is true when the identification transcript opens to user j in which it helps solving the discrete logarithm of C to base b (which is x', the user secret). However the same argument is not applicable to the case of tracing because the tracing information x for user j is in fact known to the adversary. The adversary is not required to use the same x' with the honest user in producing the signature for framing to be successful.

The Proposed Fix. It turns out that the same attack is not applicable to the pairing-based traceable signatures [25] (CPY). The reason is that the tracing information tr is of the form g^x and, although tr is known to GM, the value x is unknown and correctness of tr is implicitly checked in a signature of knowledge of x. The same idea, however, is not applicable to the original KTY scheme because the tracing mechanism in CPY requires the use of a bilinear map[5] which is not known to exists in the group of which KTY is built on. Thus, we propose another fix. That is, the tracing information tr is no longer randomly chosen. Instead, it is set to be $H(C_i)$, where $C_i = b_i^{x'}$ is known to GM during the join protocol in KTY, for some collision-resistant hash function H. The group signature will be modified so that the user will encrypt C_i under the public key g^{tr} (using El-Gamal Encryption), together with a proof-of-correctness, including the knowledge of C_i to base b_i. The corresponding Trace algorithm is also modified to include a test that $\mathrm{tr} \stackrel{?}{=} H(C_i)$ when tr is given. Indeed, this idea is employed in our construction of traceable signatures.

---

[5] Specifically, for each signature, user produces values T_4, T_5 such that the tracing agent test if $\hat{e}(\mathrm{tr}, T_4) \stackrel{?}{=} T_5$. The user also includes a proof-of-knowledge of discrete logarithm (that is, knowledge of x) of T_5 to base $\hat{e}(g, T_5)$ in the signature.

Pattern Recognition Techniques for the Classification of Malware Packers

Li Sun[1], Steven Versteeg[2], Serdar Boztaş[1], and Trevor Yann[3]

[1] School of Mathematical and Geospatial Sciences, RMIT University,
GPO Box 2476V, Melbourne, Australia
`li.sun@ca.com, serdar.boztas@ems.rmit.edu.au`
[2] CA Labs, Melbourne, Australia
`steven.versteeg@ca.com`
[3] HCL Australia, Melbourne, Australia
`tyann@hcl.in`

Abstract. Packing is the most common obfuscation method used by malware writers to hinder malware detection and analysis. There has been a dramatic increase in the number of new packers and variants of existing ones combined with packers employing increasingly sophisticated anti-unpacker tricks and obfuscation methods. This makes it difficult, costly and time-consuming for anti-virus (AV) researchers to carry out the traditional static packer identification and classification methods which are mainly based on the packer's byte signature.

In this paper[1], we present a simple, yet fast and effective packer classification framework that applies pattern recognition techniques on automatically extracted randomness profiles of packers. This system can be run without AV researcher's manual input. We test various statistical classification algorithms, including k−Nearest Neighbor, Best-first Decision Tree, Sequential Minimal Optimization and Naive Bayes. We test these algorithms on a large data set that consists of clean packed files and 17,336 real malware samples. Experimental results demonstrate that our packer classification system achieves extremely high effectiveness ($> 99\%$). The experiments also confirm that the randomness profile used in the system is a very strong feature for packer classification. It can be applied with high accuracy on real malware samples.

1 Introduction

The Internet has become an essential part of human life in the modern information society. While the Internet brings enormous convenience to users, *Mal*licious soft*ware* (Malware) which includes computer viruses, worms, Trojan horses, other malicious or unwanted software, produce a tremendous impact on users' security, reliability and privacy. This is a serious threat to the security of computer networks. In recent years, the number of malware infections has risen sharply.

Many efforts have been made to combat malware. For every malware binary, anti-virus (AV) researchers need to reverse the malware and update the AV scanner to detect

---

[1] This work was supported by CA Labs and an ARC linkage grant. Ms. Sun is currently a Ph.D. candidate sponsored by CA Labs.

R. Steinfeld and P. Hawkes (Eds.): ACISP 2010, LNCS 6168, pp. 370–390, 2010.

and counter it. Unfortunately, malware authors are aware of this. Most new malware implement various obfuscation techniques in order to disguise themselves, therefore preventing successful analysis and thwarting the detection by AV scanners.

Packing is a favorite obfuscation technology used by malware. It has been reported that the majority of malware in the Wildlist, (the list of current in-the-wild viruses [1]), is runtime packed [2, 3, 4]. The packing technique which is informally discussed in this section will be formally defned later on in this paper.

Packing malware has several benefts for the attacker. Firstly, it usually reduces its file size, thus allowing easy transfer on the net. Secondly, it increases the AV scanner's scanning time as runtime packers require additional work from the scanner, such as checking the file format and the code, unpacking, etc. Thirdly, it is more resistant to AV scanners due to the compression and encryption it employs. Lastly, the packer's complexity can be enhanced almost without limit by applying various new armoring techniques as described in Section 2.2. *The quick analysis and identification of a complex packer is a very challenging task to the AV researchers.*

Packer classification is an important step in malware analysis, if it can be designed so that it gives a reliable approach to detect packed files. The objective of packer classification is to quickly detect and identify the packer, allowing AV researchers to easily and correctly unpack the file and retrieve original payload for further malware detection and analysis. An efficient and effective packer classification system can yield benefits to both back-end AV researchers and real time anti-malware engines, running live on the client machine.

However, in recent years, the traditional packer identification technique based on signature scanning has been confounded by counter-counter attacks by packer and malware developers. An increasing number of anti-unpacking tricks and obfuscation methods are now being applied to packers. The targets include both existing common packers and the newly emerging ones, with the aim of providing a hard shell for malware.

As the number of new packers keeps growing, *a reliable computer-based packer classification scheme would facilitate the identification and characterizing process and reduce cost significantly.* In this paper, an automatically generated randomness profile based scheme is proposed as a replacement for the inefficient manually created signatures, which is the current approach, to the packer classification problem.

The main contributions of this paper are

- We develop a complete packer classification system that runs automatically, without the need to manually reverse engineer every packer variation. To the best of our knowledge, this is the first published system that achieves high accuracy on real malware.
- We refine Ebringer et al. [5]'s sliding window randomness test algorithm to produce randomness feature set of the packer and conduct experiments to give guidance for optimal parameter settings.
- We evaluate the effectiveness of four pattern recognition algorithms for classifying packers using the randomness profile information.
- We carry out large scale testing on real malware. The tests cover various packers with different levels of sophistication.

The remainder of the paper is organized as follows. Section 2 provides research background on packers, existing classification methods and pattern recognition techniques. It also discusses related work. Section 3 gives the implementation of our system. Section 4 describes experimental setting and reports evaluation results of different classification algorithms. Section 5 discusses future directions and concludes the paper.

2   Background

2.1   Packing

A *packer* is an executable program that takes an executable file or dynamic link library (DLL), *compresses* and/or *encrypts* its contents and then packs it into a new executable file. The packed files discussed in this paper are in Portable Executable (PE) format [6].

Assuming the original executable file contains already known malicious code, the signature based AV scanner should be able to detect it. However, the appearance of the malicious code is changed after packing due to the compression and/or encryption. Therefore, the packed file will thwart malware detection as no signature match will be found. The analysis and detection of malware can only be undertaken after the file is unpacked, i.e. the compressed/encrypted original file has been decompressed/decrypted completely.

A packed file contains two basic components. The first part is a number of data blocks which form the compressed and/or encrypted original executable file. The second part is an unpacking stub which can dynamically recover the original executable file on the fly. When the packed file is running, the unpacking stub is executed firstly to unpack the original executable code and then transfers the control to the original file. The execution of the original file is mostly unchanged and starts from its original entry point (OEP) with no runtime performance penalties (after the unpacking has been completed.)

2.2   Packer Evolution

As the variety of packing programs grows, their level of sophistication also increases. This is because that anti-unpacker tricks continually evolve and are implemented quickly by malware writers in a range of packers, from long-existing packers to modern new packers, with the goal of protecting malware. The main types of anti-unpacker tricks are listed below and most tricks have been described in Ferrie's papers [7, 8, 9, 10, 11]. Basically the harder/more the tricks that the packer employ, the higher the level of sophistication of the packer.

- *Anti-dumping* mainly alters the process memory of the running process to hinder further analysis on the dumped memory. The alteration is mainly applied on some useful information, such as PE header, imports, entry point codes, etc.
- *Anti-debugging* prevents AV researchers from using a debugger easily. The debugger is the most common tool used to trace the execution of malware in action.
- *Anti-emulating* attacks the software-based environment such as an emulator or a virtual machine, e.g., VMware [12]. Such an environment is essential to safely execute/monitor malicious behavior.
- *Anti-intercepting* thwarts page-level interception which stops the packer execution of newly written pages.

A lot of existing packers can be easily modified as their source codes are freely available. Moreover, most newly emerging packers can be customized. For example, as shown in Figure 1, Themida, one of most well-known sophisticated packers, provides various protection options and the user is allowed to configure the protection level of the packer. Consequently, the number and the complexity of new packer strains and variants are dramatically increased.

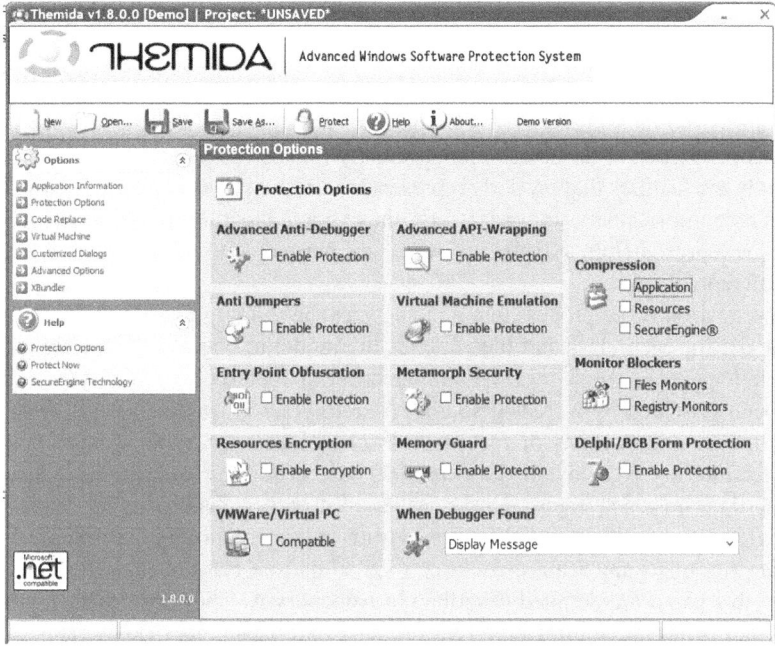

Fig. 1. Sample packer protection options of Themida

2.3   Traditional Signature Based Packer Classification

The traditional packer classification approach is mainly based on matching the packer's byte signature. Byte-signature based packer scanners, such as PEiD [13] and pefile [14], use a signature database to determine if a binary contains packed-code. A packer signature is typically a distinctive set of bytes which occurs at the entry point or in sections in a PE file. In this approach, the incoming packer is checked against the database of the signatures for known packers. If there is an exact match, the packer is considered being used and the name of the packer is also identified.

The byte-signature based packer classification method is effective at detecting known packers, however, the large diversity of packers, and the number of different variants of each packer, severely undermines the effectiveness of classical signature-based detection. Besides, this approach is expensive as the signature detection and updating need to be performed accurately by AV experts.

2.4 Pattern Recognition

Pattern recognition techniques have recently been used in anti-malware community, mainly for the purpose of identifying new or unknown malware [15, 16, 17, 18, 19, 20, 21, 22, 23]. Pattern recognition aims to recognize a particular class from a measurement vector. Different pattern classes with different measurement vectors correspond to different points in measurement space and patterns with similar appearance tend to cluster together. Therefore, a mapping relationship can be established from the measurement space into the decision space. There are two essential technologies involved in a pattern recognition system, namely feature extraction and classification.

Feature extraction retrieves the common features (patterns) among a set of objects. A feature is the measurement of a property of an object. A feature set describes the essential properties of an object using a greatly reduced number of parameters. However, only the features that properly represent the original object can lead to a satisfying pattern recognition result. In other words, the extracted features of objects in each class should be represented in a distinctive way. This permits a set of objects to be classified into different classes.

Classification is the process that first analyzes the training set, develops a classification model for each class and then applies the model to classify the testing set based on their features. A *training set* is a collection of records for which the class label have been provided by a trusted source. A *testing set* is used to verify the accuracy of the model and consists of records with class labels that you want to predict. A classification model (also called a *classifier*) can be built as follows: given a set of N training data in which each record consists of a pair: a feature vector of n features $\mathbf{x} = \mathbf{x}_1, \mathbf{x}_2, \ldots, \mathbf{x}_n$ and the associated "truth" class y_j, produce a relationship f : $\mathbf{x} \rightarrow y_j$ that maps any feature vector $\mathbf{x} \in X$ to its true class $y_j \in y$. Four classifiers used in this paper are detailed described in Appendix A. They are Naive Bayes (NB), Sequential Minimal Optimization (SMO), $k-$Nearest Neighbor (kNN) and Best-first Decision Tree (BFTree).

2.5 Related Work

The closest work to this research was conducted by Perdisci et al. [24]. They applied various pattern recognition techniques to classify executables into two categories, packed and non-packed. Nine features are combined together for classification, namely number of standard and non-standard sections, number of executable sections, number of readable/writable/executable sections, number of entries in the PE file's Import Address Table (IAT), PE header entropy, code section entropy, data section entropy and file entropy. The system achieved very high accuracy (above 95%) using NB, J48 decision tree, bagged, kNN or Multi Layer Perceptron (MLP) classifiers. However, this approach is not able to detect what family of packers a packed file belongs to.

Other than packer classification, there have been a few recent attempts to use pattern recognition techniques for automated malware detection [15, 16, 17, 18, 19]. Most of them only classified files between malicious and benign, but not by family of malware.

In early attempts, Tesauro et al. [15, 16] developed a neural network for virus detection. The system is specially designed for the detection of boot sector viruses using feature trigrams. The *trigrams* are three byte strings. They are selected in the feature set if they appear frequently in viral boot sectors but infrequently in uninfected software and definitely do not appear in legitimate ones. In the experiments, 200 viral boot sectors and 100 legitimate boot sectors were used as the data set in which half of them were the training set and the other half were the test set. Using a classification threshold of 0.5, performance on the test set was typically 80-85% for the viral boot sectors and 100% for the legitimate boot sectors. The classifier has also been incorporated into the IBM anti-virus product and has caught approximately 75% of new boot sector viruses.

Schultz et al. [19] used data mining methods to detect malware. They used three types of features, binary profiles of DLLs, strings and sequences of n adjacent bytes (also called n-grams), and paired each feature with a single learning algorithm. That is, a rule-base classifier was applied to the binary profiling; string data was used to fit a naive Bayes; and an ensemble of multi- NB classifiers is used on the n-grams data. In the latter, the n-grams data were partitioned into six parts firstly, then each classifier was trained on each partition of the data. The experiments were carried on a data set which contained 3301 malicious programs and 1001 clean programs. Among them, 38 of the malicious programs and 206 of the benign programs were in the Windows Portable Executable (PE) format. The results showed that naive Bayes with strings achieved the best accuracy than others. However, their experiments did not provide a fair comparison among the classifiers as different features were used for different classifiers. Moreover, different training sets were used to training different classifiers.

Using the same ideas, MECS [18] extracted byte sequences from the executables, converted these into n-grams, and constructed several classifiers: kNN, NB, support vector machines, decision trees J48, boosted NB, boosted J48 and boosted SVM. Muazzam tried other features [17]. Instead of using fixed length instructions or n-gram features, the author used Vector Space Model [25, 26, 27] to extract variable length instruction sequence as the primary classification feature and applied an array of classification models, including logistic regression, neural network, decision tree, SVM, Bagging and random forest.

Several researchers addressed the issue of email classification with the aid of machine learning techniques [20, 21, 22, 23]. Cohen [20] used a rule-based algorithm to classify email into folders based on the text of messages. Sahami et al. [21] employed a naive Bayes technique to the problem of junk E-mail filtering. Androutsopoulos et al. [22, 23] also used this approach to classify spam emails and legitimate ones. They compared two classifiers, naive Bayes and k-nearest neighbor. Both algorithms achieved very high classification accuracy but k-nearest neighbor with $k = 2$ slightly outperformed others.

3 Methodology

In this section, we present our approach to the packer classification problem by analysing the performance of various statistical classifiers, as a replacement for signature matching approaches. Firstly, it extracts a unique feature set, randomness profile, from each packed file. Then it maps the randomness feature vectors into an n−dimensional vector space in which various learning algorithms can be applied.

3.1 Feature Extraction

Feature extraction retrieves the 'characteristic' of the packer which represents the packer in a distinctive way. This permits the packed file to be compared with the candidate packers.

Ebringer et al. developed a randomness test [5] that preserves local detail of the packer. The randomness test measures the amount of "randomness" in different parts of a sample executable program. It was noted by authors that the randomness distribution of each packer family exhibits a distinctive pattern which suggests a kind of signal of each packer. In this paper, we investigate whether packer's randomness profile contains sufficient information to classify packers with high accuracy.

In order to extract the best features from packed files, we employ a refined version of the sliding window randomness test with trunk pruning method [5]. Compare with the original sliding window randomness test (see Appendix B), the window size and skip size used in this research are set to the same value. That is, there is no overlapping windows. Therefore, no repeat information is used in the feature set. All parameters are determined empirically (see Appendix C and D). Both window size and skip size are set to 32 and the pruning size is 50.

3.2 Classification

Classification refers to the way the packed file is examined and assigned to a predefined class. In this research, we first evaluate Ebringer et al.'s randomness signature scanning technique [5] on a large malware data set and then apply pattern recognition techniques to classify packers.

In a randomness signature based packer classification system, a packer signature is simply calculated as the average value of a set of randomness profiles of training files pre-classified as this packer. During the identification process, the distance between the test file and each packer signature is measured. The shorter the distance, the more likely the file is packed with this packer.

Instead of manual identification of signatures, a randomness signature can be automatically created from packed files. However, as stated in [5], although the preliminary packer classification results on a small clean data set are good, the results on real malware samples were of insufficient accuracy. This might be due to the fact that they used a small data set in the experiments. Besides, each tested packer contains only one version of this specific packer. They suggested that using larger data set might improve the performance.

Therefore, we firstly repeated their experiment using a large malware data set described in Table 4. As shown in Table 1 and Table 2, the performance of the system using large data set is still unsatisfactory. the average true positive rate in three tests ($n = 30, 40$ and 50) are all below 90%, while in the $n = 50$ test, NSPACK and PETITE only achieve 40.63% and 65.13% respectively.

Table 1. Performance of the sliding window algorithm using malware sample data set, where the window size $w = 32$ and the skip size is 32, the distance measure is *Cosine* measure and the pruning method is *Trunk*. The pruning size are in the range $30 - 50$.

Pruning size	Total files	TP rate
30	17336	85.68%
40	17336	85.27%
50	17336	85.17%

Table 2. Detailed performance of the sliding window algorithm using malware sample data set, where the window size $w = 32$ and the skip size is 32, the distance measure is *Cosine* measure and the pruning method is *Trunk*. The pruning size is 50.

Packer	Total files	TP rate
FSG	5105	99.53%
NSPACK	256	40.63%
PECOMPACT	1058	75.80%
PETITE	152	65.13%
UPACK	834	98.68%
UPX	9931	79.12%

The low accuracy in the results obtained by the randomness signature approach motivated us to investigate the pattern recognition techniques described in Section 2.4 for packer classification using the extracted randomness profile. In the remainder of this paper, we evaluate four statistical classifiers, namely Naive Bayes, Sequential Minimal Optimization, $k-$Nearest Neighbor and Best-first Decision Tree. These classifiers are selected since they are relatively fast. This is very important for a client-side AV scanner which needs to scan millions of files in a short user-tolerable time frame.

4 Experiments and Results

4.1 Data Sets

Experiments have been carried out on a large set of real malware samples. There are two types of data sets, namely a malware sample data set and a mixed data set. The malware sample data set only contains real malware samples which have reliable predefined class. The mixed data set has both packed clean files and real malware samples. The packers used in this data set are mixed with low complex packers and sophisticated packers. Below are the detailed descriptions of these two data sets.

Malware Samples Preparation. All malware samples have been prepared by an independent third party, Computer Associates (CA) Threat Management Team in Melbourne, Australia. To construct the data set, real malware downloaded over January and February 2009 by CA are collected. Each file is scanned by three AV scanners, Microsoft, Kaspersky and CA, for packer labeling. In addition, CA's VET engine and anti-virus Arclib Archive Library are used to determine whether the file can be unpacked. Files that reported by Microsoft, Kaspersky or CA, or that can be unpacked by either the VET engine or Arclib are identified as packed file. As a result of this method, we got a total of 103,392 packed files. The top five packers detected are UPX, ASPACK, FSG, UPACK and NSIS installer. They comprise a total of 91.35% of packed files.

For the collection of 103,392 packed files, each file is assigned as having been packed by one packer. This is done by combining the packer scanning results of all three scanners. The packer name is set if it is identified by any scanner and is 100% confirmed if it is identified by all three scanners. If there is conflicting information, the result taken is the one given by the two scanners which agree. If all three scanners disagree, PEiD is further applied to identify packer. PEiD is a byte-signature based packer scanner [13] which is supported by a large number of packer signatures. However, it is so popular that many packers start to use fake signatures to hide from PEiD detection. Therefore, PEiD's scanning results are not reliable and are only used by us for confirming information.

All scanners contain a different packer signature schema. For some packers, some scanners might only provide the packer family name without the version information. When collecting the version information, if any scanner obtains the version detail, this information is used. If there is no version information, PEiD is used to retrieve the version detail.

Table 3 lists four packer detection results extracted from our database. The first file is detected as Aspack by all three scanners. PEiD also confirms the packer. Therefore, it belongs to the Aspack family. Though Kaspersky doesn't provide the version information, all other scanners indicate it is Aspack 2.12. The second file is detected as Aspack by CA and Kaspersky, and CA provides its version number,namely version 2.0, while Kaspersky doesn't. So PEiD is further used to confirm that it is Aspack 2.0. CA identifies the third file as PC Shrinker while the other two do not have any scanning result. In this case, PEiD is used. It not only confirms the packer family, but also gives out the version 0.71. In the last example, there is information conflict between the scanning results. CA says Petite 2.1 and Microsoft gives Petite 2.3. Again, PEiD is used. Its result is Petite 2.1 or 2.2. All results are adjusted and Petite 2.1 is finally assigned to this file.

Table 3. Determination of packer name for malware samples

No	CA	Microsoft	Kaspersky	PEiD	Family	Detail
1	ASPack 2.12	ASPack v2.12	ASPack	ASPack v2.12	ASPack	ASPack 2.12
2	ASPack 2.0	NULL	ASPack	ASPack v2.001	ASPack	ASPack 2.0
3	PC Shrinker	NULL	NULL	PC Shrinker v0.71	PC Shrinker	PC Shrinker 0.71
4	Petite 2.1	Petite 2.3	NULL	Petite v2.1 (2)	Petite	Petite 2.1

Malware Sample Data Set. As discussed in the previous section, the packer names assigned to the samples are not 100% precise, especially when there is conflicting information between the scanning results from different scanners. In order to get a reliable training data set for the packer classification experiment, two criteria are used to select the packers and files in the malware sample set. These two constraints are:

- *Only confirmed cases are used.* In other words, the file's packer name has been identified as same by all three scanners, i.e., the first file in Table 3.
- *Only packers with a sufficient number of confirmed cases are chosen.* In this paper, each packer should have more than 100 confirmed packed files to be chosen.

According to the selection conditions above, 6 packers of 17,336 files, for file sizes ranging from 2 − 6880 KB are chosen from the sample collection described in Section 4.1. The details of the set is presented in Table 4. As the top packer in the collection, UPX has 39,799 confirmed packed files. However, to balance the distribution of the sample set, only 9,931 randomly selected samples from these files are used. Note that each packer contains samples with different versions. For example, packer NsPack has cross versions of 2.x, 2.9, 3.4, 3.5, 3.6 and 3.7.

Table 4. Data set one: malware sample set

Packer	Versions	Total Number
FSG	1.33 and 2.0	5,105
NSPACK	2.x, 2.9, 3.4, 3.5, 3.6 and 3.7	256
PECOMPACT	2.xx	1,058
PETITE	2.1 and 2.2	152
UPACK	0.2x-0.3x	834
UPX	UPX, UPX(LZMA), UPX(Delphi), 2.90, 2.92(LZMA), 2.93 and 3.00	9,931
		17,336

Mixed Sample Data Set. The above malware sample data set consists of six popular packers. Though each packer contains cross version samples, these packer's complexity is relatively low. In order to assess the robustness of our classification system, the system capability of classifying a wide range of packers that have not only different variants but also different levels of sophistication, we also create the mixed sample data set.

One problem is that our sample collection does not have a sufficient number of reliable samples of the sophisticated packers. To address this problem, the selection conditions used for malware sample data set are relaxed. The new selection criteria are:

- The file's packer name is identified by two scanners, or is identified by one scanner and confirmed by PEiD.
- Packers with a sufficient number of classified cases (more than 100) are chosen.

466 files of two packers, Asprotect and Mew, that match the above criteria have been added into the data set. In addition, a popular and sophisticated packer, Themida, is chosen for this data set. As most samples of Themida in the database contain conflicting packer information, Themida packed clean files are used instead. 117 clean files in the UnxUtils binaries [28] are packed with Themida v1.8.0.0 demo version. Figure 1 shows that Themida provides various protection options and the user is allowed to configure the protection level of the packer. Consequently, the number and the

complexity of Themida variants are dramatically increased. In this paper, six different combinations of packing options are evenly applied on these files.

The details of the set is presented in Table 5. Though this data set is not as reliable as the previous malware sample set, the experimental results on this data set will still provide an overall score of the system effectiveness.

Table 5. Data set two: mixed sample set

Packer	Versions	Total Number
ASPROTECT	unknown, 1.2 and 1.23	205
FSG	1.33 and 2.0	5,105
MEW	11 and 11 SE 1.2	261
NSPACK	2.x, 2.9, 3.4, 3.5, 3.6 and 3.7	256
PECOMPACT	2.xx	1,058
PETITE	2.1 and 2.2	152
THEMIDA	v1.8.0.0 with 6 option sets	117
UPACK	0.2x-0.3x	834
UPX	UPX, UPX(LZMA), UPX(Delphi), 2.90, 2.92(LZMA), 2.93 and 3.00	9,931
		17,919

4.2 Evaluation Metrics

When comparing the performance of different classification techniques, it is important to assess how well a classification model is able to correctly predict records to the actual classes. Several metrics are conventionally in use to numerically quantify classification effectiveness performance.

To introduce the metrics, let us define that for a class y_j, a record is *positive* if it is predicted to belong this specific class and is *negative* if it is predicted to belong other classes. Suppose that for a test set with n records, the set of positive records and negative records for the class are known (for example, as the result of human judgment), and P and N are the number of positive records and negative records respectively, $n = P + N$. Using four important counts defined below, $P = TP + FN$ and $N = FP + TN$.

- *TP* represents the true positives which is the number of positive records correctly identified as specific class.
- *FP* represents the false positives, the number of negative records which do not belong to the class but were incorrectly identified as it.
- *TN* represents the true negatives which refers the number of negative records correctly identified as other classes.
- *FN* represents the false negatives, that is the number of positive records which belong to the class but were incorrectly identified as other classes.

The *accuracy* is the percentage of test set records that are correctly identified by the classifier. That is,

$$Accuracy = \frac{TP + TN}{n} = \frac{TP + TN}{P + N} \tag{1}$$

Accuracy provides an overall performance of the effectiveness. However, this measure has one limitation. Suppose that a test set contains a large number of negative records and very small number of positive records, and we use a classifier which labels every class as negative (no matter what the input data). That is, TN is very high and TP is very low. Despite the classifier being very primitive, it will achieve a very high classification accuracy on this data set.

The *true positive rate (TPrate)* and *false positive rate (FPrate)* are introduced to measure the proportion of the positive records that are correctly identified and the proportion of the negative records that are incorrectly identified, respectively. For each class, they are calculated as

$$TPrate = \frac{TP}{P} = \frac{TP}{TP + FN} \quad \text{and} \quad FPrate = \frac{FP}{N} = \frac{FP}{FP + TN}. \tag{2}$$

Two other fundamental ways to measure classification effectiveness are *precision* and *recall*. Precision is the proportion of records classified as positive which are classified correctly, and recall is the proportion of positive records that have been correctly identified. So, for each class, the precision is defined as

$$Precision = \frac{TP}{TP + FP} \tag{3}$$

and the recall is

$$Recall = \frac{TP}{TP + FN} \tag{4}$$

4.3 Pattern Recognition Results

As described before, in the feature extraction process, each packed file is passed to the sliding window randomness test with a window size 32 bytes (256 bits). There is no overlap between windows, i.e., the skip size a is 32. Then the feature vector is constructed by extracting low randomness values in the range of 30 to 50 from the output using the Trunk pruning method.

Classification was carried out using the Weka 3.6.0 (Waikato Environment for Knowledge Analysis) machine learning package, developed by University of Waikato [29, 30, 31]. All selected statistical classifiers, NB, SMO, kNN (called IBk in WEKA) and BFTree are implemented in Weka. In the experiments, all classifiers used the default settings defined by Weka.

In the tests, 10-fold cross validation [32] is used. For each data set, the whole set is randomly partitioned into ten equal-size subsets. There are a total of 10 runs. During each run, one subset is used for testing and the other nine subsets are used for training. Therefore, each vector is used as a test sample exactly once.

Results of the Malware Sample Data Set. Experiments were firstly carried on the malware sample data set with feature vector size (pruning size) 50. The performance, in terms of effectiveness and efficiency, of various statistical classifiers are listed in

Table 6. All these classifiers work very well with high positive rate ($> 93\%$) and low false positive rate. This provides very strong evidence that the randomness profile plays a significant role in a packer classification system.

Among all classifiers, the kNN classifier achieves the best overall performance. Its TP rate is 99.6% and FP rate is only 0.1%. Moreover, it takes least time to build a model on training data (*Model building time* in Table 6).

Table 6. Comparison of statistical classifiers. The feature vector contains 50 points.

Classifier	TP rate	FP rate	Model building time (s)
Bayes.NaiveBayes	93.9%	1.1%	0.91
Functions.SMO	98.9%	0.8%	72.11
Lazy.kNN(k=1)	99.6%	0.1%	0.02
Trees.BFTree	99.3%	0.5%	16.45

Two other sets of experiments are used to determine the k values used in the kNN classifier and the size of the extracted feature vector. Table 7 shows that $k = 1$ outperforms other two k values, 3 and 5. Table 8 shows feature vectors of 30-50 points all achieve high effectiveness (TP $> 99\%$) while feature vector of 50 points yields the best result.

Table 7. Comparison of kNN with different k values. The feature vector contains 50 points.

k	TP rate	FP rate	Model building time (s)
1	99.6%	0.1%	0.02
3	99.5%	0.2%	0.02
5	99.4%	0.3%	0.02

Table 8. Comparison of kNN (k = 1) with different feature vector size

Vector size	TP rate	FP rate	Precision	Recall
30	99.4%	0.3%	99.4%	99.4%
40	99.6%	0.1%	99.6%	99.6%
50	99.6%	0.1%	99.8%	99.7%

Results of the Mixed Sample Data Set. The above results of malware sample data set shows that our packer classification system can achieve extremely effective performance using the pattern recognition techniques. Through the experiments, it is proved that this novel packer classification technique works well with different packer variants. However, it was unknown whether the conclusions drawn in previous sections can be applied to packers with different level of sophistication. To address this, an experiment was run on the mixed sample data set. This data set contains not only packed clean files and malware samples, but also lowly complex packers and highly complex packers.

50 randomness values are extracted from the file to construct the feature vector. In the classification process, the kNN (k=1) classifier is used. The results in Table 9 show that sophisticated packers, such as Asprotect and Themida, can also be effectively classified by applying the pattern recognition techniques on packer's randomness profile. As shown, the average TP rate is 99.4%. Among 9 packers, the TP rates of Upack and FSG obtain nearly perfect while all other packers achieve more than 90%.

Table 9. Detailed accuracy by class using the mixed sample data set. The classifier is kNN with $k = 1$ and the feature vector contains 50 points.

Packer	TP rate	FP rate	Precision	Recall
ASPROTECT	92.7%	0.1%	95.5%	92.7%
FSG	99.9%	0.0%	99.9%	99.9%
MEW	99.6%	0.0%	100.%	99.6%
NSPACK	91.0%	0.1%	90.7%	91.0%
PECOMPACT	98.5%	0.1%	98.2%	98.5%
PETITE	98.0%	0.0%	98.7%	98.0%
THEMIDA	92.3%	0.1%	86.4%	92.3%
UPACK	100.%	0.0%	99.4%	100.%
UPX	99.7%	0.3%	99.8%	99.7%
Weighted Avg	99.4%	0.2%	99.5%	99.4%

5   Conclusions and Future Work

This paper has discussed packers and presented a *fast yet effective* packer classification system which applies pattern recognition techniques. In this approach, the low randomness profile of the packer is extracted and then passed to a statistical classifier.

Our work demonstrates that the randomness profile combined with strong pattern recognition algorithms can be used to produce a highly accurate packer classification system on real life data. Such a system identifies the packer automatically and therefore is essential to keeping up with the accelerating growth in packer varieties.

The system has been tested on a large data set, including clean packed files and more than $17,000$ malware samples from the wild. The data set has a wide coverage of packers since that files are packed by not only different versions of packers, but also packers of different complexity. We evaluated four popular fast statistical classifiers, namely Naive Bayes, Sequential Minimal Optimization, $k-$Nearest Neighbor and Best-first Tree. All four classifiers were extremely effective, three of the four algorithms achieved an average true positive rate of around 99% or above, Naive Bayes was the lowest, with a true positive rate of around 94%. The $k-$Nearest Neighbor classifier with $k = 1$ obtains the best overall performance. Its true positive rate is 99.6% while false positive rate is 0.1%. Moreover, it is the fastest classifier.

The system also reveals that the low randomness profile of the packed file, normally produced by the PE header and unpacking stub, contains important packer's information. Thus it is very useful in distinguishing between families of packers.

Following the very encouraging preliminary results described here, there are several other promising steps which can be undertaken. Our future work will focus on the exploration of different statistical classifiers' performance as the main tool in a packer classification system. Several other classifiers can be added to the list, such as Random Forest, other Bayesian methods, Bagging, etc. Furthermore, we can apply various multi-classifier algorithms [33] or rank the output of different classifiers instead of choosing the best one among them.

It is still unknown whether there are a set of attributes more important than others in the extracted profile. We need to make enhancements to the feature extraction algorithm to select most important features from the randomness profile. Useful packer features other than the randomness profile, such as PE header information, string information, can also be explored and incorporated into the feature vector to improve the system's effectiveness. Moreover,new pruning methods that consider not only low randomness values but also certain sections can be developed. For example, we can only retain the lowest values in the code section.

Acknowledgements

The authors wish to thank Dr. Tim Ebringer for his important contributions during discussions related to this work.

References

1. The WildList Organization International: WildList, http://www.wildlist.org/
2. Brosch, T., Morgenstern, M.: Runtime Packers: The hidden problem? Black Hat USA (2006), http://www.blackhat.com/presentations/bh-usa-06/BH-US-06-Morgenstern.pdf
3. Bustamante, P.: Mal(ware)formation Statistics (2007), http://research.pandasecurity.com/malwareformation-statistics/
4. Morgenstern, M., Marx, A.: Runtime Packer Testing Experiences. In: 2nd International CARO Workshop (2008), www.datasecurity-event.com/uploads/runtimepacker.ppt
5. Ebringer, T., Sun, L., Boztaş, S.: A Fast Randomness Test that Preserves Local Detail. In: Proceedings of 18th Virus Bulletin International Conference, pp. 34–42 (2008)
6. Pietrek, M.: An In-depth Look into the Win32 Portable Executable File Format (2002), http://msdn.microsoft.com/msdnmag/issue/02/02/PE/print.asp
7. Ferrie, P.: Anti-unpacker Tricks Current. In: 2nd International CARO Workshop (2008), http://www.datasecurity-event.com/uploads/unpackers.pdf
8. Ferrie, P.: Anti-unpacker Tricks 2 Part One. Virus Bulletin, 4–8 (December 2008)
9. Ferrie, P.: Anti-unpacker Tricks 2 Part Two. Virus Bulletin, 4–9 (January 2009)
10. Ferrie, P.: Anti-unpacker Tricks 2 Part Three. Virus Bulletin, 4–9 (Febuary 2009)
11. Ferrie, P.: Anti-unpacker Tricks 2 Part Tour. Virus Bulletin, 4–7 (March 2009)
12. VMware workstation, http://www.vmware.com/products/ws/
13. PEiD, http://www.peid.info/
14. Carrera, E.: pefile, http://code.google.com/p/pefile/
15. Kephart, J.O., Sorkin, G.B., Arnold, W.C., Chess, D.M., Tesauro, G.J., White, S.R.: Biologically Inspired Defenses against Computer Viruses. In: Proceedings of the Fourteenth International Joint Conference on Artificial Intelligence, pp. 985–996 (1995)
16. Tesauro, G.J., Kephart, J.O., Sorkin, G.B.: Neural Networks for Computer Virus Recognition. IEEE Expert 11(4), 5–6 (1996)
17. Siddiqui, M.A.: Data Mining Methods for Malware Detection. Master's thesis, University of Central Florida, Orlando (2008)
18. Kolter, J.Z., Maloof, M.A.: Learning to Detect and Classify Malicious Executables in the Wild. JMLR 7, 2699–2720 (2006)

19. Schultz, M.G., Eskin, E., Zadok, E., Stolfo, S.J.: Data Mining Methods for Detection of New Malicious Executables. In: Proceedings of the IEEE Symposium on Security and Privacy, pp. 38–49 (2001)
20. Cohen, W.W.: Learning Rules that Classify E-mail. In: Proceedings of the AAAI Spring Symposium on Machine Learning in Information Access, pp. 18–25 (1996)
21. Sahami, M., Dumais, S., Heckerman, D., Horvitz, E.: A Bayesian Approach to Filtering Junk E-mail. AAAI Technical Report WS-98-05, pp. 55–62 (1998)
22. Androutsopoulos, I., Paliouras, G., Karkaletsis, V., Sakkis, G., Spyropoulos, C.D., Stamatopoulos, P.: Learning to Filter Spam E-mail: A Comparison of a Naive Bayesian and a Memory-based Approach. In: Proceedings of Workshop on Machine Learning and Textual Information Access, 4th European Conference on Principles and Practice of Knowledge Discovery in Databases (PKDD), pp. 1–13 (2000)
23. Androutsopoulos, I., Koutsias, J., Chandrinos, K.V., Spyropoulos, C.D.: An Experimental Comparison of Naive Bayesian and Keyword-based Anti-spam Filtering with Encrypted Personal Messages. In: Proceedings of the 23rd Annual International ACM SIGIR Conference on Research and Development in Information Retrieval, pp. 160–167 (2000)
24. Perdisci, R., Lanzi, A., Lee, W.: Classification of Packed Executables for Accurate Computer Virus Detection. Pattern Recognition Letters 29(14), 1941–1946 (2008)
25. Salton, G., McGill, M.J.: Introduction to Modern Information Retrieval. McGraw-Hill Book Co., New York (1983)
26. Frakes, W.B., Baeza-Yates, R.: Information Retrieval: Data Structures and Algorithms. Prentice Hall, Englewood Cliffs (1992)
27. van Rijsbergen, C.J.: Information Retrieval, Butterworths (1979)
28. Syring, K.M.: GNU Utilities for Win32 (2004), http://unxutils.sourceforge.net/
29. Witten, I.H., Frank, E.: Data Mining: Practical Machine Learning Tools and Techniques, 2nd edn. Morgan Kaufmann, San Francisco (2005)
30. Holmes, G., Donkin, A., Witten, I.H.: Weka: A Machine Learning Workbench. In: Proceedings of 2nd Australia and New Zealand Conference on Intelligent Information Systems, Brisbane, Australia (1994)
31. Weka, http://www.cs.waikato.ac.nz/~ml/weka/
32. Kohavi, R.: A Study of Cross-Validation and Bootstrap for Accuracy Estimation and Model Selection. In: IJCAI, pp. 1137–1145 (1995)
33. Chou, Y.Y., Shapiro, L.G.: A Hierarchical Multiple Classifier Learning Algorithm. In: Proceedings of 15th International Conference on Pattern Recognition (ICPR 2000), vol. 2, pp. 2152–2155 (2000)
34. Tan, P.N., Steinbach, M., Kumar, V.: Introduction to Data Mining. Pearson Education, Inc., London (2006)
35. Zhang, H.: The Optimality of Naive Bayes. In: FLAIRS Conf. (2004)
36. Aha, D.W., Kibler, D., Albert, M.K.: Instance-based Learning Algorithms. Machine Learning 6(1), 37–66 (1991)
37. Burges, C.J.C.: A Tutorial on Support Vector Machines for Pattern Recognition. Data Mining and Knowledge Discovery 2, 121–167 (1998)
38. Platt, J.C.: Sequential Minimal Optimization: A Fast Algorithm for Training Support Vector Machines. Microsoft Research (1998)
39. Quinlan, J.R.: Induction of Decision Trees. Machine Learning 1(1), 81–106 (1986)
40. Shi, H.J.: Best-first Decision Tree Learning. Master's thesis, The University of Waikato (2007)
41. Quinlan, J.R.: C4.5: Programs for Machine Learning. Morgan Kaufmann, San Francisco (1993)
42. Breiman, L., Friedman, J., Stone, C.J., Olshen, R.A.: Classification and Regression Trees. Wadsworth, Monterey (1984)

A   Pattern Recognition Algorithms

A.1   The Naive Bayes (NB) Algorithm

The NB classifier [34] uses a statistical approach to the problem of pattern recognition. The Bayes rule is the fundamental idea behind a NB classifier. For a feature vector x with n attributes $\mathbf{x} = \mathbf{x}_1, \mathbf{x}_2, \ldots, \mathbf{x}_n$ and a class variable y_j, let $P(\mathbf{x}|y_j)$ be the class-conditional probability for the feature vector x whose distribution depends on the class y_j. Then $P(y_j|\mathbf{x})$, the *posteriori* probability that feature vector x belongs to class y_j can be computed from $P(\mathbf{x}|y_j)$ by Bayes rule:

$$P(y_j|\mathbf{x}) = \frac{P(\mathbf{x}|y_j) \times P(y_j)}{P(\mathbf{x})} \tag{5}$$

The NB algorithm applies "naive" conditional independence assumptions which states that all n features $\mathbf{x}_1, \mathbf{x}_2, \ldots, \mathbf{x}_n$ of the feature vector x are all conditionally independent of one another, given y_j. The value of this assumption is that it dramatically simplifies the representation of $P(\mathbf{x}|y_j)$, and the problem of estimating it from the training data. In this case,

$$\begin{aligned} P(\mathbf{x}|y_j) &= P(\mathbf{x}_1 \ldots \mathbf{x}_n|y_j) \\ &= \prod_{i=1}^{n} P(\mathbf{x}_i|y_j) \end{aligned} \tag{6}$$

and equation (5) becomes

$$P(y_j|\mathbf{x}) = \frac{P(y_j) \prod_{i=1}^{n} P(\mathbf{x}_i|y_j)}{P(\mathbf{x})} \tag{7}$$

In a classification system, the feature vector x belongs to the class y_j with the highest probability $P(y_j|\mathbf{x})$. Since $P(\mathbf{x})$ is always constant for every class y_j, it is sufficient to choose the class that maximizes the numerator in (7), $P(y_j) \prod_{i=1}^{n} P(\mathbf{x}_i|y_j)$. In other words,

$$y_{max} = \arg\max_{y_m} P(y_j) \prod_{i=1}^{n} P(\mathbf{x}_i|y_j) \tag{8}$$

The NB classifier is easy to implement and can be trained very efficiently in a supervised learning setting, computation time varies approximately linearly with the number of training samples. Despite the apparent over-simplified assumptions of independence, the NB classifier often competes well with more sophisticated classifiers [35].

A.2   The $k-$Nearest Neighbor (kNN) Algorithm

k-Nearest Neighbor [36] is amongst the simplest of all machine learning algorithms. The idea behind it is quite straightforward. To classify the test packed file, the system firstly finds the k training files which are the most similar to the attributes of the test file.

These training files are called *Nearest Neighbors* as they have the shortest distance to the test file. Then the test file is categorized based on the category of its nearest neighbors. In the case where neighbors belong to more than one class, the test is assigned to the majority class of its nearest neighbors.

As shown in Figure 2, the test file (green cross) can be classified as the first class of the red plus sign or the second class of the blue minus sign. If set k to 1, the test file will be classified as the class of red plus sign since it is inside the inner circle. Similarly, if set k to 3, the test file will be classified as the class of the blue minus sign as there are two blue minus signs but only one red plus sign inside the outer circle.

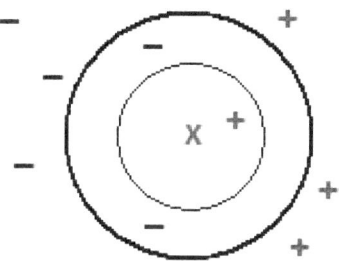

Fig. 2. Examples of kNN algorithm

A.3   The Sequential Minimal Optimization (**SMO**) Algorithm

SMO is a fast implementation of Support Vector Machines (SVM) [37]. Given data of two classes as two sets of feature vectors in an $n-$dimensional space, a SVM constructs an optimal hyperplane that separates a set of one class instances from a set of other class instances and maximizes the margin between the two data sets. That is to say if two parallel hyperplanes are constructed, one on each side of the hyperplane and passing through the nearest data point in each data class, the distance between the parallel hyperplanes needs to be as far apart as possible while still separating the data into two classes.

Training a SVM is slow due to solving a very large quadratic programming (QP) optimization problem. SMO [38] decomposes the large QP problem of SVM into QP sub-problems, which can be solved analytically and thus avoids using an entire numerical QP as an inner loop. In addition, SMO requires no extra matrix storage at all therefore the amount of memory required is linear in training set size. It allows SMO to handle very large training sets.

A.4   The Best-First Decision Tree (**BFTree**) Algorithm

The BFTree algorithm is a decision tree [39] that maps from attributes of an item to conclusions about its target class. In a tree that describes a set of packed files, each internal node represents a test on a file feature, each branch from a node corresponds to a possible outcome of the test, and each terminal node contains a packer class prediction. In each step of tree expansion, the best-first top-down strategy is applied [40]. i.e. the "best" node which maximally reduces the impurity (e.g. information [41] and Gini index [42]) among all nodes available for splitting, is added to the tree first. This partitioning of the feature space is recursively executed until all nodes are non-overlapping or a specific number of expansions is reached. For the latter, pruning methods are used to decrease the noise and variability in the data and therefore to achieve better performance.

B   Sliding Window Randomness Test

As described in [5], there are four steps involved in a sliding window version of local randomness test (see Algorithm 1). Firstly, for each file, all bytes are counted and a byte-frequency histogram is built. Secondly, using the global bytes information, the Huffman tree is constructed for the entire file by inserting bytes into the tree in the order of increasing frequency. Thirdly, a "length-encoding" array $e_{B_0}, \ldots, e_{B_{255}}$ is constructed where B_i is the corresponding byte. The entries of the array give the distance to the root of the tree for each element. Thus e_{B_0} gives the distance to the root for the byte B_0, e_{B_1} gives the distance to the root of the byte B_1, and so on. This distance is also the number of bits needed to encode this byte, in the prefix-free Huffman code. At the end, we set a window size w and a skip size a, so that there are a total $\lceil \frac{n-w}{a} \rceil$ windows (indexed by $1, 2, \ldots, \lceil \frac{n-w}{a} \rceil$) for the whole file. The randomness value r_i in each window is calculated as $\sum_{j=a(i-1)+1}^{a(i-1)+w+1} e_{b_j}$, i.e., as the total code length of the corresponding data. At the end, the r_i are scaled so that the minimum value is zero and the maximum value is one.

Data   : The packed file in the form of bytes $b_1, \ldots, b_j, \ldots, b_n$ where $b_j \in \{B_0, B_1, \ldots, B_{255}\}$, a window size w and a skip size a

Result : An array of $\lceil \frac{n-w}{a} \rceil$ samples of the randomness, ranging from 0.0 to 1.0

begin

1. Build a byte-frequency histogram for all bytes $B_0, B_1, \ldots, B_{255}$ in the entire file ;

2. Construct the Huffman tree by inserting bytes into the tree in the order of their frequency ;

3. Construct an array $e_{B_0}, \ldots, e_{B_{255}}$ containing the encoding length for each of the input bytes.

　　for i from 1 to $\lceil \dfrac{n-w}{a} \rceil$ **do**

　　　Set the randomness value $r_i \longleftarrow \displaystyle\sum_{j=a(i-1)+1}^{a(i-1)+w+1} e_{b_j}$;

　　endfor

4. **for** i from 1 to s **do**
　　　Rescale r_i between 0.0 and 1.0, where $\min(r_i) = 0.0$ and $\max(r_i) = 1.0$;
　　endfor

end

Algorithm 1. The sliding window algorithm: generate randomness measurements for a file, output proportional to file length. This is a revised version of Algorithm 2 in [5].

C   Determination of Window Size w and Skip Size a

Two sets of experiments have been initially carried out to determine the window size w and the skip size a for the sliding window algorithm. The data set used in the experiments comprises of a total of 708 packed clean files of six packers. For each packer, each file in the UnxUtils binaries [28] is packed with this packer. The selected collection contains 118 executable files whose file size ranged from 3 to 1058 KB, though most files (116 out of 118) are in the range $3 - 191KB$. Six packers used are FSG 2.0, Mew 11, Morphine 2.7, RLPack 1.19, Upack 0.399 and UPX 2.03w.

As shown in [5], the Cosine measure combined with Trunk pruning gives the best performance. We therefore use this combination in the following experiments to find the best parameter settings. Five files are removed from the data set as they produce more than n same smallest randomness values, which are all rescaled to 0. As the result, there are total 703 files in each experiment.

In each set of experiments, the total number of original bytes used for the feature vector remains roughly the same. In other words, similar features of the file are used for packer classification. For example, when test the window size w, the skip size is set to $\frac{w}{2}$ and the pruning size n will vary as w changes. Consider a run with $w = 32, a = 16$ and $n = 100$, the number of bytes from the file used is are around $100 \times 16 + 32 = 1632$. As another example, if w is set to 16, then $a = 8$ and n will be set to 200 ($200 \times 8 + 16 = 1616$).

The results are illustrated in Table 10 and 11. For w, the true positive rate of window size of 8 and 16 are slightly higher than 32. However, windows of small size carries less flexible information than big size does. Besides, feature extraction efficiency is also a factor when selecting the window size. The smaller the window size, the more randomness outputs are generated and the more time it takes. Therefore, 32 is a suitable window size used in the algorithm for packer classification purposes. If the experimental results are examined with respect to the skip size a, it is noted that when using a similar amount of information for comparison, the performance of different systems are close. In Table 11, if windows do not overlap, that is, $a = w = 32$, the system achieves slightly high TP rate than others. In this case, the total number of outputs of the randomness test and the extracted feature vector size are the smallest, so the system performs most efficiently as both of the feature extraction process, including the scaling and pruning, and the classification process is fast.

Table 10. Determination of the window size w, where the skip size $a = w/2$, the distance measure is *cosine* measure and the pruning method is *Trunk*. The pruning size varies with different window size so that that the extracted features used for comparison are roughly same.

Window size	Skip size	Pruning size	Total files	TP rate
8	4	400	703	98.29%
16	8	200	703	98.29%
32	16	100	703	98.15%
64	32	50	703	97.29%

Table 11. Determination of the skip size a, where the window size $w = 32$, the distance measure is *cosine* measure and the pruning method is *Trunk*. The pruning size varies with different skip size so that the extracted features used for comparison are roughly same.

Skip size	Pruning size	Total files	TP rate
2	800	703	98.00%
4	400	703	98.15%
8	200	703	98.15%
12	134	703	98.15%
16	100	703	98.15%
20	80	703	98.00%
32	50	703	98.29%

D   Determination of Pruning Size n

This set of experiments is used to determine how detailed the information used for classification should be. This depends on the feature vector size (pruning size n). If n is small, the classification takes less time building the model and classifying the file. However if n is too small, there may not be sufficient information to distinguish between packers. If n is too large, the classification process is slow and also information noise generated by compressed/encrypted data will effect system performance. Therefore, experiments are run with various pruning size n in the range of $30 - 70$ and other settings given in Table 12. Thus, the information used are around $1 - 2$ KB. Table 12 suggests that n should be set between $30 - 50$.

Table 12. Determination of the pruning size n, where the window size $w = 32$, the skip size $a = 32$, the distance measure is *cosine* measure and the pruning method is *Trunk*

Pruning size	Total files	TP rate
30	703	98.57%
40	703	98.57%
50	703	98.29%
60	703	95.87%
70	703	96.01%

Repelling Sybil-Type Attacks in Wireless Ad Hoc Systems*

Marek Klonowski, Michał Koza, and Mirosław Kutyłowski

Institute of Mathematics and Computer Science, Wrocław University of Technology
{Michal.Koza,Marek.Klonowski,Miroslaw.Kutylowski}@pwr.wroc.pl

Abstract. We consider ad hoc wireless networks and adversaries that try to gain control over the network by Sybil attacks, that is by emulating more physical nodes that are really under his control. We present the first defense method that works for the case when the adversary controls more than one device and these devices have some prior agreement on strategy executed and share preloaded secrets.

1 Introduction

One of the most difficult problems for ad hoc networks are Sybil-type attacks [6]: a selfish adversary controlling some of the devices may change the identities of these devices and/or create and emulate virtual devices with new identities, and thereby disproportionately large influence on a network. Indeed, it would help the adversary to gain a higher share of the communication channel and to perform more tasks in the network if the distinguished nodes are chosen at random.

Preventing such attacks might be difficult, especially if the devices are communicating in a wireless way only and the network is dynamic. In this case, there is no direct physical control over the devices and the adversary can launch an attack without risking of being caught. Moreover, at present it seems unrealistic to build a global system registering all identities and authorization mechanisms (like a PKI system) for small devices. Apart from necessity to keep the devices as simple as possible, this is due to the fact that the main purpose of an ad hoc network is to work without consulting any central system and to survive even in the hardest circumstances. So it does not help much, if the identities are assigned to the devices under strict control - the adversary may clone them and use the same identity at different locations.

Problem statement. We assume that an ad hoc network has to initialize itself, in the sense that a common list of identities is generated so that:

- the number of identities on the list does not exceed the number of physical devices in the network,
- each honest device gets its ID on the list.

---

* Supported by funds from Polish Ministry of Science and Higher Education – grant No. N N206 257335.

R. Steinfeld and P. Hawkes (Eds.): ACISP 2010, LNCS 6168, pp. 391–402, 2010.

The network consists of N honest devices and M devices under control of an adversary. However, there is no prior knowledge about the network and its participants. In particular, N and M are unknown, it is also unknown which devices are under control of the adversary. We only know that $N_{min} \leq N$ and $M + N \leq N_{max}$, where N_{min}, N_{max} are known parameters.

If the procedure succeeds, then it prevents the adversary from registering more identities than M. We are interested in a solution for an ad hoc system. All mechanisms deployed must not break the principle that the network must work on its own and cannot depend on any external help (such as external verification of identities claimed).

Network model. An ad hoc network consists of physical devices communicating via a radio channel. Time is divided into discrete slots and devices are synchronized enough to cooperate in time slots. Each device has its own identity, however, nothing prevents a dishonest device to change its identity (Sybil attack) or to emulate many physical stations with different identities.

The devices have no prior knowledge about other devices in the network and can learn about their presence only by exchanging messages.

Devices are in the signal range of each other, i.e. a message sent by one station can be received by any other station, provided that is not jammed by other messages sent at the same time. If two or more stations are transmitting at the same time each listening station receives a noise and no message can be recognized. However, noise can be distinguished from silence observed when no station transmits. Hence, we consider the single-hop model with collision detection, which models a wide class of real systems and is intensively studied in the literature.

In a single slot a device can either transmit or listen; simultaneous transmitting and listening is impossible (as it is the case for standard small size devices with a single antenna - see for instance 802.11 standard [1]). In particular, a device does not know, if its message was correctly transmitted or jammed.

Even if we focus on weak devices, we assume that devices are capable of computing a value of a one-way hash function and have an access to stochastically independent pseudo-random number generators. Computational limitations of each device is described by parameter a defined as follows. Parameter a is such a number that if we assume that some device is given value y such that $H(x) = y$, and such part of value x – that a bits of x are unknown, provided that for each missing bit b_i we have $0.5^{\frac{N_{min}}{N_{max}}} \leq Pr[b_i = 1] \leq 0.5$, then probability that device will find exact value of x during one verification procedure execution is less than $\frac{1}{n^2}$. Parameter a depends on station computational power and influences length of verification message.

Adversary model. A malicious adversary can gain control over some number of devices in the network and access all data stored in these devices. He can coordinate the actions of these devices by means of a preloaded strategy or shared secrets. However, during protocol execution the adversarial devices have exactly the same capabilities and limitations as honest devices. In particular, they can communicate only via the radio channel shared with the honest devices. This scenario describes, among others, the case wherein the adversary infects some devices with malcode.

The adversary attempts to emulate more devices than he really has at his disposal. In this way the adversary attempts to improve his chances to gain access to the shared radio channel or to be elected to play some role in the network. This works, since most algorithms perform a kind of random choice from the set of all participating devices.

In the considered scenario, the adversary does not try to block the network communication, but he tries to gain advantage in an unfair way. We do not consider energy complexity thus we have to assume that adversary can always block the network e.g. transmitting continuously.

Previous and related work. The problem of Sybil attacks, described in the seminal paper [6], is not limited to ad hoc networks. Similar attacks were considered in various settings, mainly for P2P systems based on DHT paradigm. Some ideas for other distributed systems can be found for example in [5, 14]. Various countermeasures has been proposed including certification, auditing and resource testing. Many of these solutions are surveyed in [3]. Most of proposed countermeasures cannot be applied to protect systems considered in our paper due to their peculiarities– their ad hoc nature (i.e no possibility of auditing, no secrets established a priori) and limited resources resulting in lack of public key cryptography.

The paper [13] suggests a method of protecting a sensor network from the Sybil attack based on the assumption that a single device is not able to broadcast on two different frequencies. Presented approach is very innovative, however it cannot be applied to the model discussed in our paper. In particular it requires several channels and a different adversary model. Moreover, the model investigated in [13] does not assume collisions, what is essential in networks considered in our paper. Generally speaking our model and assumptions seems to be more suitable for systems of small and extremely weak devices.

Other papers considering Sybil attack in the context of ad hoc networks are [4, 14]. However, none of proposed solutions seems to be applicable to our problem statement. For example, the first paper is focused on mobile devices with multi-hop graph of connections. Also methods from the second paper based on pre distributing secrets require some pre deployment phase not assumed in considered model.

Our paper is motivated by the solution presented in [7], where fairness of a leader election protocols were investigated. Apart from some negative results, the paper presents a protocol that provides fair leader election (i.e. each device has the same probability of becoming a leader) for the network model considered in this paper. However, this solution works only if the adversary controls exactly one station. If the adversary is controlling even two stations, then the defense method fails completely.

General model of the network and communication between nodes assumed in our paper appears in different contexts in vast body of works. Some notable examples include [2, 8, 10, 11, 12, 15].

Our contribution. We propose a probabilistic algorithm that solves the stated problem of initialization of an ad hoc network. Thereby, we solve the fundamental problem that exist for the solution from [7] – namely, our protocol works even against adversary with multiple devices.

As in the papers [7, 13], we make advantage of particular properties of communication model in wireless systems and combine them with some lightweight cryptographic tools. To the best of our knowledge this is the first solution to such stated problem.

2   Algorithm Description

Our algorithm consists of two phases:

1. each device declares its identity (we admit that a cheating device declares more than one identity),
2. it is checked that each identity corresponds to a different physical device; if cheaters are detected, they are eliminated from the further procedure and Phase 2 is restarted.

In more detail, the algorithm has the following structure (each subprocedure is described in the following subsections):

Algorithm 1. High level description

---
  registering identities
  repeat
      commitment to random seeds for PRNG
      verification procedure
      revealing seeds of PRNG and removing cheaters
  until no cheater detected

Registering identities.   Many techniques can be applied here. For the sake of completeness we sketch one of them. During the first phase we use $2N_{max}$ time slots. A device D willing to register chooses $1 \leq j \leq N_{max}$ uniformly at random, transmits its identity in slot j and listens during the remaining slots $1 \leq i \leq N_{max}$. So D knows all devices that succeeded to transmit their identities except itself. Then D forms a vector of length N_{max}, which contains a 1 on position i, iff D has heard an identifier in slot i, and a zero otherwise. D transmits this vector in slot $N_{max} + j$. D can check if it has transmitted in slot j without collision by inspecting any of such vectors transmitted by the devices that it heard during the first N_{max} slots. Assuming that at least two devices succeeded (which occurs with high probability since at least $N_{min} > 2$ devices participate in the protocol) each station learns if it has succeeded to broadcast its identifier.

If any collision has been detected, then there are stations that are not registered yet. In this case the next similar phase is started. However, now only one device transmits the vector indicating in which slots successful transmission has occurred. Namely, this is the device with the smallest identifier from those that came through in the first phase. The phases are repeated until no unregistered device is left. Let us assume that finally there are n identities registered, and the ith identity is ID_i.

One of the key properties of the above procedure is that eventually all devices get registered and the adversary cannot prohibit any honest device from registering. However, a dishonest party may register more than one identity.

Commitment subprocedure. The goal of this subprocedure is to commit to pseudo-random choices made during verification subprocedure. Without this an adversarial device could adapt its behavior to the events observed and transmit information to other adversarial devices by appropriate choice of transmitting pattern jeopardizing verification.

Let H be a secure hash function. First, for i such that $0 \leq i \leq n-1$, device controlling ID_i generates at random a secret r_i, computes $c_i = H(r_i)$ and broadcasts c_i in the ith time slot. Within the next n slots the devices broadcast the values c_i, $0 \leq i \leq n-1$. Afterward, each device computes $x = H(c_0 \oplus \ldots \oplus c_{n-1})$. Finally, ID_i computes $s_i := H(r_i||x)$ as the seed for the pseudo-random number generator for the verification subprocedure.

Verification subprocedure. This subprocedure consists of $2n-1$ consecutive *trials* T_t where $t \in \{0, \ldots, 2n-2\}$. Each trial is dedicated to one identity, namely trial t has to examine identity $ID_{t \bmod n}$. So each identity (except for ID_{n-1}) has two dedicated trials: for ID_i, $(0 \leq i \leq n-1)$ these are T_i and T_{n+i} (see Fig. 1). This way each identity's last trial takes place after at least one trial dedicated to each other identity. This way it will be possible to verify each identity against all others (see Fig. 2).

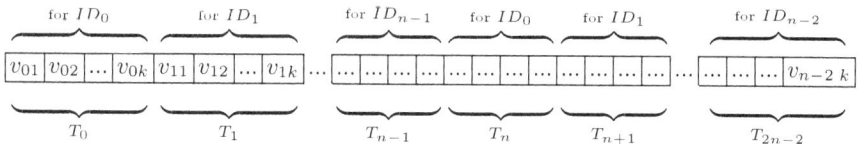

Fig. 1. Trials assignment

Trial. Each trial consists of k transmission slots. In each slot of trial T_t, device $ID_{t \bmod n}$ transmits a *verification message* in a form described below. The other devices may cause transmission collisions in these slots in a pseudo-random way with transmitting probability $1 - \sqrt[n]{0.5}$. ID_i determines in which slots to transmit using a PRNG with seed s_i.

Since there are n devices registered, the final probability that there is no collision in a given slot equals 0.5. The outcome of trial T_t is described by a k-bit jamming pattern P_t. It contains a 1 on position j, iff there is a collision in slot j of T_t.

The collisions discussed above is the main mechanism preventing emulation of more ID's than physical devices. Namely, observe that for a given identity some device has to transmit in all k slots and is unaware of the jamming pattern. Every other device should know exactly the jamming pattern: if it does not transmit in slot j, it can listen and detect a collision, if it transmits, then it knows for sure that there is a collision. So, if a device sends verification messages on behalf of another ID, it is unable to gain knowledge of the jamming pattern of this trial. Since a verification message depends on all preceding jamming patterns, lack of knowledge will be eventually detected (for details see the discussion in the subsequent sections).

Verification message v_{ts}. For constructing verification messages we use a function $F : (0,1)^k \times \mathbb{N} \times \mathbb{N} \mapsto \mathbb{N}$ having certain properties of a hash function (e.g SHA-256 can be used). Namely, given y, t and s it is infeasible to calculate x such that $F(x,t,s) = y$, even if all but a bits of x are known. Moreover, for given t, s and $y = F(x,t,s)$, and all but a bits of x, it must be infeasible to calculate y' such that $y' = F(x,t',s')$ for some $t' \neq t$ or $s' \neq s$.

Let $F_{ts}(P_i)$ stand for $F(P_i,t,s)$. We define v_{ts} as a concatenation of values of F for all preceding jamming patterns, except the jamming pattern P_{t-n} of the same device for $t \geq n$:

$$v_{ts} = \begin{cases} 0 & \text{if } t = 0, \\ F_{ts}(P_0)||\ldots||F_{ts}(P_{t-1}) & \text{if } t < n, \\ F_{ts}(P_0)||\ldots||F_{ts}(P_{t-n-1})||F_{ts}(P_{t-n+1})||\ldots||F_{ts}(P_{t-1}) & \text{if } t \geq n. \end{cases} \quad (1)$$

Referring to all verification messages in a trial we will use notation v_t.

Each device should know all the jamming patterns for all those trials which were not dedicated to its identity. Thus each ID_u can deconcatenate v_{ts} and verify $F_{ts}(P_j)$ values for all $j \neq u$ calculating them itself. So in fact each device should be able to verify correctness of the verification message except for the part dedicated to its own pattern.

Adversarial devices which emulate more than one identity will be unable to calculate correct verification messages.

To see it we consider a simple example depicted on Fig. 2.

We assume that ID_1 and ID_3 are simulated by the same device. This example does not reflect all possible strategies of the adversary, but shows the general idea of the verification.

Trial 1: ID_0 $\longrightarrow$ transmits v_0 OK
Trial 2: ID_1 $\longrightarrow$ transmits v_1 OK
Trial 3: ID_2 $\longrightarrow$ transmits v_2 OK
Trial 4: ID_3 $\longrightarrow$ unable to calculate v_3 WRONG

$\vdots$

Trial n: ID_{n-1} $\longrightarrow$ transmits v_{n-1} OK
Trial n+1: ID_0 $\longrightarrow$ transmits v_n OK
Trial n+2: ID_1 $\longrightarrow$ unable to calculate v_{n+1} WRONG
Trial n+3: ID_2 $\longrightarrow$ transmits v_{n+2} OK
Trial n+4: ID_3 $\longrightarrow$ unable to calculate v_{n+3} WRONG

$\vdots$

Trial 2n-1: ID_{n-2} $\longrightarrow$ transmits v_{2n-2} OK

Fig. 2. Exemplary verification subprocedure. We assume that ID_1 and ID_3 are simulated by the same physical device. Notice that "device ID_1" can compose correct verification message at trial T_1. "Device ID_3" does not know P_1 so it cannot broadcast correct verification messages and is immediately recognized as a fake device. However, after "device ID_3" has transmitted, "device ID_1" is also in trouble: it lacks information about jamming pattern P_3 it was supposed to have and is no more able to calculate the verification message. So "device ID_1" fails in trial T_{n+1}.

Pseudocode. We can now summarize the verification subprocedure in the form of the following pseudocode:

Algorithm 2. Verification subprocedure

---

for t=0 to 2n-2 **do**
    All stations calculate v_{ts} according to Eqn. 1
    for s=1 to k **do**
        if ID_i such that $(i = t \bmod n)$ **then**
            calculate and send v_{ts}
        else
            Jam slot with probability $1 - \sqrt[n]{0.5}$
            Update information about current jamming pattern $P_t[s]$
            check v_{ts} transmitted by $ID_{t \bmod n}$
        end if
    end for
    if there are any mistakes **then**
        each station mark $ID_{t \bmod n}$ as cheater.
    end if
end for

Revealing seeds. Finally, the seeds r_i are revealed. Then each device checks if the observed jamming patterns agree with the patterns that are computed from seeds s_i. If it is detected that ID_i was supposed to jam at a given slot, but the verification message came through, then ID_i is declared as a cheater.

If any cheater has been detected during this and the previous subprocedure, we restart without the cheaters at commitment subprocedure.

3   Algorithm Analysis

In this section we show that even sophisticated strategies of the adversary do not help him to avoid detection of cheating.

First let us observe that before verification subprocedure starts, the pattern of jamming by each ID is fixed and committed to by strings s_i. So in particular, the devices (may be virtual ones) registered by the adversary must behave in a predefined way and cannot transmit any knowledge about observed jamming patterns to other adversarial devices.

Also, if there is at least one honest device, no device can determine x and therefore influence its broadcasting schedule in advance. Since the numbers r_i are not revealed at first, it is also impossible to determine at the end of commitment subprocedure the future behavior of the honest devices. On the other hand, adversarial devices can choose their strings r_i in some way agreed in advance, so we have to assume that they know each other transmission patterns. Consequently, they can share their duties and this is not true that each adversarial ID must be implemented by a single adversarial device. This complicates considerably the analysis below.

3.1   Effectiveness of Cheater Detection

No honest device can be eliminated. First let us observe that honest devices cannot be declared as cheaters. Indeed, an identity can be eliminated only for being caught on being silent in a slot in its dedicated trial, transmitting wrong verification message, not jamming at some moments (i.e. when no other is suppose to jam) or cheating with randomness seed. All those factors depend only on given device behavior and cannot be influenced in any way by the adversary. So a device acting according to the protocol cannot be found guilty of cheating.

Eliminating cheaters. As no honest ID can be declared as cheater, the main issue is to show that in case of cheating it becomes detected with a high probability. Procedure is then executed anew without the detected cheaters.

Theorem 1 (Non-detection probability). *If there are more ID's declared than the number of physical devices, then with probability at least $1 - \frac{1}{n^2}$ at least one cheater is detected.*

Proof. To prove this theorem we show that adversarial devices are unable to calculate correctly all verification messages within the first $n - 1$ trials of the verification subprocedure. The only case when the adversary looses some information is when an adversarial device sends a verification message and no adversarial device jams it (according to the committed schedule) - the status of this slot remains hidden for the adversarial device sending the verification message. First we have to show that there are many such slots.

Lemma 1. *Assume that among n registered identities m identities have been declared by adversarial devices.*

Let $p = 1 - q$ denote the probability that an identity jams in a single slot of a trial. Let $L > 0$ be an arbitrary constant. Then for

$$k > \frac{\sqrt{\log(n^4)(4L + \log(n^4))} + 2L + \log(n^4)}{2q^m} \tag{2}$$

probability that less than L slots out of k slots of a trial have not been jammed by the adversarial identities is less than $\frac{1}{n^2}$.

Proof. Probability that a given slot is not jammed by any of the adversarial identities is q^m. Since jamming different slots can be modeled as events independent stochastically, the number of slots in the trial not being jammed by the adversary is a random variable with binomial distribution $Bin(k, q^m)$. We use use the following inequality:

Fact 1. *If $X \sim Bin(n, p)$, then for every $t > 0$*

$$\Pr[X \le np - t] \le \exp\left(-\frac{t^2}{2np}\right).$$

This is a variant of a Chernoff bound adjusted to binomial distribution case (see for example [9]). Note that in our case $n \cdot p$ is substituted by $q^m \cdot k$. If we substitute $t = \sqrt{q^m \cdot k \log(n^4)}$ and then solve inequality $q^m k > L + t$, we get

$$\Pr[X \le F] \le \exp\left(-\log(n^2)\right) = \frac{1}{n^2}$$

for $F < L$ as demanded.   □

By Lemma 1, for

$$k > \frac{\sqrt{\log(n^4)(4Ma + \log(n^4))} + 2Ma + \log(n^4)}{2(\sqrt[n]{0.5})^m}$$

in each trial there are $L > M \cdot a$ slots not jammed by the adversary. The state of the channel in each of these slots is unknown to adversarial device that transmits the verification message in them. We focus our attention on these slots. For each trial we select adversarial devices that transmitted in more than a of those slots and call them *owners* of the identity assigned to this trial. Since there are M adversarial devices, we can see that for each trial we can assign at least one owner. We will use notation $D_j \overset{t}{\rightsquigarrow} ID_i$ to state that device D_j is the owner of identity ID_i in trial T_t. We can state the following fact:

Fact 2 (Owners problems). *Assume that* $D_j \overset{i}{\leadsto} ID_{i \bmod n}$. *Then the probability that D_j can calculate v_{ts} for $(t > i$ and $t \neq i \bmod n)$ is smaller than $\frac{1}{n^2}$.*

This fact results from the assumption that the probability of guessing missing a bits of information is smaller than $\frac{1}{n^2}$ and construction of verification message v_{ts} which contains $H_{ts}(P_i)$ for all $t > i$, $t \neq i \bmod n$.

We can see that in the first $M+1$ trials dedicated to the adversarial identities there must be at least one adversarial device D_j such that $D_j \overset{i_1}{\leadsto} ID_{i_1}$ and $D_j \overset{i_2}{\leadsto} ID_{i_2}$ for some $i_1 \neq i_2$. From the Fact 2 we can see that chance that D_j will succeed to correct verification messages is smaller than $\frac{1}{n^2}$, which completes the proof of Theorem 1.   ∎

3.2 Adversary's Optimal Strategy

Adversary's goal is to go through verification algorithm with as many ID's as possible. In the previous subsection we have seen that in a single verification some cheaters are detected with a high probability. However, in principle if this procedure is repeated many times, then may be it can succeed with some cheated identities. We argue that after the first loop execution the expected number of adversarial identities is even less than the number of devices held by the adversary.

We will look closer at the issue of assigning owners to identities. Note that each identity can have multiple owners. Also in both of its dedicated trials an identity may have different owners. Assume now that there are M adversarial devices that have registered m identities $ID_{a_1}, \ldots, ID_{a_m}$. A good measure of adversary's efficiency is the expected value of number of his identities which successfully go through verification procedure $E[X] = \sum_{i=a_1}^{a_m} \bar{p}(i)$ where $\bar{p}(i)$ is probability that ID_i will successfully go through the verification procedure. Using Fact 2, we can estimate $\bar{p}(i)$ from above by $p(i)$ defined as follows:

Definition 1

$$
p(i) = \begin{cases}
0 & \text{if} \quad (\neg \exists\, D_j : D_j \overset{i}{\leadsto} ID_i) \vee (\neg \exists\, D_j : D_j \overset{i+n}{\leadsto} ID_i) \\
\frac{1}{n^2} & \text{if} \quad \exists D_j, i', t, t' : (i' \neq i) \wedge (t' < t) \wedge (D_j \overset{t}{\leadsto} ID_i) \wedge (D_j \overset{t'}{\leadsto} ID_{i'}) \\
1 & \text{if} \quad \forall t \forall D_j : [t = i \bmod n\ \&\ D_j \overset{t}{\leadsto} ID_i] \Rightarrow \\
& \quad \Rightarrow \neg \exists t' : (t' < t) \wedge (t' \neq i \bmod n) \wedge D_j \overset{t'}{\leadsto} ID_{t' \bmod n}
\end{cases}
$$

It is obvious that if adversary acts according to protocol (so in particular has only one identity per device), then for each ID_{a_i} we have $p(a_i) = 1$ and thus $E[X] = M$. We will show that otherwise $E[X] < M - 1 + \frac{1}{n}$. This means that acting according to the protocol is the optimal strategy for the adversary.

Theorem 2 (Optimal adversary behavior). *If any adversarial device decides to become owner of more than one identity, then $E[X] < (M-1) + \frac{1}{n}$.*

Proof. Let us define Ψ_1 as the set of those adversarial identities ID_i for which $p(i) = 1$. Since $\frac{m}{n^2} < \frac{1}{n}$ it is obvious that if $|\Psi_1| < M$, then $E[X] < M - 1 + \frac{1}{n}$. So it remains to show that if any of adversarial devices decides to become owner of more than one identity, then $|\Psi_1| < M$.

We can divide each verification procedure into two phases: $\mathcal{P}_1 = \{T_i : 0 \le i \le n-1\}$ and $\mathcal{P}_2 = \{T_i : n-1 \le i \le 2n-2\}$ each containing exactly one trial dedicated to each identity. $\mathcal{P}_1$ and $\mathcal{P}_2$ overlap – both contain T_{n-1}.

Definition 2. *Let us define*

$$A = \{T_t : T_t \text{ is assigned to adversarial identity}\}$$

$$\mathcal{A}_1 = \left\{ T_t : T_t \in \mathcal{P}_1 \wedge A \wedge \forall D_j \left(D_j \overset{t}{\rightsquigarrow} ID_t \right) \Rightarrow \left(\neg \exists t' : (0 \le t' < t) \wedge D_j \overset{t'}{\rightsquigarrow} ID_{t'} \right) \right\}$$

$$\mathcal{A}_2 = \left\{ T_t : T_t \in \mathcal{P}_2 \wedge A \wedge \forall D_j \left(D_j \overset{t}{\rightsquigarrow} ID_{t \bmod n} \right) \Rightarrow \right.$$

$$\left. \Rightarrow \left(\neg \exists t' : (0 \le t' < t) \wedge (t' \ne t \bmod n) \wedge D_j \overset{t'}{\rightsquigarrow} ID_{t' \bmod n} \right) \right\}$$

We can state the following facts:

Fact 3 (Ψ_1 size bound). $|\Psi_1| \le |\mathcal{A}_1|$ *and* $|\Psi_1| \le |\mathcal{A}_2|$.

The above fact is a direct consequence of Definition 1 and observation that:

$$ID_i \in \Psi_1 \Leftrightarrow (\exists t : (t = i \bmod n) \wedge T_t \in \mathcal{A}_1) \wedge (\exists t : (t = i \bmod n) \wedge T_t \in \mathcal{A}_2) \qquad (3)$$

Fact 4 ($\mathcal{A}_1$ and $\mathcal{A}_2$ size bounds). *For* $i \in \{1, 2\}$, *the size of* $\mathcal{A}_i$ *does not exceed the number of devices that are able to calculate verification message in trial* $T_t \in P_i \cap A$.

Now it is enough to notice that if the adversary is trying to emulate more identities than devices under its control, then there is at least one adversarial device D_α that became an owner of some ID_{i_1} and ID_{i_2} such that $T_{i_1} \in \mathcal{P}_1$ and $T_{i_2} \in \mathcal{P}_1$. According to Definition 2, we can see that $D_\alpha \notin \mathcal{A}_2$ and thus $|\mathcal{A}_2| \le M - 1$, which finishes the proof. ∎

4   Final Remarks

We propose an algorithm significantly limiting Sybil-type attacks in a classical model of ad hoc sensor networks of devices. Our solution can be added as a additional subprocedure constituting an additional security layer in the network. The price we have to pay is a communication overhead. This can be partially avoided, if there is another limitation in the system, e.g. the devices have strict limits on internal memory that they can use for the sake of initialization.

References

1. IEEE 802.11 wireless local area networks,
   http://www.ieee802.org/11/ [cited February 14, 2010]
2. Bordim, J.L., Ito, Y., Nakano, K.: Randomized leader election protocols in noisy radio networks with a single transceiver. In: Guo, M., Yang, L.T., Di Martino, B., Zima, H.P., Dongarra, J., Tang, F. (eds.) ISPA 2006. LNCS, vol. 4330, pp. 246–256. Springer, Heidelberg (2006)
3. Shields, C., Levine, B.N., Boris Margolin, N.: A survey of solutions to the sybil attack (2006),
   http://prisms.cs.umass.edu/brian/bubs/levine.sybil.tr.2006.pdf
4. Levine, B.N., Piro, C., Shields, C.: Detecting the sybil attack in ad hoc networks. In: Proc. IEEE/ACM SecureComm., pp. 1–11 (2006)
5. Danezis, G., Lesniewski-Laas, C., Frans Kaashoek, M., Anderson, R.J.: Sybil-resistant dht routing. In: di Vimercati, S.d.C., Syverson, P.F., Gollmann, D. (eds.) ESORICS 2005. LNCS, vol. 3679, pp. 305–318. Springer, Heidelberg (2005)
6. Douceur, J.R.: The sybil attack. In: Druschel, P., Frans Kaashoek, M., Rowstron, A.I.T. (eds.) IPTPS 2002. LNCS, vol. 2429, pp. 251–260. Springer, Heidelberg (2002)
7. Golebiewski, Z., Klonowski, M., Koza, M., Kutylowski, M.: Towards fair leader election in wireless networks. In: Ruiz, P.M., Garcia-Luna-Aceves, J.J. (eds.) ADHOC-NOW 2009. LNCS, vol. 5793, pp. 166–179. Springer, Heidelberg (2009)
8. Hayashi, T., Nakano, K., Olariu, S.: Randomized initialization protocols for packet radio networks. In: IPPS/SPDP, p. 544. IEEE Computer Society, Los Alamitos (1999)
9. Janson, S., Luczak, T., Rucinski, A.: Random Graphs. Wiley, Chichester (2000)
10. Janson, S., Szpankowski, W.: Analysis of an asymmetric leader election algorithm. Electr. J. Comb. 4(1) (1997)
11. Metcalfe, R.M., Boggs, D.R.: Ethernet: distributed packet switching for local computer networks. Commun. ACM 19(7), 395–404 (1976)
12. Nakano, K., Olariu, S.: Randomized o (log log n)-round leader election protocols in packet radio networks. In: Chwa, K.-Y., Ibarra, O.H. (eds.) ISAAC 1998. LNCS, vol. 1533, pp. 209–218. Springer, Heidelberg (1998)
13. Newsome, J., Shi, E., Song, D.X., Perrig, A.: The sybil attack in sensor networks: analysis & defenses. In: Ramchandran, K., Sztipanovits, J., Hou, J.C., Pappas, T.N. (eds.) IPSN, pp. 259–268. ACM, New York (2004)
14. Perrig, A., Stankovic, J.A., Wagner, D.: Security in wireless sensor networks. Commun. ACM 47(6), 53–57 (2004)
15. Willard, D.E.: Log-logarithmic selection resolution protocols in a multiple access channel. SIAM J. Comput. 15(2), 468–477 (1986)

Author Index

Araragi, Tadashi   135
Au, Man Ho   352
Aumasson, Jean-Philippe   87

Barbosa, Manuel   145, 164
Bard, Gregory V.   19
Biasse, Jean-François   233
Birkett, James   282
Boyd, Colin   300
Boztaş, Serdar   370
Buldas, Ahto   318

Cao, Zhenfu   216
Chaabouni, Rafik   336
Chen, Xiaofeng   200

Etrog, Jonathan   74

Farshim, Pooya   145, 164
Feng, Min   216
Fleischmann, Ewan   117
Fujisaki, Eiichiro   182

González Nieto, Juan Manuel   300
Gorantla, M. Choudary   300
Gorski, Michael   117

Henricksen, Matt   53

Jacobson Jr., Michael J.   233

Käsper, Emilia   87
Kim, Kwangjo   200
Kiyomoto, Shinsaku   53
Klonowski, Marek   391
Knudsen, Lars Ramkilde   87
Koza, Michał   391
Kunihiro, Noboru   248
Kutyłowski, Mirosław   391

Li, Chao   1
Li, Jin   200
Li, Ruilin   1
Ling, San   37
Lipmaa, Helger   336
Liu, Peng   216
Lucks, Stefan   117

Matusiewicz, Krystian   87
Mu, Yi   352

Nguyen, Phuong Ha   37
Niitsoo, Margus   318
Nishide, Takashi   135
Nishimaki, Ryo   182

Ødegård, Rune   87

Paterson, Kenneth G.   264
Peyrin, Thomas   87

Qu, Longjiang   1

Rechberger, Christian   104
Robshaw, Matthew J.B.   74

Sakurai, Kouichi   135
Schläffer, Martin   87
Shao, Jun   216
Shelat, Abhi   336
Silvester, Alan K.   233
Stebila, Douglas   264, 282
Sun, Bing   1
Sun, Li   370
Susilo, Willy   200, 352

Tanaka, Keisuke   182
Tanaka, Toshiaki   53
Tian, Haibo   200

Versteeg, Steven   370

Wang, Huaxiong   37
Wei, Lei   37
Wong, Kenneth Koon-Ho   19

Yann, Trevor   370
Yap, Wun She   53
Yian, Chee Hoo   53

Zhang, Fangguo   200
Zhu, Bin   216
Zhu, Huafei   135

GPSR Compliance

*The European Union's (EU) General Product Safety Regulation (GPSR)
is a set of rules that requires consumer products to be safe and our
obligations to ensure this.*

*If you have any concerns about our products, you can contact us on
ProductSafety@springernature.com*

In case Publisher is established outside the EU, the EU authorized
representative is:

Springer Nature Customer Service Center GmbH
Europaplatz 3
69115 Heidelberg, Germany

Batch number: 09473985

Printed by Printforce, the Netherlands